The Concise Oxford Dictionary of

Mathematics

FIFTH EDITION

CHRISTOPHER CLAPHAM

JAMES NICHOLSON

Christopher Clapham wrote the first and second editions of this dictionary. Until 1993 he was Senior Lecturer in Mathematics at the University of Aberdeen. His publications include *Introduction to Abstract Algebra* and *Introduction to Mathematical Analysis*.

James Nicholson has a mathematics degree from Cambridge, and taught at Harrow School for twelve years before becoming Head of Mathematics at Belfast Royal Academy in 1990. He lives in Belfast, but now works mostly with the School of Education at Durham University. He is the author of two A-level Statistics texts, two GCSE Mathematics revision guides, and a contributing author for a number of other mathematics textbooks.

 SEE WEB LINKS

For recommended web links for this title, visit www. oxfordreference.com/page/math when you see this sign.

OXFORD
UNIVERSITY PRESS

Great Clarendon Street, Oxford, OX2 6DP,
United Kingdom

Oxford University Press is a department of the University of Oxford.
It furthers the University's objective of excellence in research, scholarship,
and education by publishing worldwide. Oxford is a registered trade mark of
Oxford University Press in the UK and in certain other countries

© Christopher Clapham 1990, 1996
© Christopher Clapham and James Nicholson 2005, 2009, 2014

The moral rights of the authors have been asserted

Database right Oxford University Press (maker)

First edition 1990
Second edition 1996
Third edition 2005
Fourth edition 2009
Fifth edition 2014

Impression: 8

British Library Cataloguing in Publication Data

Data available

Library of Congress Control Number: 2013957561

ISBN 978-0-19-967959-1

Printed in Great Britain by
Clays Ltd., Elcograf S.p.A.

Links to third party websites are provided by Oxford in good faith and
for information only. Oxford disclaims any responsibility for the materials
contained in any third party website referenced in this work.

Contents

Contributors

C. Chatfield, BSc, PhD
R. Cheal, BSc
J. B. Gavin, BSc, MSc
University of Bath

J. R. Pulham, BSc, PhD
University of Aberdeen

D. P. Thomas, BSc, PhD
University of Dundee

Preface to Second Edition*

This dictionary is intended to be a reference book that gives reliable definitions or clear and precise explanations of mathematical terms. The level is such that it will suit, among others, sixth-form pupils, college students and first-year university students who are taking mathematics as one of their courses. Such students will be able to look up any term they may meet and be led on to other entries by following up cross-references or by browsing more generally.

The concepts and terminology of all those topics that feature in pure and applied mathematics and statistics courses at this level today are covered. There are also entries on mathematicians of the past and important mathematics of more general interest. Computing is not included. The reader's attention is drawn to the appendices which give useful tables for ready reference.

Some entries give a straight definition in an opening phrase. Others give the definition in the form of a complete sentence, sometimes following an explanation of the context. An asterisk is used to indicate words with their own entry, to which cross-reference can be made if required.

This edition is more than half as large again as the first edition. A significant change has been the inclusion of entries covering applied mathematics and statistics. In these areas, I am very much indebted to the contributors, whose names are given on page iii. I am most grateful to these colleagues for their specialist advice and drafting work. They are not, however, to be held responsible for the final form of the entries on their subjects. There has also been a considerable increase in the number of short biographies, so that all the major names are included. Other additional entries have greatly increased the comprehensiveness of the dictionary.

The text has benefited from the comments of colleagues who have read different parts of it. Even though the names of all of them will not be given, I should like to acknowledge here their help and express my thanks.

Christopher Clapham

* The formatting for cross-referencing has been changed since the second edition, and the above text has been modified to be consistent with the new format.

Preface to Third Edition

Since the second edition was published the content and emphasis of applied mathematics and statistics at sixth-form, college and first-year university levels has changed considerably. This edition includes many more applied statistics entries as well as dealing comprehensively with the new decision and discrete mathematics courses, and a large number of new biographies on 20th-century mathematicians. I am grateful to the Headmaster and Governors of Belfast Royal Academy for their support and encouragement to take on this task, and to Louise, Joanne, and Laura for transcribing my notes.

James Nicholson

Preface to Fourth Edition

Since the third edition was published there has been a dramatic increase in both access to the internet and the amount of information available. The major change to this edition is the introduction of a substantial number of web links, many of which contain dynamic or interactive illustrations related to the definition.

James Nicholson

Preface to Fifth Edition

The fifth edition introduces much fuller coverage of mathematical logic and of topology, and has extended the appendix on trigonometry and increased the number of appendices to include summaries of geometry, algebra, and probability distributions, along with a description of the Millennium Prize problems.

James Nicholson

A The number 10 in hexadecimal notation.

a- Prefix meaning 'not'. For example, an asymmetric figure is one which possesses no symmetry, which is not symmetrical.

abacus A counting device consisting of rods on which beads can be moved so as to represent numbers.

(⊕) SEE WEB LINKS
• A description of how one abacus works.

Abel, Niels Henrik (1802–29) Norwegian mathematician who, at the age of 19, proved that the general equation of degree greater than 4 cannot be solved algebraically. In other words, there can be no formula for the roots of such an equation similar to the familiar formula for a quadratic equation. He was also responsible for fundamental developments in the theory of algebraic functions. He died in some poverty at the age of 26, just a few days before he would have received a letter announcing his appointment to a professorship in Berlin.

abelian group Suppose that G is a *group with the operation $\circ$. Then G is abelian if the operation $\circ$ is commutative; that is, if, for all elements a and b in G, $a \circ b = b \circ a$.

Abel's Limit Theorem For a *convergent series $\{a_k\}$, the limit assigned by the Abel summation method exists and is equal to the sum of the series.

Abel's partial summation formula For two arbitrary sequences $\{a_k\}$ and $\{b_k\}$ with $A_k = \sum_{r=1}^{k} a_r$, the result that $\sum_{r=j}^{k} a_r b_r = \sum_{r=j}^{k} A_j(b_j - b_{j+1}) + A_k b_{k+1} - A_{j-1} b_j$. This has some similarities with the formula for *integration by parts.

Abel summation A method of computing the sum of a possibly *divergent series of *complex numbers $\{a_k\}$ as the limit, as z approaches 1 from below, of the *power series $\sum(a_k z^k)$, assuming the series $\{a_k\}$ has *radius of convergence $= 1$.

Abel's test A test for the convergence of an infinite series which states that if $\sum a_n$ is a convergent sequence, and $\{b_n\}$ is monotically decreasing, i.e. $b_{n+1} \leq b_n$ for all n, then $\sum a_n b_n$ is also convergent.

above Greater than. The limit of a function at a from above is the limit of $f(x)$ as $x \to a$ for values of $x > a$. It is of particular importance when $f(x)$ has a discontinuity at a, i.e. where the limits from above and from below do not coincide. It can be written as $f(a+)$ or $\lim_{x \to a+} f(x)$.

abscissa The x-coordinate in a Cartesian coordinate system in the plane.

absolute address In spreadsheets a formula which is to appear in a number of cells may wish to use the contents of another cell or cells. Since the relative position of those cells will be different each time the formula appears in a new location, the spreadsheet syntax allows an absolute address to be specified, identifying the actual row and column for each cell. When a formula is copied and pasted to another cell, a cell reference using an absolute address will remain unchanged. A formula can contain a mixture of absolute and *relative addresses.

absolute error *See* ERROR.

absolute frequency The number of occurrences of an event. For example, if a die is rolled 20 times and 4 sixes are observed the absolute frequency of sixes is 4 and the *relative frequency is 4/20.

absolutely continuous function A stronger condition than *continuous function or *uniformly continuous function. It says that a function f on an interval I (of the real line) is absolutely continuous on I if, for every positive number ε, there exists a positive number δ so that for every finite sequence of *pairwise disjoint subintervals of I such that $\sum_k |y_k - x_k| < \delta$ then $\sum_k |f(y_k) - f(x_k)| < \varepsilon$.

absolutely convergent series A series $\{a_n\}$ is said to be absolutely convergent if $\sum_{n=1}^{\infty} |a_n|$ is *convergent. For example, if $a_n = (-1)^{n-1} \times \frac{1}{n}$ then the series is convergent but not absolutely convergent, whereas $a_n = (-1)^{n-1} \times \frac{1}{n^2}$ is absolutely convergent.

absolutely summable $=$ ABSOLUTELY CONVERGENT.

absolute measure of dispersion $=$ MEASURE OF DISPERSION.

absolute value For any real number a, the absolute value (also called the *modulus) of a, denoted by $|a|$, is a itself if $a \geq 0$, and $-a$ if $a \leq 0$. Thus $|a|$ is positive except when $a = 0$. The following properties hold:

(i) $|ab| = |a||b|$.
(ii) $|a + b| \leq |a| + |b|$.

(iii) $|a - b| \geq ||a| - |b||$.

(iv) For $a > 0$, $|x| \leq a$ if and only if $-a \leq x \leq a$.

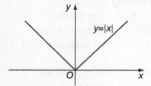

absorbing set A subset S of a *vector space X is an absorbing set if for any point x in X then tx lies in S (if $t \in \mathbf{R}$ is small enough and positive). An alternative name is a 'radial set', which reflects the notion that the set S will generate all elements of X in the sense that any $x \in X$ can be written as $x = \alpha s$ for some $s \in S$ and some real α.

absorbing state *See* RANDOM WALK.

absorption laws For all sets A and B (subsets of some *universal set), $A \cap (A \cup B) = A$ and $A \cup (A \cap B) = A$. These are the absorption laws.

abstract algebra The area of mathematics concerned with algebraic structures, such as *groups, *rings and *fields, involving sets of elements with particular operations satisfying certain axioms. The purpose is to derive, from the set of axioms, general results that are then applicable to any particular example of the algebraic structure in question. The theory of certain algebraic structures is highly developed; in particular, the theory of vector spaces is so extensive that its study, known as *linear algebra, would probably no longer be classified as abstract algebra.

abstraction The process of making a general statement which summarizes what can be observed in particular instances. For example, we can say that $x^2 < x$ for $0 < x < 1$ and $x^2 > x$ for $x < 0$ or $x > 1$. Mathematical theorems are essentially abstraction of concepts to a higher level.

abundant number An integer that is smaller than the sum of its positive divisors, not including itself, For example, 12 is divisible by 1, 2, 3, 4 and 6, and $1 + 2 + 3 + 4 + 6 = 16 > 12$.

acceleration Suppose that a particle is moving in a straight line, with a point O on the line taken as origin and one direction taken as positive. Let x be the *displacement of the particle at time t. The acceleration of the particle is equal to $\ddot{x}$ or d^2x/dt^2, the *rate of change of the *velocity with respect to t. If the velocity is positive (that is, if the particle is moving in the positive direction), the acceleration is positive when the particle is speeding up and negative when it is slowing down. However, if the velocity is negative, a positive acceleration means that the particle is slowing down and a negative acceleration means that it is speeding up.

In the preceding paragraph, a common convention has been followed, in which the unit vector **i** in the positive direction along the line has been suppressed. Acceleration is in fact a vector quantity, and in the 1-dimensional case above it is equal to $\ddot{x}\mathbf{i}$.

When the motion is in two or three dimensions, vectors are used explicitly. The acceleration **a** of a particle is a vector equal to the rate of change of the velocity **v** with respect to t. Thus $\mathbf{a} = d\mathbf{v}/dt$. If the particle has *position vector **r**, then $\mathbf{a} = d^2\mathbf{r}/dt^2 = \ddot{\mathbf{r}}$. When Cartesian coordinates are used, $\mathbf{r} = x\mathbf{i} + y\mathbf{j} + z\mathbf{k}$, and then $\ddot{\mathbf{r}} = \ddot{x}\mathbf{i} + \ddot{y}\mathbf{j} + \ddot{z}\mathbf{k}$.

If a particle is travelling in a circle with constant speed, it still has an acceleration because of the changing direction of the velocity. This acceleration is towards the centre of the circle and has magnitude $\frac{v^2}{r}$ where v is the speed of the particle and r is the radius of the circle.

Acceleration has the dimensions LT^{-2}, and the SI unit of measurement is the metre per second per second, abbreviated to 'm s^{-2}'.

acceleration–time graph A graph that shows acceleration plotted against time for a particle moving in a straight line. Let $v(t)$ and $a(t)$ be the velocity and acceleration, respectively, of the particle at time t. The acceleration-time graph is the graph $y = a(t)$, where the t-axis is horizontal and the y-axis is vertical with the positive direction upwards. With the convention that any area below the horizontal axis is negative, the area under the graph between $t = t_1$ and $t = t_2$ is equal to $v(t_2) - v(t_1)$. (Here a common convention has been followed, in which the unit vector **i** in the positive direction along the line has been suppressed. The velocity and acceleration of the particle are in fact vector quantities equal to $v(t)\mathbf{i}$ and $a(t)\mathbf{i}$, respectively.)

acceptance region *See* HYPOTHESIS TESTING.

acceptance sampling A method of quality control where a sample is taken from a batch and a decision whether to accept the batch is made on the basis of the quality of the sample. The most simple method is to have a straight accept/reject criterion, but a more sophisticated approach is to take another sample if the evidence from the existing sample, or a set of samples, is not clearly indicating whether the batch should be accepted or rejected. One of the main advantages of this approach is reducing the cost of taking samples to satisfy quality control criteria.

accessibility In a *connected graph, accessibility is a measure of the sum of total distances from a *node to all other nodes. The most accessible node is the one with the smallest sum of distances.

accumulation point For a set S in a *topological space X, a point x (which is in X but not necessarily in S) is an accumulation point if every *neighbourhood of x in X also contains an element of S distinct from x. Also called cluster or limit point.

accuracy A measure of the precision of a numerical quantity, usually given to n *significant figures (where the proportional accuracy is the important aspect) or n *decimal places (where the absolute accuracy is more important).

accurate (correct) to n decimal places Rounding a number with the accuracy specified by the number of *decimal places given in the rounded value. So $e = 2.71828 \ldots = 2.718$ to three decimal places and $= 2.72$ to two decimal places. $\sqrt{86.56} = 9.30076$ is 9.30 correct to two decimal places. Where a number of quantities are being measured and added or subtracted, using values correct to the same number of decimal places ensures that they have the same degree of accuracy. However, if the units are changed, for example between centimetres and metres, then the accuracy of the measurements will be different if the same number of decimal places is used in the measurements.

accurate (correct) to n significant figures Rounding a number with the accuracy specified by the number of *significant figures given in the rounded value. So $e = 2.71828 \ldots = 2.718$ to four significant figures and $= 2.72$ to three significant figures. $e^{-3} = 0.049787 \ldots = 0.0498$ correct to three significant figures. Rounding to the same number of significant figures ensures all the measurements have about the same proportionate accuracy. If the units are changed, for example between centimetres and metres, then the accuracy of the measurements will not be changed if the same number of significant figures is used in the measurements.

Achilles paradox The paradox which arises from considering how overtaking takes place. Achilles gives a tortoise a head start in a race. To overtake, he must reach the tortoise's initial position, then where the tortoise had moved to, and so on *ad infinitum*. The conclusion that he cannot overtake because he has to cover an infinite sum of well-defined non-zero distances is false, hence the paradox.

acre An imperial unit of surface area, which is 4 840 square yards. This is the area of a furlong (220 yards) by a chain (22 yards) which used to be standard units of measurement in the UK. A square mile contains 640 acres. In the metric system a *hectare is approximately 0.4 acre.

action limits The outer limits set on a *control chart in a production process. If the observed value falls outside these limits, then action will be taken, often resetting the machine. For the means of samples of sizes n in a process with standard deviation σ with target mean μ, the action limits will be set at $\mu \pm 3.09 \dfrac{\sigma}{\sqrt{n}}$.

active constraint An inequality such as $y + 2x \geq 13$ is said to be active at a point on the boundary, i.e. where equality holds, for example (6, 1) and (0, 13).

activity networks (edges as activities) The networks used in *critical path analysis where the edges (arcs) represent activities to be performed. Paths in the network represent the precedence relations between the activities, and *dummy activities are required to link paths where common activities appear, but the paths are at least partly independent. While this is a complication, each activity appears on only one edge and the sequence of activities needed is easier to follow than when the vertices are used to represent the activities. Once the activity network has been constructed from the precedence table, the *critical path algorithm (edges as activities) can be applied.

activity networks (vertices as activities) The networks used in *critical path analysis where the vertices (nodes) represent activities to be performed. The edges (arcs) coming out from any vertex X join X to any vertex Y whose activity cannot start until X has been completed, and the edge is labelled with the time taken for activity X. Note that there will often be more than one such activity, in which case each edge will carry the same time. The construction of an activity network requires a listing of the activities, their duration, and the precedence relations which identify which activities are dependent on the prior completion of other activities. With this structure, *dummy activities are not needed, but the same activity will be represented by more than one edge when more than one other activity depends on its prior completion, and the sequence of activities is less easy to follow than the alternative structure *activity networks (edges as activities).

acute angle An angle that is less than a *right angle. An *acute-angled triangle is one all of whose angles are acute.

adders (in combinatorial circuits) The half-adder and full adders are sections of circuits which use a system of logic gates to add binary digits, by using a combination for which the truth table output is identical to the output required by the binary addition.

((())) SEE WEB LINKS
• An article demonstrating how a simple adder works.

addition (of complex numbers) Let the complex numbers z_1 and z_2, where $z_1 = a + bi$ and $z_2 = c + di$, be represented by the points P_1 and P_2 in the *complex plane. Then $z_1 + z_2 = (a + c) + (b + d)i$, and $z_1 + z_2$ is represented in the complex plane by the point Q such that OP_1QP_2 is a parallelogram; that is, such that $\overrightarrow{OQ} = \overrightarrow{OP}_1 + \overrightarrow{OP}_2$. Thus, if the complex number z is associated with the *directed line segment $\overrightarrow{OP}$, where P represents z, then the addition of complex numbers corresponds exactly to the addition of the directed line segments.

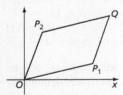

addition (of directed line segments) *See* ADDITION (of vectors).

addition (of matrices) Let $\mathbf{A}$ and $\mathbf{B}$ be $m \times n$ matrices, with $\mathbf{A} = [a_{ij}]$ and $\mathbf{B} = [b_{ij}]$. The operation of addition is defined by taking the SUM $\mathbf{A} + \mathbf{B}$ to be the $m \times n$ matrix $\mathbf{C}$, where $\mathbf{C} = [c_{ij}]$ and $c_{ij} = a_{ij} + b_{ij}$. The sum $\mathbf{A} + \mathbf{B}$ is not defined if $\mathbf{A}$ and $\mathbf{B}$ are not of the same order. This operation $+$ of addition on the set of all $m \times n$ matrices is *associative and *commutative.

addition (of vectors) Given vectors $\mathbf{a}$ and $\mathbf{b}$, let $\overrightarrow{OA}$ and $\overrightarrow{OB}$ be *directed line segments that represent $\mathbf{a}$ and $\mathbf{b}$, with the same initial point O. The sum of $\overrightarrow{OA}$ and $\overrightarrow{OB}$ is the directed line segment $\overrightarrow{OC}$, where $OACB$ is a parallelogram, and the SUM $\mathbf{a} + \mathbf{b}$ is defined to be the vector $\mathbf{c}$ represented by $\overrightarrow{OC}$. This is called the parallelogram law. Alternatively, the sum of vectors $\mathbf{a}$ and $\mathbf{b}$ can be defined by representing $\mathbf{a}$ by a directed line segment $\overrightarrow{OP}$ and $\mathbf{b}$ by $\overrightarrow{PQ}$ where the final point of the first directed line segment is the initial point of the second. Then $\mathbf{a} + \mathbf{b}$ is the vector represented by $\overrightarrow{OQ}$. This is called the triangle law. Addition of vectors has the following properties, which hold for all $\mathbf{a}$, $\mathbf{b}$ and $\mathbf{c}$:

(i) $\mathbf{a} + \mathbf{b} = \mathbf{b} + \mathbf{a}$, the commutative law.
(ii) $\mathbf{a} + (\mathbf{b} + \mathbf{c}) = (\mathbf{a} + \mathbf{b}) + \mathbf{c}$, the associative law.
(iii) $\mathbf{a} + \mathbf{0} = \mathbf{0} + \mathbf{a} = \mathbf{a}$, where $\mathbf{0}$ is the zero vector.
(iv) $\mathbf{a} + (-\mathbf{a}) = (-\mathbf{a}) + \mathbf{a} = \mathbf{0}$, where $-\mathbf{a}$ is the negative of $\mathbf{a}$.

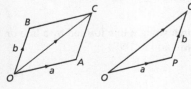

addition formula *See* COMPOUND ANGLE FORMULA.

addition law If A and B are two events then the addition law states that the $\Pr(A \text{ or } B) = \Pr(A) + \Pr(B) - \Pr(A \text{ and } B)$, or using set notation $P(A \cup B) = P(A) + P(B) - P(A \cap B)$. In the special case where A and B are *mutually exclusive events this reduces to $P(A \cup B) = P(A) + P(B)$.

addition modulo *n* *See* MODULO N, ADDITION AND MULTIPLICATION.

additive function A function for which $f(x+y) = f(x) + f(y)$. For example, $f(x) = 3x$ is an additive function since $f(x+y) = 3(x+y) = f(x) + f(y)$ but $g(x) = \sqrt{x}$ is not additive since $\sqrt{x+y}$ is not in general equal to $\sqrt{x} + \sqrt{y}$.

additive group A *group with the operation $+$, called addition, may be called an additive group. The operation in a group is normally denoted by addition only if it is *commutative, so an additive group is usually *abelian.

additive identity The identity element under an operation of addition, usually denoted by 0, so $a + 0 = 0 + a = a$.

additive inverse *See* INVERSE ELEMENT.

adherent point A point of the *closure of a set.

ad infinitum Repeating infinitely many times.

adj Abbreviation for *adjoint.

adjacency matrix For a *simple graph G, with n vertices $v_1, v_2, \ldots, v_n$, the adjacency matrix **A** is the $n \times n$ matrix $[a_{ij}]$ with $a_{ij} = 1$, if v_i is joined to v_j, and $a_{ij} = 0$, otherwise. The matrix **A** is *symmetric and the diagonal entries are zero. The number of ones in any row (or column) is equal to the *degree of the corresponding vertex. An example of a graph and its adjacency matrix **A** is shown in the figure.

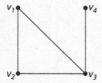

$$\mathbf{A} = \begin{bmatrix} 0 & 1 & 1 & 0 \\ 1 & 0 & 1 & 0 \\ 1 & 1 & 0 & 1 \\ 0 & 0 & 1 & 0 \end{bmatrix}$$

adjacent angles A pair of angles on a straight line formed by a line or half-line meeting it. Adjacent angles will add to $180°$.

adjacent edges A pair of edges in a graph joined by a common vertex.

adjacent side The side of a right-angled triangle between the right angle and the given angle.

adjacent vertices A pair of vertices in a *graph joined by a common edge.

adjoint The adjoint of a square matrix $\mathbf{A}$, denoted by adj $\mathbf{A}$, is the transpose of the matrix of cofactors of $\mathbf{A}$. For $\mathbf{A} = [a_{ij}]$, let A_{ij} denote the *cofactor of the entry a_{ij}. Then the matrix of cofactors is the matrix $[A_{ij}]$ and adj $\mathbf{A} = [A_{ij}]^T$. For example, a 3×3 matrix $\mathbf{A}$ and its adjoint can be written

$$\mathbf{A} = \begin{bmatrix} a_{11} & a_{12} & a_{13} \\ a_{21} & a_{22} & a_{23} \\ a_{31} & a_{32} & a_{33} \end{bmatrix}, \quad \text{adj}\,\mathbf{A} = \begin{bmatrix} A_{11} & A_{21} & A_{31} \\ A_{12} & A_{22} & A_{32} \\ A_{13} & A_{23} & A_{33} \end{bmatrix}.$$

In the 2×2 case, a matrix $\mathbf{A}$ and its adjoint have the form

$$\mathbf{A} = \begin{bmatrix} a & b \\ c & d \end{bmatrix}, \quad \text{adj}\,\mathbf{A} = \begin{bmatrix} d & -b \\ -c & a \end{bmatrix}.$$

The adjoint is important because it can be used to find the *inverse of a matrix. From the properties of cofactors, it can be shown that $\mathbf{A}$ adj $\mathbf{A} =$ (det $\mathbf{A}$)$\mathbf{I}$. It follows that, when det? $\mathbf{A} \neq 0$, the inverse of $\mathbf{A}$ is $(1/\text{det}\,\mathbf{A})$ adj $\mathbf{A}$.

adjoint equation For a system of *linear differential equations $y' = Ay$, where A is a matrix with *adjoint A^*, the adjoint equation is $z' = -A^*z$.

adjugate $=$ ADJOINT.

aerodynamic drag A body moving through the air, such as an aeroplane flying in the Earth's atmosphere, experiences a force due to the flow of air over the surface of the body. The force is the sum of the aerodynamic drag, which is tangential to the flight path, and the lift, which is normal to the flight path.

affine geometry A geometry in which some properties are preserved by *parallel projection from one plane to another. However, others are not, and in particular *Euclid's third and fourth axioms do not hold.

affine hull The affine hull of a set S is the smallest *affine set containing S. This is equivalent to the intersection of all affine sets containing S.

affine manifold (affine subspace) A subset of a *vector space which contains all lines between points of the subset.

affine plane An *affine geometry of two dimensions.

affine set Contains all the points on the line through any two distinct points of a set S.

affine space For any vectors **a**, **b** in X a *vector space over the *field K, and elements p, q in a non-empty set S, define the addition $p + $ **a** such that

(i) $p + \mathbf{0} = p$,
(ii) $(p + \mathbf{a}) + \mathbf{b} = p + (\mathbf{a} + \mathbf{b})$.

If, for any q in S, there exists a unique vector **a** in X such that $q = p + \mathbf{a}$, then S is an affine space.

In an affine space, you can add a vector to a point to get another point and you can get a vector by subtracting two points, but you cannot add points.

affine span The smallest *affine manifold which contains a given subset of a vector space.

affine subspace *See* AFFINE MANIFOLD.

affine transformation A *transformation which preserves *collinearity and therefore the straightness and parallel nature of lines, and the ratios of distances.

a fortiori From the Latin, meaning 'with even stronger reason', this is often used to refer to a general statement from which another follows immediately; for example, since 11 is a prime number, it is *a fortiori* not divisible by 5.

aggregate Census returns are used to construct aggregate statistics concerning a wide range of characteristics such as economic, social, health, education, often grouped by geographical region, gender, age, etc. The name derives from the way they are compiled as counts from a large number of individual census returns. Other examples of aggregate statistics include indices such as the *retail price index.

agree If $f(x)$ and $g(x)$ are defined on a set S, and $f(x) = g(x)$ for all $x \in S$, then f and g agree on the set S.

air resistance The resistance to motion experienced by an object moving through the air caused by the flow of air over the surface of the object. It is a force that affects, for example, the speed of a drop of rain or of a parachutist falling towards the Earth's surface. As well as depending on the nature of the object, air resistance depends on the speed of the object. Possible *mathematical models are to assume that the magnitude of the air resistance is proportional to the speed or to the square of the speed.

Airy function The differential equation $\phi'' - t\phi = 0$ has the Airy function $\phi(t) = \frac{1}{\pi} \int_0^\infty \cos\left(tx + \frac{x^3}{3}\right) dx$ as its solution.

Aitken's method (in numerical methods) If an iterative formula $x_{r+1} = f(x_r)$ is to be used to solve an equation, Aitken's method of accelerating convergence uses the initial value and the first two values obtained by the formula to calculate a better approximation than the iterative formula would produce. This can then be used as a new starting point from which to repeat the process until the required accuracy has been reached. While this is computationally intensive, it is the sort of process which spreadsheets handle very easily.

If x_0, x_1, x_2 are the initial value and the first two iterations and $\Delta x_r = x_{r+1} - x_r$, $\Delta^2 x_r = \Delta x_{r+1} - \Delta x_r$ are the forward differences then

$x_4 = x_3 - \dfrac{(\Delta x_2)^2}{\Delta^2 x_1}$. More generally this will be expressed as

$x_{r+1} = x_r - \dfrac{(\Delta x_{r-1})^2}{\Delta^2 x_{r-2}}$.

aleph Any infinite *cardinal number, usually denoted by the Hebrew letter ℵ. *See also* TRANSFINITE NUMBER.

aleph-null The smallest infinite cardinal number. The cardinality of any set which can be put in one-to-one correspondence with the set of natural numbers. Such sets are said to be *countable or *denumerable. One of the apparent paradoxes in number theory is that the set of rational numbers between 0 and 1, the set of all rational numbers, and the set of natural numbers all have the same cardinality. The symbol $\aleph_0$ is used.

algebra The area of mathematics related to the general properties of arithmetic. Relationships can be summarized by using variables, usually denoted by letters $x, y, n, \ldots$ to stand for unknown quantities, whose value (s) may be determined by solving the resulting equations. *See also* ABSTRACT ALGEBRA and LINEAR ALGEBRA.

Algebra, Fundamental Theorem of *See* FUNDAMENTAL THEOREM OF ALGEBRA.

algebraic closure The extension of a given set to include all the roots of polynomials with coefficients in the given set. The smallest algebraically closed set of numbers is **C**, the set of complex numbers, since the very simple equation $x^2 + 1 = 0$ has a complex solution.

algebraic curve A curve that can be represented as a polynomial equation. The degree of the curve is the order of the polynomial.

algebraic geometry The area of mathematics related to the study of *geometry by algebraic methods.

algebraic number A real number that is the root of a *polynomial equation with integer coefficients. All *rational numbers are algebraic, since a/b is the root of the equation $bx - a = 0$. Some *irrational numbers are algebraic; for example, $\sqrt{2}$ is the root of the equation $x^2 - 2 = 0$. An irrational number that is not algebraic (such as π) is called a *transcendental number.

algebraic structure The term used to describe an abstract concept defined as consisting of certain elements with operations satisfying given axioms. Thus, a *group or a *ring or a *field is an algebraic structure. The purpose of the definition is to recognize similarities that appear in different contexts within mathematics and to encapsulate these by means of a set of axioms.

algebraic system A set together with the *operations and *relations defined on that set.

algebraic topology The area of mathematics which uses *abstract algebra to study topology.

algebra of sets The set of all subsets of a *universal set E is closed under the binary operations $\cup$ (*union) and $\cap$ (*intersection) and the unary operation′ (*complementation). The following are some of the properties, or laws, that hold for subsets A, B and C of E:

   (i) $A \cup (B \cup C) = (A \cup B) \cup C$ and $A \cap (B \cap C) = (A \cap B) \cap C$, the associative properties.

  (ii) $A \cup B = B \cup A$ and $A \cap B = B \cap A$, the commutative properties.

 (iii) $A \cup \varnothing = A$ and $A \cap \varnothing = \varnothing$, where $\varnothing$ is the *empty set.

 (iv) $A \cup E = E$ and $A \cap E = A$.

  (v) $A \cup A = A$ and $A \cap A = A$.

 (vi) $A \cap (B \cup C) = (A \cap B) \cup (A \cap C)$ and $A \cup (B \cap C) = (A \cup B) \cap (A \cup C)$, the distributive properties.

 (vii) $A \cup A' = E$ and $A \cap A' = \varnothing$.

(viii) $E' = \varnothing$ and $\varnothing' = E$.

 (ix) $(A')' = A$.

  (x) $(A \cup B)' = A' \cap B'$ and $(A \cap B)' = A' \cup B'$, De Morgan's laws.

The application of these laws to subsets of E is known as the algebra of sets. Despite some similarities with the algebra of numbers, there are important and striking differences.

algorithm A precisely described routine procedure that can be applied and systematically followed through to a conclusion.

algorithmic complexity The number of steps required in an algorithm depends on the process and on the number of items being operated on. An algorithm where the number of steps increases in direct proportion to the number of items is said to have linear complexity, so if there are three times as many items the algorithm will have three times as many steps. If the algorithm was of quadratic complexity then there would have been nine times as many steps.

aliquant part A number or expression which is not an exact divisor of a given number or expression. For example, 2 is an aliquant part of any odd number, and $x+1$ is an aliquant part of x^2+1.

aliquot part A number or expression which is an exact divisor of a given number or expression, and is usually required to be a proper divisor. For example, 2 is an aliquot part of any even number larger than 2, and $x+1$ is an aliquot part of x^2-1.

al-Khwārizmī *See under* KHWĀRIZMĪ, MUHAMMAD IBN MŪSĀ AL-.

almost all (almost everywhere) Holding for all values except on a set of *zero measure. The most striking example of a set of zero measure is the set of rational numbers, so if $f(x)=a$ when x is rational, and $f(x)=b$ when x is irrational, and $g(x)=b$ everywhere, then f and g agree almost everywhere.

almost discrete space A space which is *discrete except on a set of *zero measure.

almost disjoint Two sets are almost disjoint if their intersection is finite (or small in some other defined sense). A collection of subsets is said to be almost disjoint if they are *pairwise almost disjoint.

almost surely = ALMOST EVERYWHERE in a probability measure.

almost uniform convergence A sequence of functions which has uniform convergence except on a set of arbitrarily small measure has almost uniform convergence. This is weaker than the convergence being uniform *almost everywhere, which would require the exception set to have zero measure.

alternate angles *See* TRANSVERSAL.

alternating group The *subgroup of the *symmetric group, S_n, containing all the *even permutations of n objects. It has order $n!/2$. For $n > 4$, it is the only proper, * *normal subgroup* of S_n apart from the *empty set. The alternating group is a *simple group.

alternating series A series in which the sign alternates between positive and negative. So any series in the form $a_n = (-1)^n \, p_n$ or $(-1)^{n-1} \, p_n$ where all $p_n > 0$ is alternating.

alternating series test The result that an *alternating series will always be *convergent if the terms decrease *monotonically to zero in *absolute value. It is attributed to Gottfried *Leibniz.

alternative hypothesis *See* HYPOTHESIS TESTING.

altitude A line through one vertex of a triangle and perpendicular to the opposite side. The three altitudes of a triangle are concurrent at the *orthocentre.

altitude (in astronomy) The angle above the horizon that a star is located at. *See also* AZIMUTH.

ambiguous case The case where two sides of a triangle are known, and an acute angle which is not the angle between the known sides. There can often be two possible triangles which satisfy all the given information, hence the name.

If you know the length of AB, and of BC, and the acute angle at A, you can construct the triangle as follows. Draw AB. From A draw a line making the required angle to AB. Place the point of a compass at B and draw an arc of a circle with radius equal to the required length of BC. Where the line from A and the arc intersect is the position of C. Unless the angle at C

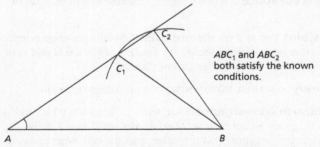

ABC_1 and ABC_2 both satisfy the known conditions.

is exactly $90°$ or the given information is inconsistent, there will be two different points of intersection and hence two possible triangles.

(((⊕))) SEE WEB LINKS

• An animation showing the construction of the two possible solutions.

amicable numbers A pair of numbers with the property that each is equal to the sum of the positive divisors of the other. (For the purposes of this definition, a number is not included as one of its own divisors.) For example, 220 and 284 are amicable numbers because the positive divisors of 220 are 1, 2, 4, 5, 10, 11, 20, 22, 44, 55 and 110, whose sum is 284, and the positive divisors of 284 are 1, 2, 4, 71 and 142, whose sum is 220.

These numbers, known to the Pythagoreans, were used as symbols of friendship. The amicable numbers 17 296 and 18 416 were found by *Fermat, and a list of 64 pairs was produced by *Euler. In 1867, a sixteen-year-old Italian boy found the second smallest pair, 1184 and 1210, overlooked by Euler. More than 600 pairs are now known. It has not been shown whether or not there are infinitely many pairs of amicable numbers.

amp Abbreviation and symbol for *amplitude.

amplitude Suppose that $x = A \sin(\omega t + \alpha)$, where A (> 0), ω and α are constants. This may, for example, give the displacement x of a particle, moving in a straight line, at time t. The particle is thus oscillating about the origin O. The constant A is the amplitude, and gives the maximum distance in each direction from O that the particle attains.

The term may also be used in the case of *damped oscillations to mean the corresponding coefficient, even though it is not constant. For example, if $x = 5e^{-2t} \sin 3t$ the oscillations are said to have amplitude $5e^{-2t}$, which tends to zero as t tends to infinity.

analog device A measuring instrument of a continuous quantity which shows the measurement on a continuous scale. For example, the hands of a clock (provided that the second hand moves smoothly).

analysis The area of mathematics generally taken to include those topics that involve the use of limiting processes. Thus *differential calculus and *integral calculus certainly come under this heading. Besides these, there are other topics, such as the summation of infinite series, which involve 'infinite' processes of this sort. The *Binomial Theorem, a theorem of algebra, leads on into analysis when the index is no longer a positive integer, and the study of sine and cosine, which begins as trigonometry, becomes analysis when the power series for the functions are derived. The term 'analysis' has also come to be used to indicate a rather more rigorous approach to the topics of calculus, and to the foundations of the real number system.

analysis of variance A general procedure for partitioning the overall variability in a set of data into components due to specified causes and random variation. It involves calculating such quantities as the 'between-groups sum of squares' and the 'residual sum of squares', and dividing by the *degrees of freedom to give so-called 'mean squares'. The results are usually presented in an ANOVA table, the name being derived from the opening letters of the words 'analysis of variance'. Such a table provides a concise summary from which the influence of the *explanatory variables can be estimated and hypotheses can be tested, usually by means of *F-tests.

(((⊕))) SEE WEB LINKS

• A description of the analysis and its interpretation in a medical context (subscription).

analytic (of a complex function) A single-valued function which is differentiable at every point of its domain. If the function is not differentiable everywhere it may be said to be analytic at points where it is differentiable.

analytic geometry The area of mathematics relating to the study of *coordinate geometry.

analytic proof Proof by algebraic reasoning or analysis only, without dependence on axioms, especially geometrical axioms. *See also* SYNTHETIC PROOF.

anchor ring = TORUS.

and *See* CONJUNCTION.

angle (between lines in space) Given two lines in space, let $\mathbf{u}_1$ and $\mathbf{u}_2$ be vectors with directions along the lines. Then the angle between the lines, even if they do not meet, is equal to the angle between the vectors $\mathbf{u}_1$ and $\mathbf{u}_2$ (*see* ANGLE (between vectors), with the directions of $\mathbf{u}_1$ and $\mathbf{u}_2$ chosen so that the angle θ satisfies $0 \leq \theta \leq \pi/2$ (θ in radians), or $0 \leq \theta \leq 90$ (θ in degrees). If l_1, m_1, n_1 and l_2, m_2, n_2 are *direction ratios for directions along the lines, the angle θ between the lines is given by

$$\cos \theta = \frac{|l_1 l_2 + m_1 m_2 + n_1 n_2|}{\sqrt{l_1^2 + m_1^2 + n_1^2}\sqrt{l_2^2 + m_2^2 + n_2^2}}.$$

angle (between lines in the plane) In coordinate geometry of the plane, the angle α between two lines with gradients m_1 and m_2 is given by

$$\tan \theta = \frac{m_1 - m_2}{1 + m_1 m_2}$$

This is obtained from the formula for $\tan(A - B)$. In the special cases when $m_1 m_2 = -1$ or when m_1 or m_2 is infinite, it has to be interpreted as follows: if $m_1 m_2 = -1$, $\theta = 90°$; if m_i is infinite, $\theta = 90° - |\tan^{-1} m_j|$ where $j \neq i$; and if m_1 and m_2 are both infinite, $\theta = 0$.

angle (between planes) Given two planes, let $\mathbf{n}_1$ and $\mathbf{n}_2$ be vectors *normal to the two planes. Then a method of obtaining the angle between the planes is to take the angle between $\mathbf{n}_1$ and $\mathbf{n}_2$ (*see* ANGLE (between vectors)), with the directions of $\mathbf{n}_1$ and $\mathbf{n}_2$ chosen so that the angle θ satisfies $0 \leq \theta \leq \pi/2$ (θ in radians), or $0 \leq \theta \leq 90$ (θ in degrees).

angle (between vectors) Given vectors $\mathbf{a}$ and $\mathbf{b}$, let $\overrightarrow{OA}$ and $\overrightarrow{OB}$ be *directed line segments representing $\mathbf{a}$ and $\mathbf{b}$. Then the angle θ between the vectors $\mathbf{a}$ and $\mathbf{b}$ is the angle $\angle AOB$, where θ is taken to satisfy $0 \leq \theta \leq \pi$ (θ in radians), or $0 \leq \theta \leq 180$ (θ in degrees). It is given by

$$\cos \theta = \frac{\mathbf{a} \cdot \mathbf{b}}{|\mathbf{a}||\mathbf{b}|}.$$

angle of depression The angle a line makes below a plane, usually the horizontal, especially for a line of sight, or the direction of projection of a projectile.

angle of elevation The angle a line makes above a plane, usually the horizontal, especially for a line of sight, or the direction of projection of a projectile.

angle of friction The angle λ such that $\tan \lambda = \mu_s$, where μ_s is the coefficient of static friction. Consider a block resting on a horizontal plane, as shown in the figure. In the limiting case when the block is about to move to the right on account of an applied force of magnitude P, $N = mg$, $P = F$ and $F = \mu_s N$. Then the *contact force, whose components are N and F, makes an angle λ with the vertical.

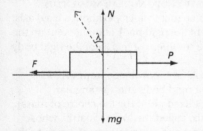

angle of inclination *See* INCLINED PLANE.

angle of projection The angle that the direction in which a particle is projected makes with the horizontal. Thus it is the angle that the initial velocity makes with the horizontal.

angular Relating to or measured in angles.

angular acceleration Suppose that the particle P is moving in the plane, in a circle with centre at the origin O and radius r_0. Let (r_0, θ) be the *polar coordinates of P. At an elementary level, the angular acceleration may be defined to be $\ddot{\theta}$.

At a more advanced level, the angular acceleration $\boldsymbol{\alpha}$ of the particle P is the vector defined by $\boldsymbol{\alpha} = \dot{\boldsymbol{\omega}}$, where $\boldsymbol{\omega}$ is the *angular velocity. Let $\mathbf{i}$ and $\mathbf{j}$ be unit vectors in the directions of the positive x- and y-axes and let $\mathbf{k} = \mathbf{i} \times \mathbf{j}$. Then, in the case above of a particle moving along a circular path, $\boldsymbol{\omega} = \dot{\theta}\mathbf{k}$ and $\boldsymbol{\alpha} = \ddot{\theta}\mathbf{k}$. If $\mathbf{r}$, $\mathbf{v}$ and $\mathbf{a}$ are the position vector, velocity and acceleration of P, then

$$\mathbf{r} = r_0\mathbf{e}_r, \qquad \mathbf{v} = \dot{\mathbf{r}} = r_0\dot{\theta}\mathbf{e}_\theta, \qquad \mathbf{a} = \ddot{\mathbf{r}} = -r_0\dot{\theta}^2\mathbf{e}_r + r_0\ddot{\theta}\mathbf{e}_\theta.$$

where $\mathbf{e}_r = \mathbf{i}\cos\theta + \mathbf{j}\sin\theta$ and $\mathbf{e}_\theta = -\mathbf{i}\sin\theta + \mathbf{j}\cos\theta$ (*see* CIRCULAR MOTION). Using the fact that $\mathbf{v} = \boldsymbol{\omega} \times \mathbf{r}$, it follows that the acceleration $\mathbf{a} = \boldsymbol{\alpha} \times \mathbf{r} + \boldsymbol{\omega} \times (\boldsymbol{\omega} \times \mathbf{r})$.

angular frequency The constant ω in the equation $\ddot{x} = -\omega^2 x$ for *simple harmonic motion. In certain respects ωt, where t is the time, acts like an angle. The angular frequency ω is usually measured in radians per second. The *frequency of the oscillations is equal to $\omega/2\pi$.

angular measure There are two principal ways of measuring angles: by using *degrees, in more elementary work, and by using *radians, essential in more advanced work, in particular, when calculus is involved.

angular momentum Suppose that the particle P of mass m has position vector $\mathbf{r}$ and is moving with velocity $\mathbf{v}$. Then the angular momentum $\mathbf{L}$ of P about the point A with position vector $\mathbf{r}_A$ is the vector defined by $\mathbf{L} = (\mathbf{r} - \mathbf{r}_A) \times m\mathbf{v}$. It is the *moment of the *linear momentum about the point A. *See also* CONSERVATION OF ANGULAR MOMENTUM.

Consider a rigid body rotating with angular velocity $\boldsymbol{\omega}$ about a fixed axis, and let $\mathbf{L}$ be the angular momentum of the rigid body about a point on the fixed axis. Then $\mathbf{L} = I\boldsymbol{\omega}$, where I is the *moment of inertia of the rigid body about the fixed axis.

To consider the general case, let $\boldsymbol{\omega}$ and $\mathbf{L}$ now be column vectors representing the angular velocity of a rigid body and the angular momentum of the rigid body about a fixed point (or the centre of mass). Then $\mathbf{L} = \mathbf{I}\boldsymbol{\omega}$, where $\mathbf{I}$ is a 3×3 matrix, called the inertia matrix, whose elements involve the moments of inertia and the *products of inertia of the rigid body relative to axes through the fixed point (or centre of mass).

The rotational motion of a rigid body depends on the angular momentum of the rigid body. In particular, the rate of change of the angular momentum about a fixed point (or centre of mass) equals the sum of the moments of the forces acting on the rigid body about the fixed point (or centre of mass).

angular speed The magnitude of the *angular velocity.

angular velocity Suppose that the particle P is moving in the plane, in a circle with centre at the origin O and radius r_0. Let (r_0, θ) be the *polar coordinates of P. At an elementary level, the angular velocity may be defined to be $\dot{\theta}$.

At a more advanced level, the angular velocity $\boldsymbol{\omega}$ of the particle P is the vector defined by $\boldsymbol{\omega} = \dot{\theta}\mathbf{k}$, where $\mathbf{i}$ and $\mathbf{j}$ are unit vectors in the directions of the positive x- and y-axes, and $\mathbf{k} = \mathbf{i} \times \mathbf{j}$. If $\mathbf{r}$ and $\mathbf{v}$ are the position vector and velocity of P, then

$$\mathbf{r} = r_0\mathbf{e}_r, \quad \mathbf{v} = \dot{\mathbf{r}} = r_0\dot{\theta}\,\mathbf{e}_\theta,$$

where $\mathbf{e}_r = \mathbf{i} \cos \theta + \mathbf{j} \sin \theta$ and $\mathbf{e}_\theta = -\mathbf{i} \sin \theta + \mathbf{j} \cos \theta$ (*see* CIRCULAR MOTION). By using the fact that $\mathbf{k} = \mathbf{e}_r \times \mathbf{e}_\theta$, it follows that the velocity $\mathbf{v}$ is given by $\mathbf{v} = \boldsymbol{\omega} \times r$.

Consider a rigid body rotating about a fixed axis, and take coordinate axes so that the z-axis is along the fixed axis. Let (r_0, θ) be the polar coordinates of some point of the rigid body, not on the axis, lying in the plane $z = 0$. Then the angular velocity $\boldsymbol{\omega}$ of the rigid body is defined by $\boldsymbol{\omega} = \dot{\theta}\mathbf{k}$.

In general, for a rigid body that is rotating, such as a top spinning about a fixed point, the rigid body possesses an angular velocity $\boldsymbol{\omega}$ whose magnitude and direction depend on time.

annualized percentage rate (APR) The true cost of borrowing money from a lender. Introduced in 1977 to provide consumers with a 'true rate' benchmark for comparison because lenders were quoting apparently very good headline rates, but charging interest on all of the initial loan for the whole period, taking no account of the reducing level of debt. Initially mortgages were excluded from the requirement that all loan illustrations show the APR, because of the variety of fees and charges associated with mortgages. Now mortgages do have to display the APR, inclusive of all compulsory fees or charges associated with them. It represents an equivalent rate with interest payable annually in arrears.

annulus (annuli) The region between two concentric circles. If the circles have radii r and $r + w$, the area of the annulus is equal to $\pi(r + w)^2 - \pi r^2$, which equals $w \times 2\pi(r + \frac{1}{2}w)$. It is therefore the same as the area of a rectangle of width w and length equal to the circumference of the circle midway in size between the two original circles.

anonymity of survey data People will often answer personal questions differently depending on how anonymous they believe their answers will be. Since there is a high premium on gaining truthful responses, methods of ensuring anonymity or at least strict confidentiality in a survey are important.

ANOVA *See* ANALYSIS OF VARIANCE.

antecedent The clause in a conditional statement which expresses the condition. So in 'if n is divisible by 2, n is even' the antecedent is 'n is divisible by 2'. Compare with *consequent.

anti- A prefix denoting the *inverse of a function.

anticlastic A surface is said to anticlastic at a point A if it is a turning point which has opposite signs of curvature in two perpendicular directions at the point, i.e. one being a maximum and one being a minimum, giving a *saddle-point.

anticlockwise Movement in the opposite direction of timing as the hands of a clock normally take. In compass terms $N \rightarrow W \rightarrow S \rightarrow E$ is anticlockwise. In polar coordinates this is taken to be the positive direction.

antiderivative Given a *real function f, any function ϕ such that $\phi'(x) = f(x)$, for all x (in the domain of f), is an antiderivative of f. If ϕ_1 and ϕ_2 are both antiderivatives of a *continuous function f, then $\phi_1(x)$ and $\phi_2(x)$ differ by a constant. In that case, the notation

$$\int f(x) \, dx$$

may be used for an antiderivative of f, with the understanding that an arbitrary constant can be added to any antiderivative. Thus,

$$\int f(x) \, dx + c$$

where c is an arbitrary constant, is an expression that gives all the antiderivatives.

antidifferentiate The reverse process to *differentiation used as a method of *integration when not working from first principles as the limit of a sum of incremental elements.

antilog Abbreviation for *antilogarithm.

antilogarithm The antilogarithm of x, denoted by antilog x, is the number whose *logarithm is equal to x. For example, suppose that common logarithm tables are used to calculate 2.75×3.12. Then, approximately, $\log 2.75 = 0.4393$ and $\log 3.12 = 0.4942$ and $0.4393 + 0.4942 = 0.9335$. Now antilog 0.9335 is required and, from tables, the answer 8.58 is obtained. Now that logarithm tables have been superseded by calculators, the term 'antilog' is little used. If y is the number whose logarithm is x, then $\log_a y = x$. This is equivalent to $y = a^x$ (from the definition of logarithm). So, if base a is being used, $\text{antilog}_a x$ is identical with a^x; for common logarithms, $\text{antilog}_{10} x$ is just 10^x, and this notation is preferable.

antiparallel A pair of *directed lines or *vectors whose directions are *parallel but having the opposite *sense.

antipodal points Two points on a sphere that are at opposite ends of a diameter.

antiprism Normally, a convex *polyhedron with two 'end' faces that are congruent regular polygons lying in parallel planes in such a way that, with each vertex of one polygon joined by an edge to two vertices of the other polygon, the remaining faces are isosceles triangles. The term could be used for a polyhedron of a similar sort in which the end faces are not regular and the triangular faces are not isosceles, in which case the first definition would be said to give a right-regular antiprism. If the end faces are regular and the triangular faces are equilateral, the antiprism is a semi-regular polygon.

antisymmetric matrix = SKEW-SYMMETRIC MATRIX.

antisymmetric relation A binary relation $\sim$ on a set S is antisymmetric if, for all a and b in S, whenever $a \sim b$ and $b \sim a$, then $a = b$. For example, the relation $\leq$ on the set of integers is antisymmetric. (Compare this with the definition of an *asymmetric relation.)

apex (apices) *See* BASE (of a triangle) and PYRAMID.

aphelion *See* APSE.

apogee *See* APSE.

Apollonius of Perga (about 262–190 BC) Greek mathematician whose most famous work *The Conics* was, until modern times, the definitive work on the *conic sections: the ellipse, parabola and hyperbola. He proposed the idea of epicyclic motion for the planets. Euclid, Archimedes and Apollonius were pre-eminent in the period covering the third century BC known as the Golden Age of Greek mathematics.

Apollonius' circle Given two points A and B in the plane and a constant k, the locus of all points P such that $AP/PB = k$ is a circle. A circle obtained like this is an Apollonius' circle. Taking $k = 1$ gives a straight line, so either this value must be excluded or, in this context, a straight line must be considered to be a special case of a circle. In the figure, $k = 2$.

((⊕)) SEE WEB LINKS

• An interactive web page where you can alter the ratio to see how the circle changes.

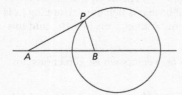

a posteriori Not able to be known without experience, i.e. the truth or otherwise cannot be argued by logic from agreed definitions. For example, *Newton's Laws of Motion describe the way bodies are observed to behave, rather than being deduced from principles in the way that the probabilities of the score when tossing a fair die are.

apothem The line, or the length of the line, joining the centre of a regular polygon to the midpoint of one of its sides. This is also the radius of the inscribed circle.

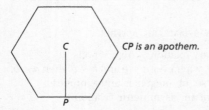

CP is an apothem.

applied The practical application of knowledge as in *applied mathematics.

applied mathematics The area of mathematics related to the study of natural phenomena, and includes mechanics of all types, probability, statistics, discrete and decision mathematics as well as the application of pure mathematics like matrices in solving real-world problems.

approximate To find a value or expression for a quantity within a specified degree of accuracy.

approximation When two quantities X and x are approximately equal, written $X \approx x$, one of them may be used in suitable circumstances in place of, or as an approximation for, the other. For example, $\pi \approx \frac{22}{7}$ and $\sqrt{2} \approx 1.414$.

approximation theory The area of numerical analysis related to finding a simpler function $f(x)$ which has approximately the same value as $F(x)$ over a specified interval. The analysis is then carried out on the more accessible $f(x)$ knowing that the difference will be small.

APR = ANNUALIZED PERCENTAGE RATE.

a priori Able to be known without experience, i.e. the truth or otherwise can be argued by logic from agreed definitions. For example, the probabilities of the score when tossing a fair die are deduced from the principle of equally likely outcomes.

a priori (in statistics) Another term for *prior probability.

apse A point in an *orbit at which the body is moving in a direction perpendicular to the radius vector. In an elliptical orbit in which the centre of attraction is at one focus there are two apses, the points at which the body is at its nearest and its furthest from the centre of attraction. When the centre of attraction is the Sun, the points at which the body is nearest and furthest are the perihelion and the aphelion. When the centre of attraction is the Earth, they are the perigee and the apogee.

apsis (apses) = APSE.

Arabic numeral *See* NUMERAL.

arbitrary constant A non-numerical symbol which is not a variable in a generalized operation. For example, $y = mx + c$ is the general equation of a straight line in two dimensions, where m and c are arbitrary constants which represent the gradient of the line and the y intercept. And $\int 2x\, dx = x^2 + c$ where c is an arbitrary constant whose value can be determined by a boundary condition.

arc (of a curve) The part of a curve between two given points on the curve. If *A* and *B* are two points on a circle, there are two arcs *AB*. When *A* and *B* are not at opposite ends of a diameter, it is possible to distinguish between the longer and shorter arcs by referring to the major arc *AB* and the minor arc *AB*.

arc (of a digraph) *See* DIGRAPH.

arc connected A *topological space in which there is a directed path (arc) between any two points is arc connected.

arccos, arccosec, arccot, arcsin, arcsec, arctan *See* INVERSE TRIGONOMETRIC FUNCTION.

arccosh, arccosech, arccoth, arcsinh, arcsech, arctanh *See* INVERSE HYPERBOLIC FUNCTION.

Archimedean property An *ordered or *normed *algebraic structure which has no infinitely large or infinitely small elements has the Archimedean property. This means that any two non-zero elements can be compared to one another, because neither is infinitesimal with respect to the other.

Archimedean solid A convex *polyhedron is called semi-regular if the faces are regular polygons, though not all congruent, and if the vertices are all alike, in the sense that the different kinds of face are arranged in the same order around each vertex. Right-regular *prisms with square side faces and (right-regular) *antiprisms whose side faces are equilateral triangles are semi-regular. Apart from these, there are thirteen semi-regular polyhedra, known as the Archimedean solids. These include the *truncated tetrahedron, the *truncated cube, the *cuboctahedron and the *icosidodecahedron.

(⊕) SEE WEB LINKS
• Links to the thirteen Archimedean solids, which can be manipulated to see the shapes fully.

Archimedean spiral A curve whose equation in polar coordinates is $r = a\theta$, where a (> 0) is a constant. In the figure, $OA = a\pi$, $OB = 2a\pi$, and so $OB = 2OA$.

(⊕) SEE WEB LINKS
• An animation showing the Archimedean spiral.

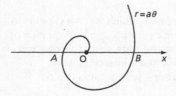

Archimedean spiral

Archimedes (287–212 BC) Greek mathematician who can be rated one of the greatest of all time. He made considerable contributions to geometry, discovering methods of finding, for example, the surface area and the volume of a sphere and the area of a segment of a parabola. His work on hydrostatics and equilibrium was also fundamental. His most fascinating work, *The Method*, was rediscovered as recently as 1906. He may or may not have shouted 'Eureka' and run naked through the streets, but he was certainly murdered by a Roman soldier, an event that marks the end of an era in mathematics.

Archimedes' method Take a circle. Computing areas or perimeters of *inscribed and *circumscribed n-sided polygons generates two sequences, the first of which all underestimate the corresponding measurement of the circle, and the second sequence are all overestimates. This allows upper and lower limits to be placed on the values of computations of polygon measurements.

arc length Let $y = f(x)$ be the graph of a function f such that f' is *continuous on $[a, b]$. The length of the arc, or arc length, of the curve $y = f(x)$ between $x = a$ and $x = b$ equals

$$\int_a^b \sqrt{1 + (f'(x))^2}\,dx.$$

Parametric form

For the curve $x = x(t)$, $y = y(t)$ ($t \in [\alpha, \beta]$), the arc length equals

$$\int_\alpha^\beta \sqrt{\left(\frac{dx}{dt}\right)^2 + \left(\frac{dy}{dt}\right)^2}\,dt.$$

Polar form

For the curve $r = r(\theta)$ ($\alpha \le \theta \le \beta$), the arc length equals

$$\int_\alpha^\beta \sqrt{r^2 + \left(\frac{dr}{d\theta}\right)^2}\,d\theta.$$

area A measure of the extent of a surface, or the part of surface enclosed by some specified boundary. For simple geometrical shapes such as rectangles, triangles, cylinders etc. the areas can be calculated by simple formulae based on their dimensions. Other more complex plane figures or 3-dimensional surfaces require the use of integration, or numerical approximations to find their areas.

area of a surface of revolution Let $y=f(x)$ be the graph of a function f such that f' is *continuous on $[a, b]$ and $f(x) \geq 0$ for all x in $[a, b]$. The area of the surface obtained by rotating, through one revolution about the x-axis, the arc of the curve $y=f(x)$ between $x=a$ and $x=b$, equals

$$2\pi \int_a^b y\sqrt{1 + \left(\frac{dy}{dx}\right)^2}\,dx, \quad \text{or} \quad 2\pi \int_a^b f(x)\sqrt{1 + (f'(x))^2}\,dx.$$

Parametric form

For the curve $x=x(t)$, $y=y(t)$ ($t \in [\alpha, \beta]$), the surface area equals

$$2\pi \int_\alpha^\beta y\sqrt{\left(\frac{dx}{dt}\right)^2 + \left(\frac{dy}{dt}\right)^2}\,dt.$$

Polar form

For the curve $r=r(\theta)$ ($\alpha \leq \theta \leq \beta$), the surface area equals

$$2\pi \int_\varepsilon^\beta r\sin\theta\sqrt{r^2 + \left(\frac{dr}{d\theta}\right)^2}\,d\theta.$$

area under a curve Suppose that the curve $y=f(x)$ lies above the x-axis, so that $f(x) \geq 0$ for all x in $[a, b]$. The area under the curve, that is, the area of the region bounded by the curve, the x-axis and the lines $x=a$ and $x=b$, equals

$$\int_a^b f(x)\,dx.$$

The definition of *integral is made precisely in order to achieve this result.

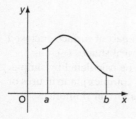

If $f(x) \leq 0$ for all x in $[a, b]$, the integral above is negative. However, it is still the case that its absolute value is equal to the area of the region bounded by the curve, the x-axis and the lines $x = a$ and $x = b$. If $y = f(x)$ crosses the x-axis, appropriate results hold. For example, if the regions A and B are as shown in the figure below, then

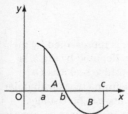

area of region $A = \int_a^b f(x) \, dx$ and area of region $B = -\int_b^c f(x) \, dx$.
It follows that

$$\int_a^c f(x) \, dx = \int_a^b f(x) \, dx + \int_b^c f(x) \, dx$$
$$= \text{area of region } A - \text{area of region } B.$$

Similarly, to find the area of the region bounded by a suitable curve, the y-axis, and lines $y = c$ and $y = d$, an equation $x = g(y)$ for the curve must be found. Then the required area equals

$$\int_c^d g(y) \, dy,$$

assuming that the curve is to the right of the y-axis, so that $g(y) \geq 0$ for all y in $[c, d]$. As before, the value of the integral is negative if $g(y) \leq 0$.

Polar areas

If a curve has an equation $r = r(\theta)$ in polar coordinates, there is an integral that gives the area of the region bounded by an arc AB of the curve and the two radial lines OA and OB. Suppose that $\angle xOA = \alpha$ and $\angle xOB = \beta$. The area of the region described equals

$$\int_\alpha^\beta \frac{1}{2} r^2 d\theta.$$

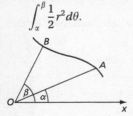

arg Abbreviation and symbol for the *argument of a complex number.

Argand, Jean Robert (1768–1822) Swiss-born mathematician who was one of several people, including Gauss, who invented a geometrical representation for complex numbers. This explains the name Argand diagram.

Argand diagram = COMPLEX PLANE.

argument Suppose that the *complex number z is represented by the point P in the *complex plane. The argument of z, denoted by arg z, is the angle θ (in radians) that OP makes with the positive real axis Ox, with the angle given a positive sense anticlockwise from Ox. As with polar coordinates, the angle θ may be taken so that $0 \leq \arg z < 2\pi$. Usually, however, the angle θ is chosen so that $-\pi < \arg z \leq \pi$. Sometimes, arg z is used to denote any of the values $\theta + 2n\pi$, where n is an integer. In that case, the particular value that lies in a certain interval, specified or understood, such as $[0, 2\pi)$ or $(-\pi, \pi]$, is called the *principal value of arg z.

Aristarchus of Samos (about 270 BC) Greek astronomer, noted for being the first to affirm that the Earth rotates and travels around the Sun. He treated astronomy mathematically and used geometrical methods to calculate the relative sizes of the Sun and Moon and their relative distances from the Earth.

Aristotelian logic The formal deductive method of logic dealing with the relations between *categorical propositions in their form as distinct from their content, especially *syllogisms.

Aristotle (384–322 BC) Greek philosopher who made important contributions to mathematics through his work on deductive logic.

arithmetic The area of mathematics relating to numerical calculations involving only the basic operations of addition, subtraction, multiplication, division and simple powers.

Arithmetic, Fundamental Theorem of *See* FUNDAMENTAL THEOREM OF ARITHMETIC.

arithmetic-geometric mean The limit of the series obtained by the *arithmetic-geometric mean iteration.

arithmetic-geometric mean inequality The arithmetic mean of a set of non-negative numbers is never less than their geometric mean. So for any $a, b \geq 0, \dfrac{a+b}{2} \geq \sqrt{ab}$. Generally, for a set of numbers $\{a_i\}, \frac{1}{n}\sum a_i \geq \sqrt[n]{\prod a_i}$ with equality if and only if all a_i take the same value.

arithmetic-geometric mean iteration If a and b are two positive real numbers, let $a_0 = a$, $b_0 = b$ and $a_{n+1} = \dfrac{a_n + b_n}{2}, b_{n+1} = \sqrt{a_n b_n}$ for $n \geq 0$. Then $\lim\limits_{n \to \infty} (a_n) = \lim\limits_{n \to \infty} (b_n)$ is the *arithmetic-geometric mean.

arithmetic mean *See* MEAN.

arithmetic progression = ARITHMETIC SERIES.

arithmetic sequence A finite or infinite sequence of terms a_1, a_2, a_3, ... with a common difference d, so that $a_2 - a_1 = d$, $a_3 - a_2 = d$, and so on. The first term is usually denoted by a. For example, 2, 5, 8, 11, ... is the arithmetic sequence with $a = 2$, $d = 3$. In such an arithmetic sequence, the n-th term a_n is given by $a_n = a + (n-1)d$.

arithmetic series A series $a_1 + a_2 + a_3 + \ldots$ (which may be finite or infinite) in which the terms form an *arithmetic sequence. Thus the terms have a common difference d, with $a_k - a_{k-1} = d$ for all $k \geq 2$. If the first term equals a, then $a_k = a + (k-1)d$. Let s_n be the sum of the first n terms of an arithmetic series, so that

$$s_n = a + (a+d) + (a+2d) + \cdots + (a + (n-1)d),$$

and let the last term here, $a + (n-1)d$, be denoted by l. Then s_n is given by the formula $s_n = \frac{1}{2}n(a+l)$. The particular case in which $a = 1$ and $d = 1$ gives the sum of the first n natural numbers:

$$\sum_{r=1}^{n} r = 1 + 2 + \cdots + n = \frac{1}{2}n(n+1).$$

arm One of the two lines forming an angle.

arrangement *See* PERMUTATION.

array An ordered collection of elements, usually numbers. A *vector is an example of a 1-dimensional array, and a *matrix is an example of a 2-dimensional array. Three or more dimensional arrays are used, but are more difficult to present on paper, though the subscript notation $a_{ij} =$ the element in the i-th row and j-th column of a matrix is easy to extend to higher dimensions. For arrays to be equal requires all corresponding elements to have the same value, and therefore equal arrays have to be the same size.

arrow paradox One of *Zeno's paradoxes which relate to the issue of whether time and space are made up of minute indivisible parts. At one of these moments in time, the arrow occupies a well-defined space. It is moving neither to where it is nor to where it is not (because no time elapses in which it can move), so there is no motion during this instant. If time is composed entirely of instants at which no motion can occur, then all motion is impossible.

Arrow's Impossibility Theorem The surprising result that when voters have three or more distinct options to choose from, there is no rank-order voting system which can aggregate the individual preferences of two or more individuals so that four apparently reasonable conditions are met.

(⊕) SEE WEB LINKS
• An illustration of the theorem.

Āryabhata (about 476–550) Indian mathematician, author of one of the oldest Indian mathematical texts. Written in verse, the *Āryabhatīya* is a summary of miscellaneous rules for calculation and mensuration. It deals with, for example, the areas of certain plane figures, values for π, and the summation of *arithmetic series. Also included is the equivalent of a table of sines, based on the half-chord rather than the whole *chord of the Greeks.

ASCII = American Standard Code for Information Interchange. A binary code representing characters used in VDUs, printers, etc.

assignment problem A problem in which things of one type are to be matched with the same number of things of another type in a way that is, in a specified sense, the best possible. For example, when n workers are to be assigned to n jobs, it may be possible to specify the value v_{ij} to the company, measured in suitable units, if the i-th worker is assigned to the j-th job. The values v_{ij} may be displayed as the entries of an $n \times n$ matrix. By introducing suitable variables, the problem of assigning workers to jobs in such a way as to maximize the total value to the company can be formulated as a *linear programming problem.

association = CORRELATION.

associative The *binary operation ∘ on a set S is associative if, for all a, b and c in S, $(a \circ b) \circ c = a \circ (b \circ c)$.

assumption A statement that is presumed to be true in the particular circumstances. Any conclusions reached are dependent on the validity of the assumption.

astroid A *hypocycloid in which the radius of the rolling circle is a quarter of the radius of the fixed circle. It has *parametric equations $x = a \cos^3 t$, $y = a \sin^3 t$, where a is the radius of the fixed circle.

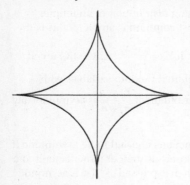

asymmetric A plane figure is asymmetric if it is neither symmetrical about a line nor symmetrical about a point.

asymmetric relation A binary relation $\sim$ on a set S is *asymmetric if, for all a and b in S, whenever $a \sim b$, then $b \sim a$ does not hold. For example, the relation $<$ on the set of integers is asymmetric. (Compare this with the definition of an *antisymmetric relation.)

asymptote A line l is an asymptote to a curve if the distance from a point P to the line l tends to zero as P tends to infinity along some unbounded part of curve. Consider the following examples:

(i) $y = \dfrac{x+3}{(x+2)(x-1)}$,   (ii) $y = \dfrac{3x^2}{x^2+x+1}$,   (iii) $y = \dfrac{x^3}{x^2+x+1}$.

Example (i) has $x = -2$ and $x = 1$ as vertical asymptotes, and $y = 0$ as a horizontal asymptote. Example (ii) has no vertical asymptotes, and $y = 3$ as a horizontal asymptote. To investigate example (iii), it can be rewritten as

$$y = x - 1 + \frac{1}{x^2 + x + 1}.$$

Then it can be seen that $y = x - 1$ is a slant asymptote, that is, an asymptote that is neither vertical nor horizontal.

asymptotically stable The qualitative theory of dynamical systems and of differential equations considers what happens to the system after a long period of time. Where a system shows periodic behaviour, it is common to be interested in what happens if a small change is made to the initial conditions. If the solution converges to the original solution, then it is asymptotically stable.

asymptotic distribution The limiting distribution of an infinite sequence of random variables. The most common example of this is the Central Limit Theorem which says that if $Y_n = \dfrac{\sum_{i=1}^{n} X_i}{n}$, where $\{X_i\}$ are a series of independent, identically distributed random variables with $E\{X_i\} = \mu$ and $Var(X_i) = \sigma^2$ then the distribution of $\dfrac{Y_n - \mu}{\sigma/\sqrt{n}}$ is asymptotically the standard normal distribution as $n \to \infty$.

asymptotic functions A pair of functions are said to be asymptotic if they approach one another infinitely closely as their arguments tend to a particular value, often infinity. This is often expressed as '$f(x)$ is asymptotic to $g(x)$ at α if $\lim_{x \to \alpha} \dfrac{f(x)}{g(x)} = 1$'.

asymptotic series A divergent series $s_n(x) = a_0 + \frac{a_1}{x} + \frac{a_2}{x^2} + \ldots + \frac{a_n}{x^n}$ is called an asymptotic series for a function $f(x)$ if $\lim_{|x| \to \infty} \{x^n[f(x) - s_n(x)]\} = 0$ for every n.

Atiyah, Sir Michael Francis FRS (1929–2019) British mathematician who made important contributions in topology, geometry, analysis, the transcendental theory of algebraic varieties, differential operators and quantum field theory. Awarded the *Fields Medal in 1966 and the Abel Prize in 2004.

atlas In *topology, a collection of *charts which cover a *manifold.

atmospheric pressure The *pressure at a point in the atmosphere due to the gravitational force acting on the air. This pressure depends on position and time and is measured by a barometer. The standard atmospheric pressure at sea level is taken to be 101 325 pascals, which is the definition of 1 atmosphere. The variations from place to place are comparatively small, but are the main cause of the wind patterns of the Earth. In general, the atmospheric pressure decreases with height.

atom In *measure theory, a set which has strictly positive measure, with the property that any subset either has equal measure or has zero measure.

attainable set The set of vertices which can be reached from a given vertex in a *digraph in one or any number of steps.

atto- Prefix used with *SI units to denote multiplication by 10^{-18}.

attractor An equilibrium state (or collection of states) to which a system evolves over time. When the system gets close enough to an attractor, it will remain close even if slightly perturbed. A system may have multiple attractors, each with its own region of attraction.

augmented matrix For a given set of m linear equations in n unknowns $x_1, x_2, \ldots x_n$,

$$a_{11}x_1 + a_{12}x_2 + \ldots + a_{1n}x_n = b_1,$$
$$a_{21}x_1 + a_{22}x_2 + \ldots + a_{2n}x_n = b_2,$$
$$\vdots$$
$$a_{m1}x_1 + a_{m2}x_2 + \ldots + a_{mn}x_n = b_m,$$

the augmented matrix is the matrix

$$\begin{bmatrix} a_{11} & a_{12} & \cdots & a_{1n} & b_1 \\ a_{21} & a_{22} & \cdots & a_{2n} & b_2 \\ \vdots & \vdots & \ddots & \vdots & \vdots \\ a_{m1} & a_{m2} & \cdots & a_{mn} & b_m \end{bmatrix}$$

obtained by adjoining to the matrix of coefficients an extra column of entries taken from the right-hand sides of the equations. The solutions of a set of linear equations may be investigated by transforming the augmented matrix to *echelon form or reduced echelon form by elementary row operations. *See* GAUSSIAN ELIMINATION; GAUSS-JORDAN ELIMINATION.

autocorrelation (serial correlation) If $\{x_i\}$ is an ordered sequence of observations then the product moment correlation coefficient between pairs (x_i, x_{i+1}) is the autocorrelation of lag 1, and between pairs (x_i, x_{i+k}) is the autocorrelation of lag k. The values of k for which the autocorrelation is not negligible can provide important information as to the underlying structure of a time series.

autocovariance Similar to *autocorrelation, but measuring the covariance between the sequences.

automatic repeat request An error control method for transmitting data that uses acknowledgements by the receiver within a prescribed period of time. If the sender has not received an acknowledgement in the time agreed, then the sender automatically repeats the transmission.

automorphic function A function $f(z)$ is *automorphic under a group of transformations if it is *analytic in a domain D except at singular points and for all transformations T in the group that if z is in D then so is $T(z)$.

automorphism A one-to-one correspondence mapping the elements of a set onto itself, so the domain and range of the function are the same. For example, $f(x) = x + 1$ is an automorphism on **R** but $g(x) = \sin x$ is not.

autonomous differential equation A differential equation where the independent variable does not explicitly occur. For example $\dfrac{dv}{dt} = -kv$ is autonomous, but $\dfrac{dv}{dt} = 3t^2 + 4t - 3$ is not.

auxiliary equation *See* LINEAR DIFFERENTIAL EQUATION WITH CONSTANT COEFFICIENTS.

average deviation = MEAN DEVIATION.

average of a rate of a quantity The constant value for which the total change in the interval would have been the same as that observed. For example, a car may vary its instantaneous speed in the duration of a journey, but the average speed = total distance travelled/time of journey.

average of a set of data Usually intended to refer to the *arithmetic mean.

average speed The total distance travelled divided by the total time taken.

average velocity The total displacement (as a vector) divided by the total time taken.

axial plane One of the planes containing two of the coordinate axes in a 3-dimensional Cartesian coordinate system. For example, one of the axial planes is the yz-plane, or (y, z)-plane, containing the y-axis and the z-axis, and it has equation $x = 0$.

axiom A statement whose truth is either to be taken as self-evident or to be assumed. Certain areas of mathematics involve choosing a set of axioms and discovering what results can be derived from them, providing proofs for the theorems that are obtained.

axiomatic set theory The presentation of set theory as comprising axioms together with rules of inference.

axiomatic system Any logical system which explicitly states *axioms from which *theorems can be deduced.

axiom of choice States that for any set of *mutually exclusive sets there is at least one set that contains exactly one element in common with each of the non-empty sets.

axis (axes) *See* COORDINATES (in the plane) and COORDINATES (in 3-dimensional space).

axis (of a cone) *See* CONE.

axis (of a cylinder) *See* CYLINDER.

axis (of a parabola) *See* PARABOLA.

axis of rotation A line about which a curve or slope is rotated to form a solid or surface.

axis of symmetry A line in which the mirror image of a curve or geometrical figure maps onto itself. The condition for this is that for each point P on the curve or figure the point P' is also, where the axis is the perpendicular bisector of PP'.

azimuth A term used in astronomy to define the bearing of a star relative, usually, to the north. Sometimes south is used as the reference direction. *See also* ALTITUDE (in astronomy).

B The number 11 in *hexadecimal notation.

Babbage, Charles (1792–1871) British mathematician and inventor of mechanical calculators. His 'analytical engine' was designed to perform mathematical operations mechanically using a number of features essential in the design of today's computers but, partly through lack of funds, the project was not completed.

back-substitution Suppose that a set of linear equations is in *echelon form. Then the last equation can be solved for the first unknown appearing in it, any other unknowns being set equal to *parameters taking arbitrary values. This solution can be substituted into the previous equation, which can then likewise be solved for the first unknown appearing in it. The process that continues in this way is back-substitution.

backward difference If $\{(x_i, f_i)\}$, $i = 0, 1, 2, \ldots$ is a given set of function values with $x_{i+1} = x_i + h$, $f_i = f(x_i)$ then the backward difference at f_i is defined by $f_i - f_{i-1} = f(x_i) - f(x_{i-1})$.

backward induction The form of *mathematical induction which shows that if the proposition $P(n)$ fails then the proposition must fail for some $P(k)$, for some $k \leq n - 1$. As with the standard form of induction, if $P(1)$ is true, this means that $P(2)$ must also be true and so on.

backward scan In an *activity network (edges as activities), the backward scan identifies the latest time for each vertex (node). Starting with the sink, work backwards through the network, for each vertex calculate the sum of edge plus total time on next vertex for all paths leaving that vertex. Put in the largest of these times onto that vertex.

Baire category In a *topological space X, a set S is (Baire) first category if it is a *countable union of *nowhere dense sets. Otherwise S is second category. *See also* MEAGRE SET.

Baire Category Theorem Every *complete metric space is a *Baire space. Every locally compact Hausdorff space is a Baire space.

Baire space A *topological space in which every intersection of any countable collection of *open *dense sets is dense. A Baire space is (Baire) second category (*See* BAIRE CATEGORY).

balanced (of a set) A subset A of a *vector space where for each element x in A, tx belongs to A if $|t| < 1$. For example, a sphere centred at the origin in Cartesian space is a balanced set.

balanced block design *See* BLOCK DESIGN.

ball A set in a *metric space containing all the points which are not more than a given distance from a fixed point. An *open ball does not include the surface where the distance is equal to the given constant and a closed ball does include the surface.

Banach, Stefan (1892–1945) Polish mathematician who was a major contributor to the subject known as functional analysis. Much subsequent work was inspired by his exposition of the theory in a paper of 1932.

Banach algebra An algebra on a *Banach space that satisfies the inequality $\|xy\| \le \|x\| . \|y\|$ for all elements x, y of the space, where $\| \ \|$ is the norm of the space.

Banach space A complete normed vector space on the real or complex numbers.

banded matrix Where the elements are zero except in a band of diagonals around the leading diagonal.

bandwidth The data transmission rate (measured in bits per second) supported by an interface or a network connection (often the internet). However, bandwidth represents the capacity of the system, and overall performance depends on other factors also, especially the *latency.

bar The superscript symbol, as in $\bar{z}$ or $\bar{x}$ is used to denote the *conjugate of a complex number or the *mean of a statistical variable.

bar chart A diagram representing the *frequency distribution for nominal or discrete data with comparatively few possible values. It consists of a sequence of bars, or rectangles, corresponding to the possible values, and the length of each is proportional to the frequency. The bars have equal widths and are usually not touching. The figure shows the kinds of vehicles recorded in a small traffic survey.

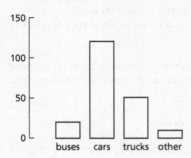

bar coding A method for automating stock control and till charging where each item carries a *Universal Product Code which is shown as a series of black bars, which are scanned by an optical reader. At many till charging points now, the customer will receive an itemized receipt showing the details of all their purchases, and the company's stock records will be automatically updated by the sale.

barrel A subset of a *normed space, or a *topological vector space, which is *closed, *convex, *absorbing, and *balanced.

barrelled space A *topological vector space in which every *barrel contains a *neighbourhood of the origin.

Barrow, Isaac (1630–77) English mathematician whose method of finding tangents, published in 1670, is essentially that now used in differential calculus. He may have been the first to appreciate that the problems of finding tangents and areas under curves are inversely related. When he resigned his chair at Cambridge, he was succeeded by *Newton, on Barrow's recommendation.

base (of an exponential function) *See* EXPONENTIAL FUNCTION TO BASE A.

base (of an isosceles triangle) *See* ISOSCELES TRIANGLE.

base (of logarithms) *See* LOGARITHMS.

base (of natural logarithms) *See* E.

base (of a pyramid) *See* PYRAMID.

base (for representation of numbers) The integer represented as 4703 in standard decimal notation is written in this way because

$$4703 = (4 \times 10^3) + (7 \times 10^2) + (0 \times 10) + 3.$$

The same integer can be written in terms of powers of 8 as follows:

$$4703 = (1 \times 8^4) + (1 \times 8^3) + (1 \times 8^2) + (3 \times 8) + 7.$$

The expression on the right-hand side is abbreviated to $(11137)_8$ and is the *representation of this number to base 8. In general, if g is an integer greater than 1, any positive integer a can be written uniquely as

$$a = c_n g^n + c_{n-1} g^{n-1} + \cdots + c_1 g + c_0,$$

where each c_i is a non-negative integer less than g. This is the representation of a to base g, and is abbreviated to $(c_n c_{n-1} \ldots c_1 c_0)_g$. Real numbers, not just integers, can also be written to any base, by using figures after a 'decimal' point, just as familiar *decimal representations of real numbers are written to base 10. *See also* BINARY REPRESENTATION, DECIMAL and HEXADECIMAL.

base (of a triangle) It may be convenient to consider one side of a triangle to be the base of the triangle. The vertex opposite the base may then be called the apex of the triangle, and the distance from the apex to the base the height of the triangle.

base angles *See* ISOSCELES TRIANGLE.

base field The *field used to define a *vector space.

baseline risk The risk at the beginning of a time period or under specific conditions.

base period The time used as a reference point for comparison of variables such as the Retail Price Index. An index of 100 is usually taken at the base period and the value at other periods is expressed proportionally, so if prices have risen 17% the index will record 117.

base unit *See* SI UNITS.

basic solution When *slack variables have been introduced into a *linear programming problem there are more equations than there are variables. If there are n more variables than equations the basic solutions are found by setting various combinations of n variables to zero, thus reducing the number of variables to the number of equations. If one of the slack variables is negative then that solution is not *feasible. For each solution, the n variables set to zero are called non-basic variables and the others are called basic variables.

basic variables *See* BASIC SOLUTION.

basis (bases) A set S of *vectors is a spanning set if any vector can be written as a linear combination of those in S. If, in addition, the vectors in S are *linearly independent, then S is a basis. It follows that any vector can be written *uniquely as a linear combination of those in a basis. In 3-dimensional space, any set of three non-coplanar vectors $\mathbf{u}$, $\mathbf{v}$, $\mathbf{w}$ is a basis, since any vector $\mathbf{p}$ can be written uniquely as $\mathbf{p} = x\mathbf{u} + y\mathbf{v} + z\mathbf{w}$. In 2-dimensional space, any set of 2 non-parallel vectors has this property and so is a basis. Any one of the vectors in a set currently being taken as a basis may be called a basis vector.

basis theorem Any *linearly independent set of n vectors is a *basis of a *vector space of finite dimension n.

basis vector *See* BASIS.

Bayes, Thomas (1702–61) English mathematician remembered for his work in probability, which led to a method of statistical inference based on the theorem that bears his name. His paper on the subject was not published until 1763, after his death.

Bayes' Theorem The following theorem for calculating a posterior probability (see PRIOR PROBABILITY):

Theorem

Let $A_1, A_2, \ldots, A_k$ be mutually exclusive events whose union is the whole sample space of an experiment and let B be an event with $\Pr(B) \neq 0$. Then

$$\Pr(A_i \mid B) = \frac{\Pr(B \mid A_i)\Pr(A_i)}{\Pr(B \mid A_1)\Pr(A_1) + \cdots + \Pr(B \mid A_k)\Pr(A_k)}.$$

For example, let A_1 be the event of tossing a double-headed coin, and A_2 the event of tossing a normal coin. Suppose that one of the coins is chosen at random so that $\Pr(A_1) = \frac{1}{2}$ and $\Pr(A_2) = \frac{1}{2}$. Let B be the event of obtaining 'heads'. Then $\Pr(B \mid A_1) = 1$ and $\Pr(B \mid A_2) = \frac{1}{2}$. So

$$\Pr(A_1 \mid B) = \frac{\Pr(B \mid A_1)\Pr(A_1)}{\Pr(B \mid A_1)\Pr(A_1) + \Pr(B \mid A_2)\Pr(A_2)}$$
$$= \frac{1 \times \frac{1}{2}}{1 \times \frac{1}{2} + \frac{1}{2} \times \frac{1}{2}} = \frac{2}{3}.$$

This says that, given that 'heads' was obtained, the probability that it was the double-headed coin that was tossed is 2/3.

Here, $\Pr(A_i)$ is a prior probability and $\Pr(A_i \mid B)$ is a posterior probability.

Bayesian An approach where a *prior distribution is modified in the light of the outcomes of an experiment.

bearing The direction of the course upon which a ship is set or the direction in which an object is sighted may be specified by giving its bearing, the angle that the direction makes with north. The angle is measured in degrees in a clockwise direction from north. For example, north-east has a bearing of 45°, and west has a bearing of 270°.

beats Suppose that a body capable of oscillating freely with *angular frequency ω is subject to an oscillatory applied force with angular frequency Ω. When ω and Ω are nearly equal, the motion appears to consist of oscillations with an amplitude that varies comparatively slowly. The beats occur as the amplitude achieves its maximum value. The effect can be heard when two notes whose frequencies are close together are sounded at the same time. Musical instruments can be tuned by listening for beats.

Bellman's principle of optimality Any part of an optimal path is itself optimal. This is one of the fundamental principles of *dynamic programming by which the length of the known optimal path is extended step by step until the complete path is known.

bell-shaped curve *See* NORMAL DISTRIBUTION.

belongs to If x is an element of a set S, then x belongs to S and this is written $x \in S$. Naturally, $x \notin S$ means that x does not belong to S.

below (of functions) Less than. The limit of a function at a from below is the limit of $f(x)$ as $x \rightarrow a$ for values of $x < a$. It is of particular importance when $f(x)$ has a discontinuity at a, i.e. where the limits from above and from below do not coincide. It can be written as $f(a-)$ or $\lim_{x \to a-} f(x)$.

bending moment The effect at a point which is the sum of all the *moments of forces acting on the structure.

Bernoulli distribution The discrete probability *distribution whose *probability mass function is given by $\Pr(X=0) = 1 - p$ and $\Pr(X=1) = p$. It is the *binomial distribution $B(1, p)$.

Bernoulli equation A differential equation which can be written in the form $y' + f(x)y = g(x)y^n$. The transformation $z = y^{1-n}$ gives $z' = \dfrac{dz}{dx} = (1-n)y^{-n}y'$. Dividing the original equation by y^n gives $y^{-n}y' + f(x)y^{1-n} = g(x)$ which reduces to $\frac{1}{1-n}z' + f(x)z = g(x)$ which is in linear form and can be solved by use of an integrating factor.

Bernoulli family A family from Basle that produced a stream of significant mathematicians, some of them very important indeed. The best known are the brothers Jacques (or James or Jakob) and Jean (or John or Johann), and Jean's son Daniel. **Jacques Bernoulli** (1654–1705) did much work on the newly developed calculus, but is chiefly remembered for his contributions to probability theory: the *Ars conjectandi* was published, after his death, in 1713. The work of **Jean Bernoulli** (1667–1748) was more definitely within calculus: he discovered *l'Hôpital's rule (*see under* H) and proposed the *brachistochrone problem. He was one of the founders of the *calculus of variations. In the next generation, **Daniel Bernoulli** (1700–82) was the member of the family whose mathematical work was mainly in hydrodynamics.

Bernoulli number Can be defined through the power series expansion of $x(1 - e^{-x}) = 1 + \frac{1}{2}x + \frac{1}{6}\left(\frac{x^2}{2!}\right) - \frac{1}{30}\left(\frac{x^4}{4!}\right) + \frac{1}{42}\left(\frac{x^6}{6!}\right) + \dots$. The Bernoulli numbers are the coefficients of $\frac{x^n}{n!}$ in this expansion, so when $n = 1$ it is $\frac{1}{2}$ and zero for all other odd n. For $n = 0, 2, 4, 6, \dots$ they are $1, \frac{1}{6}, -\frac{1}{30}, \frac{1}{42} \dots$.

Bernoulli's Theorem The *relative frequency of an event happening is $\frac{m}{n}$ where m is the number of times the event occurs in n trials. If p is the probability of the event then Bernoulli's Theorem, which is a special case of the *weak law of large numbers, says that the relative frequency must tend toward the probability as the sample size increases to infinity.

Theorem
For any $\varepsilon > 0$, $\lim_{n\to\infty} \Pr\left\{\left|\frac{m}{n} - p\right| < \varepsilon\right\} = 1$.

Bernoulli trial One of a sequence of independent experiments, each of which has an outcome considered to be success or failure, all with the same probability p of success. The number of successes in a sequence of Bernoulli trials has a *binomial distribution. The number of experiments required to achieve the first success has a *geometric distribution.

Bertrand's postulate For any integer greater than 3, there is always at least one prime between n and $2n - 2$. For example, for $n = 7$ the number 11 is prime and lies between 7 and 12.

Bessel, Friedrich Wilhelm (1784–1846) German astronomer and mathematician who made a major contribution to mathematics in the development of what are now called Bessel functions. These functions, which satisfy certain differential equations, are probably the most commonly occurring functions in physics and engineering after the elementary functions.

Bessel function Functions that provide solutions to Bessel's equation. These are of the form $J_n(z) = \sum_{r=0}^{\infty} \frac{(-1)^r}{r!\Gamma(n+r+1)}\left(\frac{z}{2}\right)^{n+2r}$, of which the simplest is the Bessel function of the first kind $J_0(z) = \sum_{r=0}^{\infty} \frac{(-1)^r}{(r!)^2}\left(\frac{z}{2}\right)^{2r}$.

Bessel's equation Second-order differential equation in the form $x^2 y'' + (x^2 - n^2)y = 0$. This occurs in physics and engineering applications.

best approximation A point of a subject in a *metric space that is closest to a given point outside the subset.

best unbiased estimator *See* ESTIMATOR.

beta distribution If X is a random variable with *probability density function given by $f(x, \alpha, \beta) = \frac{1}{B(\alpha, \beta)}x^{\alpha-1}(1-x)^{\beta-1}$ $0 \leq x \leq 1$ where $B(\alpha, \beta)$ is a *beta function and λ, v, x all > 0, then we say that X has a beta distribution with parameters α, β. The beta distribution has mean $\frac{\alpha}{\alpha + \beta}$ and variance $\frac{\alpha\beta}{(\alpha + \beta)^2(\alpha + \beta + 1)}$.

beta function The function $B(p, q) = \int_0^1 x^{p-1}(1-x)^{q-1}\, dx = \frac{\Gamma(p)\Gamma(q)}{\Gamma(p+q)}$ where $\Gamma(a)$ is the *gamma function. For integers m, n $B(m, n) = \frac{(n-1)!(m-1)!}{(m+n-1)!}$.

between-groups design An experiment designed to measure the value of a (dependent) variable for distinct groups, under each of the experimental conditions.

between-subjects design = BETWEEN-GROUPS DESIGN.

Bhāskara (1114–85) Eminent Indian mathematician who continued in the tradition of *Brahmagupta, making corrections and filling in many gaps in the earlier work. He solved examples of *Pell's equation and grappled with the problem of division by zero.

bi- Prefix denoting two, for example biannual happens twice a year.

bias A prejudice or a lack of objectivity or randomness resulting in an imbalance that makes it likely that the outcome will tend to be distorted. In statistics this is when a process contains some systematic imbalance so that, on average, the outcome of the process is not equal to the true value. Randomization techniques are employed to try and remove bias that may result from other sample estimator selection methods requiring choices to be made. *See also* SELECTION BIAS, NON-RESPONSE BIAS, RESPONSE BIAS, SELF-SELECTED SAMPLES.

biased estimator The distribution of an estimator is biased if its *expected value does not equal the population mean. Individual values returned by estimators will usually not be exactly equal to the true value, and the determination of bias is a property of the sampling distribution.

biased sample A sample whose composition is not determined only by the population from which it has been taken, but also by some property of the sampling method which has a tendency to cause an over-representation of some parts of the population. It is a property of the sampling method rather than the individual sample.

bi-conditional A logical statement in the form 'p if and only if q', meaning that p and q are *equivalent.

bifurcation The values at which the nature of an attractor changes in a dynamical system.

bifurcation theory The area of mathematics concerned with the sudden changes which occur in systems when one or more parameters are varied, particularly in the study of differential equations.

bijection A *one-to-one onto mapping, that is, a mapping that is both injective and surjective.

bijective mapping A mapping that is injective (that is, *one-to-one) and surjective (that is, *onto).

bilateral shift The linear operator T which moves every element of a sequence by one position, so if $v = T(u)$ then $v_k = u_{k-1}$ for every k.

bilateral symmetry *See* SYMMETRICAL ABOUT A LINE.

bilinear A function of two functions is bilinear if it is linear with respect to each variable independently, e.g. $f(x, y) = 3xy + x$ is bilinear but $g(x, y) = 3xy + x^2$ is not.

billion A thousand million (10^9). This is now standard usage, though in Britain billion used to mean a million million (10^{12}) while the United States billion was a thousand million.

bimodal A frequency distribution is said to be bimodal if it shows two clear peaks.

bin packing problem The scheduling problem of how to pack a number of boxes, with the same cross-section as the bins but of varying heights, using as few bins as possible. This is a situation which is a model for a range of problems. The *first-fit packing algorithm and the *first-fit decreasing packing algorithm are two standard approaches.

binary code A binary code of length n is a set of *binary words of length n which are called the codewords. For example, to send a message conveying the information as to whether a direction is north, south, east or west, a code of length 3 might be chosen, in which, say, 000 means 'north', 110 means 'south', 011 means 'east' and 101 means 'west'.

binary digit The values 0 and 1 are the only digits used in binary arithmetic, and form the basis of computer instructions, the size of which is measured in bits and bytes which are sequences of 8, 16 or 32 bits. The term 'bit' comes from **B**inary dig**IT**.

binary notation The *place value notation in base 2.

binary number A number expressed in *binary notation so the binary number $1101 = 1 \times 2^3 + 1 \times 2^2 + 0 + 1$ or 13 in base 10.

binary operation A binary operation ∘ on a set S is a rule that associates with any elements a and b of S an element denoted by $a \circ b$. If, for all a and b, the element $a \circ b$ also belongs to S, S is said to be closed under the operation ∘. This is often taken to be implied in saying that ∘ is a binary operation on S.

binary relation A formal definition of a binary relation on a set S is as a subset R of the *Cartesian product $S \times S$. Thus, it can be said that, for a given *ordered pair (a, b), either $(a, b) \in R$ or $(a, b) \notin R$. However, it is more natural to denote a relation by a symbol such as ∼ placed between the elements a and b, where ∼ stands for the words 'is related (in some way) to'. Familiar examples are normally written in this way: ' $<$ ' is a binary relation on the set of integers; '$\subseteq$' is a binary relation on the set of subsets of some set E; and 'is perpendicular to' is a binary relation on the set of straight lines in the plane. The letter 'R' may be used in this way, '$a\ R\ b$' meaning that 'a is related to b'. If this notation is used, the set $\{(a, b)|(a, b) \in S \times S$ and $aRb\}$ may be called the *graph of R. For any relation ∼, a corresponding relation $\not\sim$ can be defined that holds whenever $\sim$ does not hold.

binary representation The representation of a number to *base 2. It uses just the two *binary digits 0 and 1, and this is the reason why it is important in computing. For example, $37 = (100101)_2$, since

$$37 = (1 \times 2^5) + (0 \times 2^4) + (0 \times 2)^3 + (1 \times 2^2) + (0 \times 2) + 1.$$

Real numbers, not just integers, can also be written in binary notation, by using binary digits after a 'decimal' point, just as familiar *decimal representations of real numbers are written to base 10. For example, the number $\frac{1}{10}$, which equals 0.1 in decimal notation, has, in binary notation, a recurring representation: $(0.00011001100110011\ldots)_2$.

binary search algorithm Used to search a list to see if it contains a particular item, and to locate it if it does. The list needs to be in alphabetical order, and the binary search algorithm identifies the middle term in the list and compares it with the search item. If it is the same the search is complete, if not it identifies the half of the list in which the item would be, and the other half is discarded. The process is repeated until the item is located, or the list consists of only one item, in which case the item was not contained in the original list.

binary system System using *binary numbers.

binary tree *See* TREE.

binary word A binary word of length n is a string of n binary digits, or bits. For example, there are 8 binary words of length 3, namely, 000, 100, 010, 001, 110, 101, 011 and 111.

binding constraint = ACTIVE CONSTRAINT.

binomial An algebraic expression containing two distinct terms. For example, $3x + 1$ is a binomial expression, but $3x + 2x$ is not as it can be simplified to $5x$.

binomial coefficient The number, denoted by $\binom{n}{r}$, where n is a positive integer and r is an integer such that $0 \leq r \leq n$, defined by the formula

$$\binom{n}{r} = \frac{n(n-1)\ldots(n-r+1)}{1 \times 2 \times \ldots \times r} = \frac{n!}{r!(n-r)!}.$$

Since, by convention, $0! = 1$, we have $\binom{n}{0} = \binom{n}{n} = 1$. These numbers are called binomial coefficients because they occur as coefficients in the

*Binomial Theorem. They are sometimes denoted by ^nC_r; this arose as the notation for the number of ways of selecting r objects out of n (*see* SELECTION), but this number can be shown to be equal to the expression given above for the binomial coefficient. The numbers have the following properties:

(i) $\dbinom{n}{r}$ is an integer (this is not obvious from the definition).

(ii) $\dbinom{n}{n-r} = \dbinom{n}{r}$.

(iii) $\dbinom{n+1}{r} = \dbinom{n}{r-1} + \dbinom{n}{r}$.

(iv) $\dbinom{n}{0} + \dbinom{n}{1} + \dbinom{n}{2} + \ldots + \dbinom{n}{n} = 2^n$.

It is instructive to see the binomial coefficients laid out in the form of *Pascal's triangle.

binomial distribution The discrete probability *distribution for the number of successes when n independent experiments are carried out, each with the same probability p of success. The *probability mass function is given by $\Pr(X=r) = {}^nC_r p^r (1-p)^{n-r}$, for $r = 0, 1, 2, \ldots, n$. This distribution is denoted by $B(n, p)$ and has mean np and variance $np(1-p)$.

binomial experiment An experiment made up of a fixed number of independent Bernoulli trials. The outcomes of the experiment, where X is the *random variable counting the number of successes, follow the *binomial distribution.

binomial series (expansion) The series

$$1 + \frac{\alpha}{1!}x + \frac{\alpha(\alpha-1)}{2!}x^2 + \cdots + \frac{\alpha(\alpha-1)\ldots(\alpha-n+1)}{n!}x^n + \cdots,$$

being the *Maclaurin series for the function $(1+x)^\alpha$. In general, it is valid for $-1 < x < 1$. If α is a non-negative integer, the expansion is a finite series and so is a polynomial, and then it is equal to $(1+x)^\alpha$ for all x.

Binomial Theorem The formulae $(x+y)^2 = x^2 + 2xy + y^2$ and $(x+y)^3 = x^3 + 3x^2 y + 3xy^2 + y^3$ are used in elementary algebra. The Binomial Theorem gives an expansion like this for $(x+y)^n$, where n is any positive integer:

Theorem

For all positive integers n,

$$(x+y)^n = \sum_{r=0}^{n} \binom{n}{r} x^{n-r} y^r$$

$$= x^n + \binom{n}{1} x^{n-1} y + \binom{n}{2} x^{n-2} y^2 + \cdots + \binom{n}{r} x^{n-r} y^r + \cdots + y^n,$$

where $\binom{n}{r} = \dfrac{n!}{r!(n-r)!}$ (*see* BINOMIAL COEFFICIENT).

The following is a special case of the Binomial Theorem. It can also be seen as a special case of the *binomial series when the series is finite:

Theorem

For all positive integers n,

$$(1+x)^n = \sum_{r=0}^{n} \binom{n}{r} x^r$$

$$= 1 + \binom{n}{1} x + \binom{n}{2} x^2 + \cdots + \binom{n}{r} x^r + \cdots + x^n.$$

biometrics Individual characteristics of human beings are often sufficiently distinctive as to allow them to be used for the purposes of identification, often in combination to reduce the likelihood of an error arising where more than one individual possesses an identical combination of biometrics. Fingerprints, and DNA, have been in widespread use for some considerable time, but with the increase in computer power available now, new biometrics such as retinal scanning, facial or voice recognition, even smell are becoming used. A signature written on paper can only be analysed for certain characteristics, but digital signatures written on an electronic pad with a stylus record not only those characteristics but also biometric parameters such as the applied pressure on the stylus and the speed.

biometry The development and application of statistical methods to biological problems.

bi-orthogonal Describing two sequences $\{a_n\}$ and $\{b_n\}$ in a *Hilbert space with an *inner product $\langle \ \rangle$ such that $\langle a_m, b_n \rangle = 1$ when $m = n$ and is 0 otherwise.

bipartite graph A *graph in which the vertices can be divided into two sets V_1 and V_2, so that no two vertices in V_1 are joined and no two vertices in V_2 are joined. The *complete bipartite graph $K_{m,n}$ is the bipartite graph with m vertices in V_1 and n vertices in V_2, with every vertex in V_1 joined to every vertex in V_2.

$K_{2,3}$

Birch and Swinnerton-Dyer conjecture One of the *Millennium Prize problems (see the list of Millennium Prize problems in the appendices). As of 2013, the conjecture has been confirmed only for special cases.

Birkhoff, George David (1884–1944) American mathematician who made important contributions to the study of dynamical systems and in ergodic theory.

bisect To divide into two equal parts.

bisection method A numerical method for finding a root of an equation $f(x) = 0$. If values a and b are found such that $f(a)$ and $f(b)$ have opposite signs and f is *continuous on the interval $[a, b]$, then (by the *intermediate value theorem) the equation has a root in (a, b). The method is to bisect the interval and replace it by either one half or the other, thereby closing in on the root.

Let $c = \frac{1}{2}(a + b)$. Calculate $f(c)$. If $f(c)$ has the same sign as $f(a)$, then take c as a new value for a; if not (so that $f(c)$ has the same sign as $f(b)$, take c as a new value for b. (If it should happen that $f(c) = 0$, then c is a root and the aim of finding a root has been achieved.) Repeat this whole process until the length of the interval $[a, b]$ is less than 2ε, where ε is specified in advance. The midpoint of the interval can then be taken as an approximation to the root, and the error will be less than ε.

(((⊕))) SEE WEB LINKS

• An interactive page which demonstrates the method for finding a root of a cubic equation.

bisector The line that divides an angle into two equal angles.
See also INTERNAL BISECTOR, EXTERNAL BISECTOR and PERPENDICULAR BISECTOR.

bit Abbreviation for *binary digit, i.e. either 0 or 1. It is the basic unit on which electronic calculators and computers function.

bivariate Relating to two random variables. *See* JOINT CUMULATIVE DISTRIBUTION FUNCTION, JOINT DISTRIBUTION, JOINT PROBABILITY DENSITY FUNCTION and JOINT PROBABILITY MASS FUNCTION.

blinding (in design of experiments) In *experimental design, especially in evaluating the effectiveness of drugs or other medical interventions, there is the possibility that beneficial effects accrue because the patient psychologically expects to feel better, which will interfere with identifying any actual benefits of the treatment. In a single-blind experiment the patient does not know which treatment they have been assigned, and in a double-blind experiment neither the patient nor the doctors or others dealing with the patient know which treatment they have been assigned.

Where the nature of the treatments means they cannot be interchanged without the subject being aware, a double-dummy may be used where patients are apparently given both treatments, but one or other will be a *placebo, and the *control group will be given both as placebos.

(⊕) SEE WEB LINKS
• A full discussion of the ethics and principles of blinding in clinical trials (subscription).

block design An experimental design where experimental units with similar characteristics are grouped together as a block and treated as though they are indistinguishable. *Repeated measures designs are an example where the same individual is subject to measurement under different experimental treatments. A balanced block design requires the blocks to be the same size and each experimental treatment to be applied the same number of times. A completely balanced block design further requires each treatment to be applied the same number of times within each block.

block diagonal matrix A square matrix in which the only non-zero elements form square matrices arranged along the main diagonal. A very simple case is shown below where a 1×1 matrix and a 2×2 matrix make up a block diagonal 3×3 matrix.

$$\begin{bmatrix} 3 & -2 & 0 \\ 4 & 2 & 0 \\ 0 & 0 & -1 \end{bmatrix}$$

block multiplication Multiplication of matrices where the entries are matrices rather than single elements.

body An object in the real world idealized in a mathematical model as a *particle, a *rigid body or an *elastic body, for example.

Bohr, Niels Henrik David FRS (1885–1962) Danish mathematician and theoretical physicist whose work on the structure of atoms and on radiation won him the Nobel Prize for physics in 1922. He made further substantial contributions in the new field of *quantum mechanics where his principle of complementarity offered a physical interpretation of *Heisenberg's uncertainty principle. During the Second World War he escaped from occupied Denmark and worked in the UK and the USA on the nuclear bomb.

Bolyai, János (1802–60) Hungarian mathematician who, in a work published in 1832 but probably dating from 1823, announced his discovery of *non-Euclidean geometry. His work was independent of *Lobachevsky. He had persisted with the problem, while serving as an army officer, despite the warnings of his father, an eminent mathematician who had himself spent many years on it without success. Later, János was disheartened by lack of recognition.

Bolzano's Theorem (intermediate value theorem) *See* INTERMEDIATE VALUE THEOREM.

Bombelli, Rafael (1526–72/3) Italian mathematician who, in his book *l'Algebra*, seems to have been the first to go some way in working with complex numbers, when the square roots of negative numbers occurred in the solution of cubic equations.

Bondi, Sir Herman (1919–2005) Austrian mathematician, physicist and astronomer who worked in the UK as an academic and government scientist. Helped develop the steady state theory of the universe.

Boole, George (1815–64) British mathematician who was one of the founding fathers of mathematical logic. His major work, published in 1854, is his *Investigation of the Laws of Thought*. The kind of symbolic argument that he developed led to the study of so-called Boolean algebras, which are of current significance in computing and algebra. His work, together with that of *De Morgan and others, helped to pave the way for the development of modern formal algebra.

Boolean A variable or function which either takes the value true or false.

Boolean algebra A set of elements defined with two *binary operations (Boolean product and Boolean sum) which possess the following properties:

    (i) Both operations are commutative.
  (ii) Both operations have identity elements within the set.
 (iii) Each operation is distributive over the other.
 (iv) Each element of the set has an inverse of both operations.

bootstrapping A relatively new process of generating confidence intervals for statistics which has the advantage of not making assumptions about the underlying distribution of the population. It involves taking repeated random samples from the actual sample of observations which has been taken, and producing confidence intervals for the statistic based on the distribution of the statistic in that process. Because it involves large numbers of repetitions of the resampling process it is dependent on computer power.

bordering The process of adding an extra row and column to a *matrix, especially where the extra entries are 0 except for a 1 where the row and column meet, so the order is increased but the value of the determinant is unchanged.

Borel, Félix Edouard Justin Emile (1871–1956) French mathematician who was one of the first to study real-valued functions, with important results in *set theory and *measure theory.

Borel–Cantelli Lemma A general result in measure theory that is a particularly useful application when applied to a probability measure. If $\{A_n\}$ is an infinite sequence of *measurable sets, where the sum of the measures $\sum_{n=1}^{\infty} \mu(A_n)$ is finite, then the set of points which lie in an infinite number of the sequence of sets must have measure zero. In an infinite sequence of events in a *probability space, the lemma states that if the sum of the probabilities is finite then the probability that infinitely many of the events occur is zero, but if the sum is infinite then the probability that infinitely many of the events occur is one.

Borel measure A *measure defined on the *sigma algebra of a *topological space onto the set of real numbers. If the mapping is onto the interval [0, 1] it is a Borel probability measure.

Borel set A set obtained from repeated applications of unions and intersections of countable collections of closed or open intervals on the real line.

Born, Max FRS (1882–1970) Polish mathematician and theoretical physicist who collaborated with many of the best minds of the day including *Heisenberg, *Pauli, *Jordan, Fermi and *Dirac. As a result of those collaborations he published important work on the foundations of *quantum mechanics, and then his own studies providing a statistical interpretation of wave function, for which he was awarded the Nobel Prize for Physics in 1954.

bottleneck problems A class of constrained optimization problems involving restrictions on network flows.

bound Let S be a non-empty subset of **R**. The real number b is said to be an upper bound for S if b is greater than or equal to every element of S.

If S has an upper bound, then S is bounded above. Moreover, b is a supremum (or least upper bound) of S if b is an upper bound for S and no upper bound for S is less than b; this is written $b = \sup S$. For example, if $S = \{0.9, 0.99, 0.999, \ldots\}$ then $\sup S = 1$. Similarly, the real number c is a LOWER BOUND for S if c is less than or equal to every element of S. If S has a lower bound, then S is bounded below. Moreover, c is an infimum (or greatest lower bound) of S if c is a lower bound for S and no lower bound for S is greater than c; this is written $c = \inf S$. A set is bounded if it is bounded above and below.

It is a non-elementary result about the real numbers that any non-empty set that is bounded above has a supremum, and any non-empty set that is bounded below has an infimum.

boundary (of a surface) The boundary of a subset S of a *topological space X is the set of points which can be approached from within S and from outside S.

boundary condition A complete set of values for all variables at some instant (often the initial condition, when $t = 0$) which provides a particular solution to a *differential equation.

boundary value problem A *differential equation to be satisfied over a region together with a set of *boundary conditions.

bounded function A real *function f, defined on a domain S, is bounded (on S) if there is a number M such that, for all x in S, $|f(x)| < M$. The fact that, if f is *continuous on a closed interval $[a, b]$ then it is bounded on $[a, b]$, is a property for which a rigorous proof is not elementary (*see* CONTINUOUS FUNCTION).

bounded sequence The sequence $a_1, a_2, a_3, \ldots$ is bounded if there is a number M such that, for all n, $|a_n| < M$.

bounded set *See* BOUND.

bounded space A *metric space X with metric μ is said to be bounded if there exists some finite number d (the *diameter) such that the distance between any two points x, y in the space satisfies $\mu(x, y) \leq d$.

Bourbaki, Nicolas The pseudonym used by a group of mathematicians, of changing membership, mostly French, who since 1939 have been publishing volumes intended to build into an encyclopaedic survey of pure mathematics, the *Élements de mathématique*. Its influence is variously described as profound or baleful, but is undoubtedly extensive. Bourbaki has been the standard-bearer for what might be called the Structuralist School of modern mathematics.

Box-Jenkins model A mathematical model first proposed by Box and Jenkins in 1967 for forecasting and prediction in time series analysis based on the variable's past behaviour. It produces very accurate short-term forecasts, but requires a large amount of past data. The method involves determining what type of model is appropriate by analysing the autocorrelations and partial autocorrelations of the stationary data and comparing the patterns with the standard behaviour of the various types. The parameters of the model can then be estimated to provide the best fit to the data.

box plot A diagram constructed from a set of numerical data showing a box that indicates the middle 50% of the ranked observations together with lines, sometimes called 'whiskers', showing the maximum and minimum observations in the sample. The median is marked on the box by a line. Box plots are particularly useful for comparing several samples.

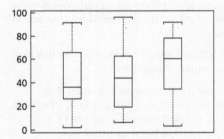

The figure shows box plots for three samples, each of size 20, drawn uniformly from the set of integers from 1 to 100.

brachistochrone Suppose that A and B are points in a vertical plane, where B is lower than A but not vertically below A. Imagine a particle starting from rest at A and travelling along a curve from A to B under the force of gravity. The curve with the property that the particle reaches B as soon as possible is called the brachistochrone (from the Greek for 'shortest time'). The straight line from A to B does not give the shortest time. The required curve is a *cycloid, vertical at A and horizontal at B. The problem was posed in 1696 by Jean * Bernoulli and his solution, together with others by * Newton, * Leibniz and Jacques Bernoulli, was published the following year.

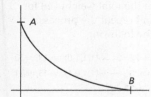

brackets The symbols used to enclose a group of symbols or numbers that are to be taken together. Brackets can be nested so that the whole contents of the inner bracket is treated as a single term in the larger bracket. Brackets can be used to change the order in which operations are to be done. For example, $7 \times 2 + 3$ will be $14 + 3 = 17$. If the sum intended is $7 \times 5 = 35$ then brackets can be put round $2 + 3$ so that the sum is taken before the product, i.e. $7 \times (2 + 3)$.

Brahmagupta (about 598–665) Indian astronomer and mathematician whose text on astronomy includes some notable mathematics for its own sake: the areas of quadrilaterals and the solution of certain *Diophantine equations, for example. Here, the systematic use of negative numbers and zero occurs for probably the first time.

branch A section of a curve with an endpoint at which it meets another branch, and where the differential has a discontinuity.

branch (of a hyperbola) The two separate parts of a *hyperbola are called the two branches.

branch and bound method (of solving the knapsack problem) This procedure constructs a branching method that terminates each branch once the constraint limit has been reached, and by allowing items to be added in an order which is determined at the start. This reduces considerably the number of combinations that have to be tried. The following simple example will be used to illustrate the detailed method.

A knapsack has maximum weight of 15, and the following items are available: A has weight (w) 5 and value (v) 10, written A (5, 10) with B (7, 11), C (4, 8) and D (4, 12).

Method. Place the items in decreasing order of value per unit weight, i.e. D, A, C, B (A and C could be reversed). Then construct a vector (x_1, x_2, x_3, x_4) where $x_i = 0$ if the item is not being taken and $x_i = 1$ if it is. Start with (0, 0, 0, 0). At each stage, if a branch has not been terminated, and there are n zeros at the end of the vector, construct n branches which change exactly one of those zeros to a one and for which the total weight does not exceed the limit so the first stage will have four branches (1, 0, 0, 0),

(0, 1, 0, 0), (0, 0, 1, 0) and (0, 0, 0, 1). Calculate the total weight (w) for each new branch created, and the value (v), and repeat the process.

For this example the process will generate the following

$$(0,0,0,0)$$
- $(1,0,0,0)\ w=4, v=12$
  - $(1,1,0,0)\ w=9, v=22$ —— $(1,1,1,0)\ w=13, v=30$
  - $(1,0,1,0)\ w=8, v=20$ —— $(1,1,0,1)\ w=15, v=31$
  - $(1,0,0,1)\ w=11, v=23$
- $(0,1,0,0)\ w=5, v=10$
  - $(0,1,1,0)\ w=9, v=18$
  - $(0,1,0,1)\ w=12, v=21$
- $(0,0,1,0)\ w=4, v=8$
  - $(0,0,1,1)\ w=11, v=19$
- $(0,0,0,1)\ w=7, v=11$

So the optimal solution is to choose items *B, C, D* with total weight 15 and value 31.

branch point A point at which two or more branches of a curve meet.

break-even point The point at which revenue begins to exceed cost. If one graph is drawn to show total revenue plotted against the number of items made and sold and another graph is drawn with the same axes to show total costs, the two graphs normally intersect at the break-even point. To the left of the break-even point, costs exceed revenue and the company runs at a loss while, to the right, revenue exceeds costs and the company runs at a profit.

bridges of Königsberg In the early 18th century, there were seven bridges in the town of Königsberg (or Kaliningrad). They crossed the different branches of the River Pregel (or Pregolya), as shown in diagrammatic form in the figure. The question was asked whether it was possible, from some starting point, to cross each bridge exactly once and return to the starting point. This prompted Euler to consider the problem in more generality and to publish what can be thought of as the first research paper in *graph theory. The original question asked, essentially, whether the graph shown is an *Eulerian graph. It can be shown that a connected graph is Eulerian if and only if every vertex has even degree, and so the answer is that it is not.

(⊕) SEE WEB LINKS

- An interactive page in which you can construct your own problem like the bridges of Königsberg.

Briggs, Henry (1561–1630) English mathematician who was responsible for the introduction of common logarithms (base 10), at one time called Briggsian logarithms. Following the publication of tables of logarithms by *Napier, Briggs consulted him and proposed an alternative definition using base 10. In 1617, the year of Napier's death, Briggs published his logarithms of the first 1000 numbers and, in 1624, tables including 30 000 logarithms to 14 decimal places.

Brouwer, Luitzen Egbertus Jan (1881–1966) Dutch mathematician, considered by many to be the founder of modern topology because of the significant theorems that he proved, mostly in the period from 1909 to 1913. He is also certainly the founder of the doctrine known as intuitionism, rejecting proofs which make use of the *principle of the excluded middle.

bubble sort algorithm Makes repeated passes through a list of numbers, and on each pass, adjacent numbers are compared and reversed if they are not in the required order. Each pass will put at least one more element into the correct position, like bubbles rising, and the process is complete when a complete set of comparisons is made which required no switching.

(⊕) SEE WEB LINKS
• A demonstration of the bubble sort in action.

Buffon's needle Suppose that a needle of length l is dropped at random onto a set of parallel lines a distance d apart, where $l < d$. The probability that the needle lands crossing one of the lines can be shown to equal $2l/\pi d$. The experiment in which this is repeated many times to estimate the value of π is called Buffon's needle. It was proposed by Georges Louis Leclerc, Comte de Buffon (1707–88).

(⊕) SEE WEB LINKS
• A simulation of Buffon's needle experiment.

Buridan's ass The paradox in logic where an ass is exactly halfway between two buckets of water and dies of thirst because it has no logical basis on which to decide to move to either one. The paradox is often constructed with a pile of hay and a bucket of water or two piles of hay, but the principle of the paradox remains the same. Although the paradox bears the name of a medieval philosopher, this is because it satirizes his philosophical arguments, and the paradox has been around at least since Aristotle.

butterfly effect *See* CHAOS.

byte A block of bits used in computing as the code for a single character.

c Abbreviation for centi- used in symbols in the metric system for one hundredth of a unit, for example cm.

C The set of *complex numbers.

C The Roman *numeral for 100, and the number 12 in *hexadecimal representation.

calculate To work out the value of a mathematical or arithmetical procedure, or the output of an *algorithm.

calculator (calculating machine) A device for performing arithmetical calculations or algebraic manipulations. The earliest example is the *abacus. Hand-held electronic calculators now have the computational powers of the early mainframe computers.

calculus *See* DIFFERENTIAL CALCULUS, INTEGRAL CALCULUS and FUNDAMENTAL THEOREM OF CALCULUS.

calculus of variations A development of calculus concerned with problems in which a function is to be determined such that some related definite integral achieves a maximum or minimum value. Examples of its application are to the *brachistochrone problem and to the problem of finding *geodesics.

cancel To eliminate terms from an expression, usually through one of the four basic arithmetical operations, to produce a simplified form. For example, $x^2 + 3x - 2 = x^2 - 2x + 8$ can be reduced $3x - 2 = -2x + 8$ by cancelling x^2 from both sides, which is effectively subtracting from both sides. $\frac{5xy}{10y} = \frac{x}{2}$, cancelling $5y$ from numerator and denominator, though care needs to be taken when division is used because the expression derived is conditional on not dividing by zero. In this case $\frac{5xy}{10y} = \frac{x}{2}$ provided $y \neq 0$.

cancellation laws Let ∘ be a *binary operation on a set S. The cancellation laws are said to hold if, for all a, b and c in S,

    (i) if $a \circ b = a \circ c$, then $b = c$,
    (ii) if $b \circ a = c \circ a$, then $b = c$.

It can be shown, for example, that in a *group the cancellation laws hold.

canonical Standard format of expression. For example, $y = mx + c$ and $ax + by + c = 0$ are canonical forms for the equation of a straight line in a plane. *See also* QUADRIC.

canonical basis The set of orthogonal unit vectors which form the simplest basis of n-dimensional Euclidean space. In 3-dimensional space the vectors $\mathbf{i}$, $\mathbf{j}$, $\mathbf{k}$ in the directions OX, OY, OZ form the canonical basis.

Cantor, Georg (Ferdinand Ludwig Philipp) (1845–1918)
Mathematician responsible for the establishment of set theory and for profound developments in the notion of the infinite. He was born in St Petersburg, but spent most of his life at the University of Halle in Germany. In 1873, he showed that the set of rational numbers is denumerable. He also showed that the set of real numbers is not. Later he fully developed his theory of infinite sets and so-called transfinite numbers. The latter part of his life was clouded by repeated mental illness.

Cantor's Diagonal Theorem The set of all subsets of any set cannot be put into one-to-one correspondence with the elements of the set. The consequence is that the power set of any set has more elements than the set had, and hence the degrees of infinite numbers which exist.

Cantor set Take the closed interval [0, 1]. Remove the open interval that forms the middle third, that is, the open interval $(\frac{1}{3}, \frac{2}{3})$. From each of the remaining intervals again remove the open interval that forms the middle third. The Cantor set is the set that remains when this process is continued indefinitely. It consists of those real numbers whose *ternary representation $(0.d_1 d_2 d_3 \ldots)_3$ has each ternary digit d_i equal to either 0 or 2.

(⊕) SEE WEB LINKS
• A visual representation of the Cantor set.

Cantor's Intersection Theorem In a *complete metric space, a nested sequence of sets $\{a_n\}$ for which the diameters of a_n decrease to zero contains a unique point of intersection.

Cantor's paradox Suppose there exists an infinite set A containing the largest possible number of elements. *Cantor's Diagonal Theorem shows that its power set has more elements than A had. This proves there is no largest cardinal number.

cap The operation ∩ (*see* INTERSECTION) is read by some as 'cap', this being derived from the shape of the symbol, and in contrast to *cup ∪.

capacity = VOLUME.

capacity (of a cut) The sum of all the maximum flows allowable across all edges intersecting the cut.

capacity (of an edge) The maximum flow possible along an edge.

Cardano, Girolamo (1501–76) Italian physician and mathematician , whose *Ars magna* contained the first published solutions of the general cubic equation and the general quartic equation. Even though these were due to Tartaglia and Cardano's assistant Ludovico Ferrari respectively, Cardano was an outstanding mathematician of the time in the fields of algebra and trigonometry.

cardinality For a finite set A, the cardinality of A, denoted by $n(A)$, is the number of elements in A. The notation $\#(A)$ or $|A|$ is also used. For subsets A, B and C of some universal set E,

(i) $n(A \cup B) = n(A) + n(B) - n(A \cap B)$,
(ii) $n(A \cup B \cup C) = n(A) + n(B) + n(C) - n(A \cap B) - n(A \cap C) - n(B \cap C) + n(A \cap B \cap C)$.

cardinal number A number that gives the number of elements in a set. If two sets can be put in one-to-one correspondence with one another they have the same cardinal number or *cardinality. For finite sets the cardinal numbers are 0, 1, 2, 3, . . . , but infinite sets require new symbols to describe their cardinality. *See* ALEPH and ALEPH-NULL.

cardioid The curve traced out by a point on the circumference of a circle rolling round another circle of the same radius. Its equation, in which a is the radius of each circle, may be taken in polar coordinates as $r = 2a(1 + \cos \theta)(-\pi < \theta \leq \pi)$. In the figure, $OA = 4a$ and $OB = 2a$.

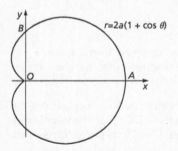

Carmichael number = PSEUDO-PRIME.

Cartesian Relating to the work of René *Descartes especially in representing geometry by an algebraic framework.

Cartesian coordinates *See* COORDINATES (in the plane) and COORDINATES (in 3-dimensional space).

Cartesian distance The distance between two points expressed in Cartesian coordinates, so the distance AB where A is (x_1, y_1, z_1) and B is (x_2, y_2, z_2) is given by $\sqrt{(x_1 - x_2)^2 + (y_1 - y_2)^2 + (z_1 - z_2)^2}$.

Cartesian plane The 2-dimensional space defined by Cartesian coordinates (x, y).

Cartesian product The Cartesian product $A \times B$, of sets A and B, is the set of all *ordered pairs (a, b), where $a \in A$ and $b \in B$. In some cases it may be possible to give a pictorial representation of $A \times B$ by taking two perpendicular axes and displaying the elements of A along one axis and the elements of B along the other axis; the ordered pair (a, b) is represented by the point with, as it were, those coordinates. In particular, if A and B are subsets of **R** and are intervals, this gives a pictorial representation as shown in the figure. Similarly, the Cartesian product $A \times B \times C$ of sets A, B and C can be defined as the set of all ordered triples (a, b, c), where $a \in A$, $b \in B$ and $c \in C$. More generally, for sets $A_1, A_2, \ldots, A_n$, the Cartesian product $A_1 \times A_2 \times \ldots \times A_n$ can be defined in a similar way.

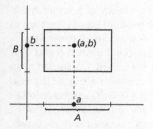

Cartesian space The 3-dimensional space defined by Cartesian coordinates (x, y, z).

cascade charts (Gantt charts) A graphical way to represent the solution following a *critical path analysis. The critical activities are drawn across the top, and non-critical activities are displayed below, showing pairs of bars where the activity is carried out as early and as late as possible without causing a delay.

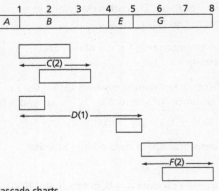

cascade charts

case control studies A study in which 'cases' that have a particular condition are compared with 'controls' that do not, to see how they differ on an explanatory variable of interest. In the same way that *matched pairs are useful in designing an experiment, matching patients by other possible explanatory variables can improve the effectiveness of the study. There are still inherent problems due to the possibility of *confounding variables, but in the world of medicine particularly there are ethical questions over the propriety of random assignation of potentially beneficial or harmful treatments.

casting out nines A way of checking the plausibility of arithmetical calculations. The sum of the digits of a sum or product equals the sum or product of the sum of the digits of the numbers used. In all cases the summing of digits is repeated until a single digit (i.e. between 0 and 9) is obtained. For example, 347 has a sum of digits of 14, then 5. Also, 514 has a sum of digits 10, then 1. Then $347 \times 514 = 178\ 358$ with a sum of digits of 32, reducing to 5 which is 5×1. However, since 178 349 (and many other numbers) also have a sum of digits which reduce to 5, this method will pick up many, but not all, errors and so does not prove a result is correct.

Catalan numbers Numbers in the form $C_n = \dfrac{1}{n+1}\dbinom{2n}{n} = \dfrac{(2n)!}{(n+1)!n!}$ for $n = 0, 1, 2, \ldots$. The first few Catalan numbers are 1, 2, 5, 14, 42 .... These numbers play an important role in problems in *combinatorial analysis.

Catalan's constant $\sum_{n=0}^{\infty}(-1)^n(2n+1)^{-2} = \frac{1}{1^2} - \frac{1}{3^2} + \frac{1}{5^2} - \frac{1}{7^2} + \ldots$. It appears in problems in *combinatorial analysis. It is not even known

whether the number is *irrational, though it is thought it may be
*transcendental.

catastrophe theory A theory providing simple models for the
behaviour of the equilibria of complex potential systems under variation
of the potential. It was developed into a theory of biological
morphogenesis by René *Thom. It obtained its name because the
transition from one stable state to another is often very rapid. It has been
proposed as a model for many situations involving a rapid change of
behaviour such as occurs when an attacking animal turns to flight or when
a stock market crashes.

categorical data = NOMINAL DATA.

categorical proposition A proposition that asserts or denies that
members of one category (the *subject) belong to another (the *predicate).
There are four main forms of categorical proposition: all P are Q; some
P are Q; some P are not Q; no P are Q.

catenary The curve in which an ideal flexible heavy rope or chain of
uniform density hangs between two points. With suitable axes, the
equation of the curve is $y = c \cosh (x/c)$ (see HYPERBOLIC FUNCTION).

(()) SEE WEB LINKS
• An interactive exploration of the shape of a catenary.

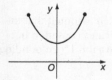

catenoid The surface generated by rotating a *catenary around its axis
of symmetry.

Cauchy, Augustin-Louis (1789–1857) One of the most important
mathematicians of the early 19th century and a dominating figure in
French mathematics. His work ranged over vast areas of mathematics, in
almost 800 papers, but he is chiefly remembered as one of the founders of
rigorous mathematical analysis. Using the definition of *limit as it is now
known, he developed sound definitions of continuity and convergence.
He was also a pioneer in the theory of functions of a complex variable.

Cauchy–Riemann equations For an *analytic function $f(z) = u + iv$ of the complex variable $z = x + iy$ the Cauchy-Riemann equations linking the real and imaginary parts of the function are $\dfrac{\partial u}{\partial x} = \dfrac{\partial v}{\partial y}, \dfrac{\partial u}{\partial y} = -\dfrac{\partial v}{\partial x}$.

Cauchy sequence A sequence $\{a_n\}$ for which the *metric $d(a_n, a_m)$, where $m > n$, satisfies $\lim\limits_{n \to \infty} d(a_n, a_m) = 0$. Cauchy sequences converge when they are defined on the set of real numbers, but do not necessarily converge on the set of rational numbers.

Cauchy's Integral Theorem For a *closed curve C and an *analytic function $f(z)$, $\int_{c} f(z).dz = 0$.

Cauchy's Lemma If G is a *finite *group and p is a *prime number that divides the *order of G, then G must contain an element of order p.

Cauchy–Schwarz inequality for integrals If $f(x)$, $g(x)$ are real functions then $\{\int [f(x)g(x)]dx\}^2 \leq \{\int [f(x)]^2 dx\}\{\int [g(x)]^2 dx\}$ if all these integrals exist.

Cauchy–Schwarz inequality for sums If a_i and b_i are real numbers, $i = 1, 2, \ldots, n$ then $\sum_{i=1}^{n} a_i b_i \leq \sqrt{(\sum_{i=1}^{n} a_i^2)(\sum_{i=1}^{n} b_i^2)}$

Cauchy's Theorem If G is a *finite *group and p is a *prime number which is a divisor of the *order of G, then G contains an element of order p. This implies that there must be a *subgroup of G whose order is p (the *cyclic group generated by the element with order p).

causation Causing or producing an effect. It is often assumed that the existence of *correlation between two variables indicates causation but this is often not the case. They may both be related to a third *confounding variable, which may be to do with time or size.

Cavalieri, Bonaventura (1598–1647) Italian mathematician known for his method of 'indivisibles' for calculating areas and volumes. In his method, an area is thought of as composed of lines and a volume as composed of areas. Here can be seen the beginnings of the ideas of integral calculus. The following theorem of his is typical of the approach: Two solids have the same volume if they have equal altitudes and if sections parallel to and at the same distance from the base have equal areas.

Cayley, Arthur (1821–95) British mathematician who contributed greatly to the resurgence of pure mathematics in Britain in the 19 century. He published over 900 papers on many aspects of geometry and algebra. He conceived and developed the theory of *matrices, and was one of the first to study abstract *groups.

Cayley–Hamilton Theorem The *characteristic polynomial $p(\lambda)$ of an $n \times n$ matrix $\mathbf{A}$ is defined by $p(\lambda) = \det(\mathbf{A} - \lambda \mathbf{I})$. The following result about the characteristic polynomial is called the Cayley-Hamilton Theorem:

Theorem

If the characteristic polynomial $p(\lambda)$ of an $n \times n$ matrix $\mathbf{A}$ is written

$$p(\lambda) = (-1)^n(\lambda^n + b_{n-1}\lambda^{n-1} + \cdots + b_1\lambda + b_0),$$

then $\mathbf{A}^n + b_{n-1}\mathbf{A}^{n-1} + \cdots + b_1\mathbf{A} + b_0\mathbf{I} = \mathbf{O}$.

Cayley Representation Theorem Every group is isomorphic to a group of *permutations.

c.d.f. = CUMULATIVE DISTRIBUTION FUNCTION.

ceiling (least integer function) The smallest integer not less than a given real number. So the ceiling function of 3.2 is 4 and of 5 is 5.

Celsius Symbol °C. The temperature scale, and the unit of measurement of temperature, which takes 0°C as the freezing point of water, and 100°C as the boiling point of water.

centesimal Hundredth or relating to hundredth parts. The *grade (grad on a scientific calculator) is an angular measure which is one-hundredth of a right angle.

centi- Prefix used with *SI units to denote multiplication by 10^{-2}.

centigrade An older term for Celsius, but one which is still in common usage.

centile = PERCENTILE. *See* QUANTILE.

central angle An angle whose vertex is the centre of a given circle.

central conic A *conic with a centre of symmetry, and thus an *ellipse or a *hyperbola. The conic with equation $ax^2 + 2hxy + by^2 + 2gx + 2fy + c = 0$ is central if and only if $ab \neq h^2$.

central difference If $\{(x_i, f_i)\}$, $i = 0, 1, 2, \ldots$ is a given set of function values with $x_{i+1} = x_i + h$, $f_i = f(x_i)$ then the central difference at f_i is defined by $\dfrac{f_{i+1} - f_{i-1}}{2} = \dfrac{f(x_{i+1}) - f(x_{i-1})}{2}$.

central difference approximation The most common numerical approximation to the derivative of a function $f(x)$ is to take the gradient of the chord joining the point and another point where x has been increased by a small amount h. The central difference approximation uses the chord joining the two points whose x values are a small amount h from the value x_0, giving $f'(x_0) \approx \dfrac{f(x_0 + h) - f(x_0 - h)}{2h}$.

central force A force acting on a particle, directed towards or away from a fixed point. The fixed point O may be called the centre of the *field of force. A central force $\mathbf{F}$ on a particle P is given by $\mathbf{F} = f(r)\mathbf{r}$, where $\mathbf{r}$ is the position vector of P and $r = |\mathbf{r}|$.

Examples are the *gravitational force $\mathbf{F} = -\{(GMm/r^3)\mathbf{r}$, and the force $\mathbf{F} = -(k(r-1)/r)\mathbf{r}$ due to an elastic string (*see* HOOKE'S LAW).

Central Limit Theorem A fundamental theorem of statistics which says that the distribution of the mean of a sequence of *random variables tends to a *normal distribution as the number in the sequence increases indefinitely. It has more general forms, but one version is the following:

Theorem

Let $X_1, X_2, X_3, \ldots$ be a sequence of independent, identically distributed random variables with mean μ and finite variance σ^2. Let

$$\overline{X}_n = \frac{X_1 + X_2 + \cdots + X_n}{n}, \qquad Z_n = \frac{(\overline{X}_n - \mu)}{\sigma/\sqrt{n}}.$$

Then, as n increases indefinitely, the distribution of Z_n tends to the standard normal distribution.

It implies, in particular, that if a reasonably large number of samples are selected from any population with finite variance, then the mean of the observations can be assumed to have a normal distribution.

(⊕) SEE WEB LINKS

• An applet that lets you define the population and then take multiple samples of different sizes to explore the behaviour of the Central Limit Theorem.

central moment *See* MOMENT ABOUT THE MEAN.

central quadric A non-*degenerate quadric with a centre of symmetry, and thus an *ellipsoid, a *hyperboloid of one sheet or a *hyperboloid of two sheets.

central vertex (of a graph) A vertex whose *eccentricity is the *radius of the graph. So it is a vertex which is as close to all the other vertices in the graph as is possible.

centre *See* CIRCLE, ELLIPSE and HYPERBOLA.

centre of curvature *See* CURVATURE.

centre of gravity When a system of particles or a rigid body with a total mass m experiences a *uniform gravitational force, the total effect on the system or body as a whole is equivalent to a single force acting at the centre of gravity. This point coincides with the *centre of mass. The centre of gravity moves in the same way as a single particle of mass m would move under the uniform gravitational force.

When the gravitational force is given by the *inverse square law of gravitation, the same is not true. However, when a particle experiences such a gravitational force due to a rigid body having spherical symmetry and total mass M, the total force on the particle is the same as that due to a single particle of mass M at the centre of the spherical body.

centre of mass Suppose that particles $P_1, \ldots, P_n$, with corresponding masses $m_1, \ldots, m_n$, have position vectors $\mathbf{r}_1, \ldots, \mathbf{r}_n$, respectively. The centre of mass (or centroid) is the point with position vector $\mathbf{r}_C$, given by

$$m\mathbf{r}_C = \sum_{i=1}^{n} m_i\mathbf{r}_i, \quad \text{where} \quad m = \sum_{i=1}^{n} m_i,$$

m being the total mass of the particles.

Now consider a rod of length l whose density at a point a distance x from one end is $\rho(x)$. Then the centre of mass is at the point a distance x_C from the end, given by

$$mx_C = \int_0^l \rho(x)x\,dx, \quad \text{where} \quad m = \int_0^l \rho(x)\,dx,$$

m being the mass of the rod.

For a lamina and for a 3-dimensional rigid body, the corresponding definitions involve double and triple integrals. In vector form, the position vector $\mathbf{r}_C$ of the centre of mass is given by

$$m\mathbf{r}_C = \int_v \rho(x)x\,dx, \quad \text{where} \quad m = \int_v \rho(\mathbf{r})\,dV,$$

where $\rho(\mathbf{r})$ is the density at the point with position vector $\mathbf{r}$, V is the region occupied by the body and m is the total mass of the body.

centre of symmetry See SYMMETRICAL ABOUT A POINT.

centrifugal force See FICTITIOUS FORCE.

centripetal force Suppose that a particle P of mass m is moving with constant speed v in a circular path, with centre at the origin O and radius r_0. Let P have polar coordinates (r_0, θ). (See CIRCULAR MOTION.) Then

$v = r_0\dot{\theta}$ and the acceleration of the particle is in the direction towards O and has magnitude $r_0\dot{\theta}^2$. It follows that, if P is acted on by a force $\mathbf{F}$, this force is in the direction towards O and has magnitude mv^2/r_0. It is called the centripetal force.

For example, when an unpowered satellite orbits the Earth at a constant speed, the centripetal force on the satellite is the gravitational force between the satellite and the Earth.

centroid = CENTRE OF MASS. *See also* CENTROID (of a triangle).

centroid (of a triangle) The geometrical definition of the centroid G of a triangle ABC is as the point at which the *medians of the triangle are concurrent. It is, in fact, 'two-thirds of the way down each median', so that, for example, if A' is the midpoint of BC, then $AG = 2GA'$. This is indeed the point at which a triangular *lamina of uniform density has its *centre of mass. It is also the centre of mass of three particles of equal mass situated at the vertices of the triangle. If A, B and C are points in the plane with Cartesian coordinates (x_1, y_1), (x_2, y_2) and (x_3, y_3), then G has coordinates $\left(\frac{1}{3}(x_1 + x_2 + x_3), \frac{1}{3}(y_1 + y_2 + y_3)\right)$.

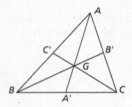

For points A, B and C in 3-dimensional space with Cartesian coordinates (x_1, y_1, z_1), (x_2, y_2, z_2) and (x_3, y_3, z_3), there is no change in the definition of the centroid G and it has coordinates

$$\left(\left(\frac{1}{3}(x_1 + x_2 + x_3), \frac{1}{3}(y_1 + y_2 + y_3), \frac{1}{3}(z_1 + z_2 + z_3)\right)\right).$$

If A, B and C have *position vectors $\mathbf{a}$, $\mathbf{b}$ and $\mathbf{c}$, then G has position vector $\frac{1}{3}(\mathbf{a} + \mathbf{b} + \mathbf{c})$.

Ceva's Theorem The following theorem, due to Giovanni Ceva (1648–1734) and published in 1678:

Theorem

Let L, M and N be points on the sides BC, CA and AB of a triangle (possibly extended). Then AL, BM and CN are concurrent if and only if

$$\frac{BL}{LC} \cdot \frac{CM}{MA} \cdot \frac{AN}{NB} = 1.$$

(Note that BC, for example, is considered to be directed from B to C so that LC, for example, is positive if LC is in the same direction as BC and negative if it is in the opposite direction; *see* MEASURE). *See* MENELAUS' THEOREM.

chain rule The following rule that gives the *derivative of the *composition of two functions: If $h(x) = (f \circ g)(x) = f(g(x))$ for all x, then $h'(x) = f'(g(x)g'(x)$. For example, if $h(x) = (x^2 + 1)^3$, then $h = f \circ g$, where $f(x) = x^3$ and $g(x) = x^2 + 1$. Then $f'(x) = 3x^2$ and $g'(x) = 2x$. So $h'(x) = 3(x^2 + 1)^2 \, 2x = 6x(x^2 + 1)^2$. Another notation can be used: if $y = f(g(x))$, write $y = f(u)$, where $u = g(x)$. Then the chain rule says that $dy/dx = (dy/du)(du/dx)$. As an example of the use of this notation, suppose that $y = (\sin x)^2$. Then $y = u^2$, where $u = \sin x$. So $dy/du = 2u$ and $du/dx = \cos x$, and hence $dy/dx = 2 \sin x \cos x$.

chance nodes (chance vertices) *See* EMV ALGORITHM.

chance variable = RANDOM VARIABLE.

change of base (of logarithms) *See* LOGARITHM.

change of coordinates (in the plane) The simplest changes from one Cartesian coordinate system to another are *translation of axes and *rotation of axes. *See also* POLAR COORDINATES for the change from Cartesian coordinates to polar coordinates, and vice versa.

change of coordinates (in 3-dimensional space) The simplest change from one Cartesian coordinate system to another is *translation of axes. *See also* CYLINDRICAL POLAR COORDINATES and SPHERICAL POLAR COORDINATES for the change from Cartesian coordinates to those coordinate systems, and vice versa.

change of observer (change of reference) Let A and B be observers in two different places. The same event will be observed differently by A and B. Knowing one observation and the relative position of A and B allows the calculation of what the other observer would see in classical mechanics, which requires time and distance to be independent of the choice of observer.

change of variable (in integration) *See* INTEGRATION.

chaos A situation in which a fully deterministic dynamical process can appear to be random and unpredictable due to the sensitive dependence of the process on its starting values and the wide range of qualitatively different behaviours available to the process. This sensitive dependence is often called the butterfly effect. A typical example of a chaotic process is that produced by iterations of the function $f(x) = \frac{1}{2}(x - 1/x)$.

characteristic The integer part of the logarithm in base 10, representing the place value, but not the digits, of the number. So the characteristic of $\log_{10} 270$ will be 2 since $270 = 2.7 \times 10^2$.

characteristic equation *See* CHARACTERISTIC POLYNOMIAL.

characteristic of a field The smallest positive whole number n such that the sum of the multiplicative identity added to itself n times equals the additive identity. If no such n exists, the field is said to have characteristic zero.

characteristic polynomial Let **A** be a square matrix. Then det $(\mathbf{A} - \lambda\mathbf{I})$ is a polynomial in λ and is called the characteristic polynomial of **A**. The equation $\det(\mathbf{A} - \lambda\mathbf{I}) = 0$ is the characteristic equation of **A**, and its roots are the *characteristic values of **A**. *See also* CAYLEY-HAMILTON THEOREM.

characteristic root = CHARACTERISTIC VALUE.

characteristic value Let **A** be a square matrix. The roots of the *characteristic equation $\det(\mathbf{A} - \lambda\mathbf{I}) = 0$ are called the characteristic values of **A**. Then λ is a characteristic value of **A** if and only if there is a non-zero vector **x** such that $\mathbf{Ax} = \lambda\mathbf{x}$. Any vector **x** such that $\mathbf{Ax} = \lambda\mathbf{x}$ is called a characteristic vector corresponding to the characteristic value λ.

characteristic vector *See* CHARACTERISTIC VALUE.

chart Another term for graph, especially statistical graphs.

chart A *neighbourhood of a point in a *manifold together with a *homeomorphism mapping it into some vector space, often Euclidean $\mathfrak{R}^n$.

Chebyshev, Pafnuty Lvovich (1821–94) Russian mathematician and founder of a notable school of mathematicians in St Petersburg. His name is remembered in results in algebra, analysis and probability theory. In number theory, he proved that, for all $n > 3$, there is at least one prime between n and $2n - 2$.

Chebyshev's inequalities Chebyshev proved a number of inequalities relating to the maximum proportion of distributions which could lie beyond a certain point: $\Pr\{|X - \mu_X| > k\sigma\} \leq \dfrac{1}{k^2}$ says that the probability a *random variable X lies more than k standard deviations from its mean is not more than $1/k^2$.

If X is a random variable and $g(X)$ is always ≥ 0 then $\text{Pr}\{g(X) \geq k\} \leq \dfrac{E\{g(X)\}}{k}$ says that for a non-negative function of a random variable, the probability the function takes a value at least k can be no more than the mean of the function divided by k.

While these inequalities are very weak statements in that most distributions do not come close to the limit specified, it is very useful sometimes to be able to identify an upper limit that it is impossible for a probability to exceed.

Chebyshev's Theorem (in number theory) For any positive integer greater than n there is always a prime between n and $2n$.

Chebyshev's Theorem (in statistics) For a random variable, whatever the distribution, with $E(X) = \mu$, $\text{Var}(X) = \sigma^2$ the proportion of values which lie within k standard deviations of the mean will be at least $1 - \frac{1}{k^2}$, i.e. $\text{Pr}\{|X - \mu| < k\sigma\} \geq 1 - \frac{1}{k^2}$.

check digit Suppose that in a *binary code of length n all the words of length n are possible codewords. Such a code is in no way an *error-detecting code. If an additional bit is added to each codeword so that the number of 1's in each new codeword is even, the new code is error-detecting. For example, from the code with codewords 00, 01, 10 and 11, the new code with codewords 000, 011, 101 and 110 would be obtained. The additional bit is called a check digit, and in this case the construction is a parity check (*see* PARITY). More complicated examples of check digits are often used for the purposes of error-detecting.

checksums A simple error detection procedure when transmitting or storing data is to include a checksum character at the end of the message based on some function of the message—for example, the number of 1s in the message. The receiver can recompute the checksum using the same algorithm on the received message. A good checksum algorithm will give a different result with high probability if the message is different in any way. However, it is only useful as protection against accidental corruption and not as a security measure against deliberate tampering.

Chinese postman problem (in graph theory) *See* ROUTE INSPECTION PROBLEM.

chi-squared contingency table test Very similar to the *chi-squared test but testing whether the characteristics used to categorize members of a sample are independent. The expected values in each cell in this case are calculated by $\dfrac{\text{Row Sum} \times \text{Column}}{\text{Grand total}}$. *See also* YATES' CORRECTION.

chi-squared distribution A type of non-negative continuous probability *distribution, normally written as the χ^2-distribution, with one parameter v called the *degrees of freedom. The distribution (whose precise definition will not be given here) is skewed to the right and has the property that the sum of independent random variables each having a χ^2-distribution also has a χ^2-distribution. It is used in the *chi-squared test for measuring goodness of fit, in tests on variance and in testing for independence in contingency tables. It has mean v and variance $2v$. Tables relating to the distribution for different values of v are available.

chi-squared test A test, normally written as the χ^2-test, to determine how well a set of observations fits a particular discrete distribution or some other given null hypothesis (*see* HYPOTHESIS TESTING). The observed frequencies in different groups are denoted by O_i, and the expected frequencies from the statistical model are denoted by E_i. For each i, the value $(O_i - E_i)^2/E_i$ is calculated, and these are summed. The result is compared with a *chi-squared distribution with an appropriate number of degrees of freedom. The number of degrees of freedom depends on the number of groups and the number of parameters being estimated. The test requires that the observations are independent and that the sample size and expected frequencies exceed minimum numbers depending on the number of groups.

chord Let A and B be two points on a curve. The straight line through A and B, or the *line segment AB, is called a chord, the word being used when a distinction is to be made between the chord AB and the arc AB.

chromatic number For a graph or map G, the maximum number of colours needed so that all regions touching one another (meeting at an edge or a vertex) are in a different colour is the chromatic number, denoted by $\chi(G)$. The *Four Colour Theorem proved that $\chi(G) \leq 4$ for all planar graphs.

Chu Shih-chieh (about AD 1300) One of the greatest of Chinese mathematicians, who wrote two major influential texts, the more important being the *Su-yuan yu-chien* ('Precious Mirror of the Four Elements'). Notable are the methods of solving equations by successive approximations and the summation of series using *finite differences. A diagram of *Pascal's triangle, as it has come to be called, known in China from before this time, also appears.

cipher (cypher) An old term for *zero. Now normally reserved for zeros which are not significant digits, but whose purpose is to give the place value. So in 0.004 02, the zeros to the left of the 4 are ciphers, but the zero between the 4 and the 2 is not.

circle The circle with centre C and radius r is the locus of all points in the plane whose distance from C is equal to r. If C has Cartesian coordinates (a, b), this circle has equation $(x - a)^2 + (y - b)^2 = r^2$. An equation of the form $x^2 + y^2 + 2gx + 2fy + c = 0$ represents a circle if $g^2 + f^2 - c > 0$, and is then an equation of the circle with centre $(-g, -f)$ and radius $\sqrt{g^2 + f^2 - c}$.

The area of a circle of radius r equals πr^2, and the length of the circumference equals $2\pi r$.

circle of convergence A circle in the *complex plane with the property that $\sum_{i=1}^{\infty} a_i(z - z_0)^i$ converges for all z within a distance $R > 0$ of z_0 and diverges for all z for which $|z - z_0| > R$. R is the radius of convergence and if the power series converges for the whole of the complex plane then R is infinite. The case where $R = 0$ is trivial since convergence only occurs when $|z - z_0| = 0$. For points on the circumference of the circle the series may either converge or diverge.

circle of curvature *See* CURVATURE.

circle theorems The following is a summary of some of the theorems that are concerned with properties of a circle:

Let A and B be two points on a circle with centre O. If P is any point on the circumference of the circle and on the same side of the chord AB as O,

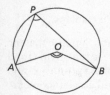 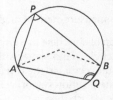

then $\angle AOB = 2\angle APB$. Hence the 'angle at the circumference' $\angle APB$ is independent of the position of P.

If Q is a point on the circumference and lies on the other side of AB from P, then $\angle AQB = 180° - \angle APB$. Hence opposite angles of a cyclic quadrilateral add up to $180°$.

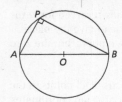

When AB is a diameter, the angle at the circumference is the 'angle in a semicircle' and is a right angle. If T is any point on the tangent at A, then $\angle APB = \angle BAT$.

Suppose now that a circle and a point P are given. Let any line through P meet the circle at points A and B. Then $PA.PB$ is constant; that is, the same

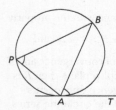

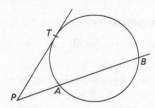

for all such lines. If P lies outside the circle and a line through P touches the circle at the point T, then $PA.PB = PT^2$.

circular argument Uses the essence of the conclusion in arriving at that conclusion.

circular function A term used to describe either of the *trigonometric functions sin and cos. Some authors also apply the term to the trigonometric function tan.

circular measure The measurement of angle size in *radians.

circular motion Motion of a particle in a circular path. Suppose that the path of the particle P is a circle in the plane, with centre at the origin O and radius r_0. Let $\mathbf{i}$ and $\mathbf{j}$ be unit vectors in the directions of the positive x- and y-axes. Let $\mathbf{r}$, $\mathbf{v}$ and $\mathbf{a}$ be the position vector, velocity and acceleration of P. If P has polar coordinates (r_0, θ), then

$$\mathbf{r} = r_0(\mathbf{i} \cos \theta + \mathbf{j} \sin \theta),$$
$$\mathbf{v} = \dot{\mathbf{r}} = r_0(-\dot{\theta}\mathbf{i} \sin \theta + \dot{\theta}\mathbf{j} \cos \theta),$$
$$\mathbf{a} = \ddot{\mathbf{r}} = r_0(-\ddot{\theta}\mathbf{i} \sin \theta - \dot{\theta}^2\mathbf{i} \cos \theta + \ddot{\theta}\mathbf{j} \cos \theta - \dot{\theta}^2\mathbf{j} \sin \theta).$$

Let $\mathbf{e}_r = \mathbf{i} \cos \theta + \mathbf{j} \sin \theta$ and $\mathbf{e}_\theta = -\mathbf{i} \sin \theta + \mathbf{j} \cos \theta$, so that $\mathbf{e}_r$ is a unit vector along OP in the direction of increasing r, and $\mathbf{e}_\theta$ is a unit vector perpendicular to this in the direction of increasing θ. Then the equations above become

$$r = r_0\mathbf{e}_r, \qquad \mathbf{v} = \dot{\mathbf{r}} = r_0\dot{\theta}\mathbf{e}_\theta, \qquad \mathbf{a} = \ddot{\mathbf{r}} = -r_0\dot{\theta}^2\mathbf{e}_r + r_0\ddot{\theta}\mathbf{e}_\theta.$$

If the particle, of mass m, is acted on by a force $\mathbf{F}$, where $\mathbf{F} = F_1\mathbf{e}_r + F_2\mathbf{e}_\theta$, then the equation of motion $m\ddot{\mathbf{r}} = \mathbf{F}$ gives $-mr_0\dot{\theta}^2 = F_1$ and $mr_0\ddot{\theta} = F_2$. If the transverse component F_2 of the force is zero, then $\dot{\theta} = $ constant and the particle has constant speed.

See also ANGULAR VELOCITY and ANGULAR ACCELERATION.

circumcentre The circumcentre of a triangle is the centre of the *circumcircle of the triangle. It is the point O, shown in the figure, at which the perpendicular bisectors of the sides of the triangle are concurrent.

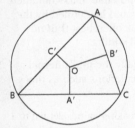

circumcircle The circumcircle of a triangle is the circle that passes through the three vertices. Its centre is at the *circumcentre.

circumference The circumference of a circle is the boundary of the circle or the length of the boundary, that is, the perimeter. The (length of the) circumference of a circle of radius r is $2\pi r$.

circumscribe Construct a geometric figure outside another so they have points in common but the circumscribed figure does not have any part of it inside the other.

circumscribing cylinder *See* ZONE.

cis The notation cis θ is sometimes used for $\cos\theta + i\sin\theta$.

class boundaries The boundaries for the classes when data is grouped. For continuous data, measurements will normally be recorded initially to some degree of accuracy, which will then allow you to determine the actual boundary of the interval. For example if intervals are $1.5 \leq x < 2.0$; $2.0 \leq x < 2.4. \ldots$ with observations recorded correct to the nearest 0.1, then the class boundaries are 1.45, 1.85, 2.45, etc. One common special case is when the data refers to ages because the boundary between recording age 22 and age 23 is actually the 23rd birthday, so age classes of $16 \leq x < 18; 18 \leq x < 23. \ldots$ would have class boundaries at 16, 18, 23, etc.

Classification Theorem for Surfaces *Topology is the study of the properties of shapes and space. This theorem states that any connected closed triangulable surface is homeomorphic to one of the standard surfaces. This is a very important theorem by virtue of the fact that the equivalence classes are huge, so there are very few of them.

🌐 SEE WEB LINKS
• A well-illustrated description of some types of surfaces.

class interval Numerical data may be *grouped by dividing the set of possible values into so-called class intervals and counting the number of observations in each interval. For example, if the possible marks obtained in a test lie between 0 and 99, inclusive, groups could be defined by the intervals 0–19, 20–39, 40–59, 60–79 and 80–99. It is often best (but not essential) to take the class intervals to be of equal widths.

class mark A value within a *class interval, usually the mid-internal value, which is used to represent the interval calculations such as the mean.

clock arithmetic Arithmetic base n where increasing values return to 1 after they reach the clock limit which is the value of n. So in clock arithmetic base 8, $5 + 6$ will come to 3.

clockwise Movement in the same direction of timing as the hands of a clock normally take. In compass terms $N \rightarrow E \rightarrow S \rightarrow W$ is clockwise.

clopen A set in a *topological space is clopen if it is both *open and *closed. This is possible because 'closed' is not defined in topology as the opposite of 'open', but rather as a property of a set whose complement is open, and a set can be open and have a complement which is also open.

closed (in graph theory) A *walk, *trail or *path which finishes at its starting point is closed.

closed (under an operation) *See* OPERATION.

closed curve A continuous plane curve that has no ends or, in other words, that begins and ends at the same point.

closed disc *See* DISC.

closed half-plane *See* HALF-PLANE.

closed half-space *See* HALF-SPACE.

closed interval The closed interval [*a*, *b*] is the set

$$\{x | x \in \mathbf{R} \text{ and } a \le x \le b\}.$$

closed set The complement of an open set in a metric space.

closed surface A surface which is *compact and without a bounding curve. Examples include a sphere and the *Klein bottle.

closure The closure of an *open set *A* is obtained by including in it all *limit points of the set *A*. For example, if *A* is the set $\{x: 1 < x < 2, x \in \mathbf{R}\}$ then the closure of A would include 1 and 2 as the limit points, giving $\{x: 1 \le x \le 2, x \in \mathbf{R}\}$.

cluster A naturally occurring subgroup, usually one which is easily accessible.

cluster point *See* ACCUMULATION POINT.

cluster sampling Where a population is geographically scattered it is reasonable to divide it into regions from which a sample is taken, and then a sample of individuals is taken from those regions only. The result is that the individuals in the final sample appear as clusters in the original population, but the costs of taking the sample are much lower than doing a full random sampling process. There are different strategies possible at both stages of the sampling process.

coaxial Having the same axis.

code *See* BINARY CODE and ERROR-CORRECTING AND ERROR-DETECTING CODE.

coded data Data which has been translated from the form in which it is collected, or the value it took originally according to some specified rule. Responses to questionnaire data may be in the form of tick boxes, or of open responses, but to make analysis easier these will often be given numerical codes. For measurements, data may be coded to standardize data from different sources in order to facilitate comparisons, or transformed, for example by taking logarithms, in order to gain greater insights into the behaviour.

codeword *See* BINARY CODE.

coding theory The area of mathematics concerned with the encryption of messages to ensure security during transmission, and with the recovery of information from corrupted data. With increasing use of the internet and other electronic communications to conduct business, this is one of the developing areas of mathematics research, for example encryption using numbers based on the product of very large primes.

codomain *See* FUNCTION and MAPPING.

coefficient *See* BINOMIAL COEFFICIENT and POLYNOMIAL.

coefficient of determination The proportion of the *variance of the *dependent variable which is explained by the model used to fit the data. For a set of data $\{x_i, y_i\}$, $i = 1, 2, 3, \ldots n$, if $\hat{y}_i$ is the value of y predicted by the model when $x = x_i$, then the unexplained variance after fitting the model is $\dfrac{\sum(y_i - \hat{y}_i)^2}{n}$ and the total variance is $\dfrac{\sum(y_i - \bar{y})^2}{n}$. The explained variance is total variance − unexplained variance. When a linear model is fitted (by the *least squares line of regression), the coefficient of determination $= \dfrac{\text{explained variance}}{\text{total variance}}$ is the square of the *correlation coefficient, i.e. $= r^2$.

( ) SEE WEB LINKS

• An interactive spreadsheet which allows you to change data values and see how r^2 and its component calculations change for a linear model.

coefficient of friction *See* FRICTION.

coefficient of kinetic friction *See* FRICTION.

coefficient of restitution A parameter associated with the behaviour of two bodies during a *collision. Suppose that two billiard balls are travelling in the same straight line and have velocities u_1 and u_2 before the collision, and velocities v_1 and v_2 after the collision. If the coefficient of restitution is e, then

$$v_2 - v_1 = -e(u_2 - u_1).$$

This formula is *Newton's law of restitution. The coefficient of restitution always satisfies $0 \leq e \leq 1$. When $e = 0$, the balls remain in contact after the collision. When $e = 1$, the collision is *elastic: there is no loss of kinetic energy.

It may be convenient to consider a collision as consisting of a deformation phase, during which the shape of each body is deformed, and a restitution phase, during which the shape of each body is completely or partially restored. Newton's law follows from the supposition that, for each body, the *impulse during restitution is e times the impulse during deformation.

coefficient of skewness *See* SKEWNESS.

coefficient of static friction *See* FRICTION.

coefficient of variation A measure of *dispersion equal to the *standard deviation of a sample divided by the mean. The value is a dimensionless quantity, not dependent on the units or scale in which the observations are made, and is often expressed as a percentage.

cofactor Let **A** be the square matrix $[a_{ij}]$. The cofactor, A_{ij}, of the entry a_{ij} is equal to $(-1)^{i+j}$ times the *determinant of the matrix obtained by deleting the i-th row and j-th column of **A**. If **A** is the 3×3 matrix shown, the factor $(-1)^{i+j}$ has the effect of introducing a $+$ or $-$ sign according to the pattern on the right:

$$\mathbf{A} = \begin{bmatrix} a_{11} & a_{12} & a_{13} \\ a_{21} & a_{22} & a_{23} \\ a_{31} & a_{32} & a_{33} \end{bmatrix} \qquad \begin{bmatrix} + & - & + \\ - & + & - \\ + & - & + \end{bmatrix}.$$

So, for example,

$$A_{12} = -\begin{vmatrix} a_{21} & a_{23} \\ a_{31} & a_{33} \end{vmatrix}, \qquad A_{31} = +\begin{vmatrix} a_{12} & a_{13} \\ a_{22} & a_{23} \end{vmatrix}.$$

for a 2×2 matrix, the pattern is:

$$\begin{bmatrix} a & b \\ c & d \end{bmatrix} \qquad \begin{bmatrix} + & - \\ - & + \end{bmatrix}.$$

So, the cofactor of a equals d, the cofactor of b equals $-c$, and so on. The following properties hold, for an $n \times n$ matrix **A**:

 (i) The expression $a_{i1}A_{i1} + a_{i2} + \cdots + a_{in}A_{in}$ has the same value for any i, and is the definition of det **A**, the determinant of **A**. This particular expression is the evaluation of det **A** by the i-th row.
 (ii) On the other hand, if $i \neq j$, $a_{i1}A_{j1} + a_{i2}A_{j1} + \cdots + a_{in}A_{in} = 0$.

Results for columns, corresponding to the results (i) and (ii) for rows, also hold.

coincident Occupying the same space or time. In particular, where geometrical shapes or functions have all points in common.

collinear Any number of points are said to be collinear if there is a straight line passing through all of them.

collision A collision occurs when two bodies move towards one another and contact takes place between the bodies. The subsequent motion is often difficult to predict. There is normally a loss of kinetic energy. A simple example is a collision between two billiard balls. The behaviour of the two balls after the collision depends upon the *coefficient of restitution.

colourable A graph or map is said to be colourable if its *chromatic number is finite. The *Four Colour Theorem shows that all planar maps are colourable.

column equivalence Let **A** and **B** be two matrices of the same size. If **A** can be transformed to **B** by carrying out a sequence of elementary matrix operations on its columns, then **A** and **B** have column equivalence.

column matrix A *matrix with exactly one column; that is, an $m \times 1$ matrix of the form

$$\begin{bmatrix} a_1 \\ a_2 \\ \vdots \\ a_m \end{bmatrix}.$$

Given an $m \times n$ matrix, it may be useful to treat its columns as individual column matrices.

column operation *See* ELEMENTARY COLUMN OPERATION.

column rank *See* RANK.

column space The *vector space defined by the columns of a matrix as *basis vectors.

column stochastic matrix *See* STOCHASTIC MATRIX.

column vector $=$ COLUMN MATRIX.

combination $=$ SELECTION.

combinatorial analysis (combinatorics) The area of mathematics concerned with counting strategies to calculate the ways in which objects can be arranged to satisfy given conditions.

combinatorial circuits Circuits and switching arrangements in electronic circuits use logic gates to control the flow of electrical pulses. Over the years, programmers have developed increasingly sophisticated methods of producing shorter *equivalent circuits. The basic logic gates represent the not, and, or connectors, with the *nand gate also used commonly to represent the combination *not and.

common denominator An integer that is exactly divisible by all the denominators of a group of fractions. Used in adding or subtracting where the first step is to express each fraction as an *equivalent fraction with a common denominator. For example, $\frac{1}{3} + \frac{7}{8} - \frac{5}{6}$ has a common denominator of 24, and the sum can be rewritten as $\frac{8}{24} + \frac{21}{24} - \frac{20}{24} = \frac{9}{24} = \frac{3}{8}$

common difference *See* ARITHMETIC SEQUENCE.

common factor (common divisor) A number or algebraic expression which is a factor of each of a group of numbers or expressions.

common fraction A term used to mean *vulgar fraction or *simple fraction. Some authors use it to mean proper fraction.

common logarithm *See* LOGARITHM.

common multiple A number or algebraic expression which is an exact multiple of each of a group of numbers or expressions. For example, 15 and 30 are common multiples of 3 and 5 and $(2x+1)(x-2)$ is a common multiple of $2x+1$ and $x-2$.

common perpendicular Let l_1 and l_2 be two straight lines in space that do not intersect and are not parallel. The common perpendicular of l_1 and l_2 is the straight line that meets both lines and is perpendicular to both.

common ratio *See* GEOMETRIC SEQUENCE.

common tangent A line which is a tangent to two or more curves.

commutative The *binary operation ∘ on a set S is commutative if, for all a and b in S, $a \circ b = b \circ a$.

commutative ring *See* RING.

commutator The element $[x, y] = x^{-1} y^{-1} xy$ for x, y in the group. It has the property that $[x, y]$ and $[y, x]$ must be commutative, because one is the inverse of the other, and therefore both products will be equal to the identity.

commute Let ∘ be a *binary operation on a set S. The elements a and b of S commute (under the operation ∘) if $a \circ b = b \circ a$. For example, multiplication on the set of all real 2×2 matrices is not commutative, but if A and B are diagonal matrices then A and B commute.

compact A *space in which any collection of *open sets whose *union is the whole space has a finite number of open sets whose union is also the whole space. *See also* FINITE INTERSECTION PROPERTY.

competitive equilibrium An equilibrium state in game theory and mathematical economics achieved when agents or players act in their own interests without cooperation. Introduced by John *Nash.

complement, complementation Let A be a subset of some *universal set E. Then the complement of A is the *difference set $E \backslash A$ (or $E - A$). It may be denoted by A' (or $\bar{A}$) when the universal set is understood or has previously been specified. Complementation (the operation of taking the complement) is a unary operation on the set of subsets of a universal set E. The following properties hold:

   (i) $E' = \emptyset$ and $\emptyset' = E$.

   (ii) For all A, $(A')' = A$.

   (iii) For all A, $A \cap A' = \emptyset$ and $A \cup A' = E$

See also RELATIVE COMPLEMENT.

complement (for angles) *See* COMPLEMENTARY ANGLES.

complementary (for probability) Events which are both *exhaustive and *mutually exclusive. So for A and B to be complementary, $P(A \cup B) = 1$ and $P(A \cap B) = 0$. It then follows that $P(B) = 1 - P(A)$.

complementary angles Two angles that add up to a right angle. Each angle is the complement of the other.

complementary function *See* LINEAR DIFFERENTIAL EQUATION WITH CONSTANT COEFFICIENTS.

complete (in logic) If a set of axioms describing a system is complete, then any true statement within that system can be deduced from the axioms.

complete graph A *simple graph in which every vertex is joined to every other. The complete graph with n vertices, denoted by K_n, is *regular of degree $n - 1$ and has $\frac{1}{2}n(n - 1)$ edges. *See also* BIPARTITE GRAPH.

K_3 K_4 K_5

completely balanced block design *See* BLOCK DESIGN.

completely normal space A *normal space X in which every subspace of X is itself a normal space.

completely regular space A *topological space X in which for every non-empty *closed subset S of X and a point p of X which is not in S there is a continuous function $f : X \rightarrow [0, 1]$ for which $f(p) = 0$ and $f(S) = 1$.

complete matching A *matching in a *bipartite graph in which all vertices are used. This requires each set to have the same number of vertices, n, and the complete matching will have n edges.

complete metric space A *metric space in which every *Cauchy sequence is *convergent. For example, the real numbers, with the usual metric.

complete quadrangle The configuration in the plane consisting of four points, no three of which are collinear, together with the six lines joining them in pairs. An example is shown in the figure.

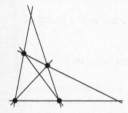

complete quadrilateral The configuration in the plane consisting of four straight lines, no three of which are concurrent, together with the six points in which they intersect in pairs. An example is shown in the figure.

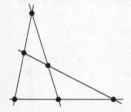

complete set of residues (modulo n) A set of n integers, one from each of the n *residue classes modulo n. Thus $\{0, 1, 2, 3\}$ is a complete set of residues modulo 4; so too are $\{1, 2, 3, 4\}$ and $\{-1, 0, 1, 2\}$.

complete solution The solution of differential equation containing a particular integral and the complementary function. The generates a family of solutions which contains all possible solutions.

complete symmetric group *See* SYMMETRIC GROUP.

completing the square Consider a numerical example: the *quadratic equation $2x^2 + 5x + 1 = 0$ can be solved by first writing it as

$$x^2 + \frac{5}{2}x = -\frac{1}{2}, \quad \text{and then} \quad \left(x + \frac{5}{4}\right)^2 = -\frac{1}{2} + \frac{25}{16} = \frac{17}{16}.$$

This step is known as completing the square: the left-hand side is made into an exact square by adding a suitable constant to both sides. The solution of the quadratic equation can then be accomplished as follows:

$$x + \frac{5}{4} = \pm\frac{\sqrt{17}}{4}, \quad \text{and so} \quad x = \frac{-5 \pm \sqrt{17}}{4}.$$

By proceeding in the same way with $ax^2 + bx + c = 0$, the standard formula for the solution of a quadratic equation can be derived.

complex analysis The area of mathematics relating to the study of complex functions.

complex conjugate = CONJUGATE (of a complex number).

complex function A function involving complex variables as either input or output, but usually both. So if $z = x + yi$, $f(z) = z^2 = x^2 - y^2 + 2xyi$ is a complex function.

complex number There is no real number x such that $x^2 + 1 = 0$. The introduction of a 'new' number i such that $i^2 = -1$ gives rise to further numbers of the form $a + bi$. A number of the form $a + bi$, where a and b are real, is a complex number. Since one may take $b = 0$, this includes all the real numbers. The set of all complex numbers is usually denoted by **C**. (The use of j in place of i is quite common.) It is assumed that two such numbers may be added and multiplied using the familiar rules of algebra, with i^2 replaced by -1 whenever it occurs. So,

$$(a + bi) + (c + di) = (a + c) + (b + d)i,$$
$$(a + bi)(c + di) = (ac - bd) + (ad + bc)i.$$

Thus the set **C** of complex numbers is closed under addition and multiplication, and the elements of this enlarged number system satisfy the laws commonly expected of numbers.

The complex number system can be put on a more rigorous basis as follows. Consider the set **R** × **R** of all ordered pairs (a, b) of real numbers (*see* CARTESIAN PRODUCT). Guided by the discussion above, addition and multiplication are defined on $\mathbf{R} \times \mathbf{R}$ by

$$(a, b) + (c, d) = (a + c, b + d),$$
$$(a, b)(c, d) = (ac - bd, ad + bc).$$

It can be verified that addition and multiplication defined in this way are associative and commutative, that the distributive law holds, that there is a zero element and an identity element, that every element has a negative and every non-zero element has an inverse (*see* INVERSE OF A COMPLEX NUMBER). This shows that $\mathbf{R} \times \mathbf{R}$ with this addition and multiplication is a *field whose elements, according to this approach, are called complex numbers. The elements of the form $(a, 0)$ can be seen to behave exactly like the corresponding real numbers a. Moreover, if the element $(0, 1)$ is denoted by i, it is reasonable to write $i^2 = -1$, since $(0, 1)^2 = -(1, 0)$. After providing this rigorous foundation, it is normal to write $a + bi$ instead of (a, b). *See also* ARGUMENT, MODULUS OF A COMPLEX NUMBER and POLAR FORM OF A COMPLEX NUMBER.

complex plane Let points in the plane be given coordinates (x, y) with respect to a Cartesian coordinate system. The plane is called the complex plane when the point (x, y) is taken to represent the *complex number $x + yi$.

component (of a compound statement) *See* COMPOUND STATEMENT.

component (of a graph) A *graph may be 'in several pieces' and these are called its components: two vertices are in the same component if and only if there is a *path from one to the other. A more precise definition can be given by defining an *equivalence relation on the set of vertices with u equivalent to v if there is a path from u to v. Then the components are the corresponding *equivalence classes.

component (of a vector) In a Cartesian coordinate system in 3-dimensional space, let $\mathbf{i}$, $\mathbf{j}$ and $\mathbf{k}$ be unit vectors along the three coordinate axes. Given a vector $\mathbf{p}$, there are unique real numbers x, y and z such that $\mathbf{p} = x\mathbf{i} + y\mathbf{j} + z\mathbf{k}$. Then x, y and z are the components of $\mathbf{p}$ (with respect to the vectors $\mathbf{i}, \mathbf{j}, \mathbf{k}$). These can be determined by using the *scalar product: $x = \mathbf{p} \cdot \mathbf{i}$, $y = \mathbf{p} \cdot \mathbf{j}$ and $z = \mathbf{p} \cdot \mathbf{k}$. (*See also* DIRECTION COSINES.)

More generally, if $\mathbf{u}$, $\mathbf{v}$ and $\mathbf{w}$ are any 3 non-coplanar vectors, then any vector $\mathbf{p}$ in 3-dimensional space can be expressed uniquely as $\mathbf{p} = x\mathbf{u} + y\mathbf{v} + z\mathbf{w}$, and x, y and z are called the components of $\mathbf{p}$ with respect to the basis $\mathbf{u}, \mathbf{v}, \mathbf{w}$. In this case, however, the components cannot be found so simply by using the scalar product.

composite A positive integer is composite if it is neither *prime, nor equal to 1; that is, if it can be written as a product hk, where the integers h and k are both greater than 1.

composition Let $f: S \to T$ and $g: T \to U$ be *mappings. With each s in S is associated the element $f(s)$ of T, and hence the element $g(f(s))$ of U. This rule gives a mapping from S to U, which is denoted by $g \circ f$ (read as 'g circle f') and is the composition of f and g. Note that f operates first, then g. Thus $g \circ f: S \to U$ is defined by $(g \circ f)(s) = g(f(s))$, and exists if and only if the domain of g equals the codomain of f. For example, suppose that $f: \mathbf{R} \to \mathbf{R}$ and $g: \mathbf{R} \to \mathbf{R}$ are defined by $f(x) = 1 - x$ and $g(x) = x/(x^2 + 1)$. Then $f \circ g: \mathbf{R} \to \mathbf{R}$ and $g \circ f: \mathbf{R} \to \mathbf{R}$ both exist, and

$$(f \circ g)(x) = 1 - \frac{x}{x^2 + 1}, \qquad (g \circ f)(x) = \frac{1 - x}{(1 - x)^2 + 1}.$$

The term 'composition' may be used for the operation $\circ$ as well as for the resulting function. The composition of mappings is associative: if $f: S \to T$, $g: T \to U$ and $h: U \to V$ are mappings,

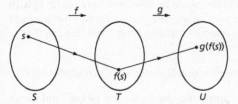

then $h \circ (g \circ f) = (h \circ g) \circ f$. This means that the mappings $h \circ (g \circ f)$ and $(h \circ g) \circ f$ have the same domain S and the same codomain V and, for all s in S, $(h \circ (g \circ f)(s) = (h \circ g) \circ f)(s)$.

compound angle formulae (in hyperbolic functions) *See* HYPERBOLIC FUNCTIONS.

compound angle formulae (in trigonometry) The trigonometric functions of the sum or difference of two angles can be expressed in terms of the functions of the individual angles.

$$\sin (A \pm B) = \sin A \cos B \pm \cos A \sin B,$$
$$\cos (A \pm B) = \cos A \cos B \pm \sin A \sin B,$$
$$\tan(A \pm B) = \frac{\tan A \pm \tan B}{1 \mp \tan A \tan B}.$$

These can be manipulated to express sums and differences of sine or cosines in alternative forms:

$$\sin A + \sin B = 2 \sin \frac{A+B}{2} \cos \frac{A-B}{2},$$

$$\sin A - \sin B = 2 \sin \frac{A-B}{2} \cos \frac{A+B}{2},$$

$$\cos A + \cos B = 2 \cos \frac{A+B}{2} \cos \frac{A-B}{2},$$

$$\cos A - \cos B = 2 \sin \frac{A+B}{2} \sin \frac{A-B}{2}.$$

compound fraction A fraction in which the numerator or denominator, or both, contain fractions. For example, $\frac{3}{4} / (1 + \frac{2}{3})$ is a compound fraction. *See* SIMPLE FRACTION.

compound interest Suppose that a sum of money P is invested, attracting interest at i per cent a year. After one year, the amount becomes $P + (i/100)P$. This equals $P(1 + i/100)$, so that adding on i per cent is equivalent to multiplying by $1 + i/100$. When interest is compounded annually, the new amount is used to calculate the interest due in the second year and so, after 2 years, the amount becomes $P(1 + i/100)^2$. After n years, the amount becomes

$$P\left(1 + \frac{i}{100}\right)^n.$$

This is the formula for compound interest. When points are plotted on graph paper to show how the amount increases, they lie on a curve that illustrates *exponential growth. This is in contrast to the straight line obtained in the case of *simple interest.

(())) **SEE WEB LINKS**
• A compound interest calculator.

compound number A quantity expressed in a mixture of units. For example, 3 metres and 25 centimetres or 3 hours and 10 minutes.

compound pendulum A pendulum consisting of a *rigid body that is free to swing about a horizontal axis. Suppose that the centre of gravity of the rigid body is a distance d from the axis, m is the mass of the body, and I is the *moment of inertia of the body about the axis. The equation of motion can be shown to give $I\ddot{\theta} = -mgd \sin \theta$. As with the *simple pendulum, this gives approximately simple harmonic motion if θ is small for all time.

compound statement A statement formed from simple statements by the use of words such as 'and', 'or', 'not', 'implies' or their corresponding symbols. The simple statements involved are the components of the compound statement. For example, $(p \wedge q) \vee (\neg r)$ is a compound statement built up from the components p, q and r.

compression A force which acts to compress a body, such as a rod in a framework.

computable Can be calculated by an algorithm. For example, the solutions of a quadratic equation are computable but the score showing on the roll of a die is not although a roll can be simulated electronically.

computation A calculation.

compute = CALCULATE.

computer Usually a digital electronic device that carries out logical and arithmetical calculations according to a very precise set of instructions contained within a program, known as software. It typically comprises a number of different components, though in laptop and hand-held computers some of these may be contained within a single unit. These components are known as computer hardware and will include input devices such as keyboard, mouse, microphone for speech recognition software, tablet and pen; a central processing unit (cpu) which carries out the actual calculations; memory storage devices such as the working memory, the computer's hard disk drive and external memory devices like floppy disk drives or CD writers; output devices such as a monitor or visual display unit (vdu) and printer and communication devices such as modems which allow computers to connect to one another or to the internet and the World Wide Web.

computer algebra Symbolic algebra manipulations carried out by a computer program. These programs are capable of solving equations, simplifying complex expressions and carrying out calculus operations. They have become sophisticated enough that they can now be found on some hand-held computers.

concave polygon Has an interior angle of more than 180°. Has the property that there are points inside the polygon which cannot be joined by a straight line without going outside the polygon.

concave up and down *See* CONCAVITY.

concavity At a point of a graph $y = f(x)$, it may be possible to specify the concavity by describing the curve as either concave up or concave down at that point, as follows,

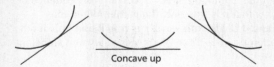

Concave up

If the second derivative $f''(x)$ exists and is positive throughout some neighbourhood of a point a, then $f'(x)$ is strictly increasing in that neighbourhood, and the curve is said to be concave up at a. At that point, the graph $y = f(x)$ and its tangent look like one of the cases shown in the first figure. If $f''(a) > 0$ and f'' is continuous at a, it follows that $y = f(x)$ is concave up at a. Consequently, if $f'(a) = 0$ and $f''(a) > 0$, the function f has a *local minimum at a. Similarly, if $f''(x)$ exists and is negative throughout some neighbourhood of a, or if $f''(a) < 0$ and f'' is continuous at a, then the graph $y = f(x)$ is concave down at a and looks like one of the cases shown in the second figure. If $f'(a) = 0$ and $f''(a) < 0$, the function f has a *local maximum at a.

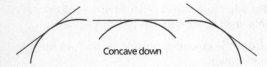

Concave down

concentric Having the same centre.

conclusion (in logic) A statement derived from starting premises by proof.

conclusion (in statistics) A decision concerning whether to accept or reject a statistical hypothesis on the basis of the evidence available.

concrete number A number which counts a specific group of objects, for example four pens. It is the first stage in the appreciation of number as an abstract concept.

concurrent Any number of lines are said to be concurrent if there is a point through which they all pass.

concyclic A number of points are concyclic if there is a circle that passes through all of them.

condition, necessary and sufficient The *implication $q \Rightarrow p$ can be read as 'if q, then p'. When this is true, it may be said that q is a sufficient condition for p; that is, the truth of the 'condition' q is sufficient to ensure the truth of p. This means that p is true if q is true. On the other hand, when the implication $p \Rightarrow q$ holds, then q is a *necessary condition for p; that is, the truth of the 'condition' q is a necessary consequence of the truth of p. This means that p is true only if q is true. When the implication between p and q holds both ways, p is true if and only if q is true, which may be written $p \Longleftrightarrow q$. Then q is a *necessary and sufficient condition for p.

conditional A statement that something [the *consequent clause] will be true provided that something else [the *antecedent clause] is true. For example 'if n is divisible by 2, n is even' is a conditional statement, in which the antecedent is 'n is divisible by 2' and the consequent is 'n is even'.

conditional distribution The probability distribution of the random variable X consisting of the conditional probabilities $\Pr(X_i \mid B)$ for each outcome X_i.

conditional equation Is valid only for certain values of a variable. For example, if a ball is thrown vertically upwards from ground level at 14.7 ms^{-1} then its height might be given by $h = 14.7\,t - 4.9\,t^2$, but only for values between 0 and 3. For other values of t the height of the ball does not correspond to the value obtained by substitution into this equation.

conditional expectation The expected value of the random variable X using the conditional distribution.

conditional probability For two events A and B, the probability that A occurs, given that B has occurred, is denoted by $\Pr(A \mid B)$, read as 'the probability of A given B'. This is called a conditional probability. Provided that $\Pr(B)$ is not zero, $\Pr(A \mid B) = \Pr(A \cap B)/\Pr(B)$. This result is often useful in the following form: $\Pr(A \cap B) = \Pr(B)\,\Pr(A \mid B)$. If A and B are *independent events, $\Pr(A \mid B) = \Pr(A)$, and this gives the product law for independent events: $\Pr(A \cap B) = \Pr(A)\,\Pr(B)$. *See also* FALSE POSITIVE.

cone In elementary work, a cone usually consists of a circle as base, a vertex lying directly above the centre of the circle, and the curved surface formed by the line segments joining the vertex to the points of the circle. The distance from the vertex to the centre of the base is the height, and the length of any of the line segments is the slant height. For a cone with base of radius r, height h and slant height l, the volume equals $\frac{1}{3}\pi r^2 h$ and the area of the curved surface equals $\pi\,rl$.

In more advanced work, a cone is the surface consisting of the points of the lines, called generators, drawn through a fixed point V, the vertex, and the points of a fixed curve, the generators being extended indefinitely in both directions. Then a right-circular cone is a cone in which the fixed curve is a circle and the vertex V lies on the line through the centre of the circle and perpendicular to the plane of the circle. The axis of a right-circular cone is the line through V and the centre of the circle, and is perpendicular to the plane of the circle. All the generators make the same angle with the axis; this is the semi-vertical angle of the cone. The right-circular cone with vertex at the origin, the z-axis as its axis, and semi-vertical angle α, has equation $x^2 + y^2 = z^2 \tan^2 \alpha$. *See also* QUADRIC CONE.

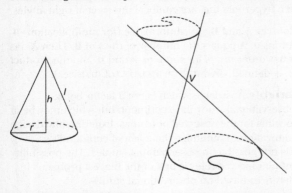

confidence interval An interval, calculated from a sample, which contains the value of a certain population *parameter with a specified probability. The end-points of the interval are the confidence limits. The specified probability is called the confidence level. An arbitrary but commonly used confidence level is 95%, which means that there is a one-in-twenty chance that the interval does not contain the true value of the parameter. For example, if $\bar{x}$ is the mean of a sample of n observations taken from a population with a normal distribution with a known standard deviation σ, then

$$\left[\bar{x} - \frac{1.96\sigma}{\sqrt{n}},\ \bar{x} + \frac{1.96\sigma}{\sqrt{n}} \right]$$

is a 95% confidence interval for the population mean μ.

confidence level *See* CONFIDENCE INTERVAL.

confidentiality (of data) It is easier to try to ensure confidentiality of data collected on sensitive topics than it is to ensure *anonymity, but surveys on sensitive issues are still unreliable because it is very difficult to be confident that the responses are truthful.

configuration A particular geometrical arrangement of points, lines, curves and, in three dimensions, planes and surfaces.

confirm (of a statistical experiment) The evidence from a statistical experiment may support belief in a hypothesis, in which case the term confirm is sometimes used, though it needs to be used carefully and in particular it needs to avoid giving the impression that the hypothesis has been proven beyond doubt.

confocal conics Two *central conics are confocal if they have the same foci. An ellipse and a hyperbola that are confocal intersect at right angles.

conformable Matrices **A** and **B** are conformable (for multiplication) if the number of columns of **A** equals the number of rows of **B**. Then **A** has order $m \times n$ and **B** has order $n \times p$, for some m, n and p, and the product **AB**, of order $m \times p$, is defined. *See* MULTIPLICATION (of matrices).

confounding variable A variable which is not a factor being considered in an observational study or experiment, but which may be at least partially responsible for the observed outcomes. Experimental design methods use randomization to minimize the effect of confounding variables, but that is not possible in observational studies. The possibility of one or more confounding variables is one of the biggest problems in trying to makes inferences based on observational studies.

congruence (modulo n) For each positive integer n, the relation of congruence between integers is defined as follows: a is congruent to b modulo n if $a - b$ is a multiple of n. This is written $a \equiv b$ (mod n). The integer n is the modulus of the congruence. Then $a \equiv b$ (mod n) if and only if a and b have the same remainder upon division by n. For example, 19 is congruent to 7 modulo 3. The following properties hold, if $a \equiv b$ (mod n) and $c \equiv d$ (mod n):

   (i) $a + c \equiv b + d$ (mod n),
   (ii) $a - c \equiv b - d$ (mod n),
   (iii) $ac \equiv bd$ (mod n).

It can be shown that congruence modulo n is an *equivalence relation and so defines a *partition of the set of integers, where two integers are in the same class if and only if they are congruent modulo n. These classes are the *residue (or congruence) classes modulo n.

congruence class = RESIDUE CLASS.

congruence equation The following are examples of congruence equations:

   (i) $x + 5 \equiv 3$ (mod 7); this has the solution $x \equiv 5$ (mod 7).

(ii) $2x \equiv 5 \pmod 4$; this has no solutions.

(iii) $x^2 \equiv 1 \pmod 8$; this has solutions $x \equiv 1,3,5$ or $7 \pmod 8$.

(iv) $x^2 + 2x + 3 \equiv 0 \pmod 6$; this has solutions $x \equiv 1$ or $3 \pmod 6$.

In seeking solutions to a congruence equation, it is necessary only to consider a *complete set of residues and find solutions in this set. The examples (i) and (ii) above are linear congruence equations. The linear congruence equation $ax \equiv b \pmod n$ has a solution if and only if (a, n) divides b, where (a, n) is the *greatest common divisor of a and n.

congruent (modulo n) *See* CONGRUENCE (modulo n).

congruent figures Two geometrical figures are congruent if they are identical in shape and size. This includes the case when one of them is a mirror image of the other, and so the three triangles shown here are all congruent to each other.

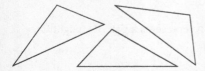

conic, conic section A curve that can be obtained as the plane section of a cone. The figure shows how an ellipse, parabola and hyperbola can be obtained.

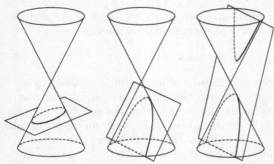

But there are other more convenient characterizations, one of which is by means of the focus and directrix property. Let F be a fixed point (the focus) and l a fixed line (the directrix), not through F, and let e be a fixed positive number (the *eccentricity). Then the locus of all points P such that the distance from P to F equals e times the distance from P to l is a curve, and any such curve is a conic. The conic is called an *ellipse if $e < 1$, a *parabola if $e = 1$ and a *hyperbola if $e > 1$. Note that a circle is certainly a conic (it is a special case of an ellipse); but it can only be obtained from the focus and directrix property by regarding it as the limiting form of an ellipse as $e \to 0$ and the directrix moves infinitely far away.

In a Cartesian coordinate system, a conic is a curve that has an equation of the second degree, that is, of the form $ax^2 + 2hxy + by^2 + 2gx + 2gy + c = 0$. This equation represents a parabola if $h^2 = ab$, an ellipse if $h^2 < ab$ and a hyperbola if $h^2 > ab$. It represents a circle if $a = b$ and $h = 0$, and a

*rectangular hyperbola if $a + b = 0$. It represents a pair of straight lines (which may coincide) if $\Delta = 0$, where

$$\Delta = \begin{vmatrix} a & h & g \\ h & b & f \\ g & f & c \end{vmatrix}.$$

The *polar equation of a conic is normally obtained by taking the origin at a focus of the conic and the direction given by $\theta = 0$ perpendicular to the directrix. Then the equation can be written $l/r = 1 + e \cos \theta$ (all θ such that $\cos \theta \neq -1/e$), where e is the eccentricity and l is another constant.

(⊕) SEE WEB LINKS
• An interactive demonstration of a conic section.

conical pendulum A pendulum in which a particle, attached by a string of constant length to a fixed point, moves in a circular path in a horizontal plane. Suppose that the string has length l and that the string makes a constant angle α with the vertical. The particle moves in a circular path in a horizontal plane a distance d below the fixed point, where $d = l \cos \alpha$. The period of the conical pendulum is equal to $2\pi \sqrt{(d/g)}$.

conjugacy class The set of all elements of a *group that are *conjugate to an element a.

conjugate (of a complex number) For any complex number z, where $z = x + yi$, its conjugate $\bar{z}$ (read as 'z bar') is equal to $x - yi$. In the *complex plane, the points representing a complex number and its conjugate are mirror images with respect to the real axis. The following properties hold:

(i) $\bar{\bar{z}} = z$; so if $z_1 = \overline{z_2}$, then $z_2 = \overline{z_1}$.

(ii) $z + \bar{z}$ is real; if $z = x + yi$, then $z + \bar{z} = 2x$.

(iii) $z\bar{z} = |z|^2$; if $z = x + yi$, then $z\bar{z} = x^2 + y^2$.

(iv) $\overline{z_1 + z_2} = \overline{z_1} + \overline{z_2}$ and $\overline{z_1 - z_2} = \overline{z_1} - \overline{z_2}$.

(v) $\overline{z_1 z_2} = \overline{z_1}\,\overline{z_2}$ and $\overline{(z_1/z_2)} = \overline{z_1}/\overline{z_2}$.

It is an important fact that if the complex number α is a root of a polynomial equation $z^n + a_1 z^{n-1} + \cdots + a_{n-1}z + n_n = 0$, where $a_1, \ldots, a_n$ are real, then $\bar{\alpha}$ is also a root of this equation.

conjugate angles Two angles that add up to four right angles.

conjugate axis *See* HYPERBOLA.

conjugate diameters If l is a diameter of a *central conic, the midpoints of the chords parallel to l lie on a straight line l', which is also a diameter. It can be shown that the midpoints of the chords parallel to l' lie on l. Then l and l' are conjugate diameters. In the case of a circle, conjugate diameters are perpendicular.

conjugate elements Elements x and y in a group G are said to be conjugate if there is an element a in G for which $y = a^{-1}xa$.

conjugate sets Subsets X and Y in a group G are said to be *conjugate if there is an element a in G for which $Y = a^{-1}Xa$.

conjugate surds Let a be a number in the form $x + \sqrt{y}$ where x and y are rational and $\sqrt{y}$ is not rational. Then $x - \sqrt{y}$ is called the conjugate of a. The conjugate surds have the property, like conjugate complex numbers, that both their product ($= x^2 - y$) and sum ($= 2x$) are rational.

conjunction If p and q are statements, then the statement 'p and q', denoted by $p \wedge q$, is the conjunction of p and q. For example, if p is 'It is raining' and q is 'It is Monday', then $p \wedge q$ is 'It is raining and it is Monday'. The conjunction of p and q is true only when p and q are both true, and so the *truth table is as follows:

p	q	$p \wedge q$
T	T	T
T	F	F
F	T	F
F	F	F

connected (of a relation) A *binary relation is connected if for all pairs of elements $x, y, x \neq y$, either $x \sim y$ or $y \sim x$. So, for example, in the set of real numbers the relation 'is greater than' is connected.

connected (of a set) A set which cannot be partitioned into two non-empty subsets so that each subset has no point in common with the set closure of the other.

connected graph A *graph in which there is a *path from any one vertex to any other. So a graph is connected if it is 'all in one piece'; that is, if it has precisely one *component.

connected surface A surface where there is a continuous path between any two points on the surface which does not cross the *boundary of the surface.

consequent The part of a conditional statement which expresses the necessary outcome if the *antecedent is true. So in 'if n is divisible by 2, n is even' the consequent is 'n is even'. *Compare with* ANTECEDENT.

conservation condition (in network flows) Nothing is allowed to build up at any vertex between a source and a sink, i.e. at each intermediate vertex the total inflow = total outflow.

conservation of angular momentum When a particle is acted on by a *central force, the *angular momentum of the particle about the centre of the field of force is constant for all time. This fact is called the principle of conservation of angular momentum.

Similarly, when the sum of the moments of the forces acting on a rigid body about a fixed point (or the centre of mass) is zero, the angular momentum of the rigid body about the fixed point (or the centre of mass) is conserved.

conservation of energy When all the forces acting on a system are *conservative forces, $E_k + E_p = $ constant, where E_k is the *kinetic energy and E_p is the *potential energy. This fact is known as the energy equation or the principle of conservation of energy.

From the equation of motion $m\,\mathbf{a} = \mathbf{F}$ for a particle with mass m moving with acceleration $\mathbf{a}$, it follows that $m\,\mathbf{a} \cdot \mathbf{v} = \mathbf{F} \cdot \mathbf{v}$ and hence $(d/dt)(\frac{1}{2}m\mathbf{v} \cdot \mathbf{v}) = \mathbf{F} \cdot \mathbf{v}$. When $\mathbf{F}$ is a conservative force, the energy equation follows by integration with respect to t.

conservation of linear momentum When the total force acting on a system is zero, the *linear momentum of the system remains constant for all time. This fact is called the principle of conservation of linear momentum.

An application is seen in the recoil of a gun when a shell is fired. Before the firing, the linear momentum of the shell and gun is zero, so the total linear momentum after the firing is also zero. Therefore the mass of the gun must be very much larger than the mass of the shell to ensure that the gun's speed of recoil is very much smaller than the speed of the shell. The principle may also be used, for example, in investigating the collision of two billiard balls.

conservative force A force or field of force is conservative if the *work done by the force as the point of application moves around any closed path is zero. It follows that for a conservative force the work done as the point of application moves from one point to another does not depend on the path taken. For example, the *uniform gravitational force, the tension in a spring satisfying Hooke's law, and the gravitational force given by the *inverse square law of gravitation are conservative forces. Only for a conservative force can *potential energy be defined.

conservative strategy In the *matrix game given by the matrix $[a_{ij}]$, suppose that the players R and C use *pure strategies. Let m_i be equal to the minimum entry in the i-th row. A maximin strategy for R is to choose the r-th row, where $m_r = \max\{m_i\}$. In doing so, R ensures that the smallest *payoff possible is as large as can be. Similarly, let M_j be equal to the maximum entry in the j-th column. A minimax strategy for C is to choose the s-th column, where $M_s = \min\{M_j\}$. These are called conservative strategies for the two players.

Now let $E(\mathbf{x}, \mathbf{y})$ be the *expectation when R and C use *mixed strategies $\mathbf{x}$ and $\mathbf{y}$. Then, for any $\mathbf{x}$, $\min_y E(\mathbf{x}, \mathbf{y})$ is the smallest expectation possible, for all mixed strategies $\mathbf{y}$ that C may use. A maximin strategy for R is a strategy $\mathbf{x}$ that maximizes $\min_y E(\mathbf{x}, \mathbf{y})$. Similarly, a minimax strategy for C is a strategy $\mathbf{y}$ that minimizes $\max_x E(\mathbf{x}, \mathbf{y})$. By the *Fundamental Theorem of Game Theory, when R and C use such strategies the expectation takes a certain value, the value of the game.

consistent A set of equations is consistent if there is a solution.

consistent (in logic) A consistent theory in logic is one which does not contain a *contradiction.

consistent estimator *See* ESTIMATOR.

constant (in physical laws) The values of quantities such as the speed of light are constant once the units of measurement are known. In certain instances units are determined to make constants 1 for simplicity. *Newton's law says that the force necessary to accelerate a body is proportional to the product of mass and acceleration. The definition of the unit of force which bears his name means that this law reduces to 'Net force = mass × acceleration' in *SI units.

constant acceleration *See* EQUATIONS OF MOTION WITH CONSTANT ACCELERATION.

constant function In real analysis, a constant function is a *real function f such that $f(x) = a$ for all x in $\mathbf{R}$, where a, the value of f, is a fixed real number.

constant matrix A matrix in which all the entries are constants. If all the entries are the same constant k, multiplication by the matrix, when it can be done, is equivalent to multiplication by the *scalar k and by a matrix of the same order in which all entries are 1.

constant of integration If ϕ is a particular *antiderivative of a *continuous function f, then any antiderivative of f differs from ϕ by a constant. It is common practice, therefore, to write

$$\int f(x)dx = \phi(x) + c,$$

where c, an arbitrary constant, is the constant of integration.

constant of proportionality *See* PROPORTION.

constant speed A particle is moving with constant speed if the magnitude of the velocity is independent of time. Thus $\mathbf{v} \cdot \mathbf{v} = \text{constant}$ and this gives $2\mathbf{v} \cdot (d\mathbf{v}/dt) = 0$. Hence there are three possibilities: $\mathbf{v} = \mathbf{0}$, or $d\mathbf{v}/dt = \mathbf{0}$, or $\mathbf{v}$ is perpendicular to $d\mathbf{v}/dt$. The third possibility shows that the velocity is not necessarily constant. One such example is when a particle is moving in a circle with constant speed, and then the acceleration is perpendicular to the velocity.

constant term *See* POLYNOMIAL.

constants Certain numbers, notably $\pi = 3.141\ 592\ 6\ldots$, $e = 2.718\ 281\ldots$ and 'the golden ratio' $= \dfrac{1+\sqrt{5}}{2} \approx 1.618\ \ldots$ occur repeatedly in the natural world. As pure numbers, their values are independent of the scales used in measurements.

constrained optimization Optimization in circumstances where constraints exist. For example, in *linear programming.

constraint A condition that causes a restriction. For example, if $f(t)$ describes the motion of a pendulum which is released at $t = 0$. Then $f(t)$ is only valid when $t \geq 0$. In a probability distribution if $p_i = \Pr\{X = x_i\}$ then the values p_i are subject to two axiomatic constraints: $p_i \geq 0$ for each i and $\sum p_i = 1$.

construct Build. In particular, many mathematical concepts and quantities are constructed from simpler concepts and quantities. For example, a *group is defined in terms of a *set, *binary operations and other constraints.

construct (figures) A geometrical shape is said to be constructed if it is drawn using only a compass and straight edge without the use of any measuring instruments.

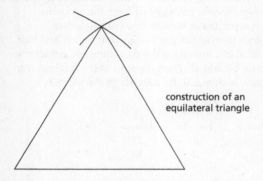

construction of an
equilateral triangle

construction with ruler and compasses *See* DUPLICATION OF THE CUBE, SQUARING THE CIRCLE and TRISECTION OF AN ANGLE.

constructivism Many cognitive psychologists now believe that humans learn most effectively where new information can be assimilated with previously held knowledge. This philosophy of learning is known as constructivism.

contact force When two bodies are in contact, each exerts a contact force on the other, the two contact forces being equal and opposite. The contact force is the sum of the frictional force, which is tangential to the surfaces at the point of contact, and the normal reaction, which is normal to the surfaces at the point of contact. If the frictional force is zero, the contact is said to be *smooth and the surfaces of the two bodies are called smooth. Otherwise, the contact is *rough and the surfaces are called rough. If the two bodies are moving, or tending to move, relative to each other, the frictional forces will act to oppose the motion. *See also* FRICTION.

contained in, contains It is tempting to say that 'x is contained in S' when $x \in S$, and also to say that 'A is contained in B' if $A \subseteq B$. To distinguish between these two different notions, it is better to say that 'x

belongs to S' and to say that 'A is included in B' or 'A is a subset of B'. However, some authors consistently say 'is contained in' for $\subseteq$. Given the same examples, it is similarly tempting to say that 'S contains x' and also that 'B contains A.' It is again desirable to distinguish between the two by saying that 'B includes A' in the second case, though some authors consistently say 'contains' in this situation. The first case is best avoided or else clarified by saying that 'S contains the element x' or 'S contains x as an element.'

contingency table A method of presenting the frequencies of the outcomes from an experiment in which the observations in the sample are categorized according to two criteria. Each cell of the table gives the number of occurrences for a particular combination of categories. A contingency table in which individuals are categorized by sex and hair colour is shown below. A final column giving the row sums and a final row giving the column sums may be added. Then the sum of the final column and the sum of the final row both equal the number in the sample.

	hair colour			
sex	black	blonde	brown	red
male	30	6	22	6
female	24	9	18	5

If the sample is categorized according to three or more criteria, the information can be presented similarly in a number of such tables.

contingent (in logic) A statement or proposition is contingent if it is neither always true nor always false. For example 'x divided by 2 is an integer' is true when $x = 2, 4, 6, \ldots$ but is not true when $x = 3, 4.3, \sqrt{7} \ldots$.

continued fraction An expression of the form $q_1 + 1/b_2$, where $b_2 = q_2 + 1/b_3$, $b_3 = q_3 + 1/b_4$, and so on, where $q_1, q_2, \ldots$ are integers, usually positive. This can be written

$$q_1 + \cfrac{1}{q_2 + \cfrac{1}{q_3 + \cfrac{1}{q_4 + \cdots}}}$$

or, in a form that is easier to print,

$$q_1 + \frac{1}{q_2 +} \frac{1}{q_3 +} \frac{1}{q_4 + \cdots}.$$

If the continued fraction terminates, it gives a rational number. The expression of any given positive rational number as a continued fraction can be found by using the *Euclidean algorithm. For example, 1274/871 is found, by using the steps which appear in the entry on the Euclidean algorithm, to equal

$$1 + \cfrac{1}{2+} \cfrac{1}{6+} \cfrac{1}{5}.$$

When the continued fraction continues indefinitely, it represents a real number that is the limit of the sequence

$$q_1, \quad q_1 + \cfrac{1}{q_2}, \quad q_1 + \cfrac{1}{q_2 + \cfrac{1}{q_3}}, \quad q_1 + \cfrac{1}{q_2 + \cfrac{1}{q_3 + \cfrac{1}{q_4}}}, \dots$$

For example, it can be shown that

$$1 + \cfrac{1}{1+} \cfrac{1}{1+} \cfrac{1}{1 + \cdots}$$

is equal to the *golden ratio, and that the representation of $\sqrt{2}$ as a continued fraction is

$$1 + \cfrac{1}{2+} \cfrac{1}{2+} \cfrac{1}{2 + \cdots}.$$

continuity correction The addition or subtraction of 0.5 when a discrete distribution, taking only integer values, is approximated by a continuous distribution, to maximize the agreement between the two distributions.

continuous data *See* DATA.

continuous function The *real function f of one variable is continuous at a if $f(x) \to f(a)$ as $x \to a$ (*see* LIMIT (of $f(x)$)). The rough idea is that, close to a, the function has values close to $f(a)$. It means that the function does not suddenly jump at $x = a$ or take widely differing values arbitrarily close to a.

A function f is continuous in an open interval if it is continuous at each point of the interval; and f is continuous on the closed interval $[a, b]$, where $a < b$, if it is continuous in the open interval (a, b) and if $\lim_{x \to a+} f(x) = f(a)$ and $\lim_{x \to b-} f(x) = f(b)$. The following properties hold:

   (i) The sum of two continuous functions is continuous.
  (ii) The product of two continuous functions is continuous.
 (iii) The quotient of two continuous functions is continuous at any point or in any interval where the denominator is not zero.
 (iv) Suppose that f is continuous at a, that $f(a) = b$ and that g is continuous at b. Then h, defined by $h(x) = (g \circ f)(x) = g(f(x))$, is continuous at a.

(v) It can be proved from first principles that the constant functions, and the function f, defined by $f(x) = x$ for all x, are continuous (at any point or in any interval). By using (i), (ii) and (iii), it follows that any *polynomial function is continuous and that any *rational function is continuous at any point or in any interval where the denominator is not zero.

The following properties of continuous functions appear to be obvious if a continuous function is thought of as one whose graph is a continuous curve; but rigorous proofs are not elementary, relying as they do on rather deep properties of the real numbers:

(vi) If f is continuous on a closed interval $[a, b]$ and η is any real number between $f(a)$ and $f(b)$, then, for some c in (a, b), $f(c) = \eta$. This is the *intermediate value theorem or property.

(vii) If f is continuous on a closed interval $[a, b]$, then f is *bounded on $[a, b]$. Furthermore, if S is the set of values $f(x)$ for x in $[a, b]$ and $M = \sup S$, then there is a ξ in $[a, b]$ such that $f(\xi) = M$ (and similarly for $m = \inf S$). It is said that 'a continuous function on a closed interval attains its bounds'.

continuous random variable *See* RANDOM VARIABLE.

continuously differentiable A function is continuously differentiable if its derivative is itself a continuous function. The function $f(x) = 0$ for $x \leq 0$ and $f(x) = x$ for $x \geq 0$ is continuous, but not continuously differentiable since $f'(x) = 0$ for $x < 0$ and $f'(x) = 1$ for $x > 0$.

continuum The set of real numbers or any interval (a, b), which can be open or closed at either end, is a continuum.

continuum hypothesis The conjecture made by Georg *Cantor that there is no set with a *cardinal number between *aleph-null which is the cardinal number of the set of *natural numbers and the cardinal number of the set of *real numbers, i.e. the continuum.

contour integral An integral $\int_C f(z)\, dz$ of a function f in the complex plane over a curve C, usually a closed curve, in the plane.

contour line A line joining points of a constant value. If $z = f(x, y)$ is a function which defines a surface and the line $y = g(x)$ has the property that $f(x, g(x))$ is constant then the line $y = g(x)$ is a contour line. Physical geography maps show height contours and weather charts show pressure isobars which are contour lines.

(⊕) SEE WEB LINKS
• An illustrated explanation of contour lines.

contractible space A *topological space X that is *homotopy equivalent to a one-point space, i.e. a space that can be continuously shrunk to a single point.

contraction mapping A *mapping, f, on a *metric space, X, which decreases the distance, d, between any two points in X, i.e. for some $a < 1$, i.e. $d\{f(x), f(y)\} \leq a \times d\{x, y\}$ for all x, y in X. *See also* FINITE INTERSECTION PROPERTY.

Contraction Mapping Theorem If f is a *contraction mapping on a non-empty *complete metric space X, then there is a unique fixed point x in X under the mapping, i.e. $f(x) = x$ for exactly one point x.

contradiction The simultaneous assertion of both the truth of a proposition and its denial. Since both cannot be true there must necessarily be a flaw in either the reasoning leading to the simultaneous assertion or in the assumptions on which the deductive reasoning is based. It is this latter situation which provides the basis for *proof by contradiction.

contraposition The logical principle, upon which *proof by contradiction is based. Let p and q be statements. If p implies not q, then q being true implies that p cannot be. For example, since all squares are rectangles, a shape which is not a rectangle cannot be a square. Here p is the statement 'is a square' and q is the statement 'is not a rectangle'.

contrapositive The contrapositive of an *implication $p \Rightarrow q$ is the implication $\neg q \Rightarrow \neg p$. An implication and its contrapositive are *logically equivalent, so that one is true if and only if the other is. So, in giving a proof of a mathematical result, it may on occasion be more convenient to establish the contrapositive rather than the original form of the theorem. For example, the theorem that if n^2 is odd then n is odd could be proved by showing instead that if n is even then n^2 is even.

control To rule out the effects of variables other than the factors the experiment wishes to explore. This may be done by ensuring certain variables are the same, either by directly controlling, for example, temperature or by matching pairs of subjects, for example by weight. Randomization is then normally used so that any *confounding variables which had not been controlled for by either of these techniques should not introduce a systematic source of *bias.

control chart A chart on which statistics from samples, taken at regular intervals, are plotted so that a production-line operator can monitor the output. The chart will normally have *warning limits and *action limits displayed on it.

SEE WEB LINKS

• Construct a control chart for your own choice of parameters.

control condition The condition of the *control group in a statistical experiment.

control group In experimental design, the control group provides a baseline assessment of any change which might happen without the treatment under investigation. For example, in looking at the effects of using a particular diet on a child's growth between 10 and 14 we need to be able to make a comparison with the growth of other children not following that regime. The control group and experimental groups should be either allocated randomly or balanced if that is possible, to provide a fair test of the treatment.

convenience sampling Where a sample is chosen by using the most conveniently available group. Data collected from such a sample are unlikely to contain much worthwhile information about a larger population.

converge (sequence) *See* LIMIT OF A SEQUENCE.

converge (series) The infinite series $a_1 + a_2 + a_3 + \ldots$ is said to converge to a limit L if for every $\varepsilon > 0$ there exists an N such that, for all $n > N, |\sum_{i=1}^{n} a_i - L| < \varepsilon$. Note that for a series to converge the sequence made up of its terms must converge to 0, though the converse is not true since $\{1, \frac{1}{2}, \frac{1}{3}, \frac{1}{4}, \ldots\}$ is a sequence which converges to 0 but $1 + \frac{1}{2} + \frac{1}{3} + \frac{1}{4} + \ldots$ is infinite.

convergence of functions There are two main types of convergence of functions: *pointwise convergence and *uniform convergence, which is a stronger condition.

converse The converse of an *implication $p \Rightarrow q$ is the implication $q \Rightarrow p$. If an implication is true, then its converse may or may not be true.

convert To change the units of a quantity or the form of expressing something. For example, an angle of $180°$ is the same as π radians and a vector of magnitude d in a given direction can be converted into Cartesian component form to enable calculations to be done more easily.

convex A plane or solid figure, such as a polygon or polyhedron, is convex if the *line segment joining any two points inside it lies wholly inside it.

convex up and **down** Some authors say that a curve is convex up when it is concave down, and convex down when it is concave up (*see* CONCAVITY).

convolution of random variables The sum of two random variables by integration or summation.

coordinate geometry The area of mathematics where geometrical relationships are described algebraically by the reference to the coordinates.

coordinates (on a line) One way of assigning coordinates to points on a line is as follows. Make the line into a *directed line by choosing one direction as the positive direction, running say from x' to x. Take a point O on the line as origin and a point A on the line such that OA is equal to the unit length. If P is any point on the line and $OP = x$, then x is the coordinate of P in this coordinate system. (Here OP denotes the *measure.) The coordinate system on the line is determined by the specified direction of the line, the origin and the given unit of length.

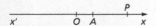

coordinates (in the plane) One way of assigning coordinates to points in the plane is as follows. Take a *directed line Ox as x-axis and a directed line Oy as y-axis, where the point O is the ORIGIN, and specify the unit length. For any point P in the plane, let M and N be points on the x-axis and y-axis such that PM is parallel to the y-axis and PN is parallel to the x-axis. If $OM = x$ and $ON = y$, then (x, y) are the coordinates of the point P in this coordinate system. The coordinate system is determined by the two directed lines and the given unit length. When the directed lines intersect at a right angle, the system is a *Cartesian, or rectangular, coordinate system and (x, y) are Cartesian coordinates of P. Normally, Ox and Oy are chosen so that an anticlockwise rotation of one right angle takes the positive x-direction to the positive y-direction.

There are other methods of assigning coordinates to points in the plane. One such is the method of *polar coordinates.

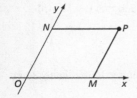

coordinates (in 3-dimensional space) One way of assigning coordinates to points in space is as follows. Take as axes three mutually perpendicular *directed lines Ox, Oy and Oz, intersecting at the point O, the origin, and forming a *right-handed system. Let L be the point where the plane through P, parallel to the plane containing the y-axis and the z-axis, meets the x-axis. Alternatively, L is the point on the x-axis such that PL is perpendicular to the x-axis. Let M and N be similarly defined points on the y-axis and the z-axis. The points L, M and N are in fact three of the vertices of the *cuboid with three of its edges along the coordinate axes and with O and P as opposite vertices. If $OL = x$, $OM = y$ and $ON = z$, then (x, y, z) are the coordinates of the point P in this *Cartesian coordinate system.

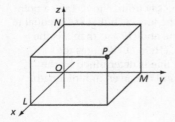

There are other methods of assigning coordinates to points in space. One is similar to that described above but using oblique axes. Others are by *spherical polar coordinates and *cylindrical polar coordinates.

coordinate system A system for identifying points on a plane or in space by their coordinates, for example Cartesian or polar coordinates.

coplanar points and lines A number of points and lines are coplanar if there is a plane in which they all lie. Three points are always coplanar: indeed, any three points that are not collinear determine a unique plane that passes through them.

coplanar vectors Let $\overrightarrow{OA}$ and $\overrightarrow{OB}$ be *directed line segments representing non-zero, non-parallel *vectors $\mathbf{a}$ and $\mathbf{b}$. A vector $\mathbf{p}$ is coplanar with $\mathbf{a}$ and $\mathbf{b}$ if $\mathbf{p}$ can be represented by a directed line segment $\overrightarrow{OP}$, where P lies in the plane determined by O, A and B. The vector $\mathbf{p}$ is coplanar with $\mathbf{a}$ and $\mathbf{b}$ if and only if there exist scalars λ and μ such that $\mathbf{p} = \lambda\mathbf{a} + \mu\mathbf{b}$.

coprime = RELATIVELY PRIME.

Coriolis force *See* FICTITIOUS FORCE.

corollary A result that follows from a theorem almost immediately, often without further proof.

correction An alteration made to the result of an observation or calculation in order to improve its accuracy. For example, if experience tells me a clock is usually 5 minutes slow I will add 5 minutes to the time it displays when I look at it, or government estimates of the number of homeless people will be higher than the recorded totals in the census, because it is known that the census will not be able to accurately record all the homeless people.

correlation Between two random variables, the correlation is a measure of the extent to which a change in one tends to correspond to a change in the other. The correlation is high or low depending on whether the relationship between the two is close or not. If the change in one corresponds to a change in the other in the same direction, there is positive correlation, and there is a negative correlation if the changes are in opposite directions. *Independent random variables have zero correlation. One measure of correlation between the random variables X and Y is the correlation coefficient ρ defined by

$$\rho = \frac{\text{Cov}(X, Y)}{\sqrt{\text{Var}(X)\,\text{Var}(Y)}}$$

(*see* *covariance and *variance). This satisfies $-1 \leq \rho \leq 1$. If X and Y are linearly related, then $\rho = -1$ or $+1$.

For a sample of n paired observations $(x_1, y_1), (x_2, y_2), \ldots (x_n, y_n)$, the (*sample) correlation coefficient is equal to

$$\frac{\sum(x_i - \bar{x})(y_i - \bar{y})}{\sqrt{\left(\sum(x_i - \bar{x})^2\right)\left(\sum(y_i - \bar{y})^2\right)}}.$$

Note that the existence of some correlation between two variables need not imply that the link between the two is one of cause and effect.

correlation matrix The $n \times n$ matrix in which $a_{ij} = \text{corr}(X_i, X_j)$ so that the correlations between all pairs of variables are contained in it. All elements a_{ii} in the leading diagonal will be 1 and the matrix will be symmetrical about that diagonal.

correspondence *See* ONE-TO-ONE CORRESPONDENCE.

corresponding angles *See* TRANSVERSAL.

corresponding sides In congruent n-sided polygons the n pairs of sides which exactly match in length. In *similar n-sided polygons, the n pairs of sides for which the ratio of the side in the large polygon to its corresponding side in the small polygon is the same in each case.

cosecant *See* TRIGONOMETRIC FUNCTION.

cosech, cosh *See* HYPERBOLIC FUNCTION.

coset If H is a *subgroup of a *group G, then for any element, x of G there is a left coset xH consisting of all the elements xh, where h is an element of H. Similarly, there is a right coset, Hx, with elements hx. *See also* NORMAL SUBGROUP.

cosine *See* TRIGONOMETRIC FUNCTION.

cosine rule *See* TRIANGLE.

cotangent *See* TRIGONOMETRIC FUNCTION.

coth *See* HYPERBOLIC FUNCTION.

count To enumerate. In a child's early development of the abstract concept of number, they will touch each one of a collection of objects in turn while saying the natural numbers (counting numbers) 1, 2, 3, ...

countable A set X is countable if there is a *one-to-one correspondence between X and a subset of the set of natural numbers. Thus a countable set is either finite or *denumerable. Some authors use 'countable' to mean denumerable.

countably infinite = DENUMERABLE.

counterexample Let $p(x)$ be a mathematical sentence involving a symbol x, so that, when x is a particular element of some universal set, $p(x)$ is a statement that is either true or false. What may be of concern is the proving or disproving of the supposed theorem that $p(x)$ is true for all x in the universal set. The supposed theorem can be shown to be false by producing just one particular element of the universal set to serve as x that makes $p(x)$ false. The particular element produced is a counterexample. For example, let $p(x)$ say that $\cos x + \sin x = 1$, and consider the supposed theorem that $\cos x + \sin x = 1$ for all real numbers x. This is demonstrably false (though $p(x)$ may be true for some values of x) because $x = \pi/4$ is a counterexample: $\cos(\pi/4) + \sin(\pi/4) \neq 1$.

counting numbers = NATURAL NUMBERS.

counts A particular statistic which simply records the number of instances of various events, such as the number of cars passing a specific position on a road, often collected by some automated process. These simple measures often then form the basis of more sophisticated statistical processes, such as analysing *contingency tables.

couple A system of forces whose sum is zero. The simplest example is a pair of equal and opposite forces acting at two different points, B and C. It can be shown that the *moment of a couple about any point A is independent of the position of A.

coupled equations A pair of interdependent equations. For example, in modelling 'predator and prey populations', coupled equations may describe the rate of increase/decrease of each group.

covariance The covariance of two random variables X and Y, denoted by $\text{Cov}(X, Y)$, is equal to $E((X - \mu_X)(Y - \mu_Y))$, where μ_X and μ_Y are the population means of X and Y respectively (*see* EXPECTED VALUE). If X and Y are *independent random variables, then $\text{Cov}(X, Y) = 0$. For computational purposes, note that $E((X - \mu_X)(Y - \mu_Y)) = E(XY) - \mu_X \mu_Y$. For a sample of n paired observations $(x_1, y_1), (x_2, y_2), \ldots, (x_n, y_n)$, the sample covariance is equal to

$$\frac{\sum (x_i - \bar{x})(y_i - \bar{y})}{n}.$$

covariance matrix *See* HYPOTHESIS TESTING.

cover A cover of a set X is a collection of sets whose union contains X as a subset.

Cox, Sir David FRS (1924–) British statistician distinguished for his work in mathematical statistics and applied probability, especially in relation to industrial and operational research. His varied contributions include the design and analysis of statistical experiments, analysis of binary data and of point stochastic processes and the development of new methods in quality control and operational research particularly in relation to queuing, congestion and renewal theory.

Cramer, Gabriel (1704–52) Swiss mathematician whose introduction to algebraic curves, published in 1750, contains the so-called *Cramer's rule. The rule was known earlier by Maclaurin.

Cramer's rule Consider a set of n linear equations in n unknowns $x_1, x_2, \ldots, x_n$, written in matrix form as $\mathbf{Ax} = \mathbf{b}$. When $\mathbf{A}$ is invertible, the set of equations has a unique solution $\mathbf{x} = \mathbf{A}^{-1}\mathbf{b}$. Since $\mathbf{A}^{-1} = (1/\det \mathbf{A}) \operatorname{adj} \mathbf{A}$, this gives the solution

$$\mathbf{x} = \frac{(\operatorname{adj} \mathbf{A})\mathbf{b}}{\det \mathbf{A}},$$

which may be written

$$x_j = \frac{b_1 A_{1j} + b_2 A_{2j} + \cdots + b_n A_{nj}}{\det \mathbf{A}} \quad (j = 1, \ldots, n),$$

using the entries of **b** and the *cofactors of **A**. This is Cramer's rule. Note that here the numerator is equal to the determinant of the matrix obtained by replacing the j-th column of **A** by the column **b**. For example, this gives the solution of

$$ax + by = h,$$
$$cx + dy = k,$$

when $ad - bc \neq 0$, as

$$x = \begin{vmatrix} h & b \\ k & d \end{vmatrix} \Big/ \begin{vmatrix} a & b \\ c & d \end{vmatrix} = \frac{hd - bk}{ad - bc},$$

$$y = \begin{vmatrix} a & h \\ c & k \end{vmatrix} \Big/ \begin{vmatrix} a & b \\ c & d \end{vmatrix} = \frac{ak - hc}{ad - bc}.$$

critical damping *See* DAMPED OSCILLATIONS.

critical events (activities) An event or activity on the *critical path.

critical path A path on an *activity network where any delay will delay the overall completion of the project.

critical path algorithm (edges as activities) A *forward scan determines the earliest time for each vertex (node) in an activity network, and a *backward scan determines the latest time for each vertex. The *critical path is any path on which the latest time = the earliest time at every vertex, i.e. on which any delays would mean the delay of the completion of the project.

critical path analysis Suppose that the vertices of a *network represent steps in a process, and the weights on the arcs represent the times that must elapse between steps. Critical path analysis is a method of determining the longest path in the network, and hence of finding the least time in which the whole process can be completed.

(((●))) SEE WEB LINKS

• An article including an example of critical path analysis.

critical point = STATIONARY POINT.

critical region *See* HYPOTHESIS TESTING.

critical value = STATIONARY VALUE.

cross-correlation The correlation between a pair of time-series variables where the values are paired by occurring at the same time.

cross-multiply Where an equation has fractions on both sides, it may be simplified by cross-multiplying. In fact both sides are multiplied by the product of the two denominators, but the denominator on each side will then cancel—so it appears that the numerator on each side has just been multiplied by the denominator on the other. For example, $\frac{x}{3} = \frac{2x-1}{4} \rightarrow 4x = 3(2x - 1)$.

cross-product = VECTOR PRODUCT.

cross ratio For a set of *coplanar points A, B, C, D, it is $\dfrac{AC \times BD}{AD \times BC}$, and for points z_1, z_2, z_3, z_4 in the *complex plane it is $\dfrac{(z_1 - z_3)(z_2 - z_4)}{(z_1 - z_4)(z_2 - z_3)}$. The definition in the complex plane can be extended to the *Riemann sphere by continuity. *See also* HARMONIC RATIO and HARMONIC RANGE.

crude data = RAW DATA.

cryptographic hash functions Any algorithm that maps data sets of variable length to a hash value which serves as a safeguard against accidental corruption or, if it is sophisticated enough, may provide some information security. *Checksums are a simple example.

cryptography The area of mathematics concerning the secure coding of information, often relying on mathematics such as prime factorizations of very large numbers.

cube A solid figure bounded by six square faces. It has eight vertices and twelve edges.

cube root *See* N-TH ROOT.

cube root of unity A complex number z such that $z^3 = 1$. The three cube roots of unity are 1, ω and ω^2, where

$$\omega = e^{2\pi i/3} = \cos\frac{2\pi}{3} + i\sin\frac{2\pi}{3} = -\frac{1}{2} + \frac{\sqrt{3}}{2}i,$$

$$\omega^2 = e^{4\pi i/3} = \cos\frac{4\pi}{3} + i\sin\frac{4\pi}{3} = -\frac{1}{2} - \frac{\sqrt{3}}{2}i.$$

Properties: (i) $\omega^2 = \bar{\omega}$ (*see* CONJUGATE), (ii) $1 + \omega + \omega^2 = 0$.

cubic (of a solid figure) Having the shape of a cube.

cubic equation A polynomial equation of degree three.

cubic polynomial A polynomial of degree three.

cuboctahedron One of the *Archimedean solids, with 6 square faces and 8 triangular faces. It can be formed by cutting off the corners of a cube to obtain a polyhedron whose vertices lie at the midpoints of the edges of the original cube. It can also be formed by cutting off the corners of an *octahedron to obtain a polyhedron whose vertices lie at the midpoints of the edges of the original octahedron.

(⊕) SEE WEB LINKS

• A cuboctahedron which can be manipulated to see its shape fully.

cuboid A *parallelepiped all of whose faces are rectangles.

cumulative distribution function For a random variable X, the cumulative distribution function (or c.d.f.) is the function F defined by $F(x) = \Pr(X \le x)$. Thus, for a discrete random variable,

$$F(x) = \sum_{xi \le x} p(x_i),$$

where p is the *probability mass function, and, for a continuous random variable,

$$F(x) = \int_{-\infty}^{x} f(t)\, dt,$$

where f is the *probability density function.

cumulative frequency The sum of the frequencies of all the values up to a given value. If the values $x_1, x_2, \ldots, x_n$, in ascending order, occur with frequencies $f_1, f_2, \ldots, f_n$, respectively, then the cumulative frequency at x_i is equal to $f_1 + f_2 + \cdots + f_i$. Cumulative frequencies may be similarly obtained for *grouped data.

cumulative frequency distribution For discrete data, the information consisting of the possible values and the corresponding *cumulative frequencies is called the cumulative frequency distribution. For *grouped data, it gives the information consisting of the groups and the corresponding cumulative frequencies. It may be presented in a table or in a diagram similar to a histogram.

cup The operation ∪ (*see* UNION) is read by some as 'cup', this being derived from the shape of the symbol, and in contrast to *cap (∩).

curl For a vector function of position $\mathbf{V}(\mathbf{r}) = V_x\mathbf{i} + V_y\mathbf{j} + V_z\mathbf{k}$, the curl of $\mathbf{V}$ is the *vector product of the operator del, $\nabla = \mathbf{i}\dfrac{\partial}{\partial x} + \mathbf{j}\dfrac{\partial}{\partial y} + \mathbf{k}\dfrac{\partial}{\partial z}$ with $\mathbf{V}$ giving curl $\mathbf{V} = \nabla \times \mathbf{V} = \mathbf{i} \times \dfrac{\partial \mathbf{V}}{\partial x} + \mathbf{j} \times \dfrac{\partial \mathbf{V}}{\partial y} + \mathbf{k} \times \dfrac{\partial \mathbf{V}}{\partial z}$ which can be written in determinant form as $\begin{vmatrix} \mathbf{i} & \mathbf{j} & \mathbf{k} \\ \dfrac{\partial}{\partial x} & \dfrac{\partial}{\partial y} & \dfrac{\partial}{\partial z} \\ V_x & V_y & V_z \end{vmatrix}$. *Compare* DIVERGENCE, GRADIENT.

curvature The rate of change of direction of a curve at a point on the curve. The Greek letter κ is used to denote curvature and

$$k = \frac{y''}{(1 + (y')^2)^{3/2}}.$$

$\rho = \frac{1}{\kappa}$ is the radius of curvature which is the radius of the circle which best fits the curve at that point, matching the position, the gradient and the second differential of the curve at that point. The centre of curvature is the centre of that best-fitting circle, known as the circle of curvature.

If κ is positive, the centre will be above the curve and if κ is negative it will be below. Since $\rho = \frac{1}{\kappa}$ is the radius of curvature, a curve which bends sharply at a point will have a small $|\rho|$ and a correspondingly large value of $|\kappa|$.

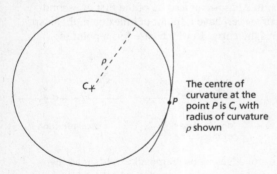

The centre of curvature at the point *P* is *C*, with radius of curvature ρ shown

curve A line which is continuously differentiable, i.e. it has no discontinuities in the value the function takes, nor in the value of the first differential.

curve sketching When a graph $y = f(x)$ is to be sketched, what is generally required is a sketch showing the general shape of the curve and the behaviour at points of special interest. The different parts of the graph do not have to be to scale. It is normal to investigate the following: *symmetry, *stationary points, intervals in which the function is always increasing or always decreasing, *asymptotes (vertical, horizontal and slant), *concavity, *points of inflexion, points of intersection with the axes, and the gradient at points of interest.

cusp A point at which two or more branches of a curve meet, and at which the limits of the tangents approaching that point along each branch coincide. There are two main characteristics used to describe cusps. In a single or simple cusp there are only two branches, and the limits of the second differentials approaching that point are different. If the branches are on opposite sides of the common tangent, it is said to be a cusp of the first kind, and if the branches are on the same side of the common tangent it is a cusp of the second kind.

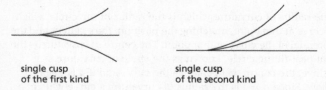

single cusp
of the first kind

single cusp
of the second kind

A double cusp or point of osculation has four branches, comprising two *continuously differentiable curves meeting at a point with a common tangent. Double cusps can also be of the first or second kind, or one or both curves can have a *point of inflexion at the cusp, so the tangent intersects the curve, in which case it is a point of osculinflection.

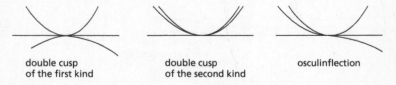

double cusp
of the first kind

double cusp
of the second kind

osculinflection

cut (in a network) A set of edges whose removal would create two separate sections with the *source in one section and the *sink in the other.

cycle (in graph theory) A *closed path with at least one edge. In a *graph, a cycle is a sequence $v_0, e_1, v_1, \ldots, e_k, v_k$ ($k \geq 1$) of alternately vertices and edges (where e_i is an edge joining v_{i-1} and v_i), with all the edges different and all the vertices different, except that $v_0 = v_k$. See HAMILTONIAN GRAPH and TREE.

cycle (in mechanics) *See* PERIOD, PERIODIC.

cycle, cyclic arrangement The arrangement of a number of objects that may be considered to be positioned in a circle may be called a cycle or cyclic arrangement. The arrangement of a, b, c and d in that order round a circle is considered the same as b, c, d and a in that order round a circle. The cycle or cyclic arrangement of a, b, c, d may be written $[a, b, c, d]$, where it is understood that $[a, b, c, d] = [b, c, d, a]$, for example, but that $[a, d, c, b]$ is different.

cyclic group Let a be an element of a *multiplicative group G. The elements a^r, where r is an integer (positive, zero or negative), form a *subgroup of G, called the subgroup generated by a. A group G is cyclic if there is an element a in G such that the subgroup generated by a is the whole of G. If G is a finite cyclic group with identity element e, the set of elements of G may be written $\{e, a, a^2, \ldots, a^{n-1}\}$, where $a^n = e$ and n is the smallest such positive integer. If G is an infinite cyclic group, the set of elements may be written $\{\ldots, a^{-2}, a^{-1}, e, a, a^2, \ldots\}$.

By making appropriate changes, a cyclic *additive group (or group with any other operation) can be defined. For example, the set $\{0, 1, 2, \ldots n - 1\}$ with addition modulo n is a cyclic group, and the set of all integers with addition is an infinite cyclic group. Any two cyclic groups of the same order are *isomorphic.

cyclic polygon A polygon whose vertices lie on a circle. From one of the *circle theorems, it follows that opposite angles of a cyclic quadrilateral add up to $180°$.

cyclic redundancy checks Error-detecting codes used in networks and storage devices to detect accidental corruption of data. Blocks of data have a check value inserted at the end of the block. Since these check values add no information to the message, they are *redundant, and the value is based on cyclic coding; hence the name.

cycling The behaviour of an *iterative method when a sequence of values which repeats itself occurs. Iterative methods hope to produce successively better approximations to the solution of an equation, but do not always succeed. Once a value recurs in the sequence it is destined to repeat (cycle) permanently.

cycloid The curve traced out by a point on the circumference of a circle that rolls without slipping along a straight line. With suitable axes, the cycloid has *parametric equations $x = a(t - \sin t)$, $y = a(1 - \cos t)$ ($t \in \mathbf{R}$), where a is a constant (equal to the radius of the rolling circle). In the figure, $OA = 2\pi a$.

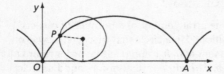

cyclotomic Relating to the n-th roots of unity. The cyclotomic equation is $z^n - 1 = 0$.

cyclotomic polynomial $\Phi_n(z)$ The polynomial whose roots are all the *primitive n-th roots of unity.

We know $z^n - 1 \equiv (z-1)(z^{n-1} + z^{n-2} + \ldots + z + 1)$,
so when n is prime $\Phi_n(z) = z^{n-1} + z^{n-2} + \ldots + z + 1$
is the cyclotomic polynomial, but when
$n = 4$, $z^4 - 1 = (z-1)(z+1)(z^2+1)$, and so $\Phi_4(z) = (z^2+1)$.

cylinder In elementary work, a cylinder, if taken, say, with its axis vertical, would be reckoned to consist of a circular base, a circular top of the same size, and the curved surface formed by the vertical line segments joining them. For a cylinder with base of radius r and height h, the volume equals $\pi r^2 h$, and the area of the curved surface equals $2\pi rh$.

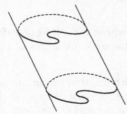

In more advanced work, a cylinder is a surface, consisting of the points of the lines, called generators, drawn through the points of a fixed curve and parallel to a fixed line, the generators being extended indefinitely in both directions. Then a right-circular cylinder is one in which the fixed curve is a circle and the fixed line is perpendicular to the plane of the circle. The axis of a right-circular cylinder is the line through the centre of the circle and perpendicular to the plane of the circle, that is, parallel to the generators.

cylindrical polar coordinates Suppose that three mutually perpendicular directed lines Ox, Oy and Oz, intersecting at the point O and forming a right-handed system, are taken as coordinate axes. For any point P, let M and N be the projections of P onto the xy-plane and the

z-axis respectively. Then $ON = PM = z$, the z-coordinate of P. Let $\rho = |\,PN\,|$, the distance of P from the z-axis, and let ϕ be the angle $\angle\,xOM$ in radians $(0 \leq \phi < 2\pi)$. Then (ρ, ϕ, z) are the cylindrical polar coordinates of P. (It should be noted that the points of the z- axis give no value for ϕ.) The two coordinates (ρ, ϕ) can be seen as polar coordinates of the point M and, as with polar coordinates, $\phi + 2k\pi$, where k is an integer, may be allowed in place of ϕ.

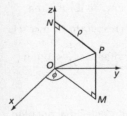

The Cartesian coordinates (x, y, z) of P can be found from (ρ, ϕ, z) by: $x = \rho \cos\phi$, $y = \rho \sin\phi$, and $z = z$. Conversely, the cylindrical polar coordinates can be found from (x, y, z) by: $\rho = \sqrt{x^2 + y^2}$, ϕ is such that $\cos\phi = x/\sqrt{x^2 + y^2}$ and $\sin\phi = y/\sqrt{x^2 + y^2}$, and $z = z$. Cylindrical polar coordinates can be useful in treating problems involving right-circular cylinders. Such a cylinder with its axis along the z-axis then has equation $\rho = $ constant.

cylindroid A *cylinder with *elliptical cross-section.

cypher *See* CIPHER.

d The symbol used to denote the differential operator. So if y is a real function of x then $y' = \dfrac{dy}{dx}$ is the differential of y with respect to x. Since it is $\dfrac{d}{dx}(y)$, when the second differential is taken, it is written as $\dfrac{d^2y}{dx^2}$ coming from $\dfrac{d}{dx}\left(\dfrac{dy}{dx}\right)$.

D The Roman *numeral for 500. The number 13 in hexadecimal notation.

damped oscillations *Oscillations in which the amplitude decreases with time. Consider the equation of motion $m\ddot{x} = -kx - c\dot{x}$, where the first term on the right-hand side arises from an elastic restoring force satisfying *Hooke's law, and the second term arises from a *resistive force. The constants k and c are positive. The form of the general solution of this *linear differential equation depends on the auxiliary equation $m\alpha^2 + c\alpha + k = 0$. When $c^2 < 4mk$, the auxiliary equation has non-real roots and damped oscillations occur. This is a case of weak damping. When $c^2 = 4mk$, the auxiliary equation has equal roots and critical damping occurs: oscillation just fails to take place. When $c^2 > 4mk$ there is strong damping: the resistive force is so strong that no oscillations take place.

(((⊕))) SEE WEB LINKS

• An applet exploring the effects of changing parameters on damped oscillations

dashpot A device consisting of a cylinder containing a liquid through which a piston moves, used for damping vibrations.

data The observations gathered from an experiment, survey or observational study. Often the data are a randomly selected sample from an underlying *population. Numerical data are discrete if the underlying population is finite or countably infinite and are continuous if the underlying population forms an interval, finite or infinite. Data are nominal if the observations are not numerical or quantitative, but are descriptive and have no natural order. Data specifying country of origin, type of vehicle or subject studied, for example, are nominal.

Note that the word 'data' is plural. The singular 'datum' may be used for a single observation.

data logging The process of automatically collecting and storing observations, typically as a sequence in time or space. For example, in hospital an electronic monitor may record a patient's temperature, heart rate, blood pressure, etc. at regular intervals and in aircraft, the flight data recorders, or 'black box', constantly monitor and record data on all the aircraft's systems.

deca- = DEKA-.

decagon A ten-sided polygon.

decahedron A solid with ten plane faces. It is not possible to construct a regular decahedron.

deceleration A slowing down. Suppose that a particle is moving in a straight line. If the particle is moving in the positive direction, then it experiences a deceleration when the acceleration is negative. Notice, however, that this is not true if the particle is moving in the negative direction (*see* ACCELERATION).

deci- Prefix used with *SI units to denote multiplication by 10^{-1}.

decidable In logic a proposition which is able to be shown either to be true or to be false is decidable.

decile *See* QUANTILE.

decimal fraction A *fraction expressed by using *decimal representation, as opposed to a *vulgar fraction. For example, $\frac{3}{4}$ is a vulgar fraction; 0.75 is a decimal fraction.

decimalize To change a measurement system to one based on multiples of 10. So in 1971 the British currency was decimalized from £1 = 20 shillings and 1 shilling = 12 pence to £1 = 100 (new) pence—the 'new' has since been dropped.

decimal places In *rounding or *truncation of a number to n decimal places, the original is replaced by a number with just n digits after the decimal point. When the rounding or truncation takes place to the left of the decimal point, a phrase such as 'to the nearest 10' or 'to the nearest 1000' has to be used. To say that $a = 1.9$ to 1 decimal place means that the exact value of a becomes 1.9 after rounding to 1 decimal place, and so $1.85 \leq a \leq 1.95$.

decimal point The separator used between the integer and fractional parts of a number expressed in *decimal representation. In the UK a dot is used, while in much of Europe a comma is used as the separator.

decimal representation Any real number a between 0 and 1 has a decimal representation, written $0.d_1\, d_2\, d_3 \ldots$, where each d_i is one of the digits $0, 1, 2, \ldots, 9$; this means that

$$a = d_1 \times 10^{-1} + d_2 \times 10^{-2} + d_3 \times 10^{-3} + \cdots.$$

This notation can be extended to enable any positive real number to be written as

$$c_n c_{n-1} \ldots c_1 c_0.d_1 d_2 d_3 \ldots$$

using, for the integer part, the normal representation $c_n c_{n-1} \ldots c_1 c_0$ to base 10 (*see* BASE). If, from some stage on, the representation consists of the repetition of a string of one or more digits, it is called a recurring or repeating decimal. For example, the recurring decimal $0.12748748748\ldots$ can be written $0.1\dot{2}74\dot{8}$, where the dots above indicate the beginning and end of the repeating string. The repeating string may consist of just one digit, and then, for example, $0.16666\ldots$ is written $0.1\dot{6}$. If the repeating string consists of a single zero, this is generally omitted and the representation may be called a terminating decimal.

The decimal representation of any real number is unique except that, if a number can be expressed as a terminating decimal, it can also be expressed as a decimal with a recurring 9. Thus 0.25 and $0.24\dot{9}$ are representations of the same number. The numbers that can be expressed as recurring (including terminating) decimals are precisely the *rational numbers.

decision analysis The branch of mathematics considering strategies to be used when decisions have to be made at stages in a process, but the outcomes resulting from the implementation of those decisions are dependent on chance.

decision nodes *See* EMV ALGORITHM.

decision theory The area of *statistics and *game theory concerned with decision making under uncertainty to maximize expected utility.

decision tree The diagram used to represent the process in a decision analysis problem. Different symbols are used to denote the different types of node or vertex. For example, decisions may be shown as rectangles, chance events as circles, and *payoffs as triangles.

decision variables The quantities to be found in *linear programming or other constrained optimization problems.

declination *See* DEPRESSION.

decompose To give a *decomposition of a number or other quantity.

decomposition The breakdown of a quantity or expression into simpler components. For example, $24 = 2 \times 2 \times 2 \times 3$ is the decomposition of 24 into prime factors, or the expression of a polynomial function as a product of factors.

decreasing function A *real function f is decreasing in or on an interval I if $f(x_1) \geq f(x_2)$ whenever x_1 and x_2 are in I with $x_1 < x_2$. Also, f is strictly decreasing if $f(x_1) > f(x_2)$ whenever $x_1 < x_2$.

decreasing sequence A sequence $a_1, a_2, a_3, \ldots$ is said to be decreasing if $a_i \geq a_{i+1}$ for all i, and strictly decreasing if $a_i > a_{i+1}$ for all i.

Dedekind, (Julius Wilhelm) Richard (1831–1916) German mathematician who developed a formal construction of the real numbers from the rational numbers by means of the so-called Dedekind cut. This new approach to *irrational numbers, contained in the very readable paper *Continuity and Irrational Numbers*, was an important step towards the formalization of mathematics. He also proposed a definition of infinite sets that was taken up by Cantor, with whom he developed a lasting friendship.

deduction The process of reasoning from axioms, premises or assumptions, in logic or in mathematics, using accepted steps of reasoning. In mathematics this includes calculation and the application of theorems not explicitly proven in the course of the reasoning, but which could be proved separately if required. The term is also used to refer to the outcome of such a reasoning process.

deficient number An integer that is larger than the sum of its positive divisors (not including itself). Any prime number must be deficient by definition but so is any power of any prime, as well as numbers like 10 whose divisors are 1, 2, 5 giving a sum of 8.

definite integral *See* INTEGRAL.

degenerate Where a family of entities can be defined in terms of parameters, and the limiting case produces an entity which is of a different nature. For example, the general quadratic function is in the form $y = \alpha x^2 + \beta x + \gamma$, and the graph is a parabola. As α decreases the curvature of the parabola steadily decreases, and the limiting case as $\alpha \to 0$ is a straight line, which is a degenerate parabola.

degenerate conic A *conic that consists of a pair of (possibly coincident) straight lines. The equation $ax^2 + 2hxy + by^2 + 2gx + 2fy + c = 0$ represents a degenerate conic if $\Delta = 0$, where

$$\Delta = \begin{vmatrix} a & h & g \\ h & b & f \\ g & f & c \end{vmatrix}.$$

degenerate quadric The *quadric with equation $ax^2 + by^2 + cz^2 + 2fyz + 2gzx + 2hxy + 2ux + 2vy + 2wz + d = 0$ is DEGENERATE if $\Delta = 0$, where

$$\Delta = \begin{vmatrix} a & h & g & u \\ h & b & f & v \\ g & f & c & w \\ u & v & w & d \end{vmatrix}.$$

The non-degenerate quadrics are the *ellipsoid, the *hyperboloid of one sheet, the *hyperboloid of two sheets, the *elliptic paraboloid and the *hyperbolic paraboloid.

degree (angular measure) The method of measuring angles in degrees dates back to Babylonian mathematics around 2000 BC. A complete revolution is divided into 360 degrees (°); a right angle measures 90°. Each degree is divided into 60 minutes (′) and each minute into 60 seconds (″). In more advanced work, angles should be measured in *radians.

degree (of a polynomial) *See* POLYNOMIAL.

degree (of a vertex of a graph) The degree of a vertex V of a *graph is the number of edges ending at V. (If loops are allowed, each loop joining v to itself contributes two to the degree of V.)

In the graph on the left, the vertices U, V, W and X have degrees 2, 2, 3 and 1. The graph on the right has vertices V_1, V_2, V_3 and V_4 with degrees 5, 4, 6 and 5.

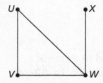
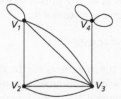

degrees of freedom (in mechanics) The number of degrees of freedom of a body is the minimum number of independent coordinates required to describe the position of the body at any instant, relative to a frame of reference. A particle in straight-line motion or circular motion has one degree of freedom. So too does a rigid body rotating about a fixed axis. A particle moving in a plane, such as a projectile, or a particle moving on a cylindrical or spherical surface has two degrees of freedom. A rigid body in general motion has six degrees of freedom.

degrees of freedom (in statistics) A positive integer normally equal to the number of independent observations in a sample minus the number of population parameters to be estimated from the sample. When the *chi-squared test is applied to a *contingency table with h rows and k columns, the number of degrees of freedom equals $(h-1)(k-1)$.

For a number of distributions, the number of degrees of freedom is required to identify which of a family of distributions is to be used. The *chi-squared distribution and *t-distribution each have a single degrees of freedom parameter, and the *F-distribution has two such parameters.

deka- Prefix used with *SI units to denote multiplication by 10.

del *See* DIFFERENTIAL OPERATOR.

Delian (altar) problem Another name for the problem of the *duplication of the cube. In 428 BC, an oracle at Delos ordered that the altar of Apollo should be doubled in volume as a means of bringing a certain plague to an end.

delta The Greek letter d, written δ, capital Δ. A small increment in the value of the variable x is usually denoted by δx. When y is a function of x, $\dfrac{\delta y}{\delta x}$ represents the average rate of change of y with respect to x at a point and the derivative of y with respect to x is defined as the limit of that ratio as $\delta x \to 0$. So $y' = \dfrac{dy}{dx} = \lim\limits_{\delta x \to 0} \dfrac{\delta y}{\delta x}$.

De Moivre, Abraham (1667–1754) Prolific mathematician, born in France, who later settled in England. In *De Moivre's Theorem, he is remembered for his use of complex numbers in trigonometry. But he was also the author of two notable early works on probability. His Doctrine of Chances of 1718, examines numerous problems and develops a number of principles, such as the notion of independent events and the product law. Later work contains the result known as *Stirling's formula and probably the first use of the normal frequency curve.

De Moivre's Theorem From the definition of *multiplication (of a complex number), it follows that $(\cos \theta_1 + i \sin \theta_1)(\cos \theta_2 + i \sin \theta_2) = \cos(\theta_1 + \theta_2) + i \sin(\theta_1 + \theta_2)$. This leads to the following result known as De Moivre's Theorem, which is crucial to any consideration of the powers z^n of a complex number z:

Theorem

For all positive integers n, $(\cos\theta + i\sin\theta)^n = \cos n\theta + i\sin n\theta$.

The result is also true for negative (and zero) integer values of n, and this may be considered as either included in or forming an extension of De Moivre's Theorem.

De Morgan, Augustus (1806-71) British mathematician and logician who was responsible for developing a more symbolic approach to algebra, and who played a considerable role in the beginnings of symbolic logic. His name is remembered in *De Morgan's laws, which he formulated. In an article of 1838, he clarified the notion of *mathematical induction.

De Morgan's laws For all sets A and B (subsets of a *universal set), $(A \cup B)' = A' \cap B'$ and $(A \cap B)' = A' \cup B'$. These are De Morgan's laws.

denominator *See* FRACTION.

dense matrix A matrix which has a high proportion of non-zero entries. *Compare* SPARSE.

dense set A set S in a *topological space X for which every point x in X either is in S or is a *limit point of S. Informally, this means that every point in X is either in S or arbitrarily close to a member of S. The set of rational numbers is dense in the space of real numbers. That every real number has a rational number within any specified distance is easy to see if you consider the decimal representation of the real number and the specified distance ε: since $\varepsilon > 0$, there exists an integer K such that $\varepsilon > 10^{-K}$, so truncating the decimal representation of the real number to K decimal places gives a rational number which is in the neighbourhood specified by ε.

density The average density of a body is the ratio of its mass to its volume. In general, the density of a body may not be constant throughout the body. The density at a point P, denoted by $\rho(P)$, is equal to the limit as $\Delta V \to 0$ of $\Delta m / \Delta V$, where ΔV is the volume of a small region containing P and Δm is the mass of the part of the body occupying that small region.

Consider a rod of length l, with density $\rho(x)$ at the point a distance x from one end of the rod. Then the mass m of the rod is given by

$$m = \int_0^l \rho(x)dx$$

In the same way, the mass of a *lamina or a 3-dimensional *rigid body, with density $\rho(\mathbf{r})$ at the point with position vector $\mathbf{r}$, is given by

$$m = \int_V \rho(\mathbf{r})dV.$$

with, as appropriate, a double or triple integral over the region V occupied by the body.

denumerable A set X is denumerable if there is a *one-to-one correspondence between X and the set of natural numbers. It can be shown that the set of rational numbers is denumerable but that the set of real numbers is not. Some authors use 'denumerable' to mean *countable.

dependent equations A set of equations where at least one of the set may be expressed as a (linear) combination of the others.

dependent events *See* INDEPENDENT EVENTS.

dependent variable In statistics, the variable which is thought might be influenced by certain other *explanatory variables. In *regression, a relationship is sought between the dependent variable and the explanatory variables. The purpose is normally to enable the value of the dependent variable to be predicted from given values of the explanatory variables.

depression The acute angle between the horizontal and a given line (*see* INCLINATION) measured positively in the direction of the downwards vertical. θ is the angle of depression of the boat from the top of the cliff. Sometimes also known as the angle of declination.

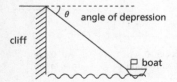

derangement A *rearrangement in which no object returns to its original place.

derivative For the *real function f, if $(f(a+h) - f(a))/h$ has a *limit as $h \to 0$, this limit is the derivative of f at a and is denoted by $f'(a)$. (The term 'derivative' may also be used loosely for the *derived function.)

Consider the graph $y = f(x)$. If (x, y) are the coordinates of a general point P on the graph, and $(x + \Delta x, y + \Delta y)$ are those of a nearby point Q on the graph, it can be said that a change Δx in x produces a change Δy in y. The quotient $\Delta y/\Delta x$ is the gradient of the chord PQ. Also, $\Delta y = f(x + \Delta x) - f(x)$. So the derivative of f at x is the limit of the quotient $\Delta y/\Delta x$ as $\Delta x \to 0$. This limit can be denoted by dy/dx, which is thus an alternative notation for $f'(x)$. The notation y' is also used.

The derivative $f'(x)$ may be denoted by $(d/dx)(\ )$, where the brackets contain a formula for $f(x)$. Some authors use the notation df/dx. The derivative $f'(a)$ gives the gradient of the curve $y = f(x)$, and hence the gradient of the tangent to the curve, at the point given by $x = a$. Suppose now, with a different notation, that x is a function of t, where t is some measurement of time. Then the derivative dx/dt, which is the *rate of change of x with respect to t, may be denoted by $\dot{x}$. The derivatives of certain common functions are given in the Table of derivatives (*Appendix 6). *See also* DIFFERENTIATION, LEFT AND RIGHT DERIVATIVE, HIGHER DERIVATIVE and PARTIAL DERIVATIVE.

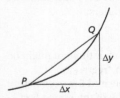

derivative test A test, usually to determine the nature of stationary points of a function, which uses one or more derivatives of the function. If $f'(\alpha) = 0$, so there is a stationary point at $x = \alpha$, the first derivative test considers the signs of $f'(x)$ at $x = \alpha^+$, α^-, i.e. whether the gradient is positive, negative, or zero as x approaches the stationary point from above and from below. The various possible combinations are shown below.

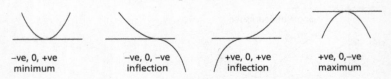

–ve, 0, +ve	–ve, 0, –ve	+ve, 0, +ve	+ve, 0, –ve
minimum	inflection	inflection	maximum

The second derivative test considers the sign of $f''(x)$ at $x = \alpha$. If it is positive, the value of $f'(x)$ is increasing as the function goes through $x = \alpha$, which requires the stationary point to be a minimum; if $f''(\alpha) < 0$ then it must be a maximum, while if $f''(\alpha) = 0$ it may be that $x = \alpha$ is a *point of inflection but this is not a sufficient condition. To determine the nature in this case, either the first derivative test may be used, or higher derivatives can be considered until the first non-zero higher derivative is found.

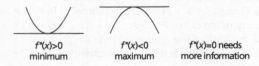

$f''(x) > 0$	$f''(x) < 0$	$f''(x) = 0$ needs
minimum	maximum	more information

derived function The function f', where $f'(x)$ is the *derivative of f at x, is the derived function of f. *See also* DIFFERENTIATION.

derived unit *See* SI UNITS.

Desargues, Girard (1591–1661) French mathematician and engineer whose work on conics and the result known as *Desargues's Theorem were to become the basis of the subject known as projective geometry. His 1639 book was largely ignored, partly because of the obscurity of the language. It was nearly 200 years later that projective geometry developed and the beauty and importance of his ideas were recognized.

Desargues's Theorem The following theorem of projective geometry:

Theorem

Consider two triangles lying in the plane or positioned in 3-dimensional space. If the lines joining corresponding vertices are concurrent, then corresponding sides intersect in points that are collinear, and conversely.

In detail, suppose that one triangle has vertices A, B and C, and the other has vertices A', B' and C'. The theorem states that if AA', BB' and CC' are concurrent, then BC and $B'C'$ intersect at a point L, CA and $C'A'$ intersect at M, and AB and $A'B'$ intersect at N, where L, M and N are collinear.

Perhaps surprisingly, the case of three dimensions is the easier to prove. The theorem can be taken as a theorem of Euclidean geometry if suitable amendments are made to cover the possibility that points of intersection do not exist because lines are parallel. It played an important role in the emergence of projective geometry.

Descartes, René (1596–1650) French philosopher and mathematician who in mathematics is known mainly for his methods of applying algebra to geometry, from which analytic geometry developed. He expounded these in *La Géométrie*, in which he was also concerned to use geometry to solve algebraic problems. Though named after him, Cartesian coordinates did not in fact feature in his work.

describe In geometry, to draw the shape of a curve or figure.

descriptive geometry The area of mathematics where 3-dimensional shapes are projected onto a plane surface so that spatial problems can be analysed graphically.

descriptive statistics The part of the subject of statistics concerned with describing the basic statistical features of a set of observations. Simple numerical summaries, using notions such as *mean, *range and *standard deviation, together with appropriate diagrams such as *histograms, are used to present an overall impression of the data.

design methods *See* EXPERIMENTAL DESIGN METHODS.

destinations (in transportation problems) *See* TRANSPORTATION PROBLEM.

detach (in logic) Given the truth of a *conditional statement, and the truth of the *antecedent clause, the *consequent clause will be true unconditionally. For example, in 'if n is divisible by 2, n is even' the antecedent is 'n is divisible by 2'. When $n = 10$, which is divisible by 2, we can detach the conditional part of the statement, leaving 'n is even' to stand alone without qualification.

determinant For the square matrix $\mathbf{A}$, the determinant of $\mathbf{A}$, denoted by det $\mathbf{A}$ or $|\mathbf{A}|$, can be defined as follows. Consider, in turn, 1×1, 2×2, 3×3, and $n \times n$ matrices.

The determinant of the 1×1 matrix $[a]$ is simply equal to a. If $\mathbf{A}$ is the 2×2 matrix below, then det $\mathbf{A} = ad - bc$, and the determinant can also be written as shown:

$$\mathbf{A} = \begin{bmatrix} a & b \\ c & d \end{bmatrix}, \quad \det \mathbf{A} = \begin{bmatrix} a & b \\ c & d \end{bmatrix}.$$

If $\mathbf{A}$ is a 3×3 matrix $[a_{ij}]$, then det $\mathbf{A}$, which may be denoted by

$$\begin{vmatrix} a_{11} & a_{12} & a_{13} \\ a_{21} & a_{22} & a_{23} \\ a_{31} & a_{32} & a_{33} \end{vmatrix},$$

is given by

$$\det \mathbf{A} = a_{11} \begin{vmatrix} a_{22} & a_{23} \\ a_{32} & a_{33} \end{vmatrix} - a_{12} \begin{vmatrix} a_{21} & a_{23} \\ a_{31} & a_{33} \end{vmatrix} + a_{13} \begin{vmatrix} a_{21} & a_{22} \\ a_{31} & a_{32} \end{vmatrix}.$$

Notice how each 2×2 determinant occurring here is obtained by deleting the row and column containing the entry by which the 2×2 determinant is multiplied. This expression for the determinant of a 3×3 matrix can be written $a_{11}A_{11} + a_{12}A_{12} + a_{13}A_{13}$, where A_{ij} is the *cofactor of a_{ij}. This is the evaluation of det $\mathbf{A}$, 'by the first row'. In fact, det $\mathbf{A}$ may be found by using evaluation by any row or column: for example, $a_{31}A_{31} + a_{32}A_{32} + a_{33}A_{33}$ is the evaluation by the third row, and $a_{12}A_{12} + a_{22}A_{22} + a_{32}A_{32}$ is the evaluation by the second column. The determinant of an $n \times n$ matrix $\mathbf{A}$ may be defined similarly, as $a_{11}A_{11} + a_{12}A_{12} + \ldots + a_{1n}A_{1n}$, and the same value is obtained using a similar evaluation by any row or column. The following properties hold:

(i) If two rows (or two columns) of a square matrix **A** are identical, then det **A** $= 0$.

(ii) If two rows (or two columns) of a square matrix **A** are interchanged, then only the sign of det **A** is changed.

(iii) The value of det **A** is unchanged if a multiple of one row is added to another row, or if a multiple of one column is added to another column.

(iv) If **A** and **B** are square matrices of the same order, then det(**AB**) = (det**A**) (det**B**).

(v) If **A** is *invertible, then det(**A**$^{-1}$) = (det**A**)$^{-1}$.

(vi) If **A** is an $n \times n$ matrix, then det k**A** $= k^n$det**A**.

In particular cases, the determinant of a given matrix may be evaluated by using operations of the kind described in (iii) to produce a matrix whose determinant is easier to evaluate.

determine Conditions which are sufficient to specify a quantity uniquely. For example, two points determine a straight line, as does one point and a gradient in two dimensions.

developable surface A surface that can be rolled out flat onto a plane without any distortion. Examples are the *cone and *cylinder. On any map of the world, distances are inevitably inaccurate because a sphere is not a developable surface.

deviation If $\{x_i\}$ is a set of observations of the random variable X then $x_i - \bar{x}$ is the deviation of the i-th observation from the mean.

df *See* DEGREES OF FREEDOM.

diagonal (of a polygon) A line joining any two vertices of a polygon that are not connected by an edge, and which does not go outside the polygon.

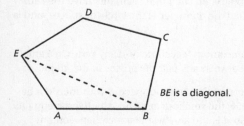

BE is a diagonal.

diagonal (of a polyhedron) A line joining any two vertices of a polyhedron that are not on a common face, and which does not go outside the polyhedron.

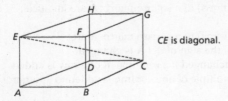

CE is diagonal.

diagonal (of a polyhedron)

diagonal entry For a square matrix $[a_{ij}]$, the diagonal entries are the entries $a_{11}, a_{22}, \ldots, a_{nn}$, which form the main diagonal.

diagonal matrix A square matrix in which all the entries not in the main diagonal are zero.

diagonally dominant matrix A *symmetric matrix where the absolute value of the entries in the main diagonal is not less than the sum of the absolute values of the rest of the row that entry appears in, i.e. $|a_{ii}| \geq \sum_{j \neq i} |a_{ij}|$ If the matrix contains complex elements, the same relationship applies with the absolute value function replaced by the modulus function.

diagram A representation of relationships or information in graphical or pictorial form. For example, statistical graphs, force diagrams in mechanics and decision trees.

diameter A diameter of a circle or *central conic is a line through the centre. If the line meets the circle or central conic at P and Q, then the line segment PQ may also be called a diameter. The term also applies in both senses to a sphere or *central quadric.

In the case of a circle or sphere, all such line segments have the same length. This length is also called the diameter of the circle or sphere, and is equal to twice the radius.

diameter (of a graph) The maximum *eccentricity of any vertex in a graph, which is the largest distance between any pair of vertices in the graph.

diameter of a set (metric space) A *metric space X with metric μ has diameter d if $\mu(x, y) \leq d$ and d is the smallest value for which the inequality holds for all pairs x, y in X. So d is the smallest distance between any two points in the metric space.

diamond See RHOMBUS.

dichotomy A splitting into two non-overlapping parts.

dichotomy paradox One of *Zeno's paradoxes, whose name stems from the basis of the argument, i.e. repeatedly splitting the distance into two parts (the dichotomy) infinitely often. The argument is that motion can never be started because that infinite process will never be completed, because there are an infinite number of milestones to reach before the destination.

die (dice) A small cube with its faces numbered from 1 to 6. When the die is thrown, the probability that any particular number from 1 to 6 is obtained on the face landing uppermost is $\frac{1}{6}$.

difference There can be some ambiguity in the way this is used, as to whether the outcome can be negative. For two individual numbers or quantities a, b the difference is usually taken as $|a - b|$ or the outcome of subtracting the smaller from the larger. From a set of paired values (a_i, b_i) the difference is usually $a_i - b_i$ as a signed value, and the term 'absolute difference' should be used if $|a - b|$ is intended. For example, in a paired test in statistics it is important to retain the information as to which variable was larger in each pair.

difference equation Let $u_0, u_1, u_2, \ldots, u_n, \ldots$ be a sequence (where it is convenient to start with a term u_0). If the terms satisfy the first-order difference equation $u_{n+1} + au_n = 0$, it is easy to see that $u_n = A(-a)^n$, where $A(=u_0)$ is arbitrary.

Suppose that the terms satisfy the second-order difference equation $u_{n+2} + au_{n+1} + bu_n = 0$. Let α and β be the roots of the quadratic equation $x^2 + ax + b = 0$. It can be shown that (i) if $\alpha \neq \beta$, then $u_n = A\alpha^n + B\beta^n$, and (ii) if $\alpha = \beta$, then $u_n = (A + Bn)\,a^n$, where A and B are arbitrary constants. The Fibonacci sequence is given by the difference equation $u_{n+2} = u_{n+1} + u_n$, with $u_0 = 1$ and $u_1 = 1$, and the above method gives

$$u_n = \frac{1}{2}\left(\frac{1 + \sqrt{5}}{2}\right)^n + \frac{1}{2}\left(\frac{1 - \sqrt{5}}{2}\right)^n.$$

Difference equations, also called recurrence relations, do not necessarily have constant coefficients like those considered above; they have similarities with differential equations. By using the notation $\Delta u_n = u_{n+1} - u_n$, difference equations can be written in terms of *finite differences. Indeed, difference equations may arise from the consideration of finite differences.

difference of two squares Since $a^2 - b^2 = (a - b)(a + b)$, any expression with the form of the left-hand side, known as the difference of two squares, can be factorized into the form of the right-hand side.

difference quotient = NEWTON QUOTIENT.

difference sequence If $\{x_i\}$ is a sequence of numbers then $\{x_{i+1} - x_i\}$ is the difference sequence, obtained by subtracting successive terms.

difference set The *difference $A \backslash B$ of sets A and B (subsets of a *universal set) is the set consisting of all elements of A that are not elements of B. The notation $A - B$ is also used. The set is represented by the shaded region of the *Venn diagram shown in the figure.

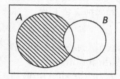

differentiable function The *real function f of one variable is differentiable at a if $(f(a+h) - f(a))/h$ has a limit as $h \to 0$; that is, if the *derivative of f at a exists. The rough idea is that a function is differentiable if it is possible to define the gradient of the graph $y = f(x)$ and hence define a tangent at the point. The function f is differentiable in an open interval if it is differentiable at every point in the interval; and f is differentiable on the closed interval $[a, b]$, where $a < b$, if it is differentiable in (a, b) and if the right derivative of f at a and the *left derivative of f at b exist.

differential calculus The part of mathematics that develops from the definition of the *derivative of a function or the gradient of a graph. The derivative is obtained as the limit of the *Newton quotient, and this is equivalent to the notion of the gradient of a graph as the limit of the gradient of a chord of the graph. From another point of view, the subject is concerned essentially with the *rate of change of one quantity with respect to another.

differential coefficient = DERIVATIVE.

differential equation Suppose that y is a function of x and that $y', y'', \ldots, y^{(n)}$ denote the *derivatives $dy/dx, d^2y/dx^2, \ldots, d^ny/dx^n$. An ordinary differential equation is an equation involving $x, y, y', y'', \ldots$. (The term 'ordinary' is used here to make the distinction from partial differential equations, which involve *partial derivatives and which will not be discussed here.) The order of the differential equation is the order n of the highest derivative $y^{(n)}$ that appears.

The problem of solving a differential equation is to find functions y whose derivatives satisfy the equation. In certain circumstances, it can be shown that a differential equation of order n has a *general solution (that is, a function y, involving n arbitrary constants) that gives all the solutions. A solution given by some set of values of the arbitrary constants is a particular solution. Here are some examples of differential equations and their general solutions, where A, B and C are arbitrary constants:

 (i) $y' - y = 3$ has general solution $y = Ae^x - 3$.
 (ii) $y' = (2x + 3y + 2)/(4x + 6y - 3)$ has general solution
      $\ln|2x + 3y| = 2y - x + C$.
 (iii) $y'' + y = 0$ has general solution $y = A \cos x + B \sin x$.
 (iv) $y'' - 2y' - 3y = e^{-x}$ has general solution $y = Ae^{-x} + Be^{3x} - \frac{1}{4}xe^{-x}$.

Example (ii) shows that it is not necessarily possible to express y explicitly as a function of x.

A differential equation of the first order (that is, of order one) can be expressed in the form $dy/dx = f(x, y)$. Whether or not it can be solved depends upon the function f. Among those that may be solvable are *separable, *homogeneous and *linear first-order differential equations. Among higher-order differential equations that may be solvable reasonably easily are *linear differential equations with constant coefficients.

differential geometry The area of mathematics which uses *differential calculus in the study of geometry. For example, to prove that the area of a circle is exactly πr^2.

differential operator Generally, any operator involving derivatives or partial derivatives. In particular, the operator del $\nabla = \mathbf{i}\dfrac{\partial}{\partial x} + \mathbf{j}\dfrac{\partial}{\partial y} + \mathbf{k}\dfrac{\partial}{\partial z}$,

where $\mathbf{i}$, $\mathbf{j}$, $\mathbf{k}$ are unit vectors in directions OX, OY, OZ, and $\dfrac{\partial}{\partial x}, \dfrac{\partial}{\partial y}, \dfrac{\partial}{\partial z}$ are

the *partial derivatives of the function with respect to x, y, z. See also CURL, DIVERGENCE and GRADIENT.

differentiation The process of obtaining the *derived function f' from the function f, where $f'(x)$ is the *derivative of f at x. See FIRST PRINCIPLES. The derivatives of certain common functions are given in the Table of derivatives (*Appendix 6), and from these many other functions can be differentiated using the following rules of differentiation:

 (i) If $h(x) = kf(x)$ for all x, where k is a constant, then $h'(x) = kf'(x)$.
 (ii) If $h(x) = f(x) + g(x)$ for all x, then $h'(x) = f'(x) + g'(x)$.
 (iii) The product rule: If $h(x) = f(x)g(x)$ for all x, then

$$h'(x) = f(x)g'(x) + f'(x)g(x).$$

(iv) The reciprocal rule: If $h(x) = 1/f(x)$ and $f(x) \neq 0$ for all x, then
$$h'(x) = -\frac{f'(x)}{\left(f(x)\right)^2}.$$

(v) The quotient rule: If $h(x) = f(x)/g(x)$ and $g(x) \neq 0$ for all x, then
$$h'(x) = \frac{g(x)f'(x) - f(x)g'(x)}{\left(g(x)\right)^2}.$$

(vi) The *chain rule: If $h(x) = (f \circ g)(x) = f(g(x))$ for all x, then
$h'(x) = f'(g(x)g'(x)$.

digit A symbol used in writing numbers in their *decimal representation or to some other *base. In decimal notation, the digits used are 0, 1, 2, 3, 4, 5, 6, 7, 8 and 9. In *hexadecimal notation, the digits are 0, 1, 2, 3, 4, 5, 6, 7, 8, 9, A, B, C, D, E and F. In *binary notation, there are just the two digits 0 and 1.

digital In numerical form. For example, a digital watch displays the time by numbers rather than the position of the hands in the traditional clock.

digital computer *See* COMPUTER.

digraph A digraph (or directed graph) consists of a number of vertices, some of which are joined by arcs, where an arc, or directed edge, joins one vertex to another and has an arrow on it to indicate its direction. The arc from the vertex u to the vertex v may be denoted by the ordered pair (u, v). The digraph with vertices u, v, w, x and arcs (u, v), (u, w), (v, u), (w, v), (w, x) is shown in the figure, on the left.

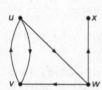
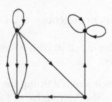

As for *graphs, and with a similar terminology, there may be multiple arcs and loops. A digraph with multiple arcs and loops is shown in the figure, on the right.

dihedral The figure formed by two half-planes and the line at which they intersect.

dihedral group The *group of *symmetries of a regular n-sided polygon; the notation D_n is often used.

Dijkstra's method *See* SHORTEST PATH ALGORITHM.

dilatation A dilatation of the plane from O with scale factor $c(\neq 0)$ is the *transformation of the plane in which the origin O is mapped to itself and a point P is mapped to the point P', where O, P and P' are collinear and $OP' = cOP$. This is given in terms of Cartesian coordinates by $x' = cx$, $y' = cy$.

dimensions In mechanics, physical quantities can be described in terms of the basic dimensions of mass M, length L and time T, using positive and negative indices. For example, the following have the dimensions given: area, L^2; velocity, LT^{-1}; force, MLT^{-2}; linear momentum, MLT^{-1}; energy, ML^2T^{-2}; and power, ML^2T^{-3}. The notation has similarities with that of SI units.

Diophantine equation An algebraic equation in one or more unknowns, with integer coefficients, for which integer solutions are required. A great variety of Diophantine equations have been studied. Some have infinitely many solutions, some have finitely many and some have no solutions. For example:

(i) $14x + 9y = 1$ has solutions $x = 2 + 9t$, $y = -3 - 14t$ (where t is any integer).
(ii) $x^2 + 1 = 2y^4$ has two solutions $x = 1$, $y = 1$ and $x = 239$, $y = 13$.
(iii) $x^3 + y^3 = z^3$ has no solutions.

See also HILBERT'S TENTH PROBLEM and PELL'S EQUATION.

Diophantus of Alexandria (about AD 250) Greek mathematician whose work displayed an algebraic approach to the solution of equations in one or more unknowns, unlike earlier Greek methods that were more geometrical. In the books of *Arithmetica* that survive, particular numerical examples of more than 100 problems are solved, probably to indicate the general methods of solution. These are mostly of the kind now referred to as *Diophantine equations.

Dirac, Paul Adrien Maurice FRS (1902–84) Born in England to a Swiss father and English mother, he was Lucasian Professor of Mathematics at Cambridge University for 37 years. He is best known for bringing together *relativity theory and *quantum mechanics and shared the Nobel prize for physics in 1933 with Erwin *Schrödinger.

direct Two variables are directly related if an increase in one is associated with an increase in the other.

directed graph = DIGRAPH.

directed line A straight line with a specified direction along the line. The specified direction may be called the positive direction and the opposite the negative direction. It may be convenient to distinguish the ends of the line by labelling them x' and x, where the positive direction runs from x' to x. Alternatively, a directed line may be denoted by Ox, where O is a point on the line and the positive direction runs towards the end x.

directed line segment If A and B are two points on a straight line, the part of the line between and including A and B, together with a specified direction along the line, is a directed line segment. Thus $\overrightarrow{AB}$ is the directed line segment from A to B, and $\overrightarrow{BA}$ is the directed line segment from B to A. *See also* VECTOR.

directed number A number with a positive or negative sign when required showing it has a direction from the origin as well as a distance from the origin on the number line.

directed ratio A *ratio of *directed numbers taking account of the sign as well as their magnitudes.

direction The orientation of a line. The line will then have two senses which a directed line will distinguish between.

direction angles The angles that are used in defining direction cosines are known as the direction angles.

direction cosines In a Cartesian coordinate system in 3-dimensional space, a certain direction can be specified as follows. Take a point P such that $\overrightarrow{OP}$ has the given direction and $|OP| = 1$. Let α, β and γ be the three angles $\angle xOP$, $\angle yOP$ and $\angle zOP$, measured in radians ($0 \leq \alpha \leq \pi$, $0 \leq \beta \leq \pi$, $0 \leq \gamma \leq \pi$). Then $\cos \alpha$, $\cos \beta$ and $\cos \gamma$ are the direction cosines of the given direction or of $\overrightarrow{OP}$. They are not independent, however, since $\cos^2 \alpha + \cos^2 \beta + \cos^2 \gamma = 1$. Point P has coordinates ($\cos \alpha$, $\cos \beta$, $\cos \gamma$) and, using the standard unit vectors $\mathbf{i}$, $\mathbf{j}$ and $\mathbf{k}$ along the coordinate axes, the position vector $\mathbf{p}$ of P is given by $\mathbf{p} = (\cos \alpha)\mathbf{i} + (\cos \beta)\mathbf{j} + (\cos \gamma)\mathbf{k}$. So the direction cosines are the components of $\mathbf{p}$. The direction cosines of the x-axis are 1, 0, 0; of the y-axis, 0, 1, 0; and of the z-axis, 0, 0, 1.

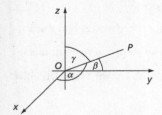

direction fields *See* TANGENT FIELDS.

direction ratios Suppose that a direction has *direction cosines cos α, cos β, cos γ. Any triple of numbers *l*, *m*, *n*, not all zero, such that $l = k \cos \alpha$, $m = k \cos \beta$, $n = k \cos \gamma$, are called direction ratios of the given direction. Since $\cos^2 \alpha + \cos^2 \beta + \cos^2 \gamma = 1$, it follows that

$$\cos\alpha = \frac{\pm l}{\sqrt{l^2 + m^2 + n^2}}, \quad \cos\beta = \frac{\pm m}{\sqrt{l^2 + m^2 + n^2}},$$

$$\cos\gamma = \frac{\pm n}{\sqrt{l^2 + m^2 + n^2}},$$

where either the + sign or the − sign is taken throughout. So any triple of numbers, not all zero, determine two possible sets of direction cosines, corresponding to opposite directions. The triple *l*, *m*, *n* are said to be direction ratios of a straight line when they are direction ratios of either direction of the line.

directly proportional *See* PROPORTION.

direct proof For a theorem that has the form $p \Rightarrow q$, a direct proof is one that supposes *p* and shows that *q* follows. Compare this with an *indirect proof.

directrix (directrices) *See* CONIC, ELLIPSE, HYPERBOLA and PARABOLA.

direct variation *See* PROPORTION.

Dirichlet, Peter Gustav Lejeune (1805–59) German mathematician who was professor at the University of Berlin before succeeding *Gauss at the University of Göttingen. He proved that in any arithmetic series *a*, $a + d$, $a + 2d, \ldots$, where *a* and *d* are relatively prime, there are infinitely many primes. He gave the modern definition of a *function. In more advanced work, he was concerned to see analysis applied to number theory and mathematical physics.

Dirichlet beta function The function $\beta(s) = \sum_{n=0}^{\infty} (-1)^n (2n+1)^{-s}$.

Dirichlet series A series in the form $\sum_{n=1}^{\infty} a_n e^{-\lambda_n z}$ where a_n and z are complex and $\{\lambda_n\}$ is a *monotonic increasing sequence of real numbers. When $\lambda_n = \log_e n$, the series reduces to $\sum_{n=1}^{\infty} a_n n^{-z}$, known as the Dirichlet L-series.

Dirichlet's test A test for *convergence of a series. If $\{a_n\}$ is a series which has bounded partial sums, i.e. $\left| \sum_{n=1}^{m} a_n \right| < K$ for all values of m, and $\{b_n\}$ is decreasing and converges to zero, i.e. $b_n < b_{n-1}$ and $\lim_{n \to \infty} b_n = 0$ then $\sum_{n=1}^{\infty} a_n n_n$ converges.

disc The circle, centre C, with coordinates (a, b) and radius r, has equation $(x-a)^2 + (y-b)^2 = r^2$. The set of points (x, y) in the plane such that $(x-a)^2 + (y-b)^2 < r^2$ forms the interior of the circle, and may be called the open disc, centre C and radius r. The closed disc, centre C and radius r, is the set of points (x, y) such that $(x-a)^2 + (y-b)^2 \leq r^2$.

disconnected graph A graph where the vertices separate into two or more distinct groups, where you cannot link a vertex in one group to a vertex in another by travelling along a series of edges.

discontinuity For a function $f(x)$ there is a discontinuity at x_0 if $f(x_0)$ is not defined or if the values of $f(x_0)$ from the right and left are different. For example, $f(x) = \dfrac{1}{1-x}$ has a discontinuity when $x = 1$, and the rectangular distribution on $[0, k]$ where $f(x) = \dfrac{1}{k}$ for $0 \leq x \leq k$ and $f(x) = 0$ elsewhere has discontinuities at $x = 0$ and $x = k$.

discontinuous function A function which is not continuous. A function only requires one discontinuity to exist to be described thus.

discrete A function or random variable is said to be discrete if it only takes values from a set of distinct values.

discrete data See DATA.

discrete (finite) Fourier transform The discrete Fourier transform of a vector $\mathbf{x} = (x_0, x_1, \ldots, x_{n-1})$ is the vector $\mathbf{y} = (y_0, y_1, \ldots, y_{n-1})$ obtained by calculating $y_k = \sum_{j=0}^{n-1} \omega^{kj} x_j$ for each $k = 0, 1, \ldots, n-1$ where

$\omega = e\left(\dfrac{-2\pi i}{n}\right) = \cos\left(\dfrac{2\pi}{n}\right) - i\sin\left(\dfrac{2\pi}{n}\right)$ This involves a large number of calculations, but the number can be reduced by using the fast Fourier transform.

(((●))) SEE WEB LINKS

• A discrete Fourier transform tool in which choosing different options from the signal menu illustrates a number of signals and the approximating function using a specified number of Fourier coefficients.

discrete metric The *metric f defined on a non-empty set X by $f(x, x) = 0; f(x, y) = 1$ if $x \neq y$.

discrete random variable See RANDOM VARIABLE.

discrete space A *topological space is said to be discrete if all the points are *isolated.

discretization The process of approximating a continuous function or relation by a discrete alternative. With the immense computational power now available, it is possible to use these methods to simulate behaviour in complex situations such as turbulence in air flow which cannot be treated analytically.

discriminant For the *quadratic equation $ax^2 + bx + c = 0$, the quantity $b^2 - 4ac$ is the discriminant. The equation has two distinct real roots, equal roots (that is, one root) or no real roots according to whether the discriminant is positive, zero or negative.

discriminatory A test is said to be discriminatory if its *power is greater than some previously specified level.

disjoint Sets A and B are disjoint if they have no elements in common; that is, if $A \cap B = \emptyset$.

disjunction See INCLUSIVE DISJUNCTION; EXCLUSIVE DISJUNCTION. The common usage of the term 'or' where the compound sentence is true if at least one of the parts is true is inclusive disjunction.

disk An open or closed ball in a metric space.

dispersion A measure of dispersion is a way of describing how scattered or spread out the observations in a sample are. The term is also applied similarly to a random variable. Common measures of dispersion are the *range, *interquartile range, *mean absolute deviation, *variance and *standard deviation. The range may be unduly affected by odd high and low values. The mean absolute deviation is difficult to work with because of the absolute value signs. The standard deviation is in the same units as the data, and it is this that is most often used. The interquartile range may be appropriate when the median is used as the measure of *location.

displacement Suppose that a particle is moving in a straight line, with a point O on the line taken as the origin and one direction along the line taken as positive. Let $|OP|$ be the distance between O and P, where P is the position of the particle at time t. Then the displacement x is equal to $|OP|$ if $\overrightarrow{OP}$ is in the positive direction and equal to $-|OP|$ if is in the negative direction. Indeed, the displacement is equal to the *measure OP.

In the preceding paragraph, a common convention has been followed, in which the unit vector $\mathbf{i}$ in the positive direction along the line has been suppressed. Displacement is in fact a vector quantity, and in the 1-dimensional case above is equal to $x\mathbf{i}$.

When the motion is in two or three dimensions, vectors are used explicitly. The displacement is a vector giving the change in the position of a particle. If the particle P moves from the point A to the point B, not necessarily along a straight line, the displacement is equal to the position vector of B relative to A, that is, the vector represented by the directed line segment $\overrightarrow{AB}$

dissection (of an interval) = PARTITION (of an interval).

dissipative force A force that causes a loss of *energy (considered as consisting of kinetic energy and potential energy). A *resistive force is dissipative because the work done by it is negative.

distance (in the complex plane) If P_1 and P_2 represent the complex numbers z_1 and z_2, the distance $|P_1P_2|$ is equal to $|z_1 - z_2|$, the *modulus of $z_1 - z_2$.

distance between two codewords The distance between two codewords in a *binary code is the number of bits in which the two codewords differ. For example, the distance between 010110 and 001100 is 3 because they differ in the second, third and fifth bits. If the distance between any two different codewords in a binary code is at least 3, the code is an *error-correcting code capable of correcting any one error.

distance between two lines (in 3-dimensional space) Let l_1 and l_2 be lines in space that do not intersect. There are two cases. If l_1 and l_2 are parallel, the distance between the two lines is the length of any line segment $N_1 N_2$, with N_1 on l_1 and N_2 on l_2 perpendicular to both lines. If l_1 and l_2 are not parallel, there are unique points N_1 on l_1 and N_2 on l_2 such that the length of the line segment $N_1 N_2$ is the shortest possible. The length $|N_1 N_2|$ is the distance between the two lines. In fact, the line $N_1 N_2$ is the *common perpendicular of l_1 and l_2.

distance between two points (in the plane) Let A and B have coordinates (x_1, y_1) and (x_2, y_2). It follows from Pythagoras' Theorem that the distance $|AB|$ is equal to $\sqrt{(x_2 - x_1)^2 + (y_2 - y_1)^2}$.

distance between two points (in 3-dimensional space) Let A and B have coordinates (x_1, y_1, z_1) and (x_2, y_2, z_2). Then the distance $|AB|$ is equal to $\sqrt{(x_2 - x_1)^2 + (y_2 - y_1)^2 + (z_2 - z_1)^2}$.

distance between two points (in n-dimensional space) *See* N-DIMENSIONAL SPACE.

distance from a point to a line (in the plane) The distance from the point P to the line l is the shortest distance between P and a point on l. It is equal to $|PN|$, where N is the point on l such that the line PN is perpendicular to l. If P has coordinates (x_1, y_1) and l has equation $ax + by + c = 0$, then the distance from P to l is equal to

$$\frac{|ax_1 + by_1 + c|}{\sqrt{a^2 + b^2}},$$

where $|ax_1 + by_1 + c|$ is the *absolute value of $ax_1 + by_1 + c$.

distance from a point to a plane (in 3-dimensional space) The distance from the point P to the plane p is the shortest distance between P and a point in p, and is equal to $|PN|$, where N is the point in p such that the line PN is normal to p. If P has coordinates (x_1, y_1, z_1) and p has equation $ax + by + cz + d = 0$, the distance from P to p is equal to

$$\frac{|ax_1 + by_1 + cz_1 + d|}{\sqrt{a^2 + b^2 + c^2}}$$

where $|ax_1 + by_1 + cz_1 + d|$ is the *absolute value of $ax_1 + by_1 + cz_1 + d$.

distance-time graph A graph that shows *displacement plotted against time for a particle moving in a straight line. Let $x(t)$ be the displacement of the particle at time t. The distance-time graph is the graph $y = x(t)$, where the t-axis is horizontal and the y-axis is vertical with the positive direction upwards. The gradient at any point is equal to the velocity of the particle at that time. (Here a common convention has been followed, in which the unit vector **i** in the positive direction along the line has been suppressed. The displacement of the particle is in fact a vector quantity equal to $x(t)\mathbf{i}$, and the velocity of the particle is a vector quantity equal to $x(t)\mathbf{i}$.)

distinct Not numerically equal.

distribution The distribution of a random variable is concerned with the way in which the probability of its taking a certain value, or a value within a certain interval, varies. It may be given by the *cumulative distribution function. More commonly, the distribution of a discrete random variable is given by its *probability mass function and that of a continuous random variable by its *probability density function.

distribution-free methods = NON-PARAMETRIC METHODS.

distribution function = CUMULATIVE DISTRIBUTION FUNCTION.

distributive Suppose that ∘ and ∗ are *binary operations on a set S. Then ∘ is distributive over ∗ if, for all a, b and c in S,

$$a \circ (b * c) = (a \circ b) * (a \circ c) \text{ and } (a * b) \circ c = (a \circ c) * (b \circ c)$$

If the two operations are multiplication and addition, 'the distributive laws' normally means those that say that multiplication is distributive over addition.

div Abbreviation for *divergence.

diverge (of a sequence) Not having a finite limit. *See* LIMIT OF A SEQUENCE.

divergence For a vector function of position $\mathbf{V}(\mathbf{r}) = V_x\mathbf{i} + V_y\mathbf{j} + V_z\mathbf{k}$, the divergence of $\mathbf{V}$ is the *scalar product of the operator

del, $\nabla = \mathbf{i}\dfrac{\partial}{\partial x} + \mathbf{j}\dfrac{\partial}{\partial y} + \mathbf{k}\dfrac{\partial}{\partial z}$ with $\mathbf{V}$ giving

div $\mathbf{V} = \nabla . \mathbf{V} = \left(\mathbf{i}\dfrac{\partial}{\partial x} + \mathbf{j}\dfrac{\partial}{\partial y} + \mathbf{k}\dfrac{\partial}{\partial z}\right) . \mathbf{V} = \dfrac{\partial V_x}{\partial x} + \dfrac{\partial V_y}{\partial y} + \dfrac{\partial V_z}{\partial z}$. *Compare* CURL, GRADIENT.

divergent An infinite series $a_1 + a_2 + a_3 + \ldots$ whose partial sums $\sum_{n=1}^{m} a_n$ do not approach a finite limit as $m \to \infty$. Either $\sum_{n=1}^{m} a_n \to \pm\infty$ as $m \to \infty$ or $\sum_{n=1}^{m} a_n$ oscillates in value. For example, $\sum_{n=1}^{m}(-2)^n$ will alternately be positive and negative, with $\left|\sum_{n=1}^{m}(-2)^n\right| \to \infty$ as $m \to \infty$, and $\sum_{n=1}^{m} i^n$ taking the values $i, -1+i, -1, 0, i, -1+i, \ldots$.

divides Let a and b be integers. Then a divides b (which may be written as $a \mid b$) if there is an integer c such that $ac = b$. It is said that a is a divisor or factor of b, that b is divisible by a, and that b is a multiple of a.

divisible *See* DIVIDES.

division (of a segment) The construction of a point which divides a line segment in specified proportions. This may be *internal division or *external division.

Division Algorithm The following theorem of elementary number theory:

Theorem

For integers a and b, with $b > 0$, there exist unique integers q and r such that $a = bq + r$, where $0 \leq r < b$.

In the division of a by b, the number q is the quotient and r is the remainder.

divisor See DIVIDES.

divisor function The number of divisors of n, including 1 and n, denoted by $d(n)$. So $d(6) = 4$ since 1, 2, 3 and 6 are divisors of 6. For any prime number p, $d(p^k) = k + 1$.

divisor of zero If in a *ring there are non-zero elements a and b such that $ab = 0$, then a and b are divisors of zero. For example, in the ring of 2×2 real matrices,

$$\begin{bmatrix} 0 & 1 \\ 0 & 0 \end{bmatrix} \begin{bmatrix} 1 & 0 \\ 0 & 0 \end{bmatrix} = \begin{bmatrix} 0 & 0 \\ 0 & 0 \end{bmatrix},$$

and so each of the matrices on the left-hand side is a divisor of zero. In the ring $\mathbf{Z}_6$, consisting of the set $\{0, 1, 2, 3, 4, 5\}$ with addition and multiplication modulo 6, the element 4 is a divisor of zero since $4.3 = 0$.

$d(n)$ The *divisor function.

dodecagon A twelve-sided polygon.

dodecahedron (dodecahedra) A *polyhedron with twelve faces, often assumed to be regular. The regular dodecahedron is one of the *Platonic solids, and its faces are regular pentagons. It has 20 vertices and 30 edges.

Dodgson, Charles Lutwidge (1832–98) British mathematician and logician, better known under the pseudonym Lewis Carroll as the author of *Alice's Adventures in Wonderland*. He was lecturer in mathematics at Christ Church, Oxford, and wrote a number of minor books on mathematics.

domain See FUNCTION and MAPPING.

dot The symbol '.' used to represent the decimal point or multiplication. It is also used to represent the differential of a function where there is no ambiguity as to what the function is being differentiated with respect to.

For example, $\dot{y}$ would mean $\dfrac{dy}{dt} = f'(t)$ if $y = f(t)$.

dot product = SCALAR PRODUCT.

double-angle formula (in hyperbolic functions) *See* HYPERBOLIC FUNCTIONS.

double-angle formula (in trigonometry) A formula in trigonometry that expresses a function of a double angle in terms of the single angle. This can be obtained from the corresponding *compound angle formulae by substituting $A = B = x$:

$$\sin(2x) = 2 \sin x \cos x$$
$$\cos(2x) = \cos^2 x - \sin^2 x$$
$$\tan(2x) = \frac{2 \tan x}{1 - \tan^2 x}.$$

double dummy *See* BLINDING.

double integral The integral of a function with respect to two variables, so if f is a function of two variables x, y, we can have $\iint f(x, y) \, dx \, dy$.

double negation The proposition that the negation of the negation of A is equivalent to the proposition A. In English, a double negative is not always precisely a double negation. For example, in hypothesis testing the conclusion 'there is not sufficient evidence to suggest that the mean is not 340' is not equivalent to saying that there is evidence to suggest that the mean is 340.

double point A point on a curve where the curve intersects itself. If the tangents are not coincident then the point is a node.

double precision *See* PRECISION.

double root *See* ROOT.

double sequence A sequence with twin indices. For example,
$$a_{n,m} = \frac{1}{n^2 + m^2}.$$

double series A series with two indices. For example, for any joint probability distribution with $p_{ij} = \Pr\{X = x_i, Y = y_j\}$,
$$E(Z = X + Y) = \sum_i \sum_j \left((x_i + y_j) \times p_{ij}\right).$$

double tangent A line which is a tangent to a curve at two distinct points. A pair of distinct tangents to the curve at a single point, as happens at a *cusp.

doubly stochastic matrix *See* STOCHASTIC MATRIX.

drag *See* AERODYNAMIC DRAG.

dummy activity (in critical path analysis) When two paths in an
*activity network (edges as activities) have a common event but are
independent, or partly independent of one another, it is necessary to
introduce a dummy activity, essentially a logical constraint, linking the two
paths at that point, with the duration of a dummy activity being zero. It is
usually shown as a dotted line, and can be labelled d_t. If R depends on P
and Q and S depends on P then the activity network would show:

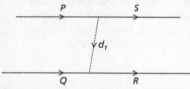

Activity networks allow at most one activity to be represented by an edge,
so if two activities Q, R have to be carried out after one activity P and before
another S a dummy activity is required as shown in the figure below.

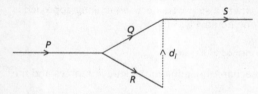

dummy variable A variable appearing in an expression is a dummy
variable if the letter being used could equally well be replaced by another
letter. For example, the two expressions

$$\int_0^1 x^2 dx \quad \text{and} \quad \int_0^1 t^2 dt$$

represent the same definite integral, and so x and t are dummy variables.
Similarly, the summation

$$\sum_{r=1}^5 r^2$$

denotes the sum $1^2 + 2^2 + 3^2 + 4^2 + 5^2$, and would still do so if the letter r
were replaced by the letter s, say; so r here is a dummy variable.

duplication of the cube One of the problems that Greek geometers
attempted (like *squaring the circle and *trisection of an angle) was to find
a construction, with ruler and pair of compasses, to obtain the side of a

cube whose volume was twice the volume of a given cube. This is equivalent to finding a geometrical construction to obtain a length of $\sqrt[3]{2}$ from a given unit length. Now constructions of the kind envisaged can only give lengths belonging to a class of numbers obtained, essentially, by addition, subtraction, multiplication, division and the taking of square roots. Since $\sqrt[3]{2}$ does not belong to this class of numbers, the duplication of the cube is impossible.

du Sautoy, Marcus (1965–) British mathematician who has done much to improve the communication of mathematical ideas through regular newspaper articles, the Royal Institution Christmas Lectures in 2006, and books. He was awarded the Berwick Prize of the London Mathematical Society in 2001 for his work on *zeta functions.

((⊕)) SEE WEB LINKS
• Marcus du Sautoy's website at Oxford University.

dynamic equilibrium A body is in dynamic equilibrium if it is moving with constant velocity and the vector sum of the forces acting on it is zero.

dynamic programming The area of mathematics relating to the study of optimization problems where a step-wise decision making approach is employed. This is often done iteratively.

((⊕)) SEE WEB LINKS
• A detailed introduction to the method with examples.

dynamics The area of mechanics relating to the study of forces and the motion of bodies.

e The number that is the base of natural logarithms. There are several ways of defining it. Probably the most satisfactory is this. First, define ln as in approach **2** to the *logarithmic function. Then define exp as the inverse function of ln (see approach **2** to the *exponential function). Then define *e* as equal to exp 1. This amounts to saying that *e* is the number that makes

$$\int_1^e \frac{1}{t}dt = 1.$$

It is necessary to go on to show that e^x and exp x are equal and so are identical as functions, and also that ln and $\log_e$ are identical functions.

The number *e* has important properties derived from some of the properties of ln and exp. For example,

$$e = \lim_{h \to 0} (1 + h)^{1/h} = \lim_{n \to \infty} \left(1 + \frac{1}{n}\right)^n.$$

Also, *e* is the sum of the series

$$1 + \frac{1}{1!} + \frac{1}{2!} + \cdots + \frac{1}{n!} + \cdots.$$

Another approach, but not a recommended one, is to make one of these properties the definition of *e*. Then exp x would be defined as e^x, ln x would be defined as its inverse function, and the properties of these functions would have to be proved.

The value of *e* is 2.718 281 83 (to 8 decimal places). The proof that *e* is *irrational is comparatively easy. In 1873, Hermite proved that *e* is *transcendental, and his proof was subsequently simplified by Hilbert.

e (in conics) The symbol for the *eccentricity of a *conic.

e (in group theory) A common notation for the *neutral element in a *group.

E The number 14 in *hexadecimal representation.

E(X) *See* EXPECTED VALUE.

Earth Our particular planet in the solar system. The Earth is often assumed to be a sphere with a radius of approximately 6400 kilometres.

A more accurate accepted value is 6371 km. A better approximation is to say that the Earth is in the shape of an oblate spheroid with an equatorial radius of 6378 km and a polar radius of 6357 km. The mass of the Earth is 5.976×10^{24} kg.

eccentricity The ratio of the distances between a point on a *conic and a fixed point (the focus) and between the point and a fixed line (the directrix). For $e = 0$ a circle is produced, for $0 < e < 1$ an ellipse is produced, for $e = 1$ a parabola is produced and for $e > 1$ the conic produced is a hyperbola.

(⊕) SEE WEB LINKS
• Two animations of ellipses with different eccentricities.

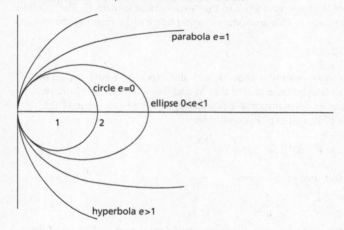

eccentricity (graph) For a vertex V in a graph, the eccentricity is the greatest distance from V to any other vertex in the graph.

echelon form Suppose that a row of a matrix is called zero if all its entries are zero. Then a matrix is in echelon form if (i) all the zero rows come below the non-zero rows, and (ii) the first non-zero entry in each non-zero row is 1 and occurs in a column to the right of the leading 1 in the row above. For example, these two matrices are in echelon form:

$$\begin{bmatrix} 1 & 6 & -1 & 4 & 2 \\ 0 & 0 & 1 & 2 & -3 \\ 0 & 0 & 0 & 1 & 5 \end{bmatrix}, \quad \begin{bmatrix} 1 & 6 & -1 & 4 & 2 \\ 0 & 1 & 2 & -3 & 5 \\ 0 & 0 & 0 & 0 & 0 \end{bmatrix}.$$

Any matrix can be transformed to a matrix in echelon form using *elementary row operations, by a method known as *Gaussian elimination. The solutions of a set of linear equations may be investigated by transforming the *augmented matrix to echelon form. Further elementary row operations may be used to transform a matrix to *reduced echelon form. A set of linear equations is said to be in echelon form if its augmented matrix is.

EDA = EXPLORATORY DATA ANALYSIS.

edge (of a graph), **edge-set** *See* GRAPH.

efficiency *See* MACHINE.

efficiency (in statistics) A statistic A is more efficient than another statistic B in estimating a parameter θ if its variance is smaller. *See* ESTIMATOR.

efficient code In computer science, efficiency is used to describe the properties of an algorithm which relate to the resources being used by the coding. The recent huge increase in computational power stems from both improvements in hardware technology and the software becoming much more efficient.

effort *See* MACHINE.

eigenvalue, eigenvector = CHARACTERISTIC VALUE, CHARACTERISTIC VECTOR.

Einstein, Albert (1879–1955) Outstanding mathematical physicist whose work was the single most important influence on the development of physics since *Newton. Born in Ulm in Germany, he lived in Switzerland and Germany before moving to the United States in 1933. He was responsible in 1905 for the Special Theory of Relativity and in 1916 for the General Theory. He made a fundamental contribution to the birth of quantum theory and had an important influence on thermodynamics. He is perhaps most widely known for his equation $E = mc^2$, quantifying the equivalence of matter and energy. He regarded himself as a physicist rather than as a mathematician, but his work has triggered off many developments in modern mathematics. Contrary to the popular image of a white-haired professor scribbling incomprehensible symbols on a blackboard, Einstein's great strength was his ability to ask simple questions and give simple answers. In that way, he changed our view of the universe and our concepts of space and time.

Eisenstein's criterion For a polynomial
$F(x) = a_n x^n + a_{n-1} x^{n-1} + \ldots + a_1 x + a_0$ with integer coefficients to be

irreducible (over the integers or rationals) it is *sufficient to find a prime number p with the following properties:

(i) p divides each a_i for $i \neq n$;
(ii) p does not divide a_n;
(iii) p^2 does not divide a_0.

For example, $x^n - 2$ is irreducible for all values of n but $x^n - 4$ is reducible as a difference of two squares when $n = 2$. Eisenstein's criterion is sufficient but it is not *necessary—there are many simple polynomials which are irreducible without the condition being met. For example, $x^2 + 1$ is irreducible.

elastic A body is said to be elastic if, after being deformed by forces applied to it, it is able to regain its original shape as soon as the deforming forces cease to act.

elastic collision A *collision in which there is no loss of kinetic energy.

(⊕) SEE WEB LINKS

• An applet allowing elastic collisions of objects of different masses and initial speeds to be explored, showing the behaviour of velocity, momentum, and kinetic energy of the two objects.

elasticity *See* MODULUS OF ELASTICITY.

elastic potential energy *See* POTENTIAL ENERGY.

elastic string A string that can be extended, but not compressed, and that resumes its natural length as soon as the forces applied to extend it are removed. How the *tension in the string varies with the *extension may be complicated. In the simplest mathematical model, it is assumed that the tension is proportional to the extension; that is, that *Hooke's law holds. A particle suspended from a fixed support by an elastic string performs *simple harmonic motion in the same way as a particle suspended by a spring does, provided that the amplitude of the oscillations is sufficiently small for the string never to go slack.

electric field A *vector field which exerts a force on any electrical charge at a given point. Electrical charge is measured in coulombs, so the field has SI units of newtons per coulomb. Electric fields which change with time influence local magnetic fields, and many advances in communications and medical technologies are based on electromagnetism.

electromagnetic field A physical field which describes the interaction between magnetic and electric fields, which will affect the behaviour of charged bodies.

electromagnetic field tensor The mathematical object that describes the *electromagnetic field of a physical system. Since the behaviour of the field is governed by Maxwell's equations, the use of tensors allows these to be written in a very concise form.

electromagnetic potentials *Maxwell's equations can be simplified considerably by defining a scalar field ϕ and a vector field $\mathbf{A}$ in terms of which $\mathbf{E}$, the electric field, and $\mathbf{B}$, the magnetic field, can be expressed. $\mathbf{A}$ and ϕ are given by $\mathbf{B} = \nabla \times \mathbf{A}$, $\mathbf{E} = -\nabla\phi - \frac{1}{c}\frac{\partial \mathbf{A}}{\partial t}$, where c is the velocity of light. $\mathbf{A}$ and ϕ are not uniquely determined by these expressions, and they are normally chosen to satisfy the extra condition

$$\nabla . \mathbf{A} + \frac{1}{c}\frac{\partial \phi}{\partial t} = 0.$$

electromagnetic radiation A form of energy which behaves like a wave as it moves. It is generated by the movement of an electrically charged body and it has both electric and magnetic field components, which oscillate in planes perpendicular to one another and to the direction the wave moves in.

electromagnetic wave The description of how *electromagnetic radiation is propagated. In a vacuum, the wave travels at the speed of light.

element An object in a set is an element of that set.

elementary column operation One of the following operations on the columns of a matrix:

   (i) interchange two columns,
  (ii) multiply a column by a non-zero scalar,
 (iii) add a multiple of one column to another column.

An elementary column operation can be produced by post-multiplication by the appropriate *elementary matrix.

elementary function Any of the following real functions: the *rational functions, the *trigonometric functions, the *logarithmic and *exponential functions, the functions f defined by $f(x) = x^{m/n}$ (where m and n are non-zero integers), and all those functions that can be obtained from these by using addition, subtraction, multiplication, division, *composition and the taking of *inverse functions.

elementary matrix A square matrix obtained from the identity matrix I by an *elementary row operation. Thus there are three types of elementary matrix. Examples of each type are:

$$
\text{(i)}
\begin{bmatrix}
1 & 0 & 0 & 0 & 0 & 0 & 0 \\
0 & 0 & 0 & 0 & 1 & 0 & 0 \\
0 & 0 & 1 & 0 & 0 & 0 & 0 \\
0 & 0 & 0 & 1 & 0 & 0 & 0 \\
0 & 1 & 0 & 0 & 0 & 0 & 0 \\
0 & 0 & 0 & 0 & 0 & 1 & 0 \\
0 & 0 & 0 & 0 & 0 & 0 & 1
\end{bmatrix},
\quad
\text{(ii)}
\begin{bmatrix}
1 & 0 & 0 & 0 & 0 & 0 & 0 \\
0 & 1 & 0 & 0 & 0 & 0 & 0 \\
0 & 0 & -3 & 0 & 0 & 0 & 0 \\
0 & 0 & 0 & 1 & 0 & 0 & 0 \\
0 & 0 & 0 & 0 & 1 & 0 & 0 \\
0 & 0 & 0 & 0 & 0 & 1 & 0 \\
0 & 0 & 0 & 0 & 0 & 0 & 1
\end{bmatrix},
$$

$$
\text{(iii)}
\begin{bmatrix}
1 & 0 & 0 & 0 & 0 & 0 & 0 \\
0 & 1 & 0 & 0 & 4 & 0 & 0 \\
0 & 0 & 1 & 0 & 0 & 0 & 0 \\
0 & 0 & 0 & 1 & 0 & 0 & 0 \\
0 & 0 & 0 & 0 & 1 & 0 & 0 \\
0 & 0 & 0 & 0 & 0 & 1 & 0 \\
0 & 0 & 0 & 0 & 0 & 0 & 1
\end{bmatrix}.
$$

The matrix (i) is obtained from $\mathbf{I}$ by interchanging the second and fifth rows, matrix (ii) by multiplying the third row by -3, and matrix (iii) by adding 4 times the fifth row to the second row. Pre-multiplication of an $m \times n$ matrix $\mathbf{A}$ by an $m \times m$ elementary matrix produces the result of the corresponding row operation on $\mathbf{A}$.

Alternatively, an elementary matrix can be seen as one obtained from the identity matrix by an *elementary column operation; and post-multiplication of an $m \times n$ matrix $\mathbf{A}$ by an $n \times n$ elementary matrix produces the result of the corresponding column operation on $\mathbf{A}$.

elementary operation Addition, subtraction, multiplication, division and finding integer roots are the elementary operations.

elementary row operation One of the following operations on the rows of a matrix:

(i) interchange two rows,
(ii) multiply a row by a non-zero scalar,
(iii) add a multiple of one row to another row.

An elementary row operation can be produced by a pre-multiplication by the appropriate *elementary matrix. Elementary row operations are applied to the *augmented matrix of a set of linear equations to transform it into *echelon form or *reduced echelon form. Each elementary row operation corresponds to an operation on the set of linear equations that does not alter the solution set of the equations.

elevation The angle between the horizontal and a straight line, measured positively in the direction of the upwards vertical.

elimination method A method of solving linear simultaneous
equations by reducing the number of variables by taking appropriate
linear combinations of the equations. For example, when $3x + 2y = 7$ (I)
and $5x - 3y = -1$ (II) then $3 \times$ (I) $+ 2 \times$ (II) $\Rightarrow 19x = 19$, so the variable y
has been eliminated, allowing the variable x to be identified as 1, and then
the substitution back into either of the original equations gives the
corresponding value for y.

ellipse A particular 'oval' shape, obtained, it could be said, by stretching
or squashing a circle. If it has length $2a$ and width $2b$, its area equals πab.

In more advanced work, a more precise definition of an ellipse is
required. One approach is to define it as a *conic with eccentricity less
than 1. Thus it is the locus of all points P such that the distance from P to a
fixed point F_1 (the focus) equals e (<1) times the distance from P to a fixed
line l_1 (the directrix). It turns out that there is another point F_2 and another
line l_2 such that the same locus would be obtained with these as focus and
directrix. An ellipse is also the conic section that results when a plane cuts
a cone in such a way that a finite section is obtained (*see* CONIC).

The line through F_1 and F_2 is the major axis, and the points V_1 and V_2
where it cuts the ellipse are the vertices. The length $|V_1 V_2|$ is the length of the
major axis and is usually taken to be $2a$. The midpoint of $V_1 V_2$ is the centre
of the ellipse. The line through the centre perpendicular to the major axis is
the minor axis, and the distance, usually taken to be $2b$, between the points
where it cuts the ellipse is the length of the minor axis. The three constants
a, b and e are related by $b^2 = a^2(1 - e^2)$ or, in another form, $e^2 = 1 - b^2/a^2$.
The eccentricity e determines the shape of the ellipse. The value $e = 0$ is
permitted and gives rise to a circle, though this requires the directrices to be
infinitely far away and invalidates the focus and directrix approach.

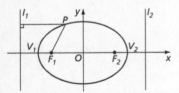

By taking a coordinate system with origin at the centre of the ellipse and
x-axis along the major axis, the foci have coordinates $(ae, 0)$ and $(-ae, 0)$,
the directrices have equations $x = a/e$ and $x = -a/e$, and the ellipse has
equation

$$\frac{x^2}{a^2} + \frac{y^2}{b^2} = 1,$$

where $a > b > 0$. When investigating the properties of an ellipse, it is a
common practice to choose this convenient coordinate system. It may be
useful to take $x = a \cos \theta$, $y = b \sin \theta$ $(0 \leq \theta < 2\pi)$ as *parametric equations.

An ellipse with its centre at the origin and its major axis, of length $2a$, along the y-axis instead has equation $y^2/a^2 + x^2/b^2 = 1$, where $a > b > 0$, and its foci are at $(0, ae)$ and $(0, -ae)$.

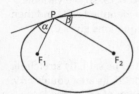

The ellipse has two important properties:

(i) If P is any point of the ellipse with foci F_1 and F_2 and length of major axis $2a$, then $|PF_1| + |PF_2| = 2a$. The fact that an ellipse can be seen as the locus of all such points is the basis of a practical method of drawing an ellipse using a string between two points.

(ii) For any point P on the ellipse, let α be the angle between the tangent at P and the line PF_1, and β the angle between the tangent at P and the line PF_2, as shown in the figure; then $\alpha = \beta$. This property is analogous to that of the parabolic reflector (*see* PARABOLA).

ellipsoid A *quadric whose equation in a suitable coordinate system is

$$\frac{x^2}{a^2} + \frac{y^2}{b^2} + \frac{z^2}{c^2} = 1.$$

The three axial planes are planes of symmetry. All non-empty plane sections are ellipses.

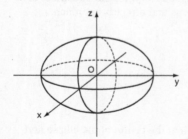

elliptic cylinder A *cylinder in which the fixed curve is an *ellipse and the fixed line to which the generators are parallel is perpendicular to the plane of the ellipse. It is a *quadric, and in a suitable coordinate system has equation

$$\frac{x^2}{a^2} + \frac{y^2}{b^2} = 1.$$

elliptic function A function defined on the complex plane for which $f(z) = f(z+a) = f(z+b)$ where a/b is not real. From this it follows that $f(z + ma + nb) = f(z)$ for all integers m, n and that the function is periodic in two distinct directions on the complex plane.

elliptic geometry *See* NON-EUCLIDEAN GEOMETRY.

elliptic integral A function $f(x)$ which can be expressed as an integral of the form $f(x) = \int_c^x R\left(u, \sqrt{P(u)}\right).du$ where R is a *rational function, P is a cubic or quartic function with no repeated roots and c is a constant. The name is because integrals in this form were first studied in connection with the arc length of an *ellipse.

elliptic paraboloid A *quadric whose equation in a suitable coordinate system is

$$\frac{x^2}{a^2} + \frac{y^2}{b^2} = \frac{2z}{c}.$$

Here the yz-plane and the zx-plane are planes of symmetry. Sections by planes $z = k$, where $k \geq 0$, are ellipses (circles if $a = b$); planes $z = k$, where $k < 0$, have no points of intersection with the paraboloid. Sections by planes parallel to the yz-plane and to the zx-plane are parabolas. Planes through the z-axis cut the paraboloid in parabolas with vertex at the origin.

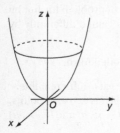

empirical Deriving from experience or observation rather than from reasoning.

empirical probability The probability that a fair die will show a four when thrown is 1/6, using an argument based on equally likely outcomes. For a die which is weighted, observations would need to be taken and an empirical probability based on their *relative frequency could be calculated.

empty set The set, denoted by $\emptyset$ with no elements in it. Consequently, its *cardinality, $n(\emptyset)$, is zero.

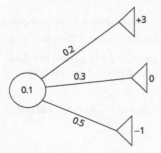

emv (expected monetary value) Where the *payoff to the player in a game is dependent on chance outcomes, the emv is the average gain the player would achieve per game in the long run. It is calculated as sum of: each payoff multiplied by the probability of that payoff.

For example

The emv is $3 \times 0.2 + 0 \times 0.3 + (-1) \times 0.5 = 0.1$ and the value can be written in the circle.

emv algorithm For a decision tree with all *payoffs and the probabilities of all chance outcomes known, determine the *emv for the chance nodes (or vertices) starting from the payoffs. At each decision node the player takes the decision which maximizes the *emv.

The notion of a *utility function develops this type of problem further by recognizing that the same amount of money may be valued differently by people in different financial circumstances.

encrypt To transform information or data into a coded form.

end-point A number defining one end of an *interval on the real line. Each of the finite intervals $[a, b]$, (a, b), $[a, b)$ and $(a, b]$ has two end-points, a and b. Each of the infinite intervals $[a, \infty)$, (a, ∞), $(-\infty, a]$ and $(-\infty, a)$ has one end-point, a.

energy Mechanics is, in general, concerned with two forms of energy: *kinetic energy and *potential energy. When there is a transference of energy of one of these forms into energy of another form such as heat or noise, there may be said to be a loss of energy.

Energy has dimensions ML^2T^{-2}, and the SI unit of measurement is the *joule.

energy equation *See* CONSERVATION OF ENERGY.

enneagon A nine-sided polygon.

entering variable *See* SIMPLEX METHOD.

entire function A complex-valued function that is *holomorphic everywhere in the complex plane. Polynomials, exponentials, and trigonometric and hyperbolic functions are all examples, while the natural logarithm and the function $f: z \rightarrow 1/z$, which is not holomorphic at zero (but is *meromorphic), are not entire functions.

entry *See* MATRIX.

enumerable = DENUMERABLE.

enumeration *See* PERMUTATIONS and SELECTIONS.

envelope A curve or surface that is tangential to every curve or surface in a family. For example, if a family of circles has radius a and centre at a distance $b > a$ from a fixed point C, the envelope will be an annulus with the two circles having radii $b + a$ and $b - a$.

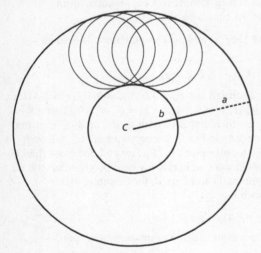

epicycloid The curve traced out by a point on the circumference of a circle rolling round the outside of a fixed circle. When the two circles have the same radius, the curve is a *cardioid.

epimorphism A *morphism $f: X \rightarrow Y$ with the property that for all morphisms g between Y and Z, $g_1 \circ f = g_2 \circ f \Rightarrow g_1 = g_2$.

epsilon The Greek letter e, written ε, commonly used to represent a small, strictly positive quantity.

epsilon–delta notation The standard notation used to define the concepts of *limits and continuity.

equality (of complex numbers) *See* EQUATING REAL AND IMAGINARY PARTS.

equality (of matrices) Matrices $\mathbf{A}$ and $\mathbf{B}$, where $\mathbf{A} = [a_{ij}]$ and $\mathbf{B} = [b_{ij}]$, are equal if and only if they have the same order and $a_{ij} = b_{ij}$ for all i and j.

equality (of sets) Sets A and B are equal if they consist of the same elements. In order to establish that $A = B$, a technique that can be useful is to show, instead, that both $A \subseteq B$ and $B \subseteq A$.

equality (of vectors) *See* VECTOR.

equals sign (equal sign) The symbol ' $=$ ' used between two expressions to indicate that they take the same value.

equate To form an *equation by joining two expressions, or an expression and a value, by an equals sign.

equating coefficients Let $f(x)$ and $g(x)$ be polynomials, and let

$$f(x) = a_n x^n + a_{n-1} x^{n-1} + \cdots + a_1 x + a_0,$$
$$g(x) = b_n x^n + b_{n-1} x^{n-1} + \cdots + b_1 x + b_0,$$

where it is not necessarily assumed that $a_n \neq 0$ and $b_n \neq 0$. If $f(x) = g(x)$ for all values of x, then $a_n = b_n$, $a_{n-1} = b_{n-1}, \ldots, a_1 = b_1$, $a_0 = b_0$. Using this fact is known as equating coefficients. The result is obtained by applying the *Fundamental Theorem of Algebra to the polynomial $h(x)$, where $h(x) = f(x) - g(x)$. If $h(x) = 0$ for all values of x (or, indeed, for more than n values of x), the only possibility is that $h(x)$ is the zero polynomial with all its coefficients zero. The method can be used, for example, to find numbers A, B, C and D such that

$$x^3 = A(x - 1)(x - 2)(x - 3) + B(x - 1)(x - 2) + C(x - 1) + D$$

for all values of x. It is often used to find the unknowns in *partial fractions.

equating real and imaginary parts Complex numbers $a + bi$ and $c + di$ are equal if and only if $a = c$ and $b = d$. Using this fact is called equating real and imaginary parts. For example, if $(a + bi)^2 = 5 + 12i$, then $a^2 - b^2 = 5$ and $2ab = 12$.

equation A statement that asserts that two mathematical expressions are equal in value. If this is true for all values of the variables involved then it is called an *identity, for example $3(x - 2) = 3x - 6$, and where it is only true for some values it is called a *conditional equation; for example $x^2 - 2x - 3 = 0$ is only true when $x = -1$ or 3, which are known as the *roots of the equation.

equation of motion An equation, based on the second of *Newton's laws of motion, that governs the motion of a particle. In vector form, the equation is $m\ddot{\mathbf{r}} = \mathbf{F}$, where $\mathbf{F}$ is the total force acting on the particle of mass m. In Cartesian coordinates, this is equivalent to the three differential equations $m\ddot{x} = F_1$, $m\ddot{y} = F_2$ and $m\ddot{z} = F_3$, where $\mathbf{F} = F_1\mathbf{i} + F_2\mathbf{j} + F_3\mathbf{k}$. In cylindrical polar coordinates, taken here to be (r, θ, z), it is equivalent to $m(\ddot{r} - r\dot{\theta}^2) = F_1$, $m(r\ddot{\theta} + 2\dot{r}\dot{\theta}) = F_2$, and $m\ddot{z} = F_3$, where now $\mathbf{F} = F_1\mathbf{e}_r + F_2\mathbf{e}_\theta + F_3\mathbf{k}$, and $\mathbf{e}_r = \mathbf{i}\cos\theta + \mathbf{j}\sin\theta$ and $\mathbf{e}_\theta = -\mathbf{i}\sin\theta + \mathbf{j}\cos\theta$ (*see* RADIAL AND TRANSVERSE COMPONENTS).

equations of motion with constant acceleration Equations relating to an object moving in a straight line with constant acceleration a. Let u be the initial velocity, v the final velocity, t the time of travel and s the displacement from the starting point. Then

$$v = u + at, \quad s = ut + \tfrac{1}{2}at^2, \quad s = \tfrac{1}{2}(u+v)t, \quad v^2 = u^2 + 2as.$$

These equations may be applied to a particle falling under *gravity near the Earth's surface. If the positive direction is taken to be downwards, then $a = g$. When a particle is projected vertically upwards from the ground, it may be appropriate to take the positive direction upwards, and then $a = -g$.

equator A circle which divides a sphere, or other body, into two symmetrical parts.

equi- A prefix denoting equality.

equiangular (of figures) A pair of figures for which the angles taken in order in each figure are equal.

equiangular (within a figure) A figure that has all angles equal.

equiangular spiral A curve whose equation in polar coordinates is $r = ae^{k\theta}$, where $a\ (>0)$ and k are constants. Let O be the origin and P be any point on the curve. The curve derives its name from the property that the angle α between OP and the tangent at P is constant. In fact, $k = \cot\alpha$. The equation can be written $r = k\theta + b$, and the curve is also called the logarithmic spiral.

• Animations of related curves, and an interactive manipulation of the spiral in Geometer's Sketchpad.

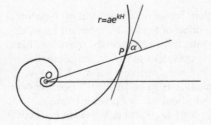

equidistant Being the same distance from one or more points or objects. The perpendicular bisector of the line segment AB is the set of points equidistant from A and B.

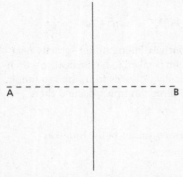

The set of points equidistant from a fixed point is a circle in two dimensions and a sphere in three dimensions.

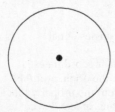

The set of points equidistant from a line segment is formed by a pair of parallel line segments with two semicircles

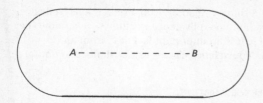

equilateral A polygon is equilateral if all its sides have the same length. In an equilateral triangle, the three angles are all equal and so each equals 60°.

equilibrium A particle is in equilibrium when it is at rest and the total force acting on it is zero for all time. These conditions are equivalent to saying that $\dot{\mathbf{r}} = \mathbf{0}$ and $\ddot{\mathbf{r}} = \mathbf{0}$ for all time, where $\mathbf{r}$ is the position vector of the particle.

Consider a rigid body experiencing a system of forces. Let F be the total force, and **M** the moment of the forces about the origin. The rigid body is in equilibrium when it is at rest and $\mathbf{F} = \mathbf{0}$ and $\mathbf{M} = \mathbf{0}$ for all time.

An equilibrium position for a body is a position in which it can be in equilibrium. A body is in stable equilibrium if, following a small change in its position, it returns to the equilibrium position. It is in unstable equilibrium if, following a small change in its position, it continues to move further from the equilibrium position. It is in neutral equilibrium if, following a small change in its position, it neither returns to nor moves further from the equilibrium position.

equipollent Two statements in logic are said to be equipollent if either one can be deduced from the other, so in that sense the statements are equivalent.

equipotent = EQUIPOLLENT.

equiprobable Events with the same probability. For example, in tossing a fair coin, the two outcomes are equiprobable. In rolling a fair die, the events 'throw an even number' and 'throw a prime number' are equiprobable.

equivalence class For an *equivalence relation $\sim$ on a set S, an equivalence class $[a]$ is the set of elements of S equivalent to a; that is to say, $[a] = \{x \mid x \in S \text{ and } a \sim x\}$. It can be shown that if two equivalence classes have an element in common, then the two classes are, as sets, equal. The collection of distinct equivalence classes having the property that every element of S belongs to exactly one of them is a *partition of S.

equivalence relation A *binary relation $\sim$ on a set S that is *reflexive, *symmetric and *transitive. For an equivalence relation $\sim$, a is said to be equivalent to b when $a \sim b$. It is an important fact that, from an equivalence relation on S, *equivalence classes can be defined to obtain a *partition of S.

equivalent *See* EQUIVALENCE RELATION.

equivalent circuits Two or more circuits which produce identical outputs for every possible input.

equivalent norms Let $\|\ \|$ and $\|\ \|'$ be two norms on a *vector space X. They are equivalent if there are positive constants a, b for which $a\|x\| \leq \|x\|' \leq b\|x\|$ for all x in X. All norms on a Euclidean space are equivalent and so give rise to the same *topology.

Eratosthenes of Cyrene (about 275–195 BC) Greek astronomer and mathematician who was the first to calculate the size of the Earth by making measurements of the angle of the Sun at two different places a known distance apart. His other achievements include measuring the tilt of the Earth's axis. He is also credited with the method known as the *sieve of Eratosthenes.

Erdos, Paul (1913–96) Hungarian mathematician who was prolific in his output, publishing more than 1500 papers, many of them jointly as he collaborated with a wide range of other mathematicians. He sought elegant and simple solutions to complex problems, which requires insight into the essential nature of the problem, as well as the technical mathematics to extract the solution. He can be regarded as the founder of *discrete mathematics and was awarded the *Wolf Prize in 1983 for contributions to number theory, combinatorics, probability, set theory and mathematical analysis.

error Let x be an approximation to a value X. According to some authors, the error is $X - x$; for example, when 1.9 is used as an approximation for 1.875, the error equals -0.025. Others define the error to be $x - X$. Whichever of these definitions is used, the error can be positive or negative. Yet other authors define the error to be $|X - x|$, the difference between the true value and the approximation, in which case the error is always greater than or equal to zero. When contrasted with *relative error, the error may be called the absolute error.

error-correcting and error-detecting code A code is said to be error-detecting if any one error in a codeword results in a word that is not a codeword, so that the receiver knows that an error has occurred. A code is error-correcting if, when any one error occurs in a codeword, it is possible to decide which codeword was intended. Certain error-correcting codes may not be able to detect errors if more than one error occurs in a codeword; other error-correcting codes can be constructed that can detect and correct more than one error in a codeword. *See also* DISTANCE BETWEEN TWO CODEWORDS.

escape speed For a celestial body, the minimum speed with which an object must be projected away from its surface so that the object does not return again because of gravity. The escape speed for the Earth is $\sqrt{2gR}$ where R is the radius of the Earth, and is approximately 11.2 kilometres per second. Similarly, the escape speed associated with a spherical celestial body of mass M and radius R is $\sqrt{2GM/R}$ where G is the *gravitational constant.

(⊕) SEE WEB LINKS

• An interactive tool allowing you to choose a rocket launch speed and then see how a rocket fired vertically and another fired horizontally (from some distance above the Earth) would behave.

escribed circle = EXCIRCLE.

essential singularity A *singularity in a complex-valued function which is neither a *pole nor a *removable singularity. This requires that $f(z)(z-a)^n$ is not differentiable at a for any integer $n > 0$.

estimate The value of an *estimator calculated from a particular sample. If the estimate is a single figure, it is a point estimate; if it is an interval, such as a *confidence interval, it is an interval estimate.

estimation The process of determining as nearly as possible the value of a population *parameter by using an *estimator. The value of an estimator found from a particular sample is called an *estimate.

estimator A *statistic used to estimate the value of a population *parameter. An estimator X of a parameter θ is consistent if the probability of the difference between the two exceeding an arbitrarily small fixed number tends to zero as the sample size increases indefinitely. An estimator X is an unbiased estimator of the parameter θ if $E(X) = \theta$, and it is a *biased estimator if not (*see* EXPECTED VALUE). The best unbiased estimator is the unbiased estimator with the minimum *variance. The relative efficiency of two unbiased estimators X and Y is the ratio $\text{Var}(Y)/\text{Var}(X)$ of their variances.

Estimators may be found in different ways, including the method of maximum likelihood (*see* LIKELIHOOD), and the method of moments (*see* MOMENT).

Euclid (about 300 BC) Outstanding mathematician of Alexandria, author of what may well be the second most influential book in Western Culture: the *Elements*. Little is known about Euclid himself, and it is not clear to what extent the book describes original work and to what extent it is a textbook. The *Elements* develops a large section of elementary geometry by rigorous logic starting from 'undeniable' axioms. It includes his proof that there are infinitely many primes, the *Euclidean algorithm, the derivation of the five *Platonic solids, and much more. It served for two millennia as a model of what pure mathematics is about.

((♦)) SEE WEB LINKS

• A fuller biography of Euclid.

Euclidean Algorithm A process, based on the *Division Algorithm, for finding the *greatest common divisor (a, b) of two positive integers a and b. Assuming that $a > b$, write $a = bq_1 + r_1$, where $0 \leq r_1 < b$. If $r_1 = 0$, the g.c.d. (a, b) is equal to b; if $r_1 \neq 0$, then $(a, b) = (b, r_1)$, so the step is repeated with b and r_1 in place of a and b. After further repetitions, the last non-zero remainder obtained is the required g.c.d. For example, for $a = 1274$ and $b = 871$, write

$$1274 = 1 \times 871 + 403,$$
$$871 = 2 \times 403 + 65,$$
$$403 = 6 \times 65 + 13,$$
$$65 = 5 \times 13.$$

and then $(1274, 871) = (871, 403) = (403, 65) = (65, 13) = 13$.

The algorithm also enables s and t to be found such that the g.c.d. can be expressed as $sa + tb$ use the equations in turn to express each remainder in this form. Thus,

$$403 = 1274 - 1 \times 871 = a - b,$$
$$65 = 871 - 2 \times 403 = b - 2(a - b) = 3b - 2a,$$
$$13 = 403 - 6 \times 65 = (a - b) - 6(3b - 2a) = 13a - 19b.$$

Euclidean construction A construction of a geometrical figure with only compasses and a straight edge, which cannot be used for measuring lengths. For example, an equilateral triangle can be constructed by drawing a line segment with the straight edge, positioning the point of the compasses at one end of the segment and drawing an arc of the circle with the same radius as the segment, and doing the same from the other end. Where the two arcs meet is the third vertex of the equilateral triangle.

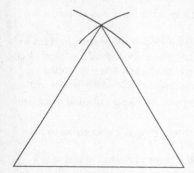

Euclidean distance (Cartesian distance) The distance between two points in Euclidean or Cartesian space. In two dimensions this is $\sqrt{(x_1 - x_2)^2 + (y_1 - y_2)^2}$ and in three dimensions it is $\sqrt{(x_1 - x_2)^2 + (y_1 - y_2)^2 + (z_1 - z_2)^2}$ with the obvious notation for the coordinates of the points.

Euclidean geometry The area of mathematics relating to the study of geometry based on the definitions and axioms set out in Euclid's book, the *Elements*.

Euclidean norm The most common *norm defined on a *vector space, given by the square root of the sum of squares of the components of the vector. Equivalent to $p = 2$ in the family of p-norms—essentially the length of a vector.

Euclidean space (Cartesian space) The number line **R**, plane $\mathbf{R}^2$ and 3-dimensional space $\mathbf{R}^3$ can be generalized to n-dimensional 'space' $\mathbf{R}^n$ with coordinates $(x_1, x_2, \ldots, x_n)$ on which the operations of addition and multiplication by a scalar have been extended in the obvious way. While $\mathbf{R}^n$ is hard to visualize for $n > 3$ it provides a very powerful framework for multivariable analysis.

Euclid numbers The *perfect numbers which are even, for example 6 and 28.

Euclid's axioms The axioms Euclid set out in his famous text, the *Elements*, are:

1. A straight line may be drawn from any point to any other point,
2. A straight line segment can be extended indefinitely at either end,

3. A circle may be described with any centre and any radius,
4. All right angles are equal,
5. If a straight line (the transversal) meets two other straight lines so that the sum of the two interior angles on one side of the transversal is less than two right angles, then the straight lines, extended indefinitely if necessary, will meet on that side of the transversal.

He also stated definitions of geometrical entities like points and lines, and five 'common notions', which are:

1. Things which are equal to the same thing are also equal to one another.
2. If equals are added to equals, the sums are also equal.
3. If equals are subtracted from equals, the remainders are also equal.
4. Things that coincide with one another are equal to one another.
5. The whole is greater than the part.

Eudoxus of Cnidus (about 380 BC) Greek mathematician and astronomer, one of the greatest of antiquity. All his original works are lost, but it is known from later writers that he was responsible for the work in Book 5 of *Euclid's Elements.* This work of his was a precise and rigorous development of the real number system in the language of his day. The significance of his sophisticated ideas was not really appreciated until the 19th century. He also developed methods of determining areas with curved boundaries.

Euler, Leonhard (1707–83) Beyond comparison, the most prolific of famous mathematicians. He was born in Switzerland but is most closely associated with the Berlin of Frederick the Great and the St Petersburg of Catherine the Great. He worked in a highly productive period when the newly developed calculus was being extended in all directions at once, and he made contributions to most areas of mathematics, pure and applied. Euler, more than any other individual, was responsible for notation that is standard today. Among his contributions to the language are the basic symbols, π, e and i, the summation notation $\sum$ and the standard function notation $f(x)$. His *Introductio in analysin infinitorum* was the most important mathematics text of the late 18th century. From the vast bulk of his work, one famous result of which he was justifiably proud is this:

$$1 + \frac{1}{2^2} + \frac{1}{3^2} + \cdots + \frac{1}{n^2} + \cdots = \frac{\pi^2}{6}.$$

Euler characteristic For a surface S this is topologically invariant, describing the shape or structure independent of how it is transformed.

It is given by the formula $e(S) = V - E + F$, where V, E, and F are the numbers of vertices, edges, and faces, respectively. For any *connected planar graph, the Euler characteristic is 2.

Eulerian graph One area of graph theory is concerned with the possibility of travelling around a *graph, going along edges in such a way as to use every edge exactly once. A *connected graph is called Eulerian if there is a sequence $v_0, e_1, v_1, \ldots, e_k, v_k$ of alternately vertices and edges (where e_i is an edge joining v_{i-1} and v_i), with $v_0 = v_k$ and with every edge of the graph occurring exactly once. Simply put, it means that 'you can draw the graph without taking your pencil off the paper or retracing any lines, ending at your starting-point'. The name arises from Euler's consideration of the problem of whether the *bridges of Königsberg could be crossed in this way. It can be shown that a connected graph is Eulerian if and only if every vertex has even degree.

Eulerian trail A trail which includes every edge of a graph.

Euler line In a triangle, the *circumcentre O, the *centroid G and the *orthocentre H lie on a straight line called the Euler line. On this line, $OG : GH = 1 : 2$. The centre of the *nine-point circle also lies on the Euler line.

Euler multiplier = MULTIPLYING FACTOR in differential equations.

Euler number Another name for e, the base of the natural logarithms.

Euler's constant Let $a_n = 1 + \frac{1}{2} + \frac{1}{3} + \cdots + (1/n) - \ln n$. This sequence has a limit whose value is known as Euler's constant, γ; that is, $a_n \to \gamma$. The value equals 0.577 215 66 to 8 decimal places. It is not known whether γ is *rational or *irrational.

Euler's formula The name given to the equation $\cos \theta + i \sin \theta = e^{i\theta}$, a special case of which gives $e^{i\pi} + 1 = 0$.

Euler's function For a positive integer n, let $\phi(n)$ be the number of positive integers less than n that are *relatively prime to n. For example, $\phi(12) = 4$, since four numbers, 1, 5, 7 and 11, are relatively prime to 12. This function ϕ, defined on the set of positive integers, is Euler's function. It can be shown that, if the prime decomposition of n is $n = p_1^{\alpha_1} p_2^{\alpha_2} \ldots p_r^{\alpha_r}$, then

$$\phi(n) = p_1^{\alpha_1 - 1} p_2^{\alpha_2 - 1} \cdots p_r^{\alpha_r - 1} (p_1 - 1)(p_2 - 1) \cdots (p_r - 1),$$
$$= n \left(1 - \frac{1}{p_1}\right) \left(1 - \frac{1}{p_2}\right) \cdots \left(1 - \frac{1}{pr}\right).$$

Euler proved the following extension of *Fermat's Little Theorem: If n is a positive integer and a is any integer such that $(a, n) = 1$, then $a^{\phi(n)} \equiv 1$ (mod n).

Euler's method The simplest numerical method for solving differential equations. If $\dfrac{dy}{dx} = f(x, y)$ and an initial condition is known, $y = y_0$ when $x = x_0$, then Euler's method generates a succession of approximations $y_{n+1} = y_n + hf(x_n, y_n)$ where $x_n = x_0 + nh$, $n = 1, 2, 3, \ldots$. This takes the known starting point, and moves along a straight line segment with horizontal distance h in the direction of the tangent at (x_0, y_0). The process is repeated from the new point (x_1, y_1) etc. If the step length h is small enough, the tangents are good approximations to the curve. The method provides a reasonably accurate estimate.

Euler's Theorem If a *planar graph G is drawn in the plane, so that no two edges cross, the plane is divided into a number of regions which may be called 'faces'. Euler's Theorem (for planar graphs) is the following:

Theorem

Let G be a connected planar graph drawn in the plane. If there are v vertices, e edges and f faces, then $v - e + f = 2$.

An application of this gives Euler's Theorem (for polyhedra):

Theorem

If a convex polyhedron has v vertices, e edges and f faces, then $v - e + f = 2$.

For particular polyhedra, it is easy to confirm the result stated in the theorem. For example, a cube has $v = 8$, $e = 12$, $f = 6$, and a tetrahedron has $v = 4$, $e = 6$, $f = 4$.

European Article Numbers (EAN) A standardized system of product bar coding introduced in 1976, as a variation on the Universal Product Code (UPC). Own-brand items often use an 8-digit EAN while many grocery products use a 13-digit EAN. All use the same system of *check digits.

evaluate To calculate the value of a function at a particular value of its independent variables.

even Divisible by two with no remainder.

even function The *real function f is an even function if $f(-x) = f(x)$ for all x (in the domain of f). Thus the graph $y = f(x)$ of an even function has the y-axis as a line of symmetry. For example, f is an even function when $f(x)$ is defined as any of the following: 5, x^2, $x^6 - 4x^4 + 1$, $1/(x^2 - 3)$, $\cos x$.

even permutation A *rearrangement of the original ordering which can be obtained by an even number of exchanges of pairs of elements.

event A subset of the *sample space relating to an experiment. For example, suppose that the sample space for an experiment in which a coin is tossed three times is {HHH, HHT, HTH, HTT, THH, THT, TTH, TTT}, and let $A =$ {HHH, HHT, HTH, THH}. Then A is the event in which at least two 'heads' are obtained. If, when the experiment is performed, the outcome is one that belongs to A, then A is said to have occurred. The *intersection $A \cap B$ of two events is the event that can be described by saying that 'both A and B occur'. The *union $A \cup B$ of two events is the event that 'either A or B occurs'. Taking the sample space as the universal set, the *complement A' of A is the event that 'A does not occur'. The probability $\Pr(A)$ of an event A is often of interest. The following laws hold:

   (i) $\Pr(A \cup B) = \Pr(A) + \Pr(B) - \Pr(A \cap B)$.

  (ii) When A and B are *mutually exclusive events, $\Pr(A \cup B) = \Pr(A) + \Pr(B)$.

 (iii) When A and B are *independent events, $\Pr(A \cap B) = \Pr(A)\Pr(B)$.

 (iv) $\Pr(A') = 1 - \Pr(A)$.

exa- Prefix used with *SI units to denote multiplication by 10^{18}.

exact The positive solution to the equation $x^2 = 3$ might be reported as 1.73, which it is correct to two *decimal places, or to three *significant figures, but the exact answer is $x = \sqrt{3}$.

exact differential If $z = f(x, y)$ is a function of two independent variables then $dz = \dfrac{\partial z}{\partial x}dx + \dfrac{\partial z}{\partial y}dy$ is the exact differential. For a function of more than two variables the exact differential will have similar partial derivative terms for each of its independent variables.

exact differential equation An equation in which the *exact differential of a function is equal to zero. If $z = f(x, y)$ then $\dfrac{\partial z}{\partial x} \cdot dx + \dfrac{\partial z}{\partial y} \cdot dy = 0$ is an exact differential equation.

exact divisor A factor of a given integer. So 3 is an exact divisor of 15.

example A particular instance of a generalized statement. A counter-example will disprove a generalized claim, but examples do not provide proof. For example, the number of chords created by joining n distinct points in a circle is 1, 2, 4, 8, 16 for $n = 1, 2, 3, 4, 5$ which are in the form 2^{n-1} but for $n = 6$, there are 31 chords produced, so the general expression cannot be 2^{n-1}. In fact it requires a quartic expression.

excentre *See* EXCIRCLE.

excircle An excircle of a triangle is a circle that lies outside the triangle and touches the three sides, two of them extended. There are three excircles. The centre of an excircle is an excentre of the triangle. Each excentre is the point of intersection of the bisector of the interior angle at one vertex and the bisectors of the exterior angles at the other two vertices.

excluded middle *See* PRINCIPLE OF THE EXCLUDED MIDDLE.

exclusive *See* MUTUALLY EXCLUSIVE.

exclusive disjunction If p and q are statements, then the statement 'p or q' where exclusive disjunction is intended, sometimes stated as 'p aut q' is denoted by $p \veebar q$, and is true only if exactly one of p, q is true. The truth table is therefore as follows:

p	q	$p \veebar q$
T	T	F
T	F	T
F	T	T
F	F	F

See also INCLUSIVE DISJUNCTION.

exhaustive A set of events in statistics whose union is the whole probability space, or a set of sets whose union is the universal set under consideration.

existential quantifier *See* QUANTIFIER.

exp The abbreviation and symbol for the *exponential function.

expand To express in an extended equivalent form. For example, $(a + b)^2 = a^2 + 2ab + b^2$.

expansion A mathematical expression written as a sum of a number of terms. Powers of *multinomials can be expanded by multiplying out the brackets, but other expansions may be derived from the *binomial series, *Taylor series, etc.

expectation (of a matrix game) Suppose that, in the *matrix game given by the $m \times n$ matrix $[a_{ij}]$, the *mixed strategies **x** and **y** for the two players are given by $\mathbf{x} = (x_1, x_2, \ldots, x_m)$, and $\mathbf{y} = (y_1, y_2, \ldots, y_n)$. The expectation $E(\mathbf{x}, \mathbf{y})$ is given by

$$E(\mathbf{x}, \mathbf{y}) = \sum_{i=1}^{m} \sum_{j=1}^{n} x_i a_{ij} y_j.$$

If $\mathbf{A} = [a_{ij}]$, and $\mathbf{x}$ and $\mathbf{y}$ are written as column matrices, $E(\mathbf{x}, \mathbf{y}) = \mathbf{x}^T \mathbf{A} \mathbf{y}$. The expectation can be said, loosely, to give the average *payoff each time when the game is played many times with the two players using these mixed strategies.

expectation (of a random variable) = EXPECTED VALUE.

expected utility In a context where the outcomes are not determined solely by the decisions an individual takes, the expected utility of a decision is the average value of the utility function weighted by the probability distribution attached to the possible outcomes of the decision.

expected value The expected value $E(X)$ of a random variable X is a value that gives the mean value of the distribution, and is defined as follows. For a discrete random variable X, $E(X) = \sum p_i x_i$, where $p_i = \Pr(X = x_i)$. For a continuous random variable X,

$$E(X) = \int_{-\infty}^{\infty} x \, f(x) \, dx,$$

where f is the *probability density function of X. The following laws hold:

(i) $E(aX + bY) = aE(X) + bE(Y)$.

(ii) When X and Y are *independent, $E(XY) = E(X)E(Y)$.

The following is a simple example. Let X be the number obtained when a die is thrown. Let $x_i = i$, for $i = 1, 2, \ldots, 6$. Then $p_i = \frac{1}{6}$, for all i. So $E(X) = \frac{1}{6} \times 1 + \frac{1}{6} \times 2 + \ldots \frac{1}{6} \times 6 = 3.5$.

experiment A statistical study in which the researchers make an intervention and measure the effect of this intervention on some outcome of interest. For example, measuring the change in reaction times of people deprived of sleep.

experimental condition In an experimental design, the distinct states under which the outcomes are to be compared are the experimental conditions. For example, different dosages of a drug may be administered to investigate the most effective treatment strategy.

experimental design (in statistics) Observational studies can suggest things for which the explanation lies with some hidden variable. For example, the performance of pupils taught in large classes is better than that of pupils in small classes, not because the large class is a more effective learning environment, but because schools operate larger classes for the more able. Experimental design methods seek to control the conditions under which observations are made so that any differences in outcome are genuinely attributable to the experimental conditions, and not to other confounding factors. Some of the most common methods are the use of *paired or matched samples, *randomization and *blind trials.

explanatory variable One of the variables that it is thought might influence the value of the *dependent variable in a statistical model.

explicit function If the dependent variable y is expressed in the form $y = f(x)$ then y is an explicit function of x. So $y = 5x + 1$ is explicit but $5x - y + 1 = 0$ is not, though it can be rearranged to be explicit.

exploratory data analysis (EDA) An approach to data analysis that sets out initially to explore the data, usually through a variety of mostly graphical techniques, to try to gain insight into the nature of the data and their underlying structure, what the important variables are and to identify outliers. The outcomes can then inform decisions as to what analyses are appropriate. The approach first gained importance with the work of *Tukey in 1977.

(()) SEE WEB LINKS

• Explanations with visual representations of many of the standard tools of exploratory data analysis.

exponent = INDEX. *See also* FLOATING-POINT NOTATION.

exponential decay Suppose that $y = Ae^{kt}$, where A (> 0) and k are constants, and t represents some measurement of time (*see* EXPONENTIAL GROWTH). When $k < 0$, y can be said to be exhibiting exponential decay. In such circumstances, the length of time it takes for y to be reduced to half its value is the same, whatever the value. This length of time, called the half-life, is a useful measure of the rate of decay. It is applicable, for example, to the decay of radioactive isotopes.

(()) SEE WEB LINKS

• An animation showing the physical decay and corresponding graph of three radioactive isotopes (subscription).

exponential distribution The continuous probability *distribution with *probability density function f given by $f(x) = \lambda \exp(-\lambda x)$, where λ is a positive parameter, and $x \geq 0$. It has mean $1/\lambda$ and variance $1/\lambda^2$. The time between events that occur randomly but at a constant rate has an exponential distribution. The distribution is skewed to the right. The following figure shows the probability density function of the exponential distribution with $\lambda = 10$.

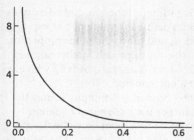

The exponential distribution
with λ = 10

exponential function The function f such that $f(x) = e^x$, or exp x, for all x in **R**. The two notations arise from different approaches described below, but are used interchangeably. Among the important properties that the exponential function has are the following:

(i) $\exp(x+y) = (\exp x)(\exp y)$, $\exp(-x) = 1/\exp x$ and $(\exp x)^r = \exp rx$. (These hold by the usual rules for indices once the equivalence of exp x and e^x has been established.)

(ii) The exponential function is the *inverse function of the *logarithmic function: $y = \exp x$ if and only if $x = \ln y$.

(iii) $\dfrac{d}{dx}(\exp x) = \exp x$.

(iv) exp x is the sum of the series $1 + \dfrac{x}{1!} + \dfrac{x^2}{2!} + \cdots + \dfrac{x^n}{n!} + \cdots$.

(v) As $n \to \infty$, $\left(1 + \dfrac{x}{n}\right)n \to \exp x$.

Three approaches can be used:

1. Suppose that the value of e has already been obtained independently. Then it is possible to define e^x, the exponential function to base e, by using approach **1** to the *exponential function to base a. Then exp x can be taken to mean just e^x. The problem with this approach is its reliance on a prior definition of e and the difficulty of subsequently proving some of the other properties of exp.

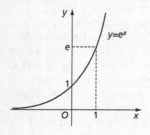

2. Define ln as in approach **2** to the logarithmic function, and take exp as its inverse function. It is then possible to define the value of e as exp 1, establish the equivalence of exp x and e^x, and prove the other properties. This is widely held to be the most satisfactory approach mathematically, but it has to be admitted that it is artificial and does not match up with any of the ways in which exp is usually first encountered.

3. Some other property of exp may be used as a definition. It may be defined as the unique function that satisfies the differential equation $dy/dx = y$ (that is, as a function that is equal to its own derivative), with $y = 1$ when $x = 0$. Alternatively, property (iv) or (v) above could be taken as the definition of exp x. In each case, it has to be shown that the other properties follow.

exponential function to base _a_ Let a be a positive number not equal to 1. The exponential function to base a is the function f such that $f(x) = a^x$ for all x in **R**. This must be clearly distinguished from what is commonly called 'the' *exponential function. The graphs $y = 2^x$ and $y = \left(\frac{1}{2}\right)^x$ illustrate the essential difference between the cases when $a > 1$ and $a < 1$. *See also* EXPONENTIAL GROWTH and EXPONENTIAL DECAY.

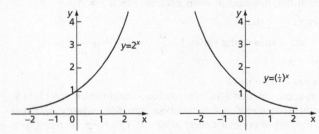

Clarifying just what is meant by a^x can be done in two ways:

1. The familiar rules for indices (*see* INDEX) give a meaning to a^x for rational values of x. For x not rational, take a sequence of rationals that approximate more and more closely to x. For example, when $x = \sqrt{2}$, such a sequence could be 1.4, 1.41, 1.414, ... Now each of the values $a^{1.4}, a^{1.41}, a^{1.414}, \ldots$ has a meaning, since in each case the index is rational. It can be proved that this sequence of values has a limit, and this limit is then taken as the definition of $a^{\sqrt{2}}$. The method is applicable for any real value of x.

2. Alternatively, suppose that exp has been defined (say by approach **2** to the *exponential function) and that ln is its *inverse function. Then the following can be taken as a definition: $a^x = \exp(x \ln a)$. This approach is less elementary, but really more satisfactory than **1**. It follows that $\ln(a^x) = x \ln a$, as would be expected, and the following can be proved:

(i) $a^{x+y} = a^x$, a^y, $a^{-x} = 1/a^x$ and $(a^x)^y = a^{xy}$.

(ii) When n is a positive integer, a^n, defined in this way, is indeed
equal to the product $a \times a \times \cdots \times a$ with n occurrences of a, and
$a^{1/n}$ is equal to $^n\sqrt{a}$.

(iii) $\dfrac{d}{dx}(a^x) = a^x \ln a$.

exponential growth When $y = Ae^{kt}$, where $A(>0)$ and k are
constants, and t represents some measurement of time, y can be said to be
exhibiting exponential growth. This occurs when $dy/dt = ky$; that is, when
the *rate of change of the quantity y at any time is proportional to the value
of y at that time. When $k>0$, then y is growing larger with x, and moreover
the rate at which y is increasing increases with x. In fact, any quantity with
exponential growth (with $k>0$) ultimately outgrows any quantity growing
linearly or in proportion to a fixed power of t. When $k<0$, the term
*exponential decay may be used.

( ) SEE WEB LINKS

• An interactive demonstration of exponential growth is included; click under the tab labelled
  'Graph'.

exponential series The series $\displaystyle\sum_{n=0}^{\infty} \frac{z^n}{n!} = 1 + z + \frac{z^2}{2} + \frac{z^3}{6} + \ldots$ which
converges for any complex number z to the *exponential function exp(z).

exponentiate To raise a number or quantity to a *power.

express To transform into equivalent terms. For example, by expanding
or factorizing an expression.

extended complex plane The set of complex numbers with a point at
infinity. The set can be denoted by $\mathbf{C}\infty$ and can be thought of as a *Riemann
sphere by means of a *stereographic projection. If a sphere is placed so that a
point S on the sphere is touching the complex plane at the origin, then S
corresponds to the point $(0,0)$ on the complex plane, which is the complex
number $z=0$. All other points on the sphere, except N which is diametrically
opposite to S on the sphere, are in a *one-to-one correspondence with points
on the complex plane through the *stereographic projection, and therefore
with a unique complex number. The point N is identified with the point at
infinity, with corresponding complex number ∞.

extended metric In some cases it is useful to allow the *metric in a metric
space to attain the value ∞, in which case it is called an extended metric.

extended real numbers The set of *real numbers, with the positive
and negative *infinite cardinals.

extension The difference $x - l$ between the actual length x of a string or spring and its natural length l. The extension of a spring is negative when the spring is compressed.

exterior *See* JORDAN CURVE THEOREM.

exterior angle (of a polygon) The angle between one side and the extension of an adjacent side of a polygon.

exterior angle (with respect to a transversal of a pair of lines) *See* TRANSVERSAL.

external bisector The bisector of the exterior angle of a triangle (or polygon) is sometimes called the external bisector of the angle of the triangle (or polygon).

external division (of a segment) Let AB be a line segment. Then the point E is the external division of AB in the ratio $1 : k$ if $\overrightarrow{AE} = k\overrightarrow{AB}$ where $\overrightarrow{AB}$ is the vector or directed line segment joining A and B.

E divides AB externally in the ratio 1:2.

external force When a system of particles or a rigid body is being considered as a whole, an external force is a force acting on the system from outside. *Compare* INTERNAL FORCE.

extrapolate To estimate a value of a quantity beyond the range already known, for example forecasting in *time series.

extrapolation Suppose that certain values $f(x_0), f(x_1), \ldots, f(x_n)$ of a function f are known, where $x_0 < x_1 < \cdots < x_n$. A method of finding from these an approximation for $f(x)$, for a given value of x that lies outside the interval $[x_0, x_n]$, is called extrapolation. Such methods are normally far less reliable than *interpolation, in which x lies between x_0 and x_n.

extreme value (of a function) The maximum or minimum value of a function.

extreme value (of a series) The first or last term of a series.

extreme value distribution The distribution of largest and smallest values in a sample. This new area of study has become an important tool in risk assessment.

extremum A point at which a function has a *turning point, i.e. at least a *local maximum or *minimum.

f Symbol for a *function, as in $f(x) = x^2 + 3$.

F The number 15 in hexadecimal notation.

face One of the plane surfaces forming a *polyhedron.

factor *See* DIVIDES.

factor analysis (in statistics) The techniques that aim to reduce the number of explanatory variables (factors) used to explain observational outcomes. New variables are constructed as combinations of the original variables with the aim of identifying a simpler model structure. For example, where a number of variables could be interpreted as measuring different aspects of a complete quantity such as intelligence, it may be possible to construct a composite variable which captures almost as much information as using all of the component variables.

factorial For a positive integer n, the notation $n!$ (read as 'n factorial') is used for the product $n(n-1)(n-2) \ldots \times 2 \times 1$. Thus $4! = 4 \times 3 \times 2 \times 1 = 24$ and $10! = 10 \times 9 \times 8 \times 7 \times 6 \times 5 \times 4 \times 3 \times 2 \times 1 = 3\ 628\ 800$. Also, by definition, $0! = 1$.

factorize To represent a number, matrix or polynomial as a product of factors. Of particular importance is the *unique factorization theorem.

factor space = QUOTIENT SPACE.

Factor Theorem The following result, which is an immediate consequence of the *remainder theorem:

Theorem

Let $f(x)$ be a polynomial. Then $x-h$ is a factor of $f(x)$ if and only if $f(h) = 0$.

The theorem is valuable for finding factors of polynomials. For example, to factorize $2x^3 + 3x^2 - 12x - 20$, look first for possible factors $x-h$, where h is an integer. Here h must divide 20. Try possible values for h, and calculate $f(h)$. It is found that $f(-2) = -16 + 12 + 24 - 20 = 0$, and so $x + 2$ is a factor. Now divide the polynomial by this factor to obtain a quadratic which it may be possible to factorize further.

Fahrenheit Symbol °F. The temperature scale, and the unit of measurement of temperature, which takes 32 °F as the freezing point of water, and 212°F as the boiling point of water. *Compare with* CELSIUS, KELVIN.

fallacy An invalid argument or a demonstrably false conclusion from plausible reasoning, giving rise to paradoxes such as *Achilles and the tortoise.

false negative In testing to determine whether a subject has a particular characteristic, especially testing whether a patient has a disease, where the test shows the characteristic is not present when it actually is.

false position (rule of false position) An iterative method for solving a non-linear equation which is similar in many respects to the *bisection method, except that the interval does not have to be bisected—any intermediate value can be chosen. For example, to solve $f(x) = x^3 + x - 3 = 0$ you find $f(1) = -1$ and $f(2) = 7$. You know there is a solution between $x = 1$ and $x = 2$, but it is likely to be closer to 1 than 2 so you might use 1.2, next with $f(x) = -0.072$. This method allows the solution to be identified more quickly than through the bisection method. Also known as trial and improvement.

false positive In testing to determine whether a subject has a particular characteristic, especially testing whether a patient has a disease, where the test shows the characteristic is present when it is not.

family A set whose elements are themselves sets may be called a family. In certain other circumstances, for example where less formal language is appropriate, the word 'family' may be used as an alternative to 'set'.

family of curves A set of similar curves which are of the same form and distinguished by the values taken by one or more parameters in their general equation. In particular, where the solution of a *differential equation is obtained, the *general solution will involve one or more constants of integration, giving rise to a family of curves. A particular member of the family may be identified as the required solution if *boundary conditions are known.

family of distributions A set of distributions which have the same general mathematical formula. A member of the family is obtained by choosing specific values for the *parameters in the formula.

farthest point A point outside a given subset of a metric space such that the distance to the nearest point in that space is greater than for any other.

fast Fourier transform *See* DISCRETE FOURIER TRANSFORM.

F-distribution A non-negative continuous *distribution formed from the ratio of the distributions of two independent random variables with *chi-squared distributions, each divided by its degrees of freedom. The mean is $\dfrac{v_2}{v_2 - 2}$ and the variance is $\dfrac{2v_2^2(v_2 + v_1 - 2)}{v_1(v_2 - 2)^2(v_2 - 4)}$, where v_1 and v_2 are the degrees of freedom of the numerator and denominator respectively. It is used to test the hypothesis that two normally distributed random variables have the same variance, and in *regression to test the relationship between an explanatory variable and the dependent variable. The distribution is skewed to the right. Tables relating to the distribution are available.

feasible A *constrained optimization problem for which the constraints can be satisfied simultaneously is said to be feasible.

feasible region *See* LINEAR PROGRAMMING.

(⊕) SEE WEB LINKS

• An interactive demonstration leading the user through steps to create the feasible region given a number of inequalities.

Feigenbaum, Mitchell (1945–   ) American mathematician instrumental in developing the mathematics of chaos theory.

femto- Prefix used with *SI units to denote multiplication by 10^{-15}.

Fermat, Pierre de (1601–65) Leading mathematician of the first half of the 17th century, remembered chiefly for his work in the theory of numbers, including *Fermat's Little Theorem and what is known as *Fermat's Last Theorem. His work on tangents was an acknowledged inspiration to *Newton in the latter's development of the calculus. Fermat introduced coordinates as a means of studying curves. Professionally he was a judge in Toulouse, and to mathematicians he is the 'Prince of Amateurs'.

Fermat point The point with the minimum total distance to the three vertices of a triangle. If the angle at any of the vertices is more than 120° then that vertex is the Fermat point, otherwise it is found by constructing an *equilateral triangle on each of the three sides of the triangle. For each side of the triangle, join the new vertex of the equilateral triangle to the vertex in

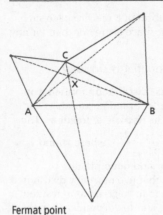

Fermat point

the original triangle which is not used in the side. These three lines intersect at the Fermat point. In the diagram X is the Fermat point.

Fermat prime A *prime of the form $2^{2^r} + 1$. At present, the only known primes of this form are those given by $r = 0$, 1, 2, 3 and 4.

Fermat's Last Theorem The statement that, for all integers $n > 2$, the equation $x^n + y^n = z^n$ has no solution in positive integers. Fermat wrote in the margin of a book that he had a proof of this, but as he never repeated the claim it is likely that he realized the incompleteness of his supposed proof. Much research was done over centuries until a proof was completed in 1995 by Andrew *Wiles.

Fermat's Little Theorem The name sometimes given to the following result:

Theorem

Let p be a prime, and let a be an integer not divisible by p. Then $a^{p-1} \equiv 1$ (mod p).

Sometimes the name is given instead to the following, which is a corollary of the preceding result:

Theorem

If p is a prime and a is any integer, then $a^p \equiv a$ (mod p).

Feuerbach's Theorem *See* NINE-POINT CIRCLE.

Feynman, Richard Phillips (1918–88) American mathematician and theoretical physicist who shared the Nobel Prize for Physics in 1965 with Scwinger and Tomonoga for their work on quantum electrodynamics.

He worked on the atomic bomb project during the Second World War, when he was already regarded as one of the leading scientists in the field at only 23. He enjoyed considerable success as an author where his intuitive grasp of fundamental physical principles allowed him to communicate to a much broader audience than most first-rate scientists. One of his best-known books is *Surely You're Joking Mr Feynman—Adventures of a Curious Character*.

Fibonacci (about 1170–1250) Pseudonym of one of the first European mathematicians to emerge after the Dark Ages. An Italian merchant by the name of Leonardo of Pisa, he was one of those who introduced the Hindu–Arabic number system to Europe. He strongly advocated this system in *Liber abaci*, published in 1202, which also contained problems including one that gives rise to the *Fibonacci numbers. Other writings of his deal with Euclidean geometry and *Diophantine equations.

Fibonacci number One of the numbers in the Fibonacci sequence 1, 1, 2, 3, 5, 8, 13, . . . , where each number after the second is the sum of the two preceding numbers in the sequence. This sequence has many interesting properties. For instance, the sequence consisting of the ratios of one Fibonacci number to the previous one, $\frac{1}{1}, \frac{2}{1}, \frac{3}{2}, \frac{5}{3}, \frac{8}{5}, \frac{13}{8}, \ldots$, has the limit τ, the *golden ratio. *See also* DIFFERENCE EQUATION and GENERATING FUNCTION.

((⊕)) SEE WEB LINKS
• A site with examples of Fibonacci numbers in nature and the relationship to the golden section.

Fibonacci sequence *See* FIBONACCI NUMBER.
((⊕)) SEE WEB LINKS
• A site with more on the Fibonacci sequence and links with Pascal's triangle.

fictitious force A force that may be assumed to exist by an observer whose *frame of reference is accelerating relative to an inertial frame of reference. Suppose, for example, that a frame of reference with origin O is rotating relative to an inertial frame of reference with the same origin. A particle P subject to a certain total force satisfies *Newton's second law of motion, relative to the inertial frame of reference. To the observer in the rotating frame, the particle appears to satisfy an equation of motion that is Newton's second law of motion with additional terms. The observer may suppose that these terms are explained by certain fictitious forces. When these forces are assumed to exist, Newton's laws appear to hold in the non-inertial frame of reference.

Consider the special case in which the rotating frame of reference has a constant angular velocity and the particle is moving in a plane perpendicular to the angular velocity of the rotating frame of reference. One fictitious force is in the direction along OP and is called the centrifugal force. This is the force outwards that is believed to exist by a rider on a roundabout. The second fictitious force is perpendicular to the path of P as seen by the observer in the rotating frame of reference and is called the Coriolis force.

To an observer standing on the Earth, which is rotating about its axis, an object such as an intercontinental missile appears to deviate from its path due to the Coriolis force. The deviation is to the right in the northern hemisphere and to the left in the southern hemisphere. This force, first described by the French mathematician and engineer, Gustave-Gaspard de Coriolis (1792–1843), also has important applications to the movement of air masses in meteorology.

Similar fictitious forces may arise whenever the observer's frame of reference is accelerating relative to an inertial frame of reference, as when a passenger in an accelerating lift witnesses a ceiling tile fall from the roof of the cabin.

field A commutative ring with identity (*see* RING) with the following additional property:

10. For each a ($\neq 0$), there is an element a^{-1} such that $a^{-1}\,a = 1$.

(The axiom numbering here follows on from that used for ring and *integral domain.) From the defining properties of a field, Axioms **1** to **8** and Axiom **10**, it can be shown that $ab = 0$ only if $a = 0$ or $b = 0$. Thus Axiom **9** holds, and so any field is an integral domain. Familiar examples of fields are the set **Q** of rational numbers, the set **R** of real numbers and the set **C** of complex numbers, each with the usual addition and multiplication. Another example is $\mathbf{Z}_p$, consisting of the set $\{0, 1, 2, \ldots, p-1\}$ with addition and multiplication modulo p, where p is a prime.

field of force A field of force is said to exist when a force acts at any point of a region of space. A particle placed at any point of the region then experiences the force, which may depend on the position and on time. Examples are gravitational, electric and magnetic fields of force.

field of integration The set of values over which a *multiple integral is defined.

Fields Medal A prize awarded for outstanding achievements in mathematics, considered by mathematicians to be equivalent to a Nobel Prize. Medals are awarded to individuals at successive International Congresses of Mathematicians, normally held at four-year intervals. The proposal was made by J. C. Fields to found two gold medals, using funds remaining after the financing of the Congress in Toronto in 1924. The first two medals were presented at the Congress in Oslo in 1936. In some instances, three or four medals have been awarded. It has been the practice to make the awards to mathematicians under the age of 40. See *Appendix 17 for a list of winners.*

SEE WEB LINKS
• The website of the awarding body of the Fields medal.

figurate numbers Numerous sequences of numbers associated with different geometrical figures were considered special by the Pythagoreans and other early mathematicians, and these are known loosely as figurate numbers. They include the *perfect squares, the *triangular numbers and the *tetrahedral numbers, and others known as the pentagonal and the hexagonal numbers.

figure (in geometry) A combination of points, lines or surfaces in a geometrical shape.

figure (in number) A *digit.

finite Not *infinite. For integers, it can be described as having a number of elements which can be put in *one-to-one correspondence with a terminating set of the *natural numbers.

finite differences Let $x_0, x_1, x_2, \ldots, x_n$ be equally spaced values, so that $x_i = x_0 + ih$, for $i = 1, 2, \ldots, n$. Suppose that the values $f_0, f_1, \ldots, f_n$ are known, where $f_i = f(x_i)$, for some function f. The first differences are defined, for $i = 0, 1, 2, \ldots, n-1$, by $\Delta f_i = f_{i+1} - f_i$. The second differences are defined by $\Delta^2 f_i = \Delta (\Delta f_i) = \Delta f_{i+1} - \Delta f_i$ and, in general, the k-th differences are defined by $\Delta^k f_i = \Delta (\Delta^{k-1} f_i) = \Delta^{k-1} f_{i+1} - \Delta^{k-1} f_i$. For a polynomial of degree n, the $(n+1)$-th differences are zero.

These finite differences may be displayed in a table, as in the following example. Alongside it is a numerical example.

x_0	f_0				1.0	1.000			
		Δf_0					0.331		
x_1	f_1		$\Delta^2 f_0$		1.1	1.331		0.066	
		Δf_1		$\Delta^3 f_0$			0.397		0.006
x_2	f_2		$\Delta^2 f_1$		1.2	1.728		0.072	
		Δf_2					0.469		
x_3	f_3				1.3	2.197			

With such tables it should be appreciated that if the values $f_0, f_1, f_2, \ldots, f_n$ are *rounded values then increasingly serious errors result in the succeeding columns.

Numerical methods using finite differences have been extensively developed. They may be used for *interpolation, as in the

*Gregory–Newton forward difference formula, for finding a polynomial that approximates to a given function, or for estimating derivatives from a table of values.

finite-dimensional A *vector space is said to be finite-dimensional if it can be defined by a finite set of linearly independent vectors.

finite element method A numerical method of solving partial differential equations with boundary conditions by considering a series of approximations which satisfy the differential equation and boundary conditions within a small region (the finite element of the name).

finite Fourier transform = DISCRETE FOURIER TRANSFORM.

finite intersection property A *space has the finite intersection property when every family of *closed subsets such that all finite subcollections have non-empty intersections means that the entire family also has a non-empty intersection. It can be shown that this property is equivalent to the space being compact.

finite population correction The *standard error of the mean based on a sample of size n assumes that the population is infinite or at least so large that the effect of withdrawing items during the sampling process has a negligible effect. If the size of the sample, n, becomes a not insignificant fraction of the population, N, then the finite population correction $\sqrt{\dfrac{N-n}{N-1}}$ is used to adjust the estimate of the standard error, which becomes $\sqrt{\dfrac{N-n}{N-1}} \times \dfrac{\sigma}{\sqrt{n}}$. It is common to use this correction if the sample size is more than 5% of the population, and it reduces the standard error, so resulting in a narrower *confidence interval for the population mean.

finite sequence See SEQUENCE.

finite series See SERIES.

first derivative A term used for the *derivative when it is being contrasted with *higher derivatives.

first derivative test See DERIVATIVE TEST.

first-fit decreasing packing algorithm In the *bin packing problem order the boxes in decreasing size, and apply the *first-fit packing algorithm. This packs the largest items first, and is more likely to produce an optimal solution than the simple first-fit method.

first-fit packing algorithm In the *bin packing problem number the bins, and always place the next box into the lowest-numbered bin it will fit into.

first-order differential equation A differential equation containing only the first differential. For example, $\frac{dy}{dx} = 5x$ and $3y\frac{dy}{dx} + 5x = 3$ are first-order differential equations.

first principles Without relying on other theorems for a result. For example, if $f(x) = x^2$, $f'(x) = 2x$ but to show this from first principles requires the following argument:

$$f'(x) = \lim_{\delta x \to 0} \frac{f(x + \delta x) - f(x)}{\delta x}$$

$$= \lim_{\delta x \to 0} \frac{x^2 + 2x\delta x + (\delta x)^2 - x^2}{\delta x}$$

$$= \lim_{\delta x \to 0} (2x + \delta x) = 2x$$

Fisher, Ronald Aylmer (1890–1962) British geneticist and statistician who established methods of designing experiments and analysing results that have been extensively used ever since. His influential book on statistical methods appeared in 1925. He developed the *t-test and the use of *contingency tables, and is responsible for the method known as *analysis of variance.

Fisher's exact test A statistical test used to examine the significance of the association between two categorical variables in a 2×2 *contingency table. The test needs the expected values in each cell to be at least 10, but does not depend on the sample characteristics.

fit Mathematical or statistical models are used to describe phenomena in the real world. The fit is the degree of correspondence between the observations and the model's predictions.

fixed point *See* TRANSFORMATION (of the plane).

fixed-point iteration To find a root of an equation $f(x) = 0$ by the method of fixed-point iteration, the equation is first rewritten in the form $x = g(x)$. Starting with an initial approximation x_0 to the root, the values $x_1, x_2, x_3, \ldots$ are calculated using $x_{n+1} = g(x_n)$. The method is said to converge if these values tend to a limit α. If they do, then $\alpha = g(\alpha)$ and so α is a root of the original equation.

(i) 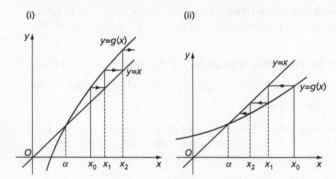 (ii)

A root of $x = g(x)$ occurs where the graph $y = g(x)$ meets the line $y = x$. It can be shown that, if $|g'(x)| < 1$ in an interval containing both the root and the value x_0, the method will converge, but not if $|g'(x)| > 1$. This can be illustrated in figures such as those shown which are for cases in which $g'(x)$ is positive. For example, the equation $x^3 - x - 1 = 0$ has a root between 1 and 2, so take $x_0 = 1.5$. The equation can be written in the form $x = g(x)$ in several ways, such as (i) $x = x^3 - 1$ or (ii) $x = (x + 1)^{1/3}$. In case (i), $g'(x) = 3x^2$, which does not satisfy $|g'(x)| < 1$ near x_0. In case (ii), $g'(x) = \frac{1}{3}(x + 1)^{-2/3}$ and $g'(1.5) \approx 0.2$, so it is likely that with this formulation the method converges.

fixed-point notation *See* FLOATING-POINT NOTATION.

fixed-point theorem Any theorem which gives conditions under which a *mapping must have a fixed point. They have been important in the development of mathematical economics.

flag A binary variable used to take some action. For example, in a conditional if...then.... else command in a computer program, or if a term is only required when certain conditions are met then including z times that term while defining $z = 1$ when the conditions are met, and zero otherwise, allows a single function to be used.

floating-point notation A method of writing real numbers, used in computing, in which a number is written as $a \times 10^n$, where $0.1 \le a < 1$ and n is an integer. The number a is called the mantissa and n is the exponent. Thus 634.8 and 0.002 34 are written as 0.6348×10^3 and 0.234×10^{-2}. (There is also a *base 2 version similar to the base 10 version just described.)

This is in contrast to fixed-point notation, in which all numbers are given by means of a fixed number of digits with a fixed number of digits after the decimal point. For example, if numbers are given by means of 8 digits with four of them after the decimal point, the two numbers above would be written (with an approximation) as 0634.8000 and 0000.0023. Integers are likely to be written in fixed-point notation; consequently, in the context of computers, some authors use 'fixed-point' to mean 'integer'.

float of an activity (in critical path analysis) The amount of time by which the start time of an activity can be varied without delaying the overall completion of the project, i.e. the latest possible time for completion—the earliest possible time for starting—the duration of the activity.

floor (greatest integer function) *See* GREATEST INTEGER FUNCTION.

fluent *See* FLUXION.

fluxion In *Newton's work on calculus, he thought of the variable x as a 'flowing quantity' or fluent. The rate of change of x was called the fluxion of x, denoted by $\dot{x}$.

focal Related to, going through or measured from the focus.

focus (foci) *See* CONIC, ELLIPSE, HYPERBOLA and PARABOLA.

foot of the perpendicular *See* PROJECTION (of a point on a line) and PROJECTION (of a point on a plane).

force In the real world, many different kinds of force are part of everyday life. A human being or an animal may use muscles to apply a force to move or try to move an object. An engine may produce a force that can be applied to turn a wheel. Commonly experienced forces are the force due to gravity, which acts throughout the region occupied by an object, forces that act between two bodies that are in contact, forces within a body that deform or restore its shape, and electrical and magnetic forces.

In a mathematical model, a force has a magnitude, a direction and a *point of application. It acts at a point and may be represented by a vector whose length is the magnitude of the force and whose direction is the direction of the force.

Force has the dimensions MLT^{-2}, and the SI unit of measurement is the *newton.

forced oscillations Oscillations that occur when a body capable of oscillating is subject to an applied force which varies with time. If the applied force is itself oscillatory, a differential equation such as $m\ddot{x} + kx = F_0 \sin(\Omega t + \varepsilon)$ may be obtained. In the solution of this equation, the *particular integral arises from the applied force. For a particular value of Ω, namely $\sqrt{k/m}$, *resonance will occur.

If the oscillations are *damped as well as forced, the complementary function part of the general solution of the differential equation tends to zero as t tends to infinity, and the particular integral arising from the applied force describes the eventual motion.

((⊕)) SEE WEB LINKS

• Animations and videos of forced oscillation and resonance.

forward difference If $\{(x_i, f_i)\}$, $i = 0, 1, 2, \ldots$ is a given set of function values with $x_{i+1} = x_i + h$, $f_i = f(x_i)$ then the forward difference operator Δ is defined by

$$\Delta f_i = f_{i+1} - f_i = f(x_{i+1}) - f(x_i).$$

forward difference formula *See* GREGORY–NEWTON FORWARD DIFFERENCE FORMULA.

forward error correction Widely used in data transmission, in unreliable communication channels, and in mass storage. It adds redundancy to the transmitted or stored information using an error-correcting code, allowing the receiver or retriever to detect a limited number of errors. This is important in situations where the storage is unique, or retransmission is expensive or impossible.

forward scan (in critical path analysis) In an *activity network (edges as activities), the forward scan identifies the earliest time for each vertex (node). Starting with the source, work forwards through the network, for each vertex calculate the sum of edge plus total time on previous vertex for all paths arriving at that vertex. Put the largest of these times onto that vertex.

Foucault pendulum A pendulum consisting of a heavy bob suspended by a long inextensible string from a fixed point, free to swing in any direction, designed to demonstrate the rotation of the Earth. In the original experiment in 1851, the French physicist Jean-Bernard-Léon Foucault (1819–68) suspended a bob of 28 kilograms by a wire 67 metres long from the dome of the Panthéon in Paris. If the experiment were set up at the north or south pole, the vertical plane of the swinging bob would appear to an observer fixed on the Earth to precess or rotate once a day. At a location of latitude λ in the northern hemisphere, the vertical plane of the swinging bob would precess in a clockwise direction with an angular speed of $\omega \sin \lambda$, where $\omega = 7.29 \times 10^{-5}$ rad s^{-1}, the angular speed of rotation of the Earth. The period of precession in Oxford, for example, would be about $30\frac{1}{2}$ hours.

foundations of mathematics The study of the logical basis for mathematics, and in particular attempts to establish an axiomatic basis upon which mathematics could be built. Euclid's geometrical text the *Elements* is one of the best-known examples, and early in the 20th century Bertrand *Russell tried to produce a unifying set of axioms for mathematics, but failed.

Four Colour Theorem It has been observed by map-makers through the centuries that any geographical map (that is, a division of the plane into regions) can be coloured with just four colours in such a way that no two neighbouring regions have the same colour. A proof of this, the Four Colour Theorem, was sought by mathematicians from about the 1850s. In 1890 Heawood proved that five colours would suffice, but it was not until 1976 that Appel and Haken proved the Four Colour Theorem itself. Initially, some mathematicians were sceptical of the proof because it relied, in an essential way, on a massive amount of checking of configurations by computer that could not easily be verified independently. However, the proof is now generally accepted and considered a magnificent achievement.

(((⊕))) SEE WEB LINKS

• A Shockwave file allowing you to create your own map and colour it in, or see an automated solution with no more than four colours.

four-group *See* KLEIN FOUR-GROUP.

Fourier, (Jean Baptiste) Joseph, Baron (1768–1830) French engineer and mathematician, best known in mathematics for his fundamental contributions to the theory of heat conduction and his study of trigonometric series. These so-called Fourier series are of immense importance in physics, engineering and other disciplines, as well as being of great mathematical interest.

Fourier analysis The use of *Fourier series and *Fourier transforms in *analysis.

Fourier coefficients The coefficients a_n and b_n of $\cos(nx)$ and $\sin(nx)$ respectively in the *Fourier series representation of a function.

$$a_n = \frac{1}{\pi} \int_0^{2\pi} f(x)\cos nx\, dx \quad n \geq 0$$

$$b_n = \frac{1}{\pi} \int_0^{2\pi} f(x)\sin nx\, dx \quad n \geq 1$$

Fourier series The infinite series $\frac{1}{2}a_0 + \sum_{n=1}^{\infty}(a_n \cos nx + b_n \sin nx)$, where a_n and b_n are the *Fourier coefficients. The Fourier series is used to decompose a waveform into component waves of different frequencies and amplitudes, allowing identification of different sources from background or random *noise in a signal.

Fourier transform The *integral transform $F(y) = \int_{-\infty}^{\infty} f(x)e^{iyx}dx$. The function F is said to be the Fourier transform of the function f. For many functions the transformation is invertible, in which case $f(x) = \frac{1}{2\pi}\int_{-\infty}^{\infty}F(y)e^{-iyx}dy$.

⊕ SEE WEB LINKS
• An illustrated discussion of the Fourier transform and its applications.

four squares theorem Any positive integer can be expressed as the sum of the squares of not more than four positive integers. Alternatively it can be expressed as the sum of squares of exactly four non-negative integers, *see* LAGRANGE'S THEOREM. So

$$1 = 1^2 + 0^2 + 0^2 + 0^2,$$
$$5 = 2^2 + 1^2 + 0^2 + 0^2,$$
$$15 = 3^2 + 2^2 + 1^2 + 1^2.$$

fourth root of unity A complex number z such that $z^4 = 1$. There are 4 fourth roots of unity and they are 1, i, -1 and $-i$. (*See* N-TH ROOT OF UNITY.)

fractal A set of points whose *fractal dimension is not an integer or, loosely, any set of similar complexity. Fractals are typically sets with infinitely complex structure and usually possess some measure of self-similarity, whereby any part of the set contains within it a scaled-down version of the whole set. Examples are the *Cantor set and the *Koch curve.

⊕ SEE WEB LINKS
• Examples of colourful fractal images.

fractal dimension One of the many extensions of the notion of dimension to objects for which the traditional concept of dimension is not appropriate. The fractal dimension may have a non-integer value. The *Koch curve has dimension $\ln 4/\ln 3 \approx 1.26$. Being between 1 and 2, this reflects the fact that the set is, as it were, too 'thick' to count as a curve and too 'thin' to count as an area. The *Cantor set has dimension $\ln 2/\ln 3$. Fractal dimension has found many practical applications in the analysis of chaotic or noisy processes (*see* CHAOS).

fraction The fraction a/b, where a and b are positive integers, was historically obtained by dividing a unit length into b parts and taking a of these parts. The number a is the numerator and the number b is the denominator. It is a proper fraction if $a < b$ and an improper fraction if $a > b$. Any fraction can be expressed as $c + d/e$, where c is an integer and d/e is a proper fraction, and in this form it is called a mixed fraction. For example, $3\frac{1}{2}$ is a mixed fraction (equal to 7/2).

fractional part For any real number x, its fractional part is equal to $x - [x]$, where $[x]$ is the *integer part of x. It may be denoted by $\{x\}$. The fractional part r of any real number always satisfies $0 \leq r < 1$.

frame (in statistics) *See* SAMPLING FRAME.

frame of reference In mechanics, a means by which an observer specifies positions and describes the motion of bodies. For example, an observer may use a Cartesian coordinate system or a *polar coordinate system. In some circumstances, it may be useful to consider two or more different frames of reference, each with its own observer. One frame of reference, its origin and axes, may be moving relative to another. The motion of a particle, for example, as it appears to one observer will be different from the motion as seen by the other observer.

A frame of reference in which *Newton's laws of motion hold is called an inertial frame (of reference). Any frame of reference that is at rest or moving with constant velocity relative to an inertial frame is an inertial frame. A frame of reference that is accelerating or rotating with respect to an inertial frame is not an inertial frame.

A frame of reference fixed on the Earth is not an inertial frame because of the rotation of the Earth. However, such a frame of reference may be assumed to be an inertial frame in problems where the rotation of the Earth has little effect.

framework (in mechanics) *See* LIGHT FRAMEWORK.

Freedman, Michael Hartley (1951–   ) American mathematician awarded the *Fields Medal in 1986 for his work on the Poincaré conjecture.

freely hinged = SMOOTHLY HINGED.

Frege, (Friedrich Ludwig) Gottlob (1848–1925) German mathematician and philosopher, founder of the subject of mathematical logic. In his works of 1879 and 1884, he developed the fundamental ideas, invented the standard notation of *quantifiers and variables, and studied the foundations of arithmetic. Not widely recognized at the time, his work was disseminated primarily through others such as Peano and Russell.

frequency (in mechanics) When *oscillations, or cycles, occur with period T, the frequency is equal to $1/T$. The frequency is equal to the number of oscillations or cycles that take place per unit time.

Frequency has the dimensions T^{-1}, and the SI unit of measurement is the *hertz.

frequency (in statistics) The number of times that a particular value occurs as an observation. In *grouped data, the frequency corresponding to a group is the number of observations that lie in that group. If numerical data are grouped by means of class intervals, the frequency corresponding to a class interval is the number of observations in that interval. *See also* RELATIVE FREQUENCY.

frequency distribution For nominal or discrete data, the information consisting of the possible values and the corresponding *frequencies is called the frequency distribution. For *grouped data, it gives the information consisting of the groups and the corresponding frequencies. It may be presented in a table or in a diagram such as a *bar chart, *histogram or *stem-and-leaf plot.

Freudenthal, Hans (1905–90) German mathematician who became one of the most important figures in mathematics education in the late 20th century. He founded the Institute for the Development of Mathematical Education in Utrecht in 1971, and was its first director. It was renamed the Freudenthal Institute in 1991 after his death.

friction Suppose that two bodies are in contact, and that the frictional force and the normal reaction have magnitudes F and N respectively (*see* CONTACT FORCE). The coefficient of static friction μ_s is the ratio F/N in the limiting case when the bodies are just about to move relative to each other. Thus if the bodies are at rest relative to each other, $F \leq \mu_s N$. The coefficient of kinetic friction μ_k is the ratio F/N when the bodies are sliding; that is, in contact and moving relative to each other. These coefficients of friction depend on the materials of which the bodies are made. Normally, μ_k is somewhat less than μ_s. *See also* ANGLE OF FRICTION.

frictional couple A *couple created by a pair of equal and opposite frictional forces. A frictional couple may occur, for example, when a rigid body rotates about an axis.

frictional force *See* CONTACT FORCE.

from above *See* RIGHT-HAND LIMIT.

from below *See* LEFT-HAND LIMIT.

from the left = FROM BELOW.

from the right = FROM ABOVE.

frontier (boundary) The set of points which are members of the closure of a set and also of the closure of its complement. For example, for an open or closed interval (a, b) or $[a, b]$, the values a and b are the frontier or boundary points.

frustum (frusta) A frustum of a right-circular cone is the part between two parallel planes perpendicular to the axis. Suppose that the planes are a distance h apart, and that the circles that form the top and bottom of the frustum have radii a and b. Then the volume of the frustum equals $\frac{1}{3}\pi h(a^2 + ab + b^2)$. Let l be the slant height of the frustum; that is, the length of the part of a generator between the top and bottom of the frustum. Then the area of the curved surface of the frustum equals $\pi (a+b)l$.

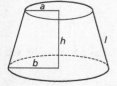

F-test A test that uses the *F-distribution.

fulcrum *See* LEVER.

full measure A set in a *measure space is said to have full measure if its complement is a set of null measure.

full rank A matrix has full rank if its rank is equal to the smaller of the number of rows and the number of columns.

full set A *compact subset S is called full if its *complement is *connected. So the closed unit disc is full, while the unit circle is not.

function A function f from S to T, where S and T are non-empty sets, is a rule that associates with each element of S (the domain) a unique element of T (the codomain). Thus it is the same thing as a *mapping. The word 'function' tends to be used when the domain S is the set **R** of real numbers, or some subset of **R**, and the codomain T is **R** (*see* REAL FUNCTION). The notation $f: S \rightarrow T$, read as "f from S to T", is used. If $x \in S$, then $f(x)$ is the image of x under f. The subset of T consisting of those elements that are images of elements of S under f, that is, the set $\{y \mid y = f(x), \text{ for some } x \text{ in } S\}$, is the range of f. If $f(x) = y$, it is said that f maps x to y, written $f: x \rightarrow y$.

If the graph of f is then taken to be $y = f(x)$, it may be said that y is a function of x. When $x = a$, $f(a)$ is the corresponding value of the function.

functional analysis The branch of mathematics concerned with the study of spaces of functions. It is largely concerned with the study of complete normed vector spaces over the set of real or complex numbers, with important applications in the study of quantum mechanics.

functionally separate Two subsets A and B of a *topological space X are functionally separate if there exists a continuous function $f: X \rightarrow [0, 1]$ such that $f(a) = 0$ for any $a \in A$ and $f(b) = 1$ for any $b \in B$.

Fundamental Theorem of Algebra The following important theorem in mathematics, concerned with the roots of polynomial equations:

Theorem

Every polynomial equation

$$a_n z^n + a_{n-1} z^{n-1} + \cdots + a_1 z + a_0 = 0,$$

where the a_i are real or complex numbers and $a_n \neq 0$, has a root in the set of complex numbers.

It follows that, if $f(z) = a_n z^n + a_{n-1} z^{n-1} + \cdots + a_1 z + a_0$, there exist complex numbers $\alpha_1, \alpha_2, \ldots, \alpha_n$ (not necessarily distinct) such that

$$f(z) = a_n(z - \alpha_1)(z - \alpha_2) \ldots (z - \alpha_n).$$

Hence the equation $f(z) = 0$ cannot have more than n distinct roots.

Fundamental Theorem of Arithmetic = UNIQUE FACTORIZATION THEOREM.

Fundamental Theorem of Calculus A sound approach to integration defines the *integral

$$\int_a^b f(x)dx$$

as the limit, in a certain sense, of a sum. That this can be evaluated, when f is continuous, by finding an *antiderivative of f, is the result embodied in the so-called Fundamental Theorem of Calculus. It establishes that integration is the reverse process to differentiation:

Theorem

If f is continuous on $[a, b]$ and ϕ is a function such that $\phi'(x) = f(x)$ for all x in $[a, b]$, then

$$\int_a^b f(x)dx = \phi(b) - \phi(a).$$

Fundamental Theorem of Game Theory The following theorem, also known as the 'Minimax Theorem', due to *Von Neumann:

Theorem

Suppose that, in a matrix game, $E(\mathbf{x}, \mathbf{y})$ is the expectation, where $\mathbf{x}$ and $\mathbf{y}$ are mixed strategies for the two players. Then

$$\max_x \min_y E(\mathbf{x}, \mathbf{y}) = \min_y \max_x E(\mathbf{x}, \mathbf{y}).$$

By using a maximin strategy, one player, R, ensures that the expectation is at least as large as the left-hand side of the equation appearing in the theorem. Similarly, by using a minimax strategy, the other player, C, ensures that the expectation is less than or equal to the right-hand side of the equation. Such strategies may be called optimal strategies for R and C. Since, by the theorem, the two sides of the equation are equal, then if R and C use optimal strategies the expectation is equal to the common value, which is called the value of the game.

For example, consider the game given by the matrix

$$\begin{bmatrix} 4 & 2 \\ 3 & 4 \end{bmatrix}.$$

if $\mathbf{x}^* = \left(\frac{1}{3}, \frac{2}{3}\right)$, it can be shown that $E(\mathbf{x}^*, \mathbf{y}) \geq 10/3$ for all $\mathbf{y}$. Also, if $\mathbf{y}^* = \left(\frac{2}{3}, \frac{1}{3}\right)$, then $E(\mathbf{x}, \mathbf{y}^*) \leq 10/3$ for all $\mathbf{x}$. It follows that the value of the game is $10/3$, and $\mathbf{x}^*$ and $\mathbf{y}^*$ are optimal strategies for the two players.

fuzzy set theory In standard set theory an element either is or is not a member of a particular set. However, in some instances, for example in pattern recognition or decision making, it is not known whether or not an element is in the set. Fuzzy set theory blurs this distinction and replaces this two-valued function with a probability distribution giving the likelihood that an element is a member of a particular set.

g *See* GRAVITY.

Galileo Galilei (1564–1642) Italian mathematician, astronomer and physicist who established the method of studying dynamics by a combination of theory and experiment. He formulated and verified by experiment the law $s = \frac{1}{2}at^2$ of uniform acceleration for falling bodies, and derived the parabolic path of a projectile. He developed the telescope and was the first to use it to make significant and outstanding astronomical observations. In later life, his support for the Copernican theory that the planets travel round the Sun resulted in conflict with the Church and consequent trial and house arrest.

Galois, Évariste (1811–32) French mathematician who had made major contributions to the theory of equations before he died at the age of 20, shot in a duel. His work developed the necessary group theory to deal with the question of whether an equation can be solved algebraically. He spent the night before the duel writing a letter containing notes of his discoveries.

Galton, Francis (1822–1911) British explorer and anthropologist, a cousin of Charles Darwin. His primary interest was eugenics. He made contributions to statistics in the areas of *regression and *correlation.

gambler's ruin *See* RANDOM WALK.

(⊕) SEE WEB LINKS
• A Java applet showing the gambler's ruin problem.

game An attempt to represent and analyse mathematically some conflict situation in which the outcome depends on the choices made by the opponents. The applications of game theory are not primarily concerned with recreational activities. Games may be used to investigate problems in business, personal relationships, military manœuvres and other areas involving decision-making. One particular kind of game for which the theory has been well developed is the *matrix game.

gamma distribution If x is a random variable with p.d.f. given by
$$f(x) = \frac{\lambda^v x^{v-1} e^{-\lambda x}}{\Gamma(v)}$$ where $\Gamma(n)$ is a *gamma function and λ, v, x all > 0, then

we say that X has a gamma distribution with parameters λ, v. When $v = 1$, $x^{v-1} = x^0 = 1$, and $\Gamma(v) = \Gamma(1) = 1$, so $f(x)$ reduces to $\lambda e^{-\lambda x}$ which is the *exponential distribution.

gamma function The function defined by $\Gamma(x) = \int_0^\infty t^{x-1} e^{-t} dt$ for $x > 0$. *Integration by parts yields $\Gamma(x+1) = x\Gamma(x)$, and $\Gamma(1) = 1$ so if n is an integer $\Gamma(n) = (n-1)!$

Gantt charts *See* CASCADE CHARTS.

Gauss, Carl Friedrich (1777–1855) German mathematician and astronomer, perhaps the greatest pure mathematician of all time. He also made enormous contributions to other parts of mathematics, physics, and astronomy. He was highly talented as a child. At the age of 18, he invented the method of least squares and made the new discovery that a 17-sided regular polygon could be constructed with ruler and compasses. By the age of 24, he was ready to publish his *Disquisitiones arithmeticae*, a book that was to have a profound influence on the theory of numbers. In this, he proved the *Fundamental Theorem of Arithmetic and the *Fundamental Theorem of Algebra. In later work, he developed the theory of curved surfaces using methods now known as differential geometry. His work on complex functions was fundamental but, like his discovery of *non-Euclidean geometry, it was not published at the time. He introduced what is now known to statisticians as the Gaussian distribution. His memoir on potential theory was just one of his contributions to applied mathematics. In astronomy, his great powers of mental calculation allowed him to calculate the orbits of comets and asteroids from limited observational data. He was the first mathematician to consider the behaviour of *knots in a formal mathematical sense.

Gaussian distribution = NORMAL DISTRIBUTION.

Gaussian elimination A particular systematic procedure for solving a set of linear equations in several unknowns. This is normally carried out by applying *elementary row operations to the augmented matrix

$$\begin{bmatrix} a_{11} & a_{12} & \cdots & a_{1n} & b_1 \\ a_{21} & a_{22} & \cdots & a_{2n} & b_2 \\ \vdots & \vdots & \ddots & \vdots & \vdots \\ a_{m1} & a_{m2} & \cdots & a_{mn} & b_m \end{bmatrix}$$

to transform it to *echelon form. The method is to divide the first row by a_{11} and then subtract suitable multiples of the first row from the subsequent rows, to obtain a matrix of the form

$$\begin{bmatrix} 1 & a'_{12} & \cdots & a'_{1n} & b'_1 \\ 0 & a'_{22} & \cdots & a'_{2n} & b'_2 \\ \vdots & \vdots & \ddots & \vdots & \vdots \\ 0 & a'_{m2} & \cdots & a'_{mn} & b'_m \end{bmatrix}.$$

(If $a_{11} = 0$, it is necessary to interchange two rows first.) The first row now remains untouched and the process is repeated with the remaining rows, dividing the second row by a'_{22} to produce a 1, and subtracting suitable multiples of the new second row from the subsequent rows to produce zeros below that 1. The method continues in the same way. The essential point is that the corresponding set of equations at any stage has the same solution set as the original. (*See also* SIMULTANEOUS LINEAR EQUATIONS.)

Gaussian function The function e^{-x^2} which has the property $\int_{-\infty}^{\infty} e^{-x^2} \, dx = \sqrt{\pi}$ which is the function underlying the *normal distribution.

Gaussian integer A *complex number whose real and imaginary parts are both integers, so $z = a + ib$ is a Gaussian integer if $a, b \in Z$.

Gaussian plane Another name for the *complex plane or Argand diagram.

Gauss–Jordan elimination An extension of the method of *Gaussian elimination. At the stage when the i-th row has been divided by a suitable value to obtain a 1, suitable multiples of this row are subtracted, not only from subsequent rows, but also from preceding rows to produce zeros both below and above the 1. The result of this systematic method is that the augmented matrix is transformed into *reduced echelon form. As a method for solving *simultaneous linear equations, Gauss–Jordan elimination in fact requires more work than Gaussian elimination followed by *back-substitution, and so it is not in general recommended.

Gauss–Markov Theorem In a *linear regression model in which the errors have zero mean, are uncorrelated, and have equal variances the best linear unbiased estimators of the coefficients are the *least squares estimators. Here, 'best' means that it has minimum variance amongst all linear unbiased estimators.

Gauss–Seidel iterative method A technique for solving a set of n linear equations in n unknowns. If the system is summarized by $\mathbf{A}\mathbf{x} = \mathbf{b}$, then taking initial values as $x_i^{(1)} = \dfrac{b_i}{a_{ii}}$, it uses the iterative relation

$$x_i^{(k)} = \frac{b_i - \sum_{j<i} a_{ij} x_j^{(k)} - \sum_{j>i} a_{ij} x_j^{(k-1)}}{a_{ii}},$$ so it uses the new values immediately they are available, unlike the *Jacobi method.

Gauss's Lemma Let $p(x)$ be a polynomial with integer coefficients. Then if $p(x)$ can be factorized using rational numbers, $p(x)$ can be factorized using only integers.

g.c.d. = GREATEST COMMON DIVISOR.

generalized eigenvalue and eigenvectors Let **A**, **B** be square matrices of the same size. Then the generalized eigenvalues (λ_i) and eigenvectors ($\mathbf{x}_i$) are the scalars and vectors for which $A\mathbf{x}_i = \lambda_i \mathbf{x}_i$ The values of λ_i are obtained by solving $\det(A - \lambda_i I) = 0$.

generalized maximum likelihood ratio test statistic The ratio of the *maximum likelihoods of the observed value under the parameters for the null and alternative hypotheses.

general relativity *See* RELATIVITY THEORY.

general solution A function containing n distinct arbitrary constants which satisfies an n-th-order differential equation is said to be a general solution. Obtained as the sum of the complementary function and a particular integral.

generating function The power series $G(x)$, where

$$G(x) = g_0 + g_1 x + g_2 x^2 + g_3 x^3 + \cdots,$$

is the generating function for the infinite sequence $g_0, g_1, g_2, g_3, \ldots$. (Notice that it is convenient here to start the sequence with a term with subscript 0). Such power series can be manipulated algebraically, and it can be shown, for example, that

$$\frac{1}{1-x} = 1 + x + x^2 + x^3 + \cdots,$$

$$\frac{1}{(1-x)^2} = 1 + 2x + 3x^2 + 4x^3 + \cdots.$$

Hence, $1/(1-x)$ and $1/(1-x)^2$ are the generating functions for the sequences $1, 1, 1, 1, \ldots$ and $1, 2, 3, 4, \ldots$, respectively.

The Fibonacci sequence $f_0, f_1, f_2, \ldots$ is given by $f_0 = 1$, $f_1 = 1$, and $f_{n+2} = f_{n+1} + f_n$. It can be shown that the generating function for this sequence is $1/(1-x-x^2)$.

The use of generating functions enables sequences to be handled concisely and algebraically. A *difference equation for a sequence can lead to an equation for the corresponding generating function, and the use of *partial fractions, for example, may then lead to a formula for the n-th term of the sequence.

*Probability and *moment generating functions are very powerful tools in statistics.

generator *See* CONE and CYLINDER.

genus The maximum number of times a surface can be cut along simple closed curves without the surface separating into disconnected parts. It is the same as the number of handles on the surface.

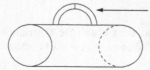

One cut can be made in the handle and still the whole of the surface is connected so the genus =1.

geodesic A curve on a surface, joining two given points, that is the shortest curve between the two points. On a sphere, a geodesic is an arc of a great circle through the two given points. This arc is unique unless the two points are *antipodal.

geometrical representation (of a vector) = REPRESENTATION (of a vector).

geometric distribution The discrete probability *distribution for the number of experiments required to achieve the first success in a sequence of independent experiments, all with the same probability p of success. The *probability mass function is given by $\Pr(X=r)=p(1-p)^{r-1}$, for $r=1$, $2,\ldots$. It has mean $1/p$ and variance $(1-p)/p^2$.

geometric mean *See* MEAN.

geometric progression = GEOMETRIC SERIES.

geometric sequence A finite or infinite sequence a_1, a_2, a_3, $\ldots$ with a common ratio r, so that $a_2/a_1=r$, $a_3/a_2=r$, $\ldots$. The first term is usually denoted by a. For example, 3, 6, 12, 24, 48, $\ldots$ is the geometric sequence with $a=3$, $r=2$. In such a geometric sequence, the n-th term a_n is given by $a_n=ar^{n-1}$.

(⊕) SEE WEB LINKS

• Stories illustrating the dramatic behaviour of geometric sequences.

geometric series A series $a_1 + a_2 + a_3 + \ldots$ (which may be finite or infinite) in which the terms form a *geometric sequence. Thus the terms have a common ratio r with $a_k/a_{k-1} = r$ for all k. If the first term a_1 equals a, then $a_k = ar^{k-1}$. Let s_n be the sum of the first n terms, so that $s_n = a + ar + ar^2 + \cdots + ar^{n-1}$. Then s_n is given (when $r \neq 1$) by the formulae

$$s_n = \frac{a(1 - r^n)}{1 - r} = \frac{a(r^n - 1)}{r - 1}.$$

If the common ratio r satisfies $-1 < r < 1$, then $r^n \to 0$ and it can be seen that $s_n \to a/(1-r)$. The value $a/(1-r)$ is called the sum to infinity of the series $a + ar + ar^2 + \ldots$. In particular, for $-1 < x < 1$, the geometric series $1 + x + x^2 + \ldots$ has sum to infinity equal to $1/(1-x)$. For example, putting $x = \frac{1}{2}$, the series $1 + \frac{1}{2} + \frac{1}{4} + \frac{1}{8} + \cdots$ as sum 2. If $x \leq -1$ or $x \geq 1$, then s_n does not tend to a limit and the series has no sum to infinity.

geometry The area of mathematics related to the study of points and figures, and their properties.

Gerschgorin's Theorem All the *characteristic values of a square complex matrix $\mathbf{A}$ lie within the circles centred on each entry in the leading diagonal a_{ii}, with radius $r_i = \sum_{j=1, j \neq i}^{n} |a_{ij}|$.

giga- Prefix used with *SI units to denote multiplication by 10^9.

given Something which is already known independently, or something which is to be used in the course of a proof. For example the epsilon-data method of proof usually states 'Given $\varepsilon > 0$, there exists a δ', by which is meant 'For any ε you choose, no matter how small, I can find a value of δ for which...'.

Gödel, Kurt (1906–78) Logician and mathematician who showed that the consistency of elementary arithmetic could not be proved from within the system itself. This result followed from his proof that any formal axiomatic system contains undecidable propositions. It undermined the hopes of those who had been attempting to determine axioms from which all mathematics could be deduced. Born in Brno, he was at the University of Vienna from 1930 until he emigrated to the United States in 1940.

Gödel's Completeness Theorem The theorem which states that in mathematical logic, if a formula is logically valid, then there is a finite formal proof of the formula based on the axioms of the system— something that a computer could be programmed to do. This is important in its own right, but is perhaps even more important as the precursor to *Gödel's Incompleteness Theorems.

Gödel's Incompleteness Theorems Taken together, the two incompleteness theorems say that it is not possible to find a set of axioms for arithmetic which are totally adequate. The first theorem says no set of axioms for arithmetic can be *consistent and *complete. So, any formal system that proves certain basic arithmetic truths must contain an arithmetical statement that is true but which cannot be proved from the axioms. The second theorem is really just a tightening-up of a particular aspect of the first incompleteness theorem. It states that if a set of axioms A is consistent then the consistency of A cannot be proved by A.

The key implication of these incompleteness theorems is that you might be able to prove all true statements about numbers (or, equivalently, about any other branch of mathematics) within a system by going outside the system to define new rules or axioms, but if you do so then you only create a larger system which will have its own unprovable statements.

Goldbach, Christian (1690–1764) Mathematician born in Prussia, who later became professor in St Petersburg and tutor to the Tsar in Moscow. *Goldbach's conjecture, for which he is remembered, was proposed in 1742 in a letter to Euler.

Goldbach's conjecture The conjecture that every even integer greater than 2 is the sum of two *primes. Neither proved nor disproved, Goldbach's conjecture remains one of the most famous unsolved problems in number theory.

golden ratio, golden rectangle *See* GOLDEN SECTION.

(🌐) SEE WEB LINKS
• Shows the golden ratio in biology and the relationship to Fibonacci numbers.

golden section A line segment is divided in golden section if the ratio of the whole length to the larger part is equal to the ratio of the larger part to the smaller part. This definition implies that, if the smaller part has unit length and the larger part has length τ, then $(\tau + 1)/\tau = \tau/1$. It follows that $\tau^2 - \tau - 1 = 0$, which gives $\tau = \frac{1}{2}(1 + \sqrt{5}) = 1.6180$, to 4 decimal places. This number τ is the golden ratio. A golden rectangle, whose sides are in this ratio, has throughout history been considered to have a particularly pleasing shape. It has the property that the removal of a square from one end of it leaves a rectangle that has the same shape.

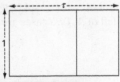

(⊕) SEE WEB LINKS
• Examples of the golden section in art and architecture.

goodness-of-fit test = CHI-SQUARED TEST.

googol A fanciful name for the number 10^{100}, written in decimal notation as a 1 followed by 100 zeros.

googolplex The number equal to the googolth power of 10, written in decimal notation as a 1 followed by a *googol of zeros.

Gosset, William Sealy (1876–1937) British industrial scientist and statistician best known for his discovery of the *t-distribution. His statistical work was motivated by his research for the brewery firm that he was with all his life. The most important of his papers, which were published under the pseudonym 'Student', appeared in 1908.

Gowers, William Timothy (1963–   ) Rouse Ball Professor of Mathematics at Cambridge University. Winner of the Fields Medal in 1998 for his research on functional analysis and combinatorics.

grade (grad function on a calculator) An angular measure which is one-hundredth of a right angle.

gradient (of a curve) The gradient of a curve at a point P may be defined as equal to the gradient of the *tangent to the curve at P. This definition presupposes an intuitive idea of what it means for a line to touch a curve. At a more advanced level, it is preferable to define the gradient of a curve by the methods of *differential calculus. In the case of a graph $y = f(x)$, the gradient is equal to $f'(x)$, the value of the derivative. The tangent at P can then be defined as the line through P whose gradient equals the gradient of the curve.

gradient (of a straight line) In coordinate geometry, suppose that A and B are two points on a given straight line, and let M be the point where the line through A parallel to the x-axis meets the line through B parallel to the y-axis. Then the gradient of the straight line is equal to MB/AM. (Notice that here MB is the *measure of $\overrightarrow{MB}$ where the line through M and B has positive direction upwards. In other words, MB equals the length $|MB|$

if B is above M, and equals $-|MB|$ if B is below M. Similarly, $AM = -|AM|$ if M is to the right of A, and $AM = -|AM|$ if M is to the left of A. Two cases are illustrated in the figure.)

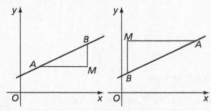

The gradient of the line through A and B may be denoted by m_{AB}, and, if A and B have coordinates (x_1, y_1) and (x_2, y_2), with $x_1 \neq x_2$, then

$$m_{AB} = \frac{y_2 - y_1}{x_2 - x_1}.$$

Though defined in terms of two points A and B on the line, the gradient of the line is independent of the choice of A and B. The line in the figure has gradient $\frac{1}{2}$.

Alternatively, the gradient may be defined as equal to $\tan \theta$, where either direction of the line makes an angle θ with the positive x-axis. (The different possible values for θ give the same value for $\tan \theta$.) If the line through A and B is vertical, that is, parallel to the y-axis, it is customary to say that the gradient is infinite. The following properties hold:

 (i) Points A, B and C are collinear if and only if $m_{AB} = m_{AC}$. (This includes the case when m_{AB} and m_{AC} are both infinite.)
 (ii) The lines with gradients m_1 and m_2 are parallel (to each other) if and only if $m_1 = m_2$. (This includes m_1 and m_2 both infinite.)
 (iii) The lines with gradients m_1 and m_2 are perpendicular (to each other) if and only if $m_1 m_2 = -1$. (This must be reckoned to include the cases when $m_1 = 0$ and m_2 is infinite and vice versa.)

gradient (grad) The vector obtained by applying the differential operator del, $\nabla = \mathbf{i}\dfrac{\partial}{\partial x} + \mathbf{j}\dfrac{\partial}{\partial y} + \mathbf{k}\dfrac{\partial}{\partial z}$, to a scalar function of position $\phi\,(\mathbf{r})$. This gives the gradient of ϕ as grad $\phi = \nabla\phi = \mathbf{i}\dfrac{\partial\phi}{\partial x} + \mathbf{j}\dfrac{\partial\phi}{\partial y} + \mathbf{k}\dfrac{\partial\phi}{\partial z}$. *See also* CURL and DIVERGENCE.

Graeco-Latin square The notion of a *Latin square can be extended to involve two sets of symbols. Suppose that one set of symbols consists of Roman letters and the other of Greek letters. A Graeco-Latin square is a square array in which each position contains one Roman letter and one Greek letter, such that the Roman letters form a Latin square, the Greek letters form a Latin square, and each Roman letter occurs with each Greek letter exactly once. An example of a 3×3 Graeco-Latin square is the following:

$A\alpha$ $B\beta$ $C\gamma$
$B\gamma$ $C\alpha$ $A\beta$
$C\beta$ $A\gamma$ $B\alpha$

Such squares are used in the design of experiments.

gram In *SI units, it is the *kilogram that is the base unit for measuring mass. A gram is one-thousandth of a kilogram.

Gram–Schmidt method If $(\mathbf{b_1}, \mathbf{b_2}, \ldots, \mathbf{b_n})$ is a set of vectors forming a *basis then an *orthonormal basis $(\mathbf{u_1}, \mathbf{u_2}, \ldots, \mathbf{u_n})$ can be constructed by $\hat{\mathbf{u}}_j = \mathbf{b_j} - \sum_{k=1}^{j-1} \left(\dfrac{\hat{\mathbf{u}}_j^T \mathbf{b_k}}{\hat{\mathbf{u}}_j^T \hat{\mathbf{u}}_j} \right) \hat{\mathbf{u}}_j, \mathbf{u_j} = \dfrac{\hat{\mathbf{u}}_j}{|\hat{\mathbf{u}}_j|}$.

graph A number of vertices (or *points or nodes), some of which are joined by edges. The edge joining the vertex U and the vertex V may be denoted by (U, V) or (V, U). The vertex-set, that is, the set of vertices, of a graph G may be denoted by $V(G)$ and the edge-set by $E(G)$. For example, the graph shown here on the left has $V(G) = \{U, V, W, X\}$ and $E(G) = \{(U, V), (U, W), (V, W), (W, X)\}$.

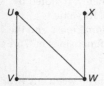
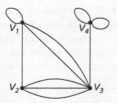

In general, a graph may have more than one edge joining a pair of vertices; when this occurs, these edges are called multiple edges. Also, a graph may have loops—a loop is an edge that joins a vertex to itself. In the other graph shown, there are 2 edges joining V_1 and V_3 and 3 edges joining V_2 and V_3; the graph also has three loops.

Normally, $V(G)$ and $E(G)$ are finite, but if this is not so, the result may also be called a graph, though some prefer to call this an *infinite graph.

graph (of a function or mapping) For a *real function f, the graph of f is the set of all pairs (x, y) in $\mathbf{R} \times \mathbf{R}$ such that $y = f(x)$ and x is in the domain of the function. For many real functions of interest, this gives a set of points

graph 206

that form a curve of some sort, possibly in a number of parts, that can be drawn in the plane. Such a curve defined by $y = f(x)$ is also called the graph of f. *See also* MAPPING.

graph (of a relation) Let R be a *binary relation on a set S, so that, when a is related to b, this is written $a \ R \ b$. The graph of R is the corresponding subset of the *Cartesian product $S \times S$, namely, the set of all pairs (a, b) such that $a \ R \ b$.

graphical solution The method of solving pairs of simultaneous equations by plotting their graphs and identifying the points of intersection. Since any point on a curve satisfies the equation producing that curve, a point of intersection of two or more lines or curves must necessarily be a solution of the equations.

graph metric The vertex set of an (undirected) *graph and the distances between vertices form a *metric space if and only if the graph is *connected. Such a metric space is known as a graph metric.

graph paper Paper printed with intersecting lines for drawing graphs or diagrams. The simplest form has equally spaced perpendicular lines, but one or both axis may use logarithmic scales, or scales derived from probability distributions.

graph theory The area of mathematics related to the study of *planar graphs and their properties.

gravitational constant The constant of proportionality, denoted by G, that occurs in the *inverse square law of gravitation. Its value is dependent on the decision to arrange for the *gravitational mass and the *inertial mass of a particle to have the same value. The dimensions of G are $L^3 \ M^{-1} \ T^{-2}$, and its value is $6.672 \times 10^{-11} \text{Nm}^2 \text{kg}^{-2}$.

gravitational force The force of attraction that exists between any two bodies, and described by the *inverse square law of gravitation. *See also* GRAVITY.

gravitational mass The parameter associated with a body that arises in the *inverse square law of gravitation.

gravitational potential energy The *potential energy associated with the gravitational force. When $\mathbf{F} = -\dfrac{GMm}{r^3}\mathbf{r}$, as in the *inverse square law of gravitation, it can be shown that the gravitational potential energy $E_p = -GMm/r + \text{constant}$. When $\mathbf{F} = -mg\mathbf{k}$, as may be assumed near the Earth's surface, $E_p = mgz + \text{constant}$.

gravity Near the Earth's surface, a body experiences the gravitational force between the body and the Earth, which may be taken to be constant. The resulting acceleration due to gravity is $-g\mathbf{k}$, where $\mathbf{k}$ is a unit vector directed vertically upwards from the Earth's surface, assumed to be a horizontal plane. The constant g, which is the magnitude of the acceleration due to gravity, is equal to GM/R^2, where G is the *gravitational constant, M is the mass of the Earth and R is the radius of the Earth. Near the Earth's surface, the value of g may be taken as 9.81 m s^{-2}, though it varies between 9.78 m s^{-2} at the equator and 9.83 m s^{-2} at one of the poles.

great circle A circle on the surface of a sphere with its centre at the centre of the sphere (in contrast to a *small circle). The shortest distance between two points on a sphere is along an arc of a great circle through the two points. This great circle is unique unless the two points are *antipodal.

greatest An element a of a set for which $a > b$ for every element $b \neq a$ in the set.

greatest common factor (**greatest common divisor**) For two non-zero integers a and b, any integer that is a divisor of both is a *common divisor. Of all the common divisors, the greatest is the greatest common divisor (or g.c.d.), denoted by (a, b). The g.c.d. of a and b has the property of being divisible by every other common divisor of a and b. It is an important theorem that there are integers s and t such that the g.c.d. can be expressed as $sa + tb$. If the prime decompositions of a and b are known, the g.c.d. is easily found: for example, if $a = 168 = 2^3 \times 3 \times 7$ and $b = 180 = 2^2 \times 3^2 \times 5$, then the g.c.d. is $2^2 \times 3 = 12$. Otherwise, the g.c.d. can be found by the *Euclidean Algorithm, which can be used also to find s and t to express the g.c.d. as $sa + tb$. Similarly, any finite set of non-zero integers $a_1, a_2, \ldots, a_n$ has a g.c.d., denoted by $(a_1, a_2, \ldots, a_n)$, and there are integers $s_1, s_2, \ldots, s_n$ such that this can be expressed as $s_1 a_1 + s_2 a_2 + \cdots + s_n a_n$.

greatest integer function (**floor**) The largest integer not greater than a given real number, so for 3.2 it would return the value 3, for -3.2 it would return -4, and for 7 it would return 7.

greatest lower bound = INFIMUM.

greatest value Let f be a *real function and D a subset of its domain. If there is a point c in D such that $f(c) \geq f(x)$ for all x in D, then $f(c)$ is the greatest value of f in D. There may be no such point: consider, for example, either the function f defined by $f(x) = 1/x$ or the function f defined by $f(x) = x$, with the open interval $(0, 1)$ as D; or the function f defined by

$f(x) = x - [x]$, with the closed interval $[0, 1]$ as D. If the greatest value does exist, it may be attained at more than one point of D.

That a *continuous function on a closed interval has a greatest value is ensured by the non-elementary theorem that such a function 'attains its bounds'. An important theorem states that a function, continuous on $[a, b]$ and *differentiable in (a, b), attains its greatest value either at a *local maximum (which is a *stationary point) or at an end-point of the interval.

Green, George (1793–1841) British mathematician who developed the mathematical theory of electricity and magnetism. In his essay of 1828, following *Poisson, he used the notion of potential and proved the result now known as Green's Theorem, which has wide applications in the subject. He had worked as a baker, and was self-taught in mathematics; he published other notable mathematical papers before beginning to study for a degree at Cambridge at the age of 40.

Gregory, James (1638–75) Scottish mathematician who studied in Italy before returning to hold chairs at St Andrews and Edinburgh. He obtained infinite series for certain trigonometric functions such as $\tan^{-1} x$, and was one of the first to appreciate the difference between convergent and divergent series. A predecessor of *Newton, he probably understood, in essence, the *Fundamental Theorem of Calculus and knew of *Taylor series forty years before Taylor's publication. He died at the age of 36.

Gregory–Newton forward difference formula Let $x_0, x_1, x_2, \ldots, x_n$ be equally spaced values, so that $x_i = x_0 + ih$, for $i = 1, 2, \ldots, n$. Suppose that the values $f_0, f_1, f_2, \ldots, f_n$ are known, where $f_i = f(x_i)$, for some function f. The Gregory–Newton forward difference formula is a formula involving *finite differences that gives an approximation for $f(x)$, where $x = x_0 + \theta h$, and $0 < \theta < 1$. It states that

$$f(x) \approx f_0 + \theta \Delta f_0 + \frac{\theta(\theta - 1)}{2!} \Delta^2 f_0 + \frac{\theta(\theta - 1)(\theta - 2)}{3!} \Delta^3 f_0 + \cdots,$$

the series being terminated at some stage. The approximation $f(x) \approx f_0 + \theta \Delta f_0$ gives the result of linear interpolation. Terminating the series after one more term provides an example of quadratic interpolation.

Grelling's paradox Certain adjectives describe themselves and others do not. For example, 'short' describes itself and so does 'polysyllabic'. But 'long' does not describe itself, 'monosyllabic' does not, and nor does 'green'. The word 'heterological' means 'not describing itself'. The German mathematician K. Grelling pointed out what is known as Grelling's paradox that results from considering whether 'heterological' describes itself or not. The paradox has some similarities with *Russell's paradox.

group An operation on a set is worth considering only if it has properties likely to lead to interesting and useful results. Certain basic properties recur in different parts of mathematics and, if these are recognized, use can be made of the similarities that exist in the different situations. One such set of basic properties is specified in the definition of a group. The following, then, are all examples of groups: the set of real numbers with addition, the set of non-zero real numbers with multiplication, the set of 2×2 real matrices with matrix addition, the set of vectors in 3-dimensional space with vector addition, the set of all bijective mappings from a set S onto itself with composition of mappings, the four numbers $1, i, -1, -i$ with multiplication. The definition is as follows: a group is a set G closed under an operation $\circ$ such that

   (i) for all a, b and c in G, $a \circ (b \circ c) = (a \circ b) \circ c$,
   (ii) there is an identity element e in G such that $a \circ e = e \circ a = a$ for all a in G,
   (iii) for each a in G, there is an inverse element a' in G such that $a \circ a' = a' \circ a = e$.

The group may be denoted by $\langle G, \circ \rangle$, or $(G, \circ)$, when it is necessary to specify the operation, but it may be called simply the group G when the intended operation is clear.

group action The group action of a group G on a non-empty set X is a *homomorphism h from G to the *symmetric group of X, with the properties that the identity element of G corresponds to the identity transformation of X and the product gh.

grouped data A set of data is said to be grouped when certain groups or categories are defined and the observations in each group are counted to give the frequencies. For numerical data, groups are often defined by means of *class intervals.

Hadamard, Jacques (1865–1963) French mathematician, known for his proof in 1896 of the *prime number theorem. He also worked, amongst other things, on the *calculus of variations and the beginnings of functional analysis.

half-angle formula Trigonometric formulae expressing a function of a half-angle in terms of the full angle or expressing a function of an angle in terms of half-angles.

$$\sin\frac{x}{2} = \pm\sqrt{\frac{1-\cos x}{2}},\ \cos\frac{x}{2} = \pm\sqrt{\frac{1+\cos x}{2}},\ \text{and if}\ t = \tan\frac{x}{2},$$

$$\sin x = \frac{2t}{1+t^2},\quad \cos x = \frac{1-t^2}{1+t^2},\quad \tan x = \frac{2t}{1-t^2}$$

half-closed An interval which includes one end but not the other, so in the form $[a, b)$ or $(a, b]$, for example $3 \le x < 7$ or $-\sqrt{2} < x \le 0$.

half-life *See* EXPONENTIAL DECAY.

half-line The part of a straight line extending from a point indefinitely in one direction.

half-open = HALF-CLOSED.

half-plane In coordinate geometry, if a line l has equation $ax + by + c = 0$, the set of points (x, y) such that $ax + by + c > 0$ forms the open half-plane on one side of l and the set of points (x, y) such that $ax + by + c < 0$ forms the open half-plane on the other side of l. When the line $ax + by + c = 0$ has been drawn, there is a useful method, if $c \ne 0$, of determining which half-plane is which: find out which of the two inequalities is satisfied by the origin. Thus the half-plane containing the origin is the one given by $ax + by + c > 0$ if $c > 0$, and is the one given by $ax + by + c < 0$ if $c < 0$.

A closed half-plane is a set of points (x, y) such that $ax + by + c \ge 0$ or such that $ax + by + c \le 0$. The use of open and closed half-planes is the basis of elementary *linear programming.

half-space Loosely, one of the two regions into which a plane divides 3-dimensional space. More precisely, if a plane has equation $ax + by + cz + d = 0$, the set of points (x, y, z) such that $ax + by + cz + d > 0$ and the set of points (x, y, z) such that $ax + by + cz + d < 0$ are open half-spaces lying on opposite sides of the plane. The set of points (x, y, z) such that $ax + by + cz + d \geq 0$, and the set of points (x, y, z) such that $ax + by + cz + d \leq 0$, are closed half-spaces.

half-turn symmetry *See* SYMMETRICAL ABOUT A POINT.

Halley, Edmond FRS (1656–1742) Published a catalogue of the stars in the southern hemisphere in 1687, and studied planetary orbits, correctly predicting the return of Halley's comet in 1758. Published the Breslau Table of Mortality in 1693 which was the foundation of actuarial science for calculating life insurance premiums and annuities.

(((∰))) SEE WEB LINKS
• Pictures of Halley's comet and related information.

Halley's method A method of solving an equation in one variable $f(x) = 0$ with a first approximation to the solution x_0 using the iteration

$$x_{n+1} = x_n - \frac{f(x_n)}{f(x_n) - \left(\frac{f(x_n)f''(x_n)}{f'(x_n)}\right)} \quad n = 0, 1, 2, \ldots \text{ This method}$$

converges faster than Newton's method, but more computation is required at each iteration, so neither method can claim to be universally more efficient.

Halmos, Paul Richard (1916–2006) Hungarian mathematician who made important contributions to functional analysis, measure theory, ergodic theory and operator theory, but is perhaps best known for his ability to communicate mathematics in a textbook style envied by many and achieved by very few.

Hamilton, William Rowan (1805–65) Ireland's greatest mathematician, whose main achievement was in the subject of geometrical optics, for which he laid a theoretical foundation that came close to anticipating quantum theory. His work is also of great significance for general mechanics. He is perhaps best known in pure mathematics for his algebraic theory of complex numbers, the invention of *quaternions and the exploitation of non-commutative algebra. A child prodigy, he could, it is claimed, speak 13 languages at the age of 13. He became Professor of Astronomy at Dublin and Royal Astronomer of Ireland at the age of 22.

Hamiltonian A function H such that a *first-order partial differential equation can be rewritten in the form $\frac{\partial u}{\partial t} = -H(t, x_1, x_2, \ldots, x_n, p_1, p_2 \ldots, p_n)$, where u, x_i, p_i are all functions of t. It is used in studying the rate of change of the conditions of sets of moving particles in both classical mechanics and in *quantum mechanics.

Hamiltonian graph In graph theory, one area of study has been concerned with the possibility of travelling around a *graph, going along edges in such a way as to visit every vertex exactly once. For this purpose, the following definitions are made. A Hamiltonian cycle is a *cycle that contains every vertex, and a graph is called *Hamiltonian if it has a Hamiltonian cycle. The term arises from Hamilton's interest in the existence of such cycles in the graph of the dodecahedron—the graph with vertices and edges corresponding to the vertices and edges of a dodecahedron.

Hamming distance The minimum *distance between any two codewords in a code. If a codeword is received which does not exist in the code, the decoder will assign a codeword of the same length with the smallest Hamming distance.

handshaking lemma The simple result that, in any *graph, the sum of the degrees of all the vertices is even. (The name arises from its application to the total number of hands shaken when some members of a group of people shake hands.) It follows from the simple observation that the sum of the degrees of all the vertices of a graph is equal to twice the number of edges. A result which follows from it is that, in any graph, the number of vertices of odd degree is even.

hardware *See* COMPUTER.

Hardy, Godfrey Harold (1877–1947) British mathematician, a leading figure in mathematics in Cambridge. Often in collaboration with J. E. Littlewood, he published many papers on prime numbers, other areas of number theory and mathematical analysis.

Hardy–Weinberg ratio The ratio of genotype frequencies that evolve when mating in a population is random. In the case where only two characteristics A and B are involved, with proportions p and $1-p$ in the population, then the ratio of the three possible pairs of genes can be derived from simple combinatorial probabilities. AA, AB and BB will occur in the ratio p^2: $2p(1-p)$: $(1-p)^2$ after a single generation.

harmonic Able to be expressed in terms of sine and cosine functions.

harmonic analysis The area of mathematics relating to the study of functions by expressing them as the sum or integral of simple trigonometric functions.

harmonic mean *See* MEAN.

harmonic progression = HARMONIC SEQUENCE.

harmonic range When the *cross ratio is −1, the four points are said to form a harmonic range.

harmonic ratio When the *cross ratio is −1.

harmonic sequence A sequence a_1, a_2, a_3, ... such that $1/a_1$, $1/a_2$, $1/a_3$, ... is an *arithmetic sequence. The most commonly occurring harmonic sequence is the sequence $1, \frac{1}{2}, \frac{1}{3}, \frac{1}{4}, \ldots$.

harmonic series A series $a_1 + a_2 + a_3 + \ldots$ in which a_1, a_2, a_3, ... is a *harmonic sequence. Often the term refers to the particular series $1 + \frac{1}{2} + \frac{1}{3} + \frac{1}{4} \ldots$, where the n-th term a_n equals $1/n$. For this series, $a_n \to 0$. However, the series does not have a sum to infinity since, if s_n is the sum of the first n terms, then $s_n \to \infty$. For large values of n, $s_n \approx \ln n + \gamma$, where γ is *Euler's constant.

Hausdorff, Felix (1868–1942) Polish mathematician who made seminal contributions in topology and the study of metric spaces. Committed suicide with his family when they were to be sent to the infamous concentration camps in the Second World War.

Hausdorff space A *topological space in which distinct points have *disjoint *neighbourhoods. More precisely, it is a *completely regular space in which any closed set S and any x not in S are *functionally separate.

Hawking, Stephen William FRS (1942–2018) British mathematician and theoretical physicist who was Lucasian Professor of Mathematics at Cambridge University. He suffered from motor neurone disease for most of his adult life, but this did not prevent him from becoming one of the most famous scientists in Britain and author of one of the bestselling science books of all time, *A Brief History of Time*, which was a popular exposition of cosmology. His main contributions relate to the understanding of the nature of black holes, where he showed that black holes can emit radiation, and that mini black holes exist with very large mass in a tiny space so that *general relativity is required to deal with their gravitational attraction and *quantum mechanics is needed to deal with their size. He was awarded the *Wolf Prize for Physics in 1988 jointly with Roger *Penrose.

Hawthorne effect The effect where participants in an experiment may respond differently than they normally would, because they know they are part of an experiment.

h.c.f. = HIGHEST COMMON FACTOR.

hectare A metric unit of surface area, of a square with side 100 metres, so 1 hectare = 10 000m^2. 1 hectare is approximately 2.5 acres.

hecto- Prefix used with *SI units to denote multiplication by 10^2.

-hedron Ending denoting a *polyhedron, for example a *dodecahedron.

height (of a triangle) *See* BASE (of a triangle).

Heisenberg, Werner Karl FRS (1901–76) German mathematician and theoretical physicist who was the first to work on what developed into *quantum mechanics for which he was awarded the Nobel Prize for Physics in 1932. He is best known for *Heisenberg's uncertainty principle.

Heisenberg's uncertainty principle The principle in *quantum mechanics stated by Werner Heisenberg in 1927 which says that it is not possible to simultaneously determine the position and momentum of a particle.

(⊕) SEE WEB LINKS
• An applet illustrating Heisenberg's uncertainty principle.

helix A curve on the surface of a (right-circular) cylinder that cuts the generators of the cylinder at a constant angle. Thus it is 'like a spiral staircase'.

(⊕) SEE WEB LINKS
• An animation of a helix.

hemi- = SEMI-. Prefix denoting half.

hemisphere One half of a sphere cut off by a plane through the centre of the sphere.

hendeca- Prefix denoting eleven.

hendecagon An eleven-sided polygon.

hepta- = SEPT-. Prefix denoting seven.

heptagon A seven-sided polygon.

hereditary property (of spaces) A *topological property is hereditary if all subspaces must share the property when the full space has it. If it is only true that all *closed subspaces must share the property, then the property is weakly hereditary.

Hermite, Charles (1822–1901) French mathematician who worked in algebra and analysis. In 1873, he proved that e is *transcendental. Also notable is his proof that the general quintic equation can be solved using elliptic functions.

Hero (Heron) of Alexandria (1st century AD) Greek scientist whose interests included optics, mechanical inventions and practical mathematics. A long-lost book, the *Metrica*, rediscovered in 1896, contains examples of mensuration, showing, for example, how to work out the areas of regular polygons and the volumes of different polyhedra. It also includes the earliest known proof of *Hero's formula.

Hero's formula *See* TRIANGLE.

Hero's method An iterative method of approximating the square root of a number. If $\sqrt{k}$ is required, and x_0 is an initial approximation, then

$$x_{n+1} = \frac{\left(x_n + \frac{k}{x_n}\right)}{2}, \quad n = 0, 1, 2, \ldots \text{ will converge to the square root of } k.$$

For example, to calculate the square root of 5, using a first approximation of 2, will give $x_1 = \frac{(2+2.5)}{2} = 2.25$, $x_2 = 2.236\,111\,11\ldots$, $x_3 = 2.236\,067$ $978\ldots$, $x_4 = 2.236\,067\,978\ldots$ and $\sqrt{5} = 2.236067978\ldots$. So this method has found the square root to considerable accuracy after only three iterations.

hertz The SI unit of *frequency, abbreviated to 'Hz'. Normally used to measure the frequency of cycles, or oscillations, one hertz is equal to one cycle per second.

heteroscedastic Essentially means different variances in some sense. So a number of variables may have difference variances. In a bivariate or multivariate context the variance of the independent variable may change as the dependent variable changes value—for example, if variability is proportional to the size of the variable.

heuristic Problem solving based on experience of working with other problems which share some characteristics of the current problem, but for which no *algorithm is known. Good heuristics can reduce the time needed to solve problems by recognizing which possible approaches are unlikely to be successful. George *Polya brought the notion of heuristics to a wide audience through his book *How To Solve It*. The second edition was published in 1957 and is still in print half a century later.

hex Abbreviation for hexadecimal representation.

hexa- = SEX-. Prefix denoting six.

hexadecimal representation The representation of a number to *base 16. In this system, 16 digits are required and it is normal to take 0, 1, 2, 3, 4, 5, 6, 7, 8, 9, A, B, C, D, E and F, where A to F represent the numbers that, in decimal notation, are denoted by 10, 11, 12, 13, 14 and 15. Then the hexadecimal representation of the decimal number 712, for example, is found by writing

$$712 = 2 \times 16^2 + 12 \times 16 + 8 = (2C8)_{16}.$$

It is particularly simple to change the representation of a number to base 2 (binary) to its representation to base 16 (hexadecimal) and vice versa: each block of 4 digits in base 2 (form blocks of 4, starting from the right-hand end) can be made to correspond to its hexadecimal equivalent. Thus

$$(101101001001101)_2 = (101\ 1010\ 0100\ 1101)_2 = (5A4D)_{16}.$$

Real numbers, not just integers, can also be written in hexadecimal notation, by using hexadecimal digits after a 'decimal' point, just as familiar *decimal representations of real numbers are obtained to base 10. Hexadecimal notation is important in computing. It translates easily into binary notation, but is more concise and easier to read.

hexagon A six-sided polygon.

hexagram The plane figure shown formed by extending the six sides of a regular hexagon to meet one another.

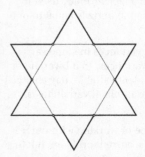

hexahedron A solid figure with six plane faces. The regular hexahedron is a *cube, *see* PLATONIC SOLIDS.

higher arithmetic = NUMBER THEORY.

higher derivative If the function f is *differentiable on an interval, its *derived function f' is defined. If f' is also differentiable, then the derived function of this, denoted by f'', is the second derived function of f; its value at x, denoted by $f''(x)$, or d^2f/dx^2, is the second derivative of f at x. (The term 'second derivative' may be used loosely also for the second derived function f''.)

Similarly, if f'' is differentiable, then $f'''(x)$ or d^3f/dx^3, the third derivative of f at x, can be formed, and so on. The n-th derivative of f at x is denoted by $f^{(n)}(x)$ or d^nf/dx^n. The n-th derivatives, for $n \geq 2$, are called the higher derivatives of f. When $y = f(x)$, the higher derivatives may be denoted by $d^2y/dx^2, \ldots, d^ny/dx^n$ or $y'', y''', \ldots, y^{(n)}$. If, with a different notation, x is a function of t and the derivative dx/dt is denoted by $\dot{x}$, the second derivative d^2x/dt^2 is denoted by $\ddot{x}$.

higher-order partial derivative Given a function f of n variables x_1, $x_2, \ldots, x_n$, the *partial derivative $\partial f/\partial x_i$, where $1 \leq i \leq n$, may also be reckoned to be a function of $x_1, x_2, \ldots, x_n$. So the partial derivatives of $\partial f/\partial x_i$ can be considered. Thus

$$\frac{\partial}{\partial x_i}\left(\frac{\partial f}{\partial x_i}\right) \quad \text{and} \quad \frac{\partial}{\partial x_j}\left(\frac{\partial f}{\partial x_i}\right) \quad \text{(for } j \neq i)$$

can be formed, and these are denoted, respectively, by

$$\frac{\partial^2 f}{\partial x_i^2} \quad \text{and} \quad \frac{\partial^2 f}{\partial x_j \partial x_i}.$$

These are the second-order partial derivatives. When $j \neq i$,

$$\frac{\partial^2 f}{\partial x_i \partial x_j} \quad \text{and} \quad \frac{\partial^2 f}{\partial x_j \partial x_i}.$$

are different by definition, but the two are equal for most 'straightforward' functions f that are likely to be met. (It is not possible to describe here just what conditions are needed for equality.) Similarly, third-order partial derivatives such as

$$\frac{\partial^3 f}{\partial x_1^3}, \qquad \frac{\partial^3 f}{\partial x_1 \partial x_2^2}, \qquad \frac{\partial^3 f}{\partial x_1 \partial x_2 \partial x_3}, \qquad \frac{\partial^3 f}{\partial x_2 \partial x_3 \partial x_1},$$

can be defined, as can fourth-order partial derivatives, and so on. Then the n-th-order partial derivatives, where $n \geq 2$, are called the higher-order partial derivatives.

When f is a function of two variables x and y, and the partial derivatives are denoted by f_x and f_y, then $f_{xx}, f_{xy}, f_{yx}, f_{yy}$ are used to denote

$$\frac{\partial^2 f}{\partial x^2}, \qquad \frac{\partial^2 f}{\partial y \partial x}, \qquad \frac{\partial^2 f}{\partial x \partial y}, \qquad \frac{\partial^2 f}{\partial y^2}$$

respectively, noting particularly that f_{xy} means $(f_x)_y$ and f_{yx} means $(f_y)_x$. This notation can be extended to third-order (and higher) partial derivatives and to functions of more variables. With the value of f at (x, y) denoted by $f(x, y)$ and the partial derivatives denoted by f_1 and f_2, the second-order partial derivatives can be denoted by f_{11}, f_{12}, f_{21} and f_{22}, and this notation can also be extended to third-order (and higher) partial derivatives and to functions of more variables.

highest common factor = GREATEST COMMON DIVISOR.

high precision *See* PRECISION.

Hilbert, David (1862–1943) German mathematician who was one of the founding fathers of 20th-century pure mathematics, and in many ways the originator of the formalist school of mathematics which has been so dominant in the pure mathematics of this century. Born at Königsberg (Kaliningrad), he became professor at Göttingen in 1895, where he remained for the rest of his life. One of his fundamental contributions to formalism was his *Grundlagen der Geometrie* (Foundations of Geometry), published in 1899, which served to put geometry on a proper axiomatic basis, unlike the rather more intuitive 'axiomatization' of *Euclid. He also made a major contribution to mathematical analysis. At the International Congress of Mathematics in 1900, he opened the new century by posing his famous list of 23 problems—problems that have kept mathematicians busy ever since and have generated a significant amount of the important work of this century. Hilbert is, for these reasons, often thought of as a thorough-going pure mathematician, but he was also the chairman of the famous atomic physics seminar at Göttingen that had a great influence on the development of quantum theory.

(((●))) SEE WEB LINKS
• The 23 mathematical problems of Hilbert.

Hilbert space A *vector space H with an *inner product $\langle \ \rangle$ for which the *norm given by $|f| = \sqrt{\langle f, f \rangle}$ makes H a *complete metric space.

Hilbert's paradox (infinite hotel paradox) The paradox stated by Hilbert illuminating the nature of infinite, but countable sets, where the number of rational numbers between zero and one is equal to the total number of rational numbers. The paradox is this: if a hotel with infinitely many rooms is full and another guest arrives, that guest can be accommodated by each existing guest moving from their current room to the room with the next highest number, leaving room 1 free for the new arrival. If an infinite number of extra guests arrived, they could be accommodated efficiently by each existing guest moving to the room whose number is twice their existing room number, leaving an infinite number of odd numbered rooms available for the new arrivals.

Hilbert's tenth problem The problem posed by Hilbert of finding an *algorithm to determine whether or not a given *Diophantine equation has solutions. It was proved by Y. Matijasevich in 1970 that no such algorithm exists.

histogram A diagram representing the *frequency distribution of data grouped by means of *class intervals. It consists of a sequence of rectangles, each of which has as its base one of the class intervals and is of a height taken so that the area is proportional to the frequency. If the class intervals are of equal lengths, then the heights of the rectangles are proportional to the frequencies. Some authors use the term 'histogram' when the data are discrete to describe a kind of *bar chart in which the rectangles are shown touching.

The figure shows a histogram of a sample of 500 observations.

SEE WEB LINKS
• An illustration of the construction of a histogram.

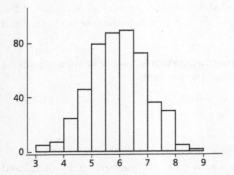

Hoffman coding = HUFFMAN CODING.

holomorphic = ANALYTIC (OF A COMPLEX FUNCTION).

homeomorphism A continuous function between *topological spaces that has a continuous inverse function. It is a mapping which preserves all the topological properties of the space, and therefore the two spaces are equivalent from a topological perspective. This forms the basis of the *Classification Theorem for Surfaces.

homogeneous Literally meaning 'of the same kind or nature'. A homogeneous population is one where the members are similar in nature. If this is not the case, then it may be more appropriate to use some form of *representative sampling, such as quota or *stratified.

homogeneous first-order differential equation A first-order differential equation $dy/dx = f(x, y)$ in which the function f, of two variables, has the property that $f(kx, ky) = f(x, y)$ for all k. Examples of such functions are

$$\frac{x^2 + 3y^2}{2x^2 - 5xy}, \qquad 1 + e^{x/y}, \qquad \frac{x}{\sqrt{x^2 + y^2}}.$$

Any such function f can be written as a function of one variable v, where $v = y/x$. The method of solving homogeneous first-order differential equations is therefore to let $y = vx$ so that $dy/dx = x\,dv/dx + v$. The differential equation for v as a function of x that is obtained is always *separable.

homogeneous linear differential equation *See* LINEAR DIFFERENTIAL EQUATION WITH CONSTANT COEFFICIENTS.

homogeneous polynomial A polynomial where all terms are of the same total degree. For example, $x^3 + 5x^2 y - y^3$ is homogeneous in the third degree.

homogeneous set of linear equations A set of m linear equations in n unknowns $x_1, x_2, \ldots, x_n$ that has the form

$$a_{11}x_1 + a_{12}x_2 + \ldots + a_{1n}x_n = 0,$$
$$a_{21}x_1 + a_{22}x_2 + \ldots + a_{2n}x_n = 0,$$
$$\vdots$$
$$a_{m1}x_1 + a_{m2}x_2 + \ldots + a_{mn}x_n = 0.$$

Here, unlike the non-homogeneous case, the numbers on the right-hand sides of the equations are all zero. In matrix notation, this set of equations can be written $\mathbf{Ax} = \mathbf{0}$, where the unknowns form a column matrix $\mathbf{x}$. Thus $\mathbf{A}$ is the $m \times n$ matrix $[a_{ij}]$, and

$$\mathbf{x} = \begin{bmatrix} x_1 \\ \vdots \\ x_n \end{bmatrix}.$$

If $\mathbf{x}$ is a solution of a homogeneous set of linear equations, then so is any scalar multiple $k\mathbf{x}$ of it. There is always the trivial solution $\mathbf{x} = \mathbf{0}$. What is generally of concern is whether it has other solutions besides this one. For a homogeneous set consisting of the same number of equations as unknowns, the matrix of coefficients **A** is a square matrix, and the set of equations has non-trivial solutions if and only if det **A** = 0.

homomorphism A *mapping between two similar algebraic structures which preserves the relational properties of elements in the two structures. So if f is a homomorphism, and $*$ and $\circ$ are corresponding operations in the two structures, $f(x * y) = f(x) \circ f(y)$.

homoscedastic Essentially means same *variance in some sense. So a number of variables may have the same variance. In a bivariate or multivariate context the variance of the independent variable may remain the same as the dependent variable changes value.

homotopy A continuous transformation of one *mapping between two spaces into another mapping between the same spaces. Another way of saying this is that a homotopy is a continuous deformation of one mapping into the other.

Hooke's law The law that says that the *tension in a spring or a stretched elastic string is proportional to the *extension. Suppose that a spring or elastic string has natural length l and actual length x. Then the tension T is given by $T = (\lambda/l)(x-l)$, where λ is the *modulus of elasticity, or $T = k(x-l)$, where k is the *stiffness.

Consider the motion of a particle suspended from a fixed support by a spring, the motion being in a vertical line through the equilibrium position of the particle. Suppose that the particle has mass m, and that the spring has stiffness k and natural length l. Let x be the length of the spring at time t. By using Hooke's law, the equation of motion $m\ddot{x} = mg - k(x - l)$ is obtained. The equilibrium position is given by $0 = mg - k(x-l)$; that is, $x = l + mg/k$. Letting $x = l + mg/k + X$ gives $\ddot{X} + \omega^2 X = 0$, where $\omega^2 = k/m$. Thus the particle performs *simple harmonic motion.

(⊕) SEE WEB LINKS
• An interactive illustration of Hooke's law.

l'Hôpital, Guillaume François Antoine, Marquis de (1661-1704) French mathematician who in 1696 produced the first textbook on differential calculus. This, and a subsequent book on analytical geometry, were standard texts for much of the 18th century. The first contains *l'Hôpital's rule, known to be due to Jean Bernouilli, who is thought to have agreed to keep the Marquis de l'Hôpital informed of his discoveries in return for financial support.

l'Hôpital's rule A rule for evaluating *indeterminate forms. One form of the rule is the following:

Theorem

Suppose that $f(x) \to 0$ and $g(x) \to 0$ as $x \to a$. Then

$$\lim_{x \to a} \frac{f(x)}{g(x)} = \lim_{x \to a} \frac{f'(x)}{g'(x)},$$

if the limit on the right-hand side exists.

For example,

$$\lim_{x \to 0} \frac{\sqrt{1+x} - 1}{x} = \lim_{x \to 0} \frac{\frac{1}{2}(1+x)^{-1/2}}{1} = \frac{1}{2}.$$

The result also holds if $f(x) \to \infty$ and $g(x) \to \infty$ as $x \to a$. Moreover, the results apply if '$x \to a$' is replaced by '$x \to +\infty$' or '$x \to -\infty$'.

Horner's method An iterative method for finding real roots of *polynomial equations, which is rather slow and cumbersome, though effective. Essentially it requires identifying the value of the root one place value at a time, starting with the largest place value, and then rewriting the equation in terms of a new variable which is the previous variable less this value. A simple example will illustrate the process. To solve $f(x) = x^2 - x - 1 = 0$, observe there is a root between 1 and 2 since $f(1) = -1$, $f(2) = 1$, so write $f_1(x) = f(x-1) = (x-1)^2 - (x-1) - 1 = x^2 - 3x + 1$, then identifying a root between 0.6 and 0.7 leads to calculating $f_2(x) = f_1(x-0.6)$ and so on until the required degree of accuracy is achieved.

Horner's rule A polynomial $f(x) = a_0 + a_1x + a_2x^2 + \ldots + a_n x^n$ can be written as $f(x) = a_0 + x(a_1 + x(a_2 + x(a_3 + \ldots x(a_{n-1} + a_n x)\ldots)))$. The method requires fewer multiplications than the standard method of working out each power of x and multiplying by the coefficient.

horsepower A unit of *power, abbreviated to 'hp', once commonly used in Britain. 1 hp $= 745.70$ watts.

Hoyles, Celia (1946–) British mathematics educationalist who was appointed as the government's Chief Adviser for Mathematics following the publication of Adrian *Smith's inquiry into the state of post-14 mathematics education in the UK, and subsequently the Director of the *National Centre for Excellence in the Teaching of Mathematics (NCETM). In 2004 she was chosen as the first recipient of the Hans Freudenthal medal, in recognition of her cumulative programme of research.

Huffman (Hoffman) coding A coding framework based on minimizing the length of binary digits necessary to transmit messages, based on the frequency with which letters and symbols appear. So 1 is used to transmit the most common letter, e, and all other characters use combinations of 0s and 1s with the property that the coding for any letter does not start with a string which identifies another character. So if n is 000, no other character starts with 000.

Huygens, Christiaan (1629–95) Dutch mathematician, astronomer and physicist. In mathematics, he is remembered for his work on pendulum clocks and his contributions in the field of dynamics. These concerned, for example, the period of oscillation of a simple pendulum and matters such as the centrifugal force in uniform circular motion.

hydrodynamics The area of mechanics which studies fluids in motion.

hydrostatics The area of mechanics which studies fluids at rest.

hyp. Abbreviation for hypotenuse or hypothesis.

Hypatia (370–415) Greek philosopher who was the head of the neoplatonist school in Alexandria. It is known that she was consulted on scientific matters, and that she wrote commentaries on works of Diophantus and Apollonius. Her brutal murder by a fanatical mob is often taken to mark the beginning of Alexandria's decline as the outstanding centre of learning.

hyperbola A *conic with eccentricity greater than 1. Thus it is the locus of all points P such that the distance from P to a fixed point F_1 (the focus) is equal to $e\,(>1)$ times the distance from P to a fixed line l_1 (the directrix). It turns out that there is another point F_2 and another line l_2 such that the same set of points would be obtained with these as focus and directrix. The hyperbola is also the conic section that results when a plane cuts a (double) cone in such a way that a section in two separate parts is obtained (*see* CONIC).

   The line through F_1 and F_2 is the transverse axis, and the points V_1 and V_2 where it cuts the hyperbola are the vertices. The length $|V_1V_2|$ is the length of the transverse axis and is usually taken to be $2a$. The

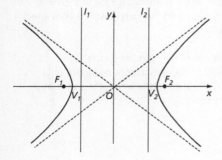

midpoint of V_1V_2 is the centre of the hyperbola. The line through the centre perpendicular to the transverse axis is the conjugate axis. It is usual to introduce b (> 0) defined by $b^2 = a^2(e^2-1)$, so that $e^2 = 1 + b^2/a^2$. It may be convenient to consider the points $(0,-b)$ and $(0, b)$ on the conjugate axis, despite the fact that the hyperbola does not cut the conjugate axis at all. The two separate parts of the hyperbola are the two *branches.

If a coordinate system is taken with origin at the centre of the hyperbola, and x-axis along the transverse axis, the foci have coordinates $(ae, 0)$ and $(-ae, 0)$, the directrices have equations $x = a/e$ and $x = -a/e$, and the hyperbola has equation

$$\frac{x^2}{a^2} - \frac{y^2}{b^2} = 1.$$

Unlike the comparable equation for an ellipse, it is not necessarily the case here that $a > b$. When investigating the properties of a hyperbola, it is normal to choose this convenient coordinate system. It may be useful to take $x = a \sec \theta$, $y = b \tan \theta (0 \leq \theta < 2\pi,\ \theta \neq \pi/2, 3\pi/2)$ as *parametric equations. As an alternative, the parametric equations $x = a \cosh t$, $y = b \sinh t$ ($t \in \mathbf{R}$) (*see* HYPERBOLIC FUNCTION) may also be used, but give only one branch of the hyperbola.

A hyperbola with its centre at the origin and its transverse axis, of length $2a$, along the y-axis instead has equation $y^2/a^2 - x^2/b^2 = 1$ and has foci at $(0, ae)$ and $(0,-ae)$.

The hyperbola

$$\frac{x^2}{a^2} - \frac{y^2}{b^2} = 1$$

has two *asymptotes, $y = (b/a)x$ and $y = (-b/a)x$. The shape of the hyperbola is determined by the eccentricity or, what is equivalent, by the ratio of b to a. The particular value $e = \sqrt{2}$ is important for this gives $b = a$. Then the asymptotes are perpendicular and the curve, which is of special interest, is a *rectangular hyperbola.

hyperbolic cylinder A *cylinder in which the fixed curve is a *hyperbola and the fixed line to which the generators are parallel is perpendicular to the plane of the hyperbola. It is a *quadric, and in a suitable coordinate system has equation

$$\frac{x^2}{a^2} - \frac{y^2}{b^2} = 1.$$

hyperbolic function Any of the functions cosh, sinh, tanh, sech, cosech and coth, defined as follows:

$$\cosh x = \tfrac{1}{2}(e^x + e^{-x}), \qquad \sinh x = \tfrac{1}{2}(e^x - e^{-x}),$$

$$\tanh x = \frac{\sinh x}{\cosh x}, \qquad \coth x = \frac{\cosh x}{\sinh x} \ (x \neq 0),$$

$$\mathrm{sech}\, x = \frac{1}{\cosh x}, \qquad \mathrm{cosech}\, x = \frac{1}{\sinh x}\,(x \neq 0).$$

The functions derive their name from the possibility of using $x = a \cosh t$, $y = b \sinh t$ ($t \in \mathbf{R}$) as *parametric equations for (one branch of) a *hyperbola. (The pronunciation of these functions causes difficulty. For instance, tanh may be pronounced as 'tansh' or 'than' (with the 'th' as in 'thing'); and sinh may be pronounced as 'shine' or 'sinch'. Some prefer to say 'hyperbolic tan' and 'hyperbolic sine'.) Many of the formulae satisfied by the hyperbolic functions are similar to corresponding formulae for the trigonometric functions, but some changes of sign must be noted. For example:

$$\cosh^2 x = 1 + \sinh^2 x,$$
$$\mathrm{sech}^2 x = 1 - \tanh^2 x,$$
$$\sinh(x + y) = \sinh x \cosh y + \cosh x \sinh y,$$
$$\cosh(x + y) = \cosh x \cosh y + \sinh x \sinh y,$$
$$\sinh 2x = 2 \sinh x \cosh x,$$
$$\cosh 2x = \cosh^2 x + \sinh^2 x.$$

Since $\cosh(-x) = \cosh x$ and $\sinh(-x) = -\sinh x$, cosh is an *even function and sinh is an *odd function. The graphs of $\cosh x$ and $\sinh x$ are shown below. It is instructive to sketch both of them, together with the graphs of e^x and e^{-x} on the same diagram.

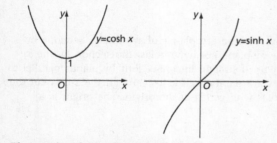

The graphs of the other hyperbolic functions are:

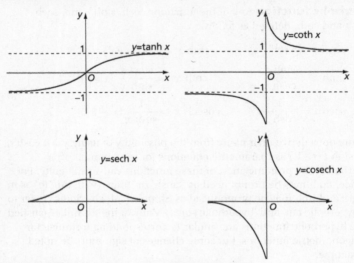

The following derivatives are easily established:

$$\frac{d}{dx}(\cosh x) = \sinh x, \quad \frac{d}{dx}(\sinh x) = \cosh x, \quad \frac{d}{dx}(\tanh x) = \operatorname{sech}^2 x.$$

See also INVERSE HYPERBOLIC FUNCTION.

hyperbolic geometry *See* NON-EUCLIDEAN GEOMETRY.

hyperbolic paraboloid A *quadric whose equation in a suitable coordinate system is

$$\frac{x^2}{a^2} - \frac{y^2}{b^2} = \frac{2z}{c}.$$

The *yz*-plane and the *zx*-plane are planes of symmetry. Sections by planes parallel to the *xy*-plane are hyperbolas, the section by the *xy*-plane itself being a pair of straight lines. Sections by planes parallel to the other axial planes are parabolas. Planes through the *z*-axis cut the paraboloid in parabolas with vertex at the origin. The origin is a *saddle-point.

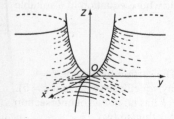

hyperbolic spiral A curve with equation $r\,\theta = k$ in polar coordinates. The figure shows the curve obtained when $k = 1$.

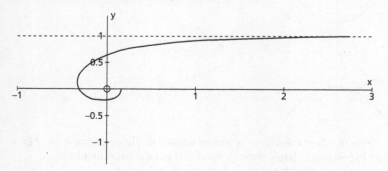

hyperboloid of one sheet A *quadric whose equation in a suitable coordinate system is

$$\frac{x^2}{a^2} + \frac{y^2}{b^2} - \frac{z^2}{c^2} = 1.$$

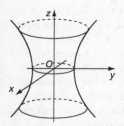

The axial planes are planes of symmetry. The section by a plane $z = k$ parallel to the xy-plane, is an ellipse (a circle if $a = b$). The hyperboloid is all in one piece or sheet. Sections by planes parallel to the other two axial planes are hyperbolas.

hyperboloid of two sheets A *quadric whose equation in a suitable coordinate system is

$$\frac{x^2}{a^2} + \frac{y^2}{b^2} - \frac{z^2}{c^2} = -1.$$

The axial planes are planes of symmetry. The section by a plane $z = k$ parallel to the xy-plane, is, when non-empty, an ellipse (a circle if $a = b$). When k lies between $-c$ and c, the plane $z = k$ has no points of intersection with the hyperboloid, and the hyperboloid is thus in two

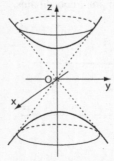

pieces or *sheets. Sections by planes parallel to the other two axial planes are hyperbolas. Planes through the z-axis cut the hyperboloid in hyperbolas with vertices at $(0, 0, c)$ and $(0, 0, -c)$.

hypercube The generalization in n dimensions of a square in two dimensions and a cube in three dimensions. A geometrical description is difficult because of the problem of visualizing more than three dimensions. But the following approach describes the hypercube most frequently encountered, which has edges of unit length.

In the plane, the points with Cartesian coordinates $(0, 0)$, $(1, 0)$, $(0, 1)$ and $(1, 1)$ are the vertices of a square. In 3-dimensional space, the eight points with Cartesian coordinates $(0, 0, 0)$, $(1, 0, 0)$, $(0, 1, 0)$, $(0, 0, 1)$, $(0, 1, 1)$, $(1, 0, 1)$, $(1, 1, 0)$ and $(1, 1, 1)$ are the vertices of a cube. So, in *n-dimensional space, the 2^n points with coordinates $(x_1, x_2, \ldots, x_n)$, where each $x_i = 0$ or 1, are the vertices of a hypercube. Two vertices are joined by an edge of the hypercube if they differ in exactly one of their coordinates. The vertices and edges of a hypercube in n dimensions form the vertices and edges of the graph known as the *n-cube.

hypergeometric distribution Suppose that, in a set of N items, M items have a certain property and the remainder do not. A sample of n items is taken from the set. The discrete probability *distribution for the number r of items in the sample having the property has *probability mass function given by $\Pr(X = r) = \dfrac{{}^{M}C_{r}\,{}^{N-M}C_{n-r}}{{}^{N}C_{n}}$, where r runs from 0 up to the smaller of M and n. This is called a hypergeometric distribution. The mean is $\dfrac{nM}{N}$, and the variance is $\dfrac{nM(N-M)\,(N-n)}{N^2(N-1)}$.

hyperplane *See* N-DIMENSIONAL SPACE.

hypocycloid The curve traced out by a point on the circumference of a circle rolling round the inside of a fixed circle. A special case is the *astroid.

hypotenuse The side of a right-angled triangle opposite the right angle.

hypothesis testing In statistics, a hypothesis is an assertion about a population. A null hypothesis, usually denoted by H_0, is a particular assertion that is to be accepted or rejected. The alternative hypothesis H_1 specifies some alternative to H_0. To decide whether H_0 is to be accepted or rejected, a significance test tests whether a sample taken from the population could have occurred by chance, given that H_0 is true.

From the sample, the *test statistic is calculated. The test partitions the set of possible values into the acceptance region and the *critical (or rejection) region. These depend upon the choice of the significance level α, which is the probability that the test statistic lies in the critical region if H_0 is true. Often α is chosen to be 5%, for example. If the test statistic falls in the critical region, the null hypothesis H_0 is rejected; otherwise, the conclusion is that there is no evidence for rejecting H_0 and it is said that H_0 is accepted.

A *Type I error occurs if H_0 is rejected when it is in fact true. The probability of a Type I error is α, so this depends on the choice of significance level. A *Type II error occurs if H_0 is accepted when it is in fact false. If the probability of a Type II error is β, the *power of the test is defined as $1-\beta$. This depends upon the choice of alternative hypothesis. The null hypothesis H_0 normally specifies that a certain parameter has a certain value. If the alternative hypothesis H_1 says that the parameter is not equal to this value, then the test is said to be two-tailed (or two-sided). If H_1 says that the parameter is greater than this value (or less than this value), then the test is one-tailed (or one-sided). *See also* P-VALUE.

i *See* COMPLEX NUMBER.

i (*i*) A unit vector, usually in the direction of the *x*-axis or a horizontal vector in the plane of a projectile's motion.

I The Roman *numeral for 1.

icosa- Prefix denoting 20.

icosahedron (icosahedra) A *polyhedron with 20 faces, often assumed to be regular. The regular icosahedron is one of the *Platonic solids, and its faces are equilateral triangles. It has 12 vertices and 30 edges.

icosidodecahedron One of the *Archimedean solids, with 12 pentagonal faces and 20 triangular faces. It can be formed by cutting off the corners of a *dodecahedron to obtain a polyhedron whose vertices lie at the midpoints of the edges of the original dodecahedron. It can also be formed by cutting off the corners of an *icosahedron to obtain a polyhedron whose vertices lie at the midpoints of the edges of the original icosahedron.

ideal A *subring *I* of a *ring *R* such that for every *a* in *R* and every *x* in *I* both *ax* and *xa* are in *I*. Where the multiplication is not commutative, it is possible that one of these conditions holds, but not the other. In these cases left ideal is used when *ax* always lies in *I*, and right ideal when *xa* always lies in *I*. If no qualification is made, the assumption is that the ideal is two-sided. The multiples of any integer form an ideal in the ring of integers.

ideal element An element added to a mathematical structure with the purpose of eliminating anomalies or exceptions. For example, $i = \sqrt{-1}$. If this element is defined as a number then the set of real numbers is extended to the set of all complex numbers, and all algebraic equations can be solved without extending the number system further.

ideal point An *ideal element, in particular a *point at infinity.

identically distributed A number of random variables are identically distributed if they have the same *distribution.

identification space = QUOTIENT SPACE.

identity An equation which states that two expressions are equal for all values of any variables that occur, such as $x^2 - y^2 = (x + y)(x - y)$ and $x(x - 1)(x - 2) = x^3 - 3x^2 + 2x$. Sometimes the symbol $\equiv$ is used instead of $=$ to indicate that a statement is an identity.

identity element *See* NEUTRAL ELEMENT.

identity function The function which maps every element to itself, i.e. $I(x) = x$ for every x.

identity mapping The identity mapping on a set S is the *mapping i_S: $S \rightarrow S$ defined by $i_S(s) = s$ for all s in S. Identity mappings have the property that if $f: S \rightarrow T$ is a mapping, then $f \circ i_S = f$ and $i_T \circ f = f$.

identity matrix The $n \times n$ identity matrix $\mathbf{I}$, or $\mathbf{I}_n$, is the $n \times n$ matrix

$$\begin{bmatrix} 1 & 0 & 0 & \ldots & 0 \\ 0 & 1 & 0 & \ldots & 0 \\ 0 & 0 & 1 & \ldots & 0 \\ \vdots & \vdots & \vdots & \ddots & \vdots \\ 0 & 0 & 0 & \ldots & 1 \end{bmatrix},$$

with each diagonal entry equal to 1 and all other entries equal to 0. It is the identity element for the set of $n \times n$ matrices with multiplication.

if and only if *See* CONDITION, NECESSARY AND SUFFICIENT.

iff Abbreviation for *if and only if.

i.i.d. = INDEPENDENT and IDENTICALLY DISTRIBUTED.

ill-conditioned A problem is said to be ill-conditioned if a small change in the input data (or coefficients in an equation) gives rise to large changes in the output data (or solutions to the equation). For example, if $f(x) = \dfrac{1}{1 - x^3}$, then $f(0.999) = 333.7, f(1.001) = -333.0$; or the solution of two almost parallel lines $y = x$; $y = (1 + \alpha)\, x + 10$: when $\alpha = 0.01$, the solution is $(-1000, -1000)$ and when $\alpha = 0.02$, the solution is $(-500, -500)$.

-illion Suffix denoting numbers of millions. Common usage now is that million, billion, trillion, quadrillion, quintillion etc. are successively multiplied by 1000. In UK billion was previously taken as a million million.

im Abbreviation and symbol for the imaginary part of a complex number.

image *See* FUNCTION and MAPPING.

imaginary axis The *y*-axis in the *complex plane. Its points represent *pure imaginary numbers.

imaginary number A *complex number where the real part is 0 so it is purely imaginary. For example, $3i$ where $i = \sqrt{-1}$.

imaginary part A *complex number z can be written $x + yi$, where x and y are real, and then y is the imaginary part. It is denoted by Im z or $\Im z$.

implication If p and q are statements, the statement 'p implies q' or 'if p then q' is an implication and is denoted by $p \Rightarrow q$. It is reckoned to be false only in the case when p is true and q is false. So its *truth table is as follows:

p	q	$p{\Rightarrow}q$
T	T	T
T	F	F
F	T	T
F	F	T

The statement that 'p implies q and q implies p' may be denoted by $p \Leftrightarrow q$, to be read as 'p if and only if q'.

implicit Where the independent variable is not the only term on one side of an equation. For example, $3x + 5y - 1 = 0$ is an implicit form while the equivalent equation $y = 0.6x + 0.2$ is explicit.

implicit differentiation When x and y satisfy a single equation, it may be possible to regard this as defining y as a function of x with a suitable domain, even though there is no explicit formula for y. In such a case, it may be possible to obtain the *derivative of y by a method, called implicit differentiation, that consists of differentiating the equation as it stands and making use of the *chain rule. For example, if $xy^2 + x^2 y^3 - 1 = 0$, then

$$x2y\frac{dy}{dx} + y^2 + x^2 3y^2 \frac{dy}{dx} + 2xy^3 = 0$$

and so, if $2xy + 3x^2 y^2 \neq 0$,

$$\frac{dy}{dx} = -\frac{y^2 + 2xy^3}{2xy + 3x^2y^2}.$$

imply To enable a conclusion to be validly inferred. For example, n is divisible by 2 implies that n is even.

improper fraction *See* FRACTION.

improper integrals There are two kinds of improper integral. The first kind is one in which the interval of integration is infinite as, for example, in

$$\int_a^\infty f(x)\,dx.$$

It is said that this integral exists, and that its value is l, if the value of the integral from a to X tends to a limit l as $X \to \infty$. For example,

$$\int_1^X \frac{1}{x^2}\,dx = 1 - \frac{1}{X}$$

and, as $X \to \infty$, the right-hand side tends to 1. So

$$\int_1^\infty \frac{1}{x^2}\,dx = 1.$$

A similar definition can be given for the improper integral from $-\infty$ to a. If both the integrals

$$\int_{-\infty}^a f(x)\,dx \quad \text{and} \quad \int_a^\infty f(x)\,dx$$

exist and have values l_1 and l_2, then it is said that the integral from $-\infty$ to ∞ exists and that its value is $l_1 + l_2$.

The second kind of improper integral is one in which the function becomes infinite at some point. Suppose first that the function becomes infinite at one of the limits of integration as, for example, in

$$\int_0^1 \frac{1}{\sqrt{x}}\,dx.$$

The function $1/\sqrt{x}$ is not bounded on the closed interval $[0, 1]$, so, in the normal way, this integral is not defined. However, the function is bounded on the interval $[\delta, 1]$, where $0 < \delta < 1$, and

$$\int_\delta^1 \frac{1}{\sqrt{x}}\,dx = 2 - 2\sqrt{\delta}.$$

As $\delta \to 0$, the right-hand side tends to the limit 2. So the integral above, from 0 to 1, is taken, by definition, to be equal to 2; in the same way, any such integral can be given a value equal to the appropriate limit, if it exists. A similar definition can be made for an integral in which the function becomes infinite at the upper limit.

Finally, an integral in which the function becomes infinite at a point between the limits is dealt with as follows. It is written as the sum of two integrals, where the function becomes infinite at the upper limit of the first and at the lower limit of the second. If both these integrals exist, the original integral is said to exist and, in this way, its value can be obtained.

impulse The impulse **J** associated with a force **F** acting during the time interval from $t = t_1$ to $t = t_2$ is the definite integral of the force, with respect to time, over the interval. Thus **J** is the vector quantity given by

$$\mathbf{J} = \int_{t_1}^{t_2} \mathbf{F}\,dt.$$

Suppose that **F** is the total force acting on a particle of mass m. From Newton's second law of motion in the form $\mathbf{F} = (d/dt)(m\mathbf{v})$, it follows that the impulse **J** is equal to the change in the *linear momentum of the particle.

If the force **F** is constant, then $\mathbf{J} = \mathbf{F}(t_2 - t_1)$. This can be stated as 'impulse = force × time'.

The concept of impulse is relevant in problems involving *collisions.

Impulse has the dimensions MLT^{-1}, and the SI unit of measurement is the newton second, abbreviated to 'N s', or the kilogram metre per second, abbreviated to 'kg m s^{-1}'.

incentre The centre of the *incircle of a triangle. It is the point at which the three internal bisectors of the angles of the triangle are concurrent.

incidence The partial overlap or intersection of two or more geometric figures. For example, a line which is a tangent to a curve is incident to the curve at at least one point.

incircle The circle that lies inside a triangle and touches the three sides. Its centre is the *incentre of the triangle.

inclination The acute angle between the horizontal and a given line (*see* DEPRESSION) measured positively in the direction of the upwards vertical. θ is the angle of inclination of the top of the cliff from the boat.

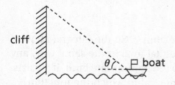

inclined plane A plane that is not horizontal. Its angle of inclination is the angle that a line of greatest slope makes with the horizontal.

include *See* SUBSET.

inclusion–exclusion principle The number of elements in the union of 3 sets is given by

$$n(A \cup B \cup C) = n(A) + n(B) + n(C) - n(A \cap B) - n(A \cap C) - n(B \cap C) + n(A \cap B \cap C),$$

where $n(X)$ is the *cardinality of the set X. This may be written

$$n(A \cup B \cup C) = \alpha_1 - \alpha_2 + \alpha_3,$$

where $\alpha_1 = n(A) + n(B) + n(C)$, $\alpha_2 = n(A \cap B) + n(A \cap C) + n(B \cap C)$ and $\alpha_3 = n(A \cap B \cap C)$. Now suppose that there are n sets instead of 3. Then the following, known as the inclusion–exclusion principle, holds:

$$n(A_1 \cup A_2 \cup \cdots \cup A_n) = \alpha_1 - \alpha_2 + \alpha_3 - \cdots + (-1)^{n-1}\alpha_n,$$

where α_i is the sum of the cardinalities of the intersections of the sets taken i at a time.

The principle is commonly used, when $A_1, A_2, \ldots, A_n$ are subsets of a universal set E, to find the number of elements not in any of the subsets. This number is equal to $n(E) - n(A_1 \cup A_2 \cup \cdots \cup A_n)$.

inclusive disjunction If p and q are statements, then the statement 'p or q', denoted by $p \vee q$, is the *disjunction of p and q. For example, if p is 'It is raining' and q is 'It is Monday' then $p \vee q$ is 'It is raining or it is Monday'. To be quite clear, notice that $p \vee q$ means 'p or q or both': the disjunction of p and q is true when *at least one of the statements p and q is true. The *truth table is therefore as follows:

p	q	$p \vee q$
T	T	T
T	F	T
F	T	T
F	F	F

This is sometimes explicitly known as 'inclusive disjunction' to distinguish it from exclusive disjunction which requires exactly one of p, q to be true.

See also EXCLUSIVE DISJUNCTION.

incompatible = INCONSISTENT.

inconsistent A set of equations is inconsistent if its solution set is empty.

increasing function A *real function f is increasing in or on an interval I if $f(x_1) \leq f(x_2)$ whenever x_1 and x_2 are in I with $x_1 < x_2$. Also, f is strictly increasing if $f(x_1) < f(x_2)$ whenever $x_1 < x_2$.

increasing sequence A sequence $a_1, a_2, a_3, \ldots$ is increasing if $a_i \leq a_{i+1}$ for all i, and strictly increasing if $a_i < a_{i+1}$ for all i.

increment The amount by which a certain quantity or variable is increased. An increment can normally be positive or negative, or zero, and is often required to be in some sense 'small'. If the variable is x, the increment may be denoted by Δx or δx.

indefinite integral *See* INTEGRAL.

independent In logic or mathematics, a set of statements, propositions or formulae where the truth of any of the statements or propositions or the value of formulae cannot be deduced from knowledge of the others.

independent equations A set of equations $e_1, e_2, \ldots, e_n$ for which $a_1 e_1 + a_2 e_2 + \cdots + a_n e_n = 0$ implies $a_1 = a_2 = \cdots = a_n = 0$. Equivalently, a set of equations which are not *linearly dependent.

independent events Two events A and B are independent if the occurrence of either does not affect the probability of the other occurring. Otherwise, A and B are dependent. For independent events, the probability that they both occur is given by the product law $\Pr(A \cap B) = \Pr(A)\Pr(B)$.

For example, when an unbiased coin is tossed twice, the probability of obtaining 'heads' on the first toss is $\frac{1}{2}$, and the probability of obtaining 'heads' on the second toss is $\frac{1}{2}$. These two events are independent. So the probability of obtaining 'heads' on both tosses is equal to $\frac{1}{2} \times \frac{1}{2} = \frac{1}{4}$.

independent random variables Two random variables X and Y are independent if the value of either does not affect the value of the other. If two discrete random variables X and Y are independent, then $\Pr(X=x, Y=y) = \Pr(X=x)\Pr(Y=y)$. If two continuous random variables X and Y, with *joint probability density function $f(x, y)$, are independent, then $f(x, y) = f_1(x) f_2(y)$, where f_1 and f_2 are the *marginal probability density functions.

independent variable (in regression) = EXPLANATORY VARIABLE.

indeterminate form Suppose that $f(x) \to 0$ and $g(x) \to 0$ as $x \to a$. Then the limit of the quotient $f(x)/g(x)$ as $x \to a$ is said to give an indeterminate form, sometimes denoted by $0/0$. It may be that the limit of $f(x)/g(x)$ can nevertheless be found by some method such as *l'Hôpital's rule.

Similarly, if $f(x) \to \infty$ and $g(x) \to \infty$ as $x \to a$, then the limit of $f(x)/g(x)$ gives an indeterminate form, denoted by ∞/∞. Also, if $f(x) \to 0$ and

$g(x) \to \infty$ as $x \to a$, then the limit of the product $f(x)g(x)$ gives an indeterminate form $0 \times \infty$.

index (indices) Suppose that a is a real number. When the product $a \times a \times a \times a \times a$ is written as a^5, the number 5 is called the index. When the index is a positive integer p, then a^p means $a \times a \times \cdots \times a$, where there are p occurrences of a. It can then be shown that

(i) $a^p \times a^q = a^{p+q}$

(ii) $a^p / a^q = a^{p-q}$ $(a \neq 0)$,

(iii) $(a^p)^q = a^{pq}$,

(iv) $(ab)^p = a^p b^p$,

where, in (ii), for the moment, it is required that $p > q$. The meaning of a^p can however be extended so that p is not restricted to being a positive integer. This is achieved by giving a meaning for a^0, for a^{-p} and for $a^{m/n}$, where m is an integer and n is a positive integer. To ensure that (i) to (iv) hold when p and q are any rational numbers, it is necessary to take the following as definitions:

(v) $a^0 = 1$,

(vi) $a^{-p} = 1/a^p$ $(a \neq 0)$,

(vii) $a^{m/n} = \sqrt[n]{a^m}$ (m an integer, n a positive integer).

Together, (i) to (vii) form the basic rules for indices.

The same notation is used in other contexts; for example, to define z^p, where z is a complex number, to define $\mathbf{A}^p$, where $\mathbf{A}$ is a square matrix, or to define g^p, where g is an element of a multiplicative group. In such cases, some of the above rules may hold and others may not.

index (in statistics) A figure used to show the variation in some quantity over a period of time, usually standardized relative to some base value. The index is given as a percentage with the base value equal to 100%. For example, the retail price index is used to measure changes in the cost of household items. Indices are often calculated by using a weighted mean of a number of constituent parts.

(⊕) SEE WEB LINKS
• Examples of calculations of index numbers.

index set A set S may consist of elements, each of which corresponds to an element of a set I. Then I is the index set, and S is said to be indexed by I. For example, the set S may consist of elements a_i, where $i \in I$. This is written $S = \{a_i | i \in I\}$ or $\{a_i, i \in I\}$.

indifference curve A contour line for a utility function. Since all points give the same value of the utility function, there is no reason to prefer one of the points on the indifference curve to any other.

indirect proof A theorem of the form $p \Rightarrow q$ can be proved by establishing instead its *contrapositive, by supposing $\neg\, q$ and showing that $\neg\, p$ follows. Such a method is called an indirect proof. Another example of an indirect proof is the method of *proof by contradiction.

indirect proportion = INVERSE PROPORTION. *See* PROPORTION.

indivisible A number or expression which is not able to be divided exactly. Since 1 divides every number exactly, it is usually taken to refer to division by any other whole number, and is equivalent to *prime in this case.

indivisible by A number or expression which is not able to be divided exactly by the stated number or expression. So 10 is indivisible by 4, though it is not indivisible since it has factors of 2 and 5.

induction *See* MATHEMATICAL INDUCTION.

inequality The following symbols have the meanings shown:

$\neq$ is not equal to,

$<$ is less than,

$\leq$ is less than or equal to,

$>$ is greater than,

$\geq$ is greater than or equal to.

An inequality is a statement of one of the forms: $a \neq b$, $a < b$, $a \leq b$, $a > b$ or $a \geq b$, where a and b are suitable quantities or expressions.

Given an inequality involving one real variable x, the problem may be to find the *solutions, that is, the values of x that satisfy the inequality, in an explicit form. Often, the set of solutions, known as the *solution set, may be given as an interval or as a union of intervals. For example, the solution set of the inequality

$$x - 2 > \frac{6}{x + 3}$$

is the set $(-4, -3) \cup (3, \infty)$.

Some authors use 'inequality' only for a statement, using one of the symbols above, that holds for all values of the variables that occur; for example, $x^2 + 2 > 2x$ is an inequality in this sense. Such authors use 'inequation' for a statement which holds only for some values of the variables involved. For them, an inequation has a solution set but an inequality does not; an inequality is comparable to an *identity.

inertia The *mass of a body is a measure of how difficult a force finds a body to move linearly, and the *moment of inertia is a measure of how difficult a *moment finds a body to rotate about an axis.

inertial frame of reference *See* FRAME OF REFERENCE.

inertial mass The inertial mass of a particle is the constant of proportionality that occurs in the statement of *Newton's second law of motion, linking the acceleration of the particle with the total force acting on the particle. Thus $ma = \mathbf{F}$, where m is the inertial mass of the particle, $\mathbf{a}$ is the acceleration and $\mathbf{F}$ is the total force acting on the particle. *See also* MASS.

inertia matrix *See* ANGULAR MOMENTUM.

inextensible string A string that has constant length, no matter how great the tension in the string.

inf Abbreviation for infimum.

infeasible A *constrained optimization problem for which the constraints cannot be satisfied simultaneously.

inference The process of coming to some conclusion about a *population based on a sample; or the conclusion about a population reached by such a process. The subject of statistical inference is concerned with the methods that can be used and the theory behind them.

infimum (infima) *See* BOUND.

infinite Not *finite. Having a size or absolute value greater than any natural number.

infinite product From an infinite sequence $a_1, a_2, a_3, \ldots$, an infinite product $a_1\, a_2\, a_3 \ldots$ can be formed and denoted by

$$\prod_{r=1}^{\infty} a_r.$$

Let P_n be the n-th *partial product, so that

$$P_n = \prod_{r=1}^{n} a_r.$$

If P_n tends to a limit P as $n \to \infty$, then P is the value of the infinite product. For example,

$$\prod_{r=2}^{\infty}\left(1-\frac{1}{r^2}\right)$$

has the value $\frac{1}{2}$, since it can be shown that $P_n = (n+1)/2n$ and $P_n \to \frac{1}{2}$.

infinite sequence *See* SEQUENCE.

infinite series *See* SERIES.

infinite set A set whose elements cannot be put into *one-to-one correspondence with a bounded sequence of the *natural numbers.

infinitesimal A variable whose limit is zero, usually involving increments of functions, and of particular use in defining differentiation. The ratio $\frac{\delta y}{\delta x}$ tends to the differential $\frac{dy}{dx}$ as $\delta x \to 0$.

infinity A value greater than any fixed bound, denoted by ∞.

inflection = INFLEXION.

inflexion *See* POINT OF INFLEXION.

information The abstraction of the content of a statement or data.

information theory The area of mathematics concerned with the transmission and processing of information, especially concerned with methods of coding, decoding, storage and retrieval and evaluating likelihoods of degree of accuracy in the process.

inhomogeneous Not *homogeneous.

initial conditions In applied mathematics, when differential equations with respect to time are to be solved, the known information about the system at a specified time, usually taken to be $t = 0$, may be called the initial conditions.

initialize Set parameters or variables at the start of an algorithm.

initial line The single axis used in *polar coordinates (from which the angle parameter is measured).

initial value The value of any physical quantity at a specified time reckoned as the starting-point, usually taken to be $t = 0$.

initial value problem Usually a *differential or *partial differential equation together with conditions specified at some initial time t_0.

injection = ONE-TO-ONE MAPPING.

injective mapping = ONE-TO-ONE MAPPING.

inner product A generalization of the *scalar product. Any product $\langle \mathbf{u}, \mathbf{v} \rangle$ of vectors which satisfies the following conditions. It must be distributive over addition, be reflexive, $\langle a\mathbf{u}, \mathbf{v} \rangle$ must equal $a\langle \mathbf{u}, \mathbf{v} \rangle$, and $\langle \mathbf{v}, \mathbf{v} \rangle = 0 \Rightarrow \mathbf{v} = \mathbf{0}$.

inscribe Construct a geometric figure inside another so they have points in common, but the inscribed figure does not have any part of it outside the other.

inscribed circle (of a triangle) = INCIRCLE.

insoluble Without a solution. So $x^2 + 1 = 0$ is insoluble within the set of real numbers, though it can be solved within the set of complex numbers.

insolvable = INSOLUBLE.

instance A particular case, often derived from a general expression by substitution of one or more parameter values.

instantaneous Occurring at a single point in time, with zero duration. For example, the velocity function of a body is the derivative of its position function with respect to time, defined by taking the limit as $\delta t \to 0$ of the average velocity $\frac{\delta v}{\delta t}$ over the interval δt.

int Symbol and abbreviation for *integral part or greatest integer function of a number.

integer One of the 'whole' numbers: $\ldots, -3, -2, -1, 0, 1, 2, 3, \ldots$. The set of all integers is often denoted by $\mathbf{Z}$. With the normal addition and multiplication, $\mathbf{Z}$ forms an *integral domain. *Kronecker said, 'God made the integers; everything else is the work of man.'

integer factorization (in cryptography) While the *unique factorization theorem states that any number can be written uniquely as a product of prime factors, there is no efficient factorization algorithm available for very large numbers. The difficulty of factorizing large composite integers, especially *semiprimes, lies at the heart of many modern secure systems of data transmission.

integer part For any real number x, there is a unique integer n such that $n \le x < n + 1$. This integer n is the integer part of x, and is denoted by $[x]$. For example, $\left[\frac{9}{4}\right] = 2$ and $[\pi] = 3$, but notice that $\left[-\frac{9}{4}\right] = -3$. In a computer language, the function **INT(X)** may convert the real number **X** into an integer by truncating. If so, **INT(9/4)** = 2 and **INT(PI)** = 3, but **INT(-9/4)** = -2. So **INT(X)** agrees with $[x]$ for $x \ge 0$ but not for $x < 0$.

The graph $y = [x]$ is shown here.

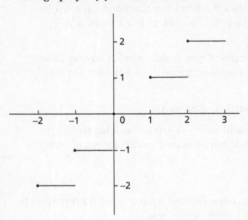

integer programming The restriction of *linear programming to cases when one or more variables can only take integer values, for example if a variable is a count.

integrable A function that has an *integral.

integral Let f be a function defined on the closed interval $[a, b]$. Take points $x_0, x_1, x_2, \ldots, x_n$ such that $a = x_0 < x_1 < x_2 < \ldots < x_{n-1} < x_n = b$, and in each subinterval $[x_i, x_{i+1}]$ take a point c_i. Form the sum

$$\sum_{i=0}^{n-1} f(c_i)(x_{i+1} - x_i);$$

that is, $f(c_0)(x_1 - x_0) + f(c_1)(x_2 - x_1) + \ldots + f(c_{n-1})(x_n - x_{n-1})$. Such a sum is called a Riemann sum for f over $[a, b]$. Geometrically, it gives the sum of the areas of n rectangles, and is an approximation to the area under the curve $y = f(x)$ between $x = a$ and $x = b$.

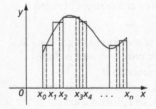

The (Riemann) integral of f over $[a, b]$ is defined to be the limit I (in a sense that needs more clarification than can be given here) of such a Riemann sum as n, the number of points, increases and the size of the subintervals gets smaller. The value of I is denoted by

$$\int_a^b f(x)\ dx, \quad \text{or} \quad \int_a^b f(t)\ dt,$$

where it is immaterial what letter, such as x or t, is used in the integral. The intention is that the value of the integral is equal to what is intuitively understood to be the area under the curve $y = f(x)$. Such a limit does not always exist, but it can be proved that it does if, for example, f is a *continuous function on $[a, b]$.

If f is continuous on $[a, b]$ and F is defined by

$$F(x) = \int_a^x f(t)dt,$$

then $F'(x) = f(x)$ for all x in $[a, b]$, so that F is an *antiderivative of f. Moreover, if an antiderivative ϕ of f is known, the integral

$$\int_a^b f(t)\ dt$$

can be easily evaluated: the *Fundamental Theorem of Calculus gives its value as $\phi(b) - \phi(a)$. Of the two integrals

$$\int_a^b f(x)\ dx \quad \text{and} \quad \int f(x)\ dx,$$

the first, with limits, is called a definite integral and the second, which denotes an antiderivative of f, is an indefinite integral.

integral calculus The subject that arose from the problem of trying to find the area of a region with a curved boundary. In general, this is calculated by a limiting process that yields gradually better approximations to the value. The *integral is defined by a limiting process based on an intuitive idea of the area under a curve; the fundamental discovery was the link that exists between this and the *differential calculus.

integral domain A commutative *ring R with identity, with the additional property that

    9. For all a and b in R, $ab = 0$ only if $a = 0$ or $b = 0$.

(The axiom numbering follows on from that used for ring.) Thus an integral domain is a commutative ring with identity with no *divisors of zero. The natural example is the set **Z** of all integers with the usual addition and multiplication. Any *field is an integral domain. Further examples of integral domains (these are not fields) are: the set $\mathbf{Z}[\sqrt{2}]$ of all real numbers of the form $a + b\sqrt{2}$, where a and b are integers, and the set $\mathbf{R}[x]$ of all polynomials in an indeterminate x, with real coefficients, each with the normal addition and multiplication.

integral equation An equation that involves *integration of functions.

integral function = ENTIRE FUNCTION.

integral part = INTEGER PART.

integral transform A relationship between two or more functions can be expressed in the form $f(x) = \int K(x, y) F(X) \, dy$ then $f(x)$ is the integral transform of $F(x)$, and $K(x, y)$ is the kernel of the transform. If $F(x)$ can be found from $f(x)$ then the transform can be inverted. Integral transformations are particularly useful where they result in simplifying equations into forms where methods of solution are much easier. For example, differential equations can be reduced to linear equations where the solution is straightforward, and when the inverse transform also exists the solution to the original problem can be obtained. *Fourier and *Laplace transformations are examples of integral transformations.

integrand The expression $f(x)$ in either of the integrals

$$\int_a^b f(x) \, dx, \qquad \int f(x) \, dx.$$

integrating factor *See* LINEAR FIRST-ORDER DIFFERENTIAL EQUATION.

integration The process of finding an *antiderivative of a given function f. 'Integrate f' means 'find an antiderivative of f'. Such an antiderivative may be called an indefinite integral of f and be denoted by

$$\int f(x) \, dx.$$

The term 'integration' is also used for any method of evaluating a definite integral. The definite integral

$$\int_a^b f(x) \, dx$$

can be evaluated if an antiderivative ϕ of f can be found, because then its value is $\phi(b) - \phi(a)$. (This is provided that a and b both belong to an interval in which f is continuous.) However, for many functions f, there is no antiderivative expressible in terms of elementary functions, and other methods for evaluating the definite integral have to be sought, one such being so-called *numerical integration.

What ways are there, then, of finding an antiderivative? If the given function can be recognized as the derivative of a familiar function, an antiderivative is immediately known. Some standard integrals are also given in the Table of integrals (*Appendix 7); more extensive tables of integrals are available. Certain techniques of integration may also be tried, among which are the following:

Change of variable

If it is possible to find a suitable function g such that the *integrand can be written as $f(g(x)) g'(x)$, it may be possible to find an indefinite integral using the change of variable $u = g(x)$; this is because

$$\int f(g(x))g'(x)\,dx = \int f(u)\,du,$$

a rule derived from the *chain rule for differentiation. For example, in the integral

$$\int 2x(x^2+1)^8\,dx,$$

let $u = g(x) = x^2 + 1$. Then $g'(x) = 2x$ (this can be written '$du = 2x\,dx$'), and, using the rule above with $f(u) = u^8$, the integral equals

$$\int (x^2+1)^8 2x\,dx = \int u^8\,du = \frac{1}{9}u^9 = \frac{1}{9}(x^2+1)^9.$$

Substitution

The rule above, derived from the chain rule for differentiation, can be written as

$$\int f(x)\,dx = \int f(g(u))g'(u)\,du.$$

It is used in this form to make the substitution $x = g(u)$. For example, in the integral

$$\int \frac{1}{(1+x^2)^{3/2}}\,dx,$$

let $x = g(u) = \tan u$. Then $g'(u) = \sec^2 u$ (this can be written '$dx = \sec^2 u\,du$'), and the integral becomes

$$\int \frac{\sec^2 u}{(1+\tan^2 u)^{3/2}}\,du = \int \frac{1}{\sec u}\,du = \int \cos u\,du = \sin u = \frac{x}{\sqrt{1+x^2}}.$$

Integration by parts The rule for integration by parts,

$$\int f(x)g'(x)dx = f(x)g(x) - \int g(x)f'(x)dx,$$

is derived from the rule for differentiating a product $f(x)g(x)$, and is useful when the integral on the right-hand side is easier to find than the integral on the left. For example, in the integral

$$\int x \cos x \, dx,$$

let $f(x) = x$ and $g'(x) = \cos x$. Then $g(x)$ can be taken as $\sin x$ and $f'(x) = 1$, so the method gives

$$\int x \cos x \, dx = x \sin x - \int \sin x \, 1 \, dx = x \sin x + \cos x.$$

See also REDUCTION FORMULA, PARTIAL FRACTIONS, and SEPARABLE FIRST-ORDER DIFFERENTIAL EQUATIONS.

integration by parts *See* INTEGRATION.

interacting variables Sometimes a second variable interacts with one of the variables in an experiment. If the experiment fails to take account of this, the reported results are unlikely to have any general applicability.

intercept *See* STRAIGHT LINE (in the plane).

interest Effectively the price attached to the use of money belonging to someone else. It is expressed as a rate, usually as a percentage per year, though often how the interest is calculated can make an important difference. For this reason, any advertisement involving interest has to show the *annualized percentage rate (APR) to allow straightforward comparisons. Car finance packages have traditionally been simple interest calculations where interest is paid on the whole principle for the whole term, so if £5000 is borrowed at 8% pa, and the finance is for 3 years, the buyer will pay $5000 \times 8/100 \times 3 = £1200$ in interest charges, and will have to pay $6200 \div 36 = £127.22$ per month for 3 years. The true rate of interest is much higher than 8% because towards the end of the 3 years most of the £5000 has already been repaid. An interest-bearing account will usually reinvest the interest (so the interest is added to the principal and subsequently interest will be paid on the larger amount), and this is known as *compound interest, though you can choose to have the interest paid into another account as a source of income.

interior (of a curve) *See* JORDAN CURVE THEOREM.

interior angle (of a polygon) The angle between two adjacent sides of a polygon that lies inside the polygon.

interior angle (with respect to a transversal of a pair of lines) *See* TRANSVERSAL.

interior of a set The set of all *interior points.

interior point A point in a *topological space contained within an open subset of the space. When the *real line is being considered, the real number x is an interior point of the set S of real numbers if there is an open interval $(x-\delta, x+\delta)$, where $\delta > 0$, included in S.

intermediate value theorem The following theorem stating an important property of *continuous functions:

Theorem

If the real function f is continuous on the closed interval $[a, b]$ and η is a real number between $f(a)$ and $f(b)$, then, for some c in (a, b), $f(c) = \eta$.

The theorem is useful for locating roots of equations. For example, suppose that $f(x) = x - \cos x$. Then f is continuous on $[0, 1]$, and $f(0) < 0$ and $f(1) > 0$, so it follows from the intermediate value theorem that the equation $f(x) = 0$ has a root in the interval $(0, 1)$.

intermediate vertex (in a network) A vertex between the *source and the *sink.

internal bisector Name sometimes given to the bisector of the interior angle of a triangle (or polygon).

internal division Let AB be a line segment; then the point E is the internal division of AB in the ratio $1: k$ if $\overrightarrow{AE} = k\,\overrightarrow{EB}$ where $\overrightarrow{AB}$ is the vector or directed line segment joining A and B. Alternatively E is the point $\dfrac{1}{1+k}$ of the way along from A to B.

internal force When a system of particles or a rigid body is being considered as a whole, an internal force is a force exerted by one particle of the system on another, or by one part of the rigid body on another. As far as the whole is concerned, the sum of the internal forces is zero. *Compare* EXTERNAL FORCE.

interpolation Suppose that the values $f(x_0), f(x_1), \ldots, f(x_n)$ of a certain function f are known for the particular values $x_0, x_1, \ldots, x_n$. A method of finding an approximation for $f(x)$, for a given value of x somewhere between these particular values, is called interpolation. If $x_0 < x < x_1$, the method known as linear interpolation gives

$$f(x) \approx f(x_0) + \frac{x - x_0}{x_1 - x_0}(f(x_1) - f(x_0)).$$

This is obtained by supposing, as an approximation, that between x_0 and x_1 the graph of the function is a straight line joining the points $(x_0, f(x_0))$ and $(x_1, f(x_1))$. More complicated methods of interpolation use the values of the function at more than two values.

interquartile range A measure of *dispersion equal to the difference between the first and third *quartiles in a set of numerical data.

intersect Two geometrical figures or curves are said to intersect if they have at least one point in common.

intersection The intersection of sets A and B (subsets of a *universal set) is the set consisting of all objects that belong to A and belong to B, and it is denoted by $A \cap B$ (read as 'A intersection B'). Thus the term 'intersection' is used for both the resulting set and the operation, a *binary operation on the set of all subsets of a universal set. The following properties hold:

    (i) For all A, $A \cap A = A$ and $A \cap \emptyset = \emptyset$.

    (ii) For all A and B, $A \cap B = B \cap A$; that is, $\cap$ is commutative.

    (iii) For all A, B and C, $(A \cap B) \cap C = A \cap (B \cap C)$; that is, $\cap$ is associative.

In view of (iii), the intersection $A_1 \cap A_2 \cap \ldots \cap A_n$ of more than two sets can be written without parentheses, and it may also be denoted by

$$\bigcap_{i=1}^{n} A_i.$$

For the intersection of two events, *see* EVENT.

interval A finite interval on the real line is a subset of **R** defined in terms of end-points a and b. Since each end-point may or may not belong to the subset, there are four types of finite interval:

    (i) the closed interval $\{x \mid x \in \mathbf{R} \text{ and } a \leq x \leq b\}$, denoted by $[a, b]$,

    (ii) the open interval $\{x \mid x \in \mathbf{R} \text{ and } a < x < b\}$, denoted by (a, b),

    (iii) the interval $\{x \mid x \in \mathbf{R} \text{ and } a \leq x < b\}$, denoted by $[a, b)$,

    (iv) the interval $\{x \mid x \in \mathbf{R} \text{ and } a < x \leq b\}$, denoted by $(a, b]$.

There are also four types of infinite interval:

(v) $\{x \mid x \in \mathbf{R}$ and $a \leq x\}$, denoted by $[a, \infty)$,
(vi) $\{x \mid x \in \mathbf{R}$ and $a < x\}$, denoted by (a, ∞),
(vii) $\{x \mid x \in \mathbf{R}$ and $x \leq a\}$, denoted by $(-\infty, a]$,
(viii) $\{x \mid x \in \mathbf{R}$ and $x < a\}$, denoted by $(-\infty, a)$.

Here ∞ (read as 'infinity') and $-\infty$ (read as 'minus infinity') are not, of course, real numbers, but the use of these symbols provides a convenient notation.

If I is any of the intervals (i) to (iv), the open interval determined by I is (a, b); if I is (v) or (vi), it is (a, ∞) and, if I is (vii) or (viii), it is $(-\infty, a)$.

interval estimate *See* ESTIMATE.

interval scale A scale of measurement where differences in value are meaningful, but not ratios of the values. Consequently the choice of zero on the scale is arbitrary. For example, temperature is an interval scale. 0 on the Fahrenheit scale does not coincide with 0 on the *Celsius or Centigrade scale, and in neither case is it meaningful to say that 10° is twice as hot as 5°.

intransitive relation A *binary relation which is not transitive for any three elements, i.e. $a \sim b$ and $b \sim c$ requires that a does not $\sim c$. Note that this is a much stronger condition than the relation being *non-transitive which simply requires that transitive does not hold for all combinations of three elements.

invalid Not valid. A conclusion which is not a necessary consequence of the conditions, so one *counterexample is sufficient to show a proposition is invalid.

invariable = CONSTANT.

invariant A property or quantity that is not changed by one or more specified operations or transformations. For example, a conic has equation

$$ax^2 + 2hxy + by^2 + 2gx + 2fy + c = 0$$

and, under a *rotation of axes, the quantity $h^2 - ab$ is an invariant. The distance between two points and the property of perpendicularity between two lines are invariants under *translations and rotations of the plane, but not under *dilatations.

An important distinction is between cases where a line is made up of invariant points, i.e. each point is mapped onto itself by the transformation, and an invariant line, where the image of any point on that line also lies on the line, but is not necessarily mapped onto itself. For example, in a reflection in the x-axis, any line perpendicular to the mirror line, i.e. in the form $x = \alpha$ is an invariant line, since (α, k) will be mapped onto $(\alpha, -k)$, but the x-axis itself is a line made up of invariant points since any point in the form $(\alpha, 0)$ will be mapped onto itself.

invariant subgroup = NORMAL SUBGROUP.

inverse correlation = NEGATIVE CORRELATION. *See* CORRELATION.

inverse element Suppose that, for the *binary operation ∘ on the set S, there is a *neutral element e. An element a' is an inverse (or inverse element) of the element a if $a \circ a' = a' \circ a = e$. If the operation is called multiplication, the neutral element is normally called the identity element and may be denoted by 1. Then the inverse a' may be called a multiplicative inverse of a and be denoted by a^{-1}, so that $aa^{-1} = a^{-1}a = 1$ (or e). If the operation is addition, the neutral element is denoted by 0, and the inverse a' may be called an additive inverse (or a negative) of a and be denoted by $-a$, so that $a + (-a) = (-a) + a = 0$. *See also* GROUP.

inverse function For a *real function f, its inverse function f^{-1} is to be a function such that if $y = f(x)$, then $x = f^{-1}(y)$. The conditions under which such a function exists need careful consideration. Suppose that f has domain S and range T, so that $f: S \rightarrow T$ is onto. The inverse function f^{-1}, with domain T and range S, can be defined (provided that f is a *one-to-one mapping), as follows. For y in T, $f^{-1}(y)$ is the unique element x of S such that $f(x) = y$.

If the domain S is an interval I and f is strictly increasing on I or strictly decreasing on I, then f is certainly one-to-one. When f is *differentiable, a sufficient condition can be given in terms of the sign of $f'(x)$. Thus, if f is continuous on an interval I and differentiable in the open interval (a, b) determined by I (*see* INTERVAL), and $f'(x) > 0$ in (a, b) (or $f'(x) < 0$ in (a, b)), then, for f (with domain I) an inverse function exists.

When an inverse function is required for a given function f, it may be necessary to restrict the domain and obtain instead the inverse function of this *restriction of f. For example, suppose that $f: \mathbf{R} \rightarrow \mathbf{R}$ is defined by $f(x) = x^2 - 4x + 5$. This function is not one-to-one. However, since $f'(x) = 2x - 4$, it can be seen that $f'(x) > 0$ for $x > 2$. Use f now to denote the function defined by $f(x) = x^2 - 4x + 5$ with domain $[2, \infty)$. The range is $[1, \infty)$ and the function $f[2, \infty) \rightarrow [1, \infty)$ has an inverse function f^{-1}: $[1, \infty) \rightarrow [2, \infty)$. A formula for f^{-1} can be found by setting $y = x^2 - 4x + 5$ and, remembering that $x \in [2, \infty)$, obtaining $x = 2 + \sqrt{y - 1}$. So, with a change of notation, $f^{-1}(x) = 2 + \sqrt{x - 1}$ for $x \geq 1$.

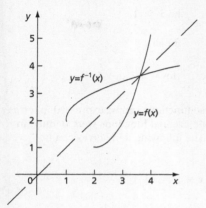

When the inverse function exists, the graphs $y = f(x)$ and $y = f^{-1}(x)$ are reflections of each other in the line $y = x$. The derivative of the inverse function can be found as follows. Suppose that f is differentiable and has inverse denoted now by g, so that if $y = f(x)$ then $x = g(y)$. Then, if $f'(x) \neq 0$, g is differentiable at y and

$$g'(y) = \frac{1}{f'(x)} = \frac{1}{f'(g(y))}.$$

When $f'(x)$ is denoted by dy/dx, then $g'(y)$ may be denoted by dx/dy and the preceding result says that, if $dy/dx \neq 0$, then

$$\frac{dx}{dy} = \frac{1}{dy/dx}.$$

This can be used safely only if it is known that the function in question has an inverse. For examples, *see* INVERSE HYPERBOLIC FUNCTION and INVERSE TRIGONOMETRIC FUNCTION.

inverse hyperbolic function Each of the *hyperbolic functions sinh and tanh is strictly increasing throughout the whole of its domain **R**, so in each case an inverse function exists. In the case of cosh, the function has to be restricted to a suitable domain (*see* INVERSE FUNCTION), taken to be $[0, \infty)$. The domain of the inverse function is, in each case, the range of the original function (after the restriction of the domain, in the case of cosh). The inverse functions obtained are: $\cosh^{-1} : [1, \infty) \to [0, \infty)$; $\sinh^{-1} : \mathbf{R} \to \mathbf{R}$; $\tanh^{-1} : (-1, 1) \to \mathbf{R}$. These functions are given by the formulae:

$$\cosh^{-1}x = \ln\left(x + \sqrt{x^2 - 1}\right), \quad \text{for } x \geq 1,$$

$$\sinh^{-1}x = \ln\left(x + \sqrt{x^2 + 1}\right), \quad \text{for all } x,$$

$$\tanh^{-1}x = \ln\sqrt{\frac{1 + x}{1 - x}}, \quad \text{for } -1 < x < 1.$$

It is not so surprising that the inverse functions can be expressed in terms of the logarithmic function, since the original functions were defined in terms of the exponential function. The following derivatives can be obtained:

$$\frac{d}{dx}(\cosh^{-1}x) = \frac{1}{\sqrt{x^2 - 1}} \quad (x \neq 1),$$

$$\frac{d}{dx}(\sinh^{-1}x) = \frac{1}{\sqrt{x^2 + 1}}, \quad \frac{d}{dx}(\tanh^{-1}x) = \frac{1}{1 - x^2}.$$

inversely proportional *See* PROPORTION.

inverse mapping Let $f: S \rightarrow T$ be a bijection, that is, a mapping that is both a *one-to-one mapping and an *onto mapping. Then a mapping, denoted by f^{-1}, from T to S may be defined as follows: for t in T, $f^{-1}(t)$ is the unique element s of S such that $f(s) = t$. The mapping $f^{-1}: T \rightarrow S$, which is also a bijection, is the inverse mapping of f. It has the property that $f \circ f^{-1} = i_T$ and $f^{-1} \circ f = i_S$, where i_S and i_T are the identity mappings on S and T, and $\circ$ denotes composition.

inverse matrix An *inverse of a square matrix $\mathbf{A}$ is a matrix $\mathbf{X}$ such that $\mathbf{AX} = \mathbf{I}$ and $\mathbf{XA} = \mathbf{I}$. (A matrix that is not square cannot have an inverse.) A square matrix $\mathbf{A}$ may or may not have an inverse, but if it has then that inverse is unique and $\mathbf{A}$ is said to be invertible. A matrix is invertible if and only if it is *non-singular. Consequently, the term 'non-singular' is sometimes used for 'invertible'.

When $\det \mathbf{A} \neq 0$, the matrix $(1/\det \mathbf{A})$ adj $\mathbf{A}$ is the inverse of $\mathbf{A}$, where adj $\mathbf{A}$ is the *adjoint of $\mathbf{A}$. For example, the 2×2 matrix $\mathbf{A}$ below is invertible if $ad - bc \neq 0$, and its inverse $\mathbf{A}^{-1}$ is as shown:

$$\mathbf{A} = \begin{bmatrix} a & b \\ c & d \end{bmatrix}, \quad \mathbf{A}^{-1} = \frac{1}{ad - bc}\begin{bmatrix} d & -b \\ -c & a \end{bmatrix}.$$

inverse of a complex number If z is a non-zero complex number and $z = x + yi$, the (*multiplicative) inverse of z, denoted by z^{-1} or $1/z$, is

$$\frac{x}{x^2 + y^2} - \frac{y}{x^2 + y^2}i.$$

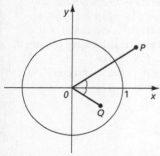

When z is written in *polar form, so that $z = re^{i\theta} = r(\cos\theta + i\sin\theta)$, where $r \neq 0$, the inverse of z is $(1/r)e^{-i\theta} = (1/r)(\cos\theta - i\sin\theta)$. If z is represented by P in the *complex plane, then z^{-1} is represented by Q, where $\angle xOQ = -\angle xOP$ and $|OP| \cdot |OQ| = 1$.

inverse square law of gravitation The following law that describes the force of attraction that exists between two particles, assumed to be isolated from all other bodies. Consider two particles P and S, with gravitational masses m and M, respectively. Let $\mathbf{r}$ be the position vector of P relative to S (that is, the vector represented by $\overrightarrow{SP}$) and let $r = |\mathbf{r}|$. Then the force $\mathbf{F}$ experienced by P is given by

$$\mathbf{F} = -\frac{GMm}{r^3}\mathbf{r},$$

where G is the *gravitational constant. Since $\mathbf{r}/r$ is a unit vector, the magnitude of the gravitational force experienced by P is inversely proportional to the square of the distance between S and P. The force experienced by S is equal and opposite to that experienced by P.

inverse trigonometric function The inverse sine function $\sin^{-1}$ is, to put it briefly, the inverse function of sine, so that $y = \sin^{-1}x$ if $x = \sin y$. Thus $\sin^{-1}\frac{1}{2} = \pi/6$ because $\sin(\pi/6) = \frac{1}{2}$. However, $\sin(5\pi/6) = \frac{1}{2}$ also, so it might be thought that $\sin^{-1}\frac{1}{2} = \pi/6$ or $5\pi/6$. It is necessary to avoid such ambiguity so it is normally agreed that the value to be taken is the one lying in the interval $[-\pi/2, \pi/2]$. Similarly, $y = \tan^{-1}x$ if $x = \tan y$, and the value y is taken to lie in the interval $(-\pi/2, \pi/2)$. Also, $y = \cos^{-1}x$ if $x = \cos y$ and the value y is taken to lie in the interval $[0, \pi]$.

A more advanced approach provides more explanation. The inverse function of a trigonometric function exists only if the original function is restricted to a suitable domain. This can be an interval I in which the function is strictly increasing or strictly decreasing (*see* INVERSE FUNCTION). So, to obtain an inverse function for $\sin x$, the function is restricted to a domain consisting of the interval $[-\pi/2, \pi/2]$; $\tan x$ is restricted to the interval $(-\pi/2, \pi/2)$; and $\cos x$ is restricted to $[0, \pi]$. The domain of the inverse function is, in each case, the range of the restricted function. Hence the following inverse functions are obtained: $\sin^{-1}: [-1, 1] \to [-\pi/2, \pi/2]$; $\tan^{-1}: \mathbf{R} \to (-\pi/2, \pi/2)$; $\cos^{-1}[-1, 1] \to [0, \pi]$. The notation arcsin, arctan and arccos, for $\sin^{-1}$, $\tan^{-1}$ and $\cos^{-1}$, is also used. The following derivatives can be obtained:

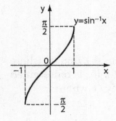 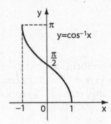

$$\frac{d}{dx}(\sin^{-1}x) = \frac{1}{\sqrt{1 - x^2}} \quad (x \neq \pm 1),$$

$$\frac{d}{dx}(\cos^{-1}x) = -\frac{1}{\sqrt{1 - x^2}} \quad (x \neq \pm 1), \quad \frac{d}{dx}(\tan^{-1}x) = \frac{1}{1 + x^2}.$$

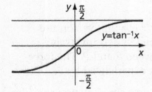

invertible matrix *See* INVERSE MATRIX.

involution A *function or *transformation which is its own inverse, that is to say applying it twice returns you to where you started. The *identity will also be an involution trivially, but more interesting ones are any reflection, rotation by 180°, multiplication by -1, taking *reciprocals, *complex conjugates, and *complementary sets.

irrational number A real number that is not *rational. A famous proof, sometimes attributed to Pythagoras, shows that $\sqrt{2}$ is irrational; the method can also be used to show that numbers such as $\sqrt{3}$ and $\sqrt{7}$ are also irrational. It follows that numbers like $1 + \sqrt{2}$ and $1/(1 + \sqrt{2})$ are irrational. The proof that e is irrational is reasonably easy, and in 1761 Lambert showed that π is irrational.

irreducible Unable to be factorized. So prime numbers are irreducible numbers and an expression such as $x^2 + 1$ is irreducible when considering only the real number system, though it can be factorized to $(x + i)(x - i)$ in the complex number system, where $i = \sqrt{-1}$.

isoclines *See* TANGENT FIELDS.

isogon A polygon with all angles equal, so that all regular polygons are isogons but not vice versa. For example, a rectangle must have four right angles but does not have to be the regular quadrilateral (a square).

A rectangle is an isogon, but not a regular polygon.

isolate To identify an interval within which only one root of an equation lies.

isolated point A *singular point which does not lie on the main curve of an equation but whose coordinates do satisfy the equation. For example, $y^2 = x^4 - x^2 = x^2(x^2 - 1)$ is not defined for $-1 < x < 0$, $0 < x < 1$, and the graph of the function looks like

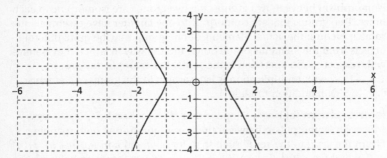

but the function is defined for $x = 0$, and the point $(0, 0)$ satisfies the equation.

isolated point (of a set) A point P is an isolated point of a set X if there exists a *neighbourhood of P which does not contain another element in X which is distinct from the point P.

isolated singularity *See* SINGULAR POINT.

isometric graph paper Has three axes, equally spaced relative to one another, to enable 3-dimensional figures to be represented on a plane. The effect of the equal spacing is to produce a grid made up of equilateral triangles.

isometry If P and Q are points in the plane, $|PQ|$ denotes the distance between P and Q. An isometry is a *transformation of the plane that preserves the distance between points: it is a transformation with the property that, if P and Q are mapped to P' and Q', then $|P'Q'| = |PQ|$. Examples of isometries are *translations, *rotations and *reflections. It can be shown that all the isometries of the plane can be obtained from translations, rotations and reflections, by composition. Two figures are *congruent if there is an isometry that maps one onto the other.

isomorphic *See* ISOMORPHISM.

isomorphic graphs Two or more graphs which have the same structure of edges joining the same vertices. Depending on where the vertices are positioned these graphs may superficially look quite different.

isomorphism (of groups) Let $\langle G, \circ \rangle$ and $\langle G', * \rangle$ be *groups, so that $\circ$ is the operation on G and $*$ is the operation on G'. An isomorphism between $\langle G, \circ \rangle$ and $\langle G', * \rangle$ is a *one-to-one onto mapping f from the set G to the set G' such that, for all a and b in G, $f(a \circ b) = f(a) * f(b)$. This means that, if f maps a to a' and b to b', then f maps $a \circ b$ to $a' * b'$. If there is an isomorphism between two groups, the two groups are isomorphic to each other. Two groups that are isomorphic to one another have essentially the same structure; the actual elements of one group may be quite different objects from the elements of the other, but the way in which they behave with respect to the operation is the same. For example, the group $1, i, -1, -i$ with multiplication is isomorphic to the group of elements $0, 1, 2, 3$ with addition modulo 4.

isomorphism (of rings) Suppose that $\langle R, +, \times \rangle$ and $\langle R', \oplus, \otimes \rangle$ are *rings. An isomorphism between them is a *one-to-one onto mapping f from the set R to the set R' such that, for all a and b in R, $f(a + b) = f(a) \oplus f(b)$ and $f(a \times b) = f(a) \otimes f(b)$. If there is an isomorphism between two rings, the rings are ISOMORPHIC to each other and, as with isomorphic groups, the two have essentially the same structure.

isoperimetric Two or more geometric figures with the same perimeter.

isoperimetric inequality If p is the perimeter of a closed curve in a plane and the area enclosed by the curve is A, then $p^2 \leq 4\pi A$, and equality is only achieved if the curve is a circle. Turning this condition round the other way, it says that for a fixed length of curve p the greatest area which can be enclosed is when the curve is a circle. The result can be generalized to surfaces in 3-dimensional space where the sphere is the most efficient shape at enclosing volume for a fixed surface area, and to higher dimensions.

isoperimetric quotient (IQ) number of a closed curve $IQ = \dfrac{4\pi A}{P^2}$ where $A =$ area enclosed by the curve, with perimeter P. The circle maximizes the area which can be enclosed by isoperimetric curves, and the IQ gives the proportion of this maximum which is actually enclosed by the curve. IQ is always ≤ 1, and equality only occurs for the circle.

isosceles trapezium A *trapezium in which the two non-parallel sides have the same length.

isosceles triangle A triangle in which two sides have the same length. Sometimes the third side (which may be called the *base) is required to be not of the same length. The two angles opposite the sides of equal length (which may be called the base angles) are equal.

iterated series A double or multiple series such as $\sum\limits_{i=1}^{\infty} \sum\limits_{j=1}^{\infty} a_{ij}$.

iteration A method uses iteration if it yields successive approximations to a required value by repetition of a certain procedure. Examples are *fixed-point iteration and *Newton's method for finding a root of an equation $f(x) = 0$.

I(x) The *identity function $I(x) = x$ for all x.

j In the notation for *complex numbers, some authors, especially engineers, use j instead of i.

***j* (*j*)** A unit vector, usually in the direction of the y-axis or a vertical vector in the plane of a projectile's motion.

J Symbol for joule.

Jacobi, Carl Gustav Jacob (1804–51) German mathematician responsible for notable developments in the theory of elliptic functions, a class of functions defined by, as it were, inverting certain integrals. Applying them to number theory, he was able to prove *Fermat's conjecture that every integer is the sum of four perfect squares. He also published contributions on *determinants and *mechanics.

Jacobian For n functions in n variables, the Jacobian is the *determinant of the *Jacobian matrix.

Jacobian matrix For m functions in n variables, $f_i (x_1, x_2, \ldots, x_n)$ $i = 1, 2, \ldots, m$, the Jacobian matrix is the m-by-n matrix whose element in the i-th row and j-th column is the *partial derivative $\frac{\partial f_i}{\partial x_j}$.

Jacobi's iterative method A technique for solving a set of n linear equations in n unknowns. If the system is summarized by $\mathbf{Ax} = \mathbf{b}$, then taking initial values as $x_i^{(1)} = \frac{b_i}{a_{ii}}$, it uses the iterative relation

$x_i^{(k)} = \frac{b_i - \sum_{j \neq i} a_{ij} x_j^{(k-1)}}{a_{ii}}$, so it computes a complete set of new values together, unlike the *Gauss–Seidel method.

Jeffreys, Sir Harold (1891–1989) British mathematician and astronomer who made important contributions in probability theory and calculus as well as our understanding of the solar system.

joint cumulative distribution function For two *random variables X and Y, the joint cumulative distribution function is the function F

defined by $F(x, y) = \Pr(X \leq x, Y \leq y)$. The term applies also to the generalization of this to more than two random variables. For two variables, it may be called the *bivariate and, for more than two, the *multivariate cumulative distribution function.

joint distribution The joint distribution of two or more *random variables is concerned with the way in which the probability of their taking certain values, or values within certain intervals, varies. It may be given by the *joint cumulative distribution function. More commonly, the joint distribution of some number of discrete random variables is given by their *joint probability mass function, and that of some number of continuous random variables by their *joint probability density function. For two variables it may be called the *bivariate and, for more than two, the *multivariate distribution.

joint probability density function For two continuous *random variables X and Y, the joint probability density function is the function f such that

$$\Pr(a \leq X \leq b, c \leq Y \leq d) = \int_a^b \int_c^d f(x,y) \; dy \; dx.$$

The term applies also to the generalization of this to more than two random variables. For two variables, it may be called the *bivariate and, for more than two, the *multivariate probability density function.

joint probability mass function For two discrete *random variables X and Y, the joint probability mass function is the function p such that $p(x_i, y_j) = \Pr(X = x_i, Y = y_j)$, for all i and j. The term applies also to the generalization of this to more than two random variables. For two variables, it may be called the *bivariate and, for more than two, the *multivariate probability mass function.

Jordan, (Marie-Ennemond-) Camille (1838–1922) French mathematician whose treatise on groups of permutations and the theory of equations drew attention to the work of Galois. In a later work on analysis, there appears his formulation of what is now called the *Jordan Curve Theorem.

Jordan curve A simple closed curve. Thus a Jordan curve is a continuous plane curve that has no ends (or, in other words, that begins and ends at the same point) and does not intersect itself.

Jordan Curve Theorem An important theorem which says that a *Jordan curve divides the plane into two regions, the interior and the exterior of the curve. The result is intuitively obvious but the proof is not elementary.

Josephus problem The story is told by the Jewish historian Josephus of a band of 41 rebels who, preferring suicide to capture, decided to stand in a circle and kill every third remaining person until no one was left. Josephus, one of the number, quickly calculated where he and his friend should stand so that they would be the last two remaining and so avoid being a part of the suicide pact. In its most general form, the Josephus problem is to find the position of the one who survives when there are n people in a circle and every m-th remaining person is eliminated.

joule The SI unit of *work, abbreviated to 'J'. One joule is the work done when the point of application of a force of one *newton moves a distance of one metre in the direction of the force.

Julia set The boundary of the set of points z_0 in the complex plane for which the application of the function $f(z) = z^2 + c$ repeatedly to the point z_0 produces a bounded sequence. The term may be used similarly for other functions as well. Julia sets are almost always *fractals, and vary enormously in structure for different values of c.

(🌐) SEE WEB LINKS

• A Java applet which allows visualizations of Mandelbrot and Julia sets.

jump The absolute value of the difference between the *left-hand and *right-hand limits of a given function, i.e. $|(f(x_+) - f(x_-))|$.

jump discontinuity A point at which a function $f(x)$ has both *left-hand and *right-hand limits but the limits are not equal. For example, $f(x) = -1$ if $x < 0$ and $f(x) = 1$ if $x > 0$, f has a jump discontinuity at $x = 0$.

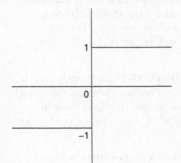

k Abbreviation for *kilo—used in symbols in the metric system for thousands of a unit, for example km.

_k_ (_k_) A unit vector, usually in the direction of the z-axis or a vertical vector in the motion described in three dimensions.

kelvin Symbol K. The temperature scale, and the unit of measurement of temperature, which takes 273.15° C as the freezing point of water, and 373.15° C as the boiling point of water. Then 0°K represents absolute zero, the lowest possible temperature, at which all thermodynamic motion has ceased.

Kelvin, Lord (1824–1907) The British mathematician, physicist and engineer, William Thomson, who, mostly in the earlier part of his career, made contributions to mathematics in the theories of electricity and magnetism and hydrodynamics.

Kendall's rank correlation coefficient *See* RANK CORRELATION.

Kepler, Johannes (1571–1630) German astronomer and mathematician, best known for his three laws of planetary motion, including the discovery that the planets travel in elliptical *orbits round the Sun. He proposed the first two laws in 1609 after observations had convinced him that the orbit of Mars was an ellipse, and the third followed ten years later. He also developed methods used by *Archimedes to determine the volumes of many solids of revolution by regarding them as composed of infinitely many infinitesimal parts.

Kepler's laws of planetary motion The following three laws about the motion of the planets round the Sun:

   (i) Each planet travels in an elliptical *orbit with the Sun at one focus.

  (ii) The line joining a planet and the Sun sweeps out equal areas in equal times.

 (iii) The squares of the times taken by the planets to complete an orbit are proportional to the cubes of the lengths of the major axes of their orbits.

These laws were formulated by Kepler, based on observations. Subsequently, *Newton formulated his *inverse square law of gravitation and his second law of motion, from which Kepler's laws can be deduced.

(⊕) SEE WEB LINKS

• Animation which illustrates all three laws.

kg Abbreviation and symbol for *kilogram.

Khwārizmī, Muhammad ibn Mūsā al- (about AD 800) Mathematician and astronomer of Baghdad, the author of two important Arabic works on arithmetic and algebra. The word 'algorithm' is derived from the occurrence of his name in the title of the work that includes a description of the Hindu–Arabic number system. The word 'algebra' comes from the title *Al-jabr wa'l muqābalah* of his other work, which contains, among other things, methods of solving quadratic equations by a method akin to *completing the square.

kilo- Prefix used with *SI units to denote multiplication by 10^3.

kilogram In *SI units, the base unit used for measuring *mass, abbreviated to 'kg'. Up until 20 May 2019, it was defined as the mass of a certain platinum–iridium cylinder kept at Sèvres near Paris; it is now defined in terms of the *Planck constant. One kilogram is approximately the mass of 1000 cm^3 of water.

kilowatt The power of 1000 *watts.

kinematics The study of the motion of bodies without reference to the forces causing the motion. It is concerned with such matters as the positions, the velocities and the accelerations of the bodies.

kinetic energy The energy of a body attributable to its motion. The kinetic energy may be denoted by E_k or T. For a particle of mass m with speed v, $E_k = \frac{1}{2}mv^2$. For a rigid body rotating about a fixed axis with angular speed ω, and with moment of inertia I about the fixed axis, $E_k = \frac{1}{2}I\omega^2$.

Kinetic energy has the dimensions $ML^2 T^{-2}$, and the SI unit of measurement is the *joule.

Kingman, Sir John Frank Charles (1939–) British mathematician who made important contributions to the theory and applications of probability and stochastic analysis, including operational research and population genetics, queuing theory, measure theory and sub-additive ergodic theory. Appointed as the Director of the Isaac Newton Institute in Cambridge in 2001, and as the first chairman of the Statistics Commission in 2000.

kite A quadrilateral that has two adjacent sides of equal length and the other two sides of equal length. If the kite *ABCD* has *AB* = *AD* and *CB* = *CD*, the diagonals *AC* and *BD* are perpendicular, and *AC* bisects *BD*.

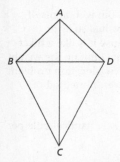

Klein, (Christian) Felix (1849–1925) German mathematician, who in his influential *Erlanger Programm* launched a unification of geometry, in which different geometries are classified by means of group theory. He is responsible for applying the terms 'elliptic' and 'hyperbolic' to describe *non-Euclidean geometries and, in topology, he is remembered in the name of the *Klein bottle.

Klein bottle The *Möbius band has just one surface and has one edge. A sphere, on the other hand, has no edges, but it has two surfaces, the outside and the inside. An example with no edges and only one surface is the Klein bottle.

Consider, in the plane, the square of all points with Cartesian coordinates (x, y) such that $-1 \leq x \leq 1$ and $-1 \leq y \leq 1$. The operation of 'identifying' the point $(x, 1)$ with the point $(x, -1)$, for all x, can be thought of as forming a cylinder in three dimensions, with two surfaces, the outside and the inside, and two edges. The operation of 'identifying' the point $(1, y)$ with $(-1, -y)$ is like forming a cylinder but including a twist, so this can be thought of as resulting in a Möbius band. The Klein bottle is constructed by doing these two operations simultaneously. This is, of course, not practically possible with a square sheet of flexible material in three dimensions, but the result is nevertheless valid mathematically and it has no edges and just one surface.

SEE WEB LINKS
• A fuller description and images of Klein bottles.

Klein four-group There are essentially two *groups with four elements. In other words, any group with four elements is *isomorphic to one of these two. One is the *cyclic group with four elements. The other is the

Klein four-group. It is *abelian, and may be written as a multiplicative group with elements e, a, b and c, where e is the identity element, $a^2 = b^2 = c^2 = e$, $ab = c$, $bc = a$ and $ca = b$.

knapsack problem The problem of packing one bin with items of different sizes, where the items have an associated value, so as to maximize the total value. The name originates because it is like a hiker packing a knapsack (an old name for rucksack) with a limit on the weight they can carry, but it acts as a model for any resource allocation problem with constraints—such as finances, time or manpower. A method of solution is the *branch and bound method.

knot A navigational measurement of speed which is 1 *nautical mile per hour.

knot (in a curve) A closed curve which does not intersect with itself, formed by looping and interweaving the curve with itself and joining the two ends together, so that it cannot be untangled to form a simple loop. A knot is topologically equivalent to a circle.

knot equivalence Two mathematical *knots are equivalent if one can be transformed into the other by a continuous deformation of Euclidean space. These deformations correspond to ways of manipulating a knotted string that do not involve cutting the string or passing the string through itself. Knots can be described in different ways, so one fundamental task in knot theory is to determine when two descriptions represent the same knot.

knot polynomial A polynomial that is associated with a *knot, with the property that if two knots are *equivalent then their polynomials are equal. This means that if the associated polynomials are different, the knots cannot be equivalent.

knot theory The study of *knots and their properties under continuous deformations in geometry and topology. It was first formally considered by *Gauss and one of his students, Listing. It has important applications in the understanding of DNA, where the normal behaviour is in the form of a double helix, but DNA often becomes knotted, which can interfere with its functions. There are other applications in chemistry, computer science, medical research, and statistical mechanics. Knot theory is continuing to develop as mathematicians look to generalize the notions by considering objects other than circles as the starting point, and knots in spaces other than 3-dimensional Euclidean space.

Koch curve Take an equilateral triangle as shown in the first diagram. Replace the middle third of each side by two sides of an equilateral triangle pointing outwards. This forms a six-pointed star, as shown in the second diagram. Repeat this construction to obtain the figure shown in the third diagram. The Koch curve, named after the Swedish mathematician Helge von Koch (1870–1924), is the curve obtained when this process is continued indefinitely. The interior of the curve has finite area, but the curve has infinite length. *See also* FRACTALS.

(⊕) SEE WEB LINKS
• Animation of a Koch curve.

Kolmogorov, Andrei Nikolaevich (1903–87) Russian mathematician who worked in a number of areas of mathematics, including probability. In this, his important contribution was to give the subject a rigorous foundation by using the language and notation of set theory.

Kolmogorov–Smirnov test A *non-parametric test for testing the null hypothesis that a given sample has been selected from a population with a specified *cumulative distribution function F. Let $x_1, x_2, \ldots, x_n$ be the sample values in ascending order, and let

$$z = \max_i \left| \frac{i}{n} - F(x_i) \right|.$$

If the null hypothesis is true, then z should be less than a value that can be obtained, for different significance levels, from tables.

Kolmogorov space A *topological space X in which every pair of distinct points are *topologically distinguishable. This is equivalent to saying that for any pair of distinct points there is an *open set which contains one of the points but not the other.

Königsberg *See* BRIDGES OF KÖNIGSBERG.

Kronecker, Leopold (1823–91) German mathematician responsible for considerable contributions to number theory and other fields. But he is more often remembered as the first to cast doubts on non-constructive existence proofs, over which he disputed with *Weierstrass and others. *See* INTEGER.

Kronecker delta The two variable function δ_{ij} that takes the value 1 when $i = j$ and the value 0 otherwise. If the elements of a square matrix are defined by the delta function, the matrix produced will be the *identity matrix.

Kronecker's Lemma If $\sum\limits_{n=1}^{\infty} \dfrac{a_n}{n}$ converges then $\dfrac{1}{N} \sum\limits_{n=1}^{N} a_n \to 0$ as $N \to \infty$.

Kruskal's algorithm (to solve the minimum connector problem) This method is very effective for a small number of vertices for which a full graph is available. Since n vertices will be joined by any *spanning tree which has $n-1$ edges, this method is a simple step-by-step procedure which leads to the one with the smallest total distance. Essentially you start with the shortest edge (and any time there is a tie for the shortest edge, choose either one at random). At each stage add in another edge which connects a new point to the connected tree being built up, without completing a cycle and adding the least distance to the total connected distance, i.e. the shortest edge which connects a new vertex without creating a cycle. Once all vertices have been connected, i.e. once $n-1$ edges have been taken, the minimum connected path will have been found.

Kruskal–Wallis test A *non-parametric test for comparisons between more than two medians, using the relative rankings within the different samples.

k-simplex In geometry, the k-simplex generalizes the notions of the triangle and tetrahedron from two and three dimensions to k-dimensional space. The k-simplex has $k + 1$ vertices and is defined as the smallest *convex set containing the given vertices.

Kuratowski closure axioms *Topology is intimately involved with the notion of limit points and how they are separated from one another. The Kuratowski closure axioms are a set of axioms that allow a topology to be defined on a set.

If A^c denotes the closure of A and $\emptyset$ denotes the empty set, the statements of the axioms and their mathematical expressions are:

(i) Every set is contained in its closure: $A \subseteq A^c$.
(ii) The closure of the closure of a set is equal to the closure of the set: $(A^c)^c = A^c$.
(iii) The closure of the union of two sets is the union of their closures: $A^c \cup B^c = (A \cup B)^c$.
(iv) The closure of the empty set is empty: $\emptyset^c = \emptyset$.

Kuratowski's Theorem A graph is *planar if and only if it does not contain the *complete graph K_5, the *bipartite graph $K_{3,3}$ or a *subdivision of K_5 or $K_{3,3}$ as a subgraph.

kurtosis Let $m_2 = E((X-\mu)^2)$ and $m_4 = E((X-\mu)^4)$, where $\mu = E(X)$ be the second and fourth moments about the mean. Then $\dfrac{m_4}{(m_2)^2}$ is the kurtosis of the distribution. For a normal distribution this has the value 3, which can be taken as a standard for comparison and hence a value of 3 is termed mesokurtic—from 'meso' meaning 'middle'. For values less than 3 the distribution is termed platykurtic and for values greater than 3 it is leptokurtic.

kW Abbreviation and symbol for *kilowatt.

L The Roman *numeral for 50.

labelling algorithm The algorithm for identifying the maximum flow possible across a network. It is a step-by-step procedure whereby any initial flow can be increased until the maximum is found. The max-flow/min-cut rule can tell you in advance what the maximum flow is, but the algorithm will find it for you anyway, as well as determining the detailed flows.

You can start with any flow, including the zero flow, where the flow on every edge is zero. Each edge should be labelled with both the current flow and the excess capacity along that edge. Choose a path from source to sink, and take the smallest of the flows along any of the edges on that path. On each edge, transfer that amount of flow from the current flow to the excess capacity. Now choose another path and repeat the process, and continue until no path can be found that does not have at least one edge with zero flow. This is often easy to see, as any cut with zero flow guarantees that no such path exists.

The excess capacities on each edge at this stage represent the maximum flow. Note that the maximum flow can often be achieved in a number of different ways, and the choice of a different order of paths from source to sink will often lead to a different network solution, but with the same maximum flow.

For example,

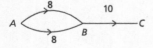

In this very simple network the maximum flow occurs when a total of 10 units flow along the two edges between A and B. However, any x satisfying $2 \leq x \leq 8$ and $10-x$ along the two edges will be a maximum flow.

Example of labelling algorithm.

For the network shown here, an initial zero flow is shown in bold, so the excess capacity currently on each edge is the capacity of the edge.

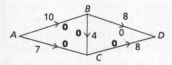

Starting with the path *ABD* the smallest flow on the route is 8, so transfer 8 to the excess capacity on both *AB* and *BD*.

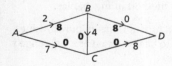

Now choose path *ACD* which has smallest flow 7, giving.

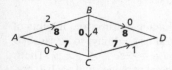

The only route left is *ABCD* with smallest flow 1, giving

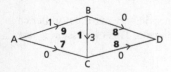

Since all routes into the sink at D have 0 excess capacity, it is obvious immediately that no further increases would be available and the maximum flow is 16, and the bold figures show a way of achieving this maximum flow. If *ABCD* had been chosen as the second path, before *ACD*, then 10 would have gone along *AB*, 2 along *BC* and 6 along *AC* instead of this solution.

lag The time difference between the pairs of observations used in calculating the *autocorrelation of a sequence. For a monthly time series the important lags are likely to be 1 (the most recent performance), 3, 6 and 12, reflecting the performance at corresponding points in the business cycle.

Lagrange, Joseph-Louis (1736–1813) Arguably the greatest, alongside *Euler, of 18th-century mathematicians. Although he was born in Turin and spent the early part of his life there, he eventually settled in Paris and is normally deemed to be French. Much of his important work was done in Berlin, where he was Euler's successor at the Academy. His work, in common with that of most important mathematicians of the time, covers the whole range of mathematics. He is known for results in number theory and algebra but is probably best remembered as a leading figure in the development of theoretical mechanics. His work *Mécanique analytique*, published in 1788, is a comprehensive account of the subject. In particular, he was mainly responsible for the methods of the *calculus of variations and the consequent Lagrangian method in mechanics.

Lagrange interpolation formula If a function has n known values $(x_1, y_1), (x_2, y_2), \ldots, (x_n, y_n)$ and the value is to be estimated at x, Lagrange's interpolation formula gives the estimate as

$$f(x) = \frac{y_1(x - x_2)\cdots(x - x_n)}{(x_1 - x_2)(x_1 - x_3)\cdots(x_1 - x_n)} + \cdots + \frac{y_n(x - x_1)\cdots(x - x_{n-1})}{(x_n - x_1)(x_n - x_2)\cdots(x_n - x_{n-1})}$$

Lagrange multiplier A method of evaluating maxima and minima of a function f where one or more exact constraints $g_i = 0$ have to be satisfied. A new function is constructed as $L = f + \lambda_1 g_1 + \lambda_2 g_2 + \cdots + \lambda_n g_n$. Then the derivatives of L with respect to the original variables and each λ_i are taken, and the *stationary points found by solving the set of simultaneous equations obtained by setting each derivative to zero.

Lagrange's Theorem (sum of four squares) Every natural number can be written as the sum of four squares of integers. This expression is not necessarily unique. For example,

$$1 = 1^2 + 0^2 + 0^2 + 0^2,$$
$$10 = 3^2 + 1^2 + 0^2 + 0^2 \text{ or } 2^2 + 2^2 + 1^2 + 1^2.$$

Lambert, Johann Heinrich (1728–77) Swiss–German mathematician, scientist, philosopher and writer on many subjects, who in 1761 proved that π is an *irrational number. He introduced the standard notation for *hyperbolic functions and appreciated the difficulty of proving the *parallel postulate.

lamina An object considered as having a 2-dimensional shape and density but having no thickness. It is used in a *mathematical model to represent an object such as a thin plate or sheet in the real world.

Lami's Theorem The following theorem in mechanics, named after Bernard Lami, or Lamy (1640–1715):

Theorem

Suppose that three forces act on a particle and that the resultant of the three forces is zero. Then the magnitude of each force is proportional to the sine of the angle between the other two forces.

In the given situation, the three forces are coplanar and the corresponding *polygon of forces is a triangle. The lengths of the sides of this triangle are proportional to the magnitudes of the three forces and the result of the theorem is obtained by an application of the *sine rule.

Langlands, Robert (1936–) A Canadian mathematician who has made significant contributions in the areas of number theory and representation theory. He was awarded the Abel Prize 2018 "for his visionary program connecting representation theory to number theory".

Laplace, Pierre-Simon, Marquis de (1749–1827) French mathematician, best known for his work on planetary motion, enshrined in his five-volume *Mécanique céleste*, and for his fundamental contributions to the theory of probability. It was Laplace who extended *Newton's gravitational theory to the study of the whole solar system. He developed the strongly deterministic view that, once the starting conditions of a closed dynamical system such as the universe are known, its future development is then totally determined. Napoleon asked him where God fitted into all this. 'I have no need of that hypothesis,' replied Laplace. He briefly held the post of Minister of the Interior under Napoleon, but seems to have got on just as well with the restored monarchy.

Laplace's equation The partial differential equation
$$\nabla^2 V = \frac{\partial^2 V}{\partial x^2} + \frac{\partial^2 V}{\partial y^2} + \frac{\partial^2 V}{\partial z^2} = 0 \text{ where } \nabla \text{ is the *differential operator del.}$$
This equation is important in potential theory.

Laplace transform An *integral transform of a function into another function of a different variable, the Laplace transform of $f(x)$ is obtained by multiplying by e^{-xy} and integrating over the range from 0 to $+\infty$ to give a function of y, i.e. $g(y) = \int_0^\infty f(x)e^{-xy}dx$. Such transforms are useful in solving differential equations.

latency Essentially the time delay observed in a data transmission system, which may mean that even when a high bandwidth is being used, transmission is slow. When a web page is being loaded, the user may see a noticeable delay as the request message is sent to the computer hosting the page (identified by the page *URL), especially if a number of requests are being received by the host server simultaneously. This is particularly noticeable at times when tickets for major sporting or entertainment events are released online at a specified time and the servers struggle to cope with the surge in demand.

latent root = CHARACTERISTIC ROOT or eigenvalue.

latent vector = CHARACTERISTIC VECTOR or eigenvector.

lateral face (surface) Any side of a geometric solid which is not the base.

Latin square A square array of symbols arranged in rows and columns in such a way that each symbol occurs exactly once in each row and once in each column. Two examples are given in the figure. Latin squares are used in the design of experiments involving, for example, trials of different kinds of seed or fertilizer.

A	B	C	D		A	B	C	D
B	A	D	C		D	C	B	A
C	D	A	B		C	D	A	B
D	C	B	A		B	A	D	C

latitude Suppose that the *meridian through a point P on the Earth's surface meets the equator at P'. (The meridian is taken here as a semicircle.) Let O be the centre of the Earth. The latitude of P is then the angle $P'OP$, measured in degrees north or south. The latitude of any point may be up to 90° north or up to 90° south. Latitude and *longitude uniquely fix the position of a point on the Earth's surface.

latus rectum The chord through the focus of a *parabola and perpendicular to the axis. For the parabola $y^2 = 4ax$, the latus rectum has length $4a$.

Laurent expansion For a function which is analytic within the annulus centred at c_0 in the complex plane, defined by $r_1 \leq |z - c_0| \leq r_2$ the Laurent expansion of $f(x)$ is $\sum_{n=-\infty}^{\infty} a_n(z - c_0)^n$ where $a_n = \dfrac{1}{2\pi i} \int f(z)(z - c_0)^{n+1} dz$ with the integral taken over the values of z for which $|z - c_0|$ is constant.

law A general theorem or principle. For example, *Newton's laws of motion. Laws are listed alphabetically by the rest of the phrase in this dictionary.

law of averages False conception that an event becomes more likely if it has been under-represented so far in a sequence of observations. For example, if a fair coin has been tossed six times, and has come up with five heads and only one tail, the 'law of averages' suggests that a tail is more likely on the next toss.

laws of large numbers (in statistics) The generic name given to a group of theorems relating to the behaviour of $\frac{S_n}{n} = \frac{1}{n}(X_1 + X_2 \ldots + X_n)$ for large n, where $\{X_i\}$ are independent random variables, usually identically distributed. For example, $\lim_{n \to \infty} \Pr\left\{\frac{S_n}{n} \to \mu\right\} = 1$ says that for a sequence of independent, identically distributed random variables, the sequence of sample means of the first n random variables will converge to the population mean μ with probability 1.

l.c.m. = LEAST COMMON MULTIPLE.

leading coefficient The coefficient of the term of highest degree in a polynomial so if $f(x) = 7x^2 - 3x + 2$ the leading coefficient of $f(x)$ is 7.

leading diagonal = MAIN DIAGONAL.

least Less than (or sometimes 'not more than') every other member of a set.

least common denominator = LEAST COMMON MULTIPLE of the *denominators. Used in adding numerical or algebraic fractions efficiently. So $\frac{5}{12} + \frac{3}{8} = \frac{10}{24} + \frac{9}{24} = \frac{19}{24}$ is simpler than using 96 as the common denominator for the equivalent fractions.

least common multiple For two non-zero integers a and b, an integer that is a multiple of both is a *common multiple. Of all the positive common multiples, the least is the least common multiple (l.c.m.), denoted by $[a, b]$. The l.c.m. of a and b has the property of dividing any other common multiple of a and b. If the prime decompositions of a and b are known, the l.c.m. is easily obtained: for example, if $a = 168 = 2^3 \times 3 \times 7$ and $b = 180 = 2^2 \times 3^2 \times 5$, then the l.c.m. is $2^3 \times 3^2 \times 5 \times 7 = 2520$. For positive integers a and b, the l.c.m. is equal to $ab/(a, b)$, where (a, b) is the *greatest common divisor (g.c.d.).

   Similarly, any finite set of non-zero integers, $a_1, a_2, \ldots, a_n$ has an l.c.m. denoted by $[a_1, a_2, \ldots, a_n]$. However, when $n > 2$ it is not true, in general, that $[a_1, a_2, \ldots, a_n]$ is equal to $a_1 a_2 \ldots a_n/(a_1, a_2, \ldots, a_n)$, where $(a_1, a_2, \ldots, a_n)$ is the g.c.d.

least squares The method of least squares is used to estimate *parameters in statistical models such as those that occur in *regression. Estimates for the parameters are obtained by minimizing the sum of the squares of the differences between the observed values and the predicted values under the model. For example, suppose that, in a case of linear regression, where $E(Y) = \alpha + \beta X$, there are n paired observations (x_1, y_1),

$(x_2, y_2), \ldots, (x_n, y_n)$. Then the method of least squares gives a and b as *estimators for α and β, where a and b are chosen so as to minimize $\sum (y_i - a - b x_i)^2$.

least squares theorem = GAUSS–MARKOV THEOREM.

least upper bound = SUPREMUM.

least value Let f be a *real function and D a subset of its domain. If there is a point c in D such that $f(c) \leq f(x)$ for all x in D, then $f(c)$ is the least value of f in D. There may be no such point: consider, for example, either the function f defined by $f(x) = -1/x$, or the function f defined by $f(x) = x$, with the open interval $(0, 1)$ as D; or the function f defined by $f(x) = [x] - x$, with the closed interval $[0, 1]$ as D. If the least value does exist, it may be attained at more than one point of D.

That a *continuous function on a closed interval has a least value is ensured by the non-elementary theorem that such a function 'attains its bounds'. An important theorem states that a function, continuous on $[a, b]$ and *differentiable in (a, b), attains its least value either at a *local minimum (which is a *stationary point) or at an end-point of the interval.

leaving variable See SIMPLEX METHOD.

Lebesgue, Henri-Léon (1875–1941) French mathematician who revolutionized the theory of integration. Building on the work of earlier French mathematicians on the notion of measure, he developed the Lebesgue integral, one of the outstanding concepts in modern analysis. It generalizes the Riemann *integral, which can deal only with continuous functions and other functions not too unlike them. His two major texts on the subject were published in the first few years of this century.

left and **right derivative** For the *real function f, if

$$\lim_{h \to 0-} \frac{f(a+h) - f(a)}{h}$$

exists, this limit is the left derivative of f at a. Similarly, if

$$\lim_{h \to 0+} \frac{f(a+h) - f(a)}{h}$$

exists, this limit is the right derivative of f at a. (See LIMIT FROM THE LEFT AND RIGHT.) The *derivative $f'(a)$ exists if and only if the left derivative and the right derivative of f at a exist and are equal. An example where the left and right derivatives both exist but are not equal is provided by the function f, where $f(x) = |x|$ for all x. At 0, the left derivative equals -1 and the right derivative equals $+1$.

left-handed system See RIGHT-HANDED SYSTEM.

left-hand limit The real number line is usually represented with values increasing from left to right, so the left-hand limit is the limit of the function from below, i.e. where $x \to a$ but takes only values less than a.

Legendre, Adrien-Marie (1752–1833) French mathematician who, with *Lagrange and *Laplace, formed a trio associated with the period of the French Revolution. He was well known in the 19th century for his highly successful textbook on the geometry of *Euclid. But his real work was concerned with calculus. He was responsible for the classification of elliptic integrals into their standard forms. The so-called *Legendre polynomials, solutions of a certain differential equation, are among the most important of the special functions. In an entirely different area, he along with *Euler conjectured and partially proved an important result in number theory known as the law of quadratic reciprocity.

Legendre polynomials The set of solutions $\{P_n(x)\}$ to *Legendre's differential equation are generated by the expansion of $\dfrac{1}{\sqrt{(1-2xt+t^2)}}$ as the coefficients of t^n.

Alternatively they can be found by RODRIGUES'S FORMULA

$$P_n(x) = \frac{1}{2^n n!} \frac{d^n}{dx^n} (x^2-1)^n.$$

Legendre's differential equation The differential equation $(1-x^2)$ $y'' - 2xy' + n(n-1)\,y = 0$ where n is a *natural number. The solutions are the *Legendre polynomials.

Leibniz, Gottfried Wilhelm (1646–1716) German mathematician, philosopher, scientist and writer on a wide range of subjects, who was, with *Newton, the founder of the calculus. Newton's discovery of differential calculus was perhaps ten years earlier than Leibniz's, but Leibniz was the first to publish his account, written independently of Newton, in 1684. Soon after, he published an exposition of integral calculus that included the *Fundamental Theorem of Calculus. He also wrote on other branches of mathematics, making significant contributions to the development of symbolic logic, a lead which was not followed up until the end of the 19th century.

Leibniz's Theorem The following result for the n-th derivative of a product:

Theorem

If $h(x) = f(x)g(x)$ for all x, the n-th derivative of h is given by

$$h^{(n)}(x) = \sum_{r=0}^{n} \binom{n}{r} f^{(r)}(x) g^{(n-r)}(x),$$

where the coefficients $\binom{n}{r}$ are *binomial coefficients.

For example, to find $h^{(8)}(x)$, when $h(x) = x^2 \sin x$, let $f(x) = x^2$ and $g(x) = \sin x$. Then $f'(x) = 2x$ and $f''(x) = 2$; and $g^{(8)}(x) = \sin x$, $g^{(7)}(x) = -\cos x$ and $g^{(6)}(x) = -\sin x$. So

$$h^{(8)}(x) = x^2 \sin x + \binom{8}{1} 2x(-\cos x) + \binom{8}{2} 2(-\sin x)$$
$$= x^2 \sin x - 16x \cos x - 56 \sin x.$$

lemma A mathematical theorem proved primarily for use in the proof of a subsequent theorem.

length (of a binary code) *See* BINARY CODE.

length (of a binary word) *See* BINARY WORD.

length (of a line segment) The length $|AB|$ of the *line segment AB, and the length $|\overrightarrow{AB}|$ of the *directed line segment, is equal to the distance between A and B. It is equal to zero when A and B coincide, and is otherwise always positive.

length (of a sequence) *See* SEQUENCE.

length (of a series) *See* SERIES.

length (of a vector) *See* VECTOR.

length of an arc The length of the straight line segment obtained by straightening the arc without stretching it. Except where the curve forms part of a known shape, the arc length has to be calculated by integration, and the form depends on the coordinates used:

in Cartesian coordinates: $\displaystyle\int_a^b \sqrt{1 + \left(\frac{dy}{dx}\right)^2}\, dx$

in polar coordinates: $\displaystyle\int_u^v \sqrt{1 + r^2 \left(\frac{d\theta}{dr}\right)^2}\, dr$ for the length between $r = u$ and $r = v$ or if the integration is carried out for a range of angles, it is in the

form $\displaystyle\int_\alpha^\beta \sqrt{r^2 + \left(\frac{dr}{d\theta}\right)^2}\, d\theta$

in parameterized form $(x(t), y(t))$ it will be $\displaystyle\int_{t_1}^{t_2} \sqrt{\left(\frac{dx}{dt}\right)^2 + \left(\frac{dy}{dt}\right)^2}\, dt$.

length of the major and minor axis *See* ELLIPSE.

Leonardo da Vinci (1452–1519) Italian polymath best known for his magnificent paintings such as the *Mona Lisa* and the *Last Supper*, Leonardo was also prolific as an architect, inventor, scientist and mathematician where he was particularly interested in geometry, mechanics, aerodynamics and hydrodynamics.

(⊕) SEE WEB LINKS
• An online exhibition about da Vinci.

Leonardo of Pisa *See* FIBONACCI.

leptokurtic *See* KURTOSIS.

lever A rigid bar or rod free to rotate about a point or axis, called the fulcrum, normally used to transmit a force at one point to a force at another. In diagrams the fulcrum is usually denoted by a small

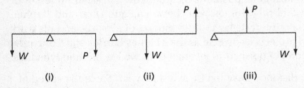

(i)   (ii)   (iii)

triangle. In the three examples shown in the figure, the lever is being used as a machine in which the effort is the force **P** and the load is **W**. In each case, the mechanical advantage can be found by using the *principle of moments. Everyday examples corresponding to the three kinds of lever shown are (i) an oar as used in rowing, (ii) a stationary wheelbarrow and (iii) the jib of a certain kind of crane.

l'Hôpital, l'Hôpital's rule *See under* H.

liar paradox The paradox that arises from considering the statement 'This statement is false.'

Lie, (Marius) Sophus (1842–99) Norwegian mathematician responsible for advances in differential equations and differential geometry, and primarily remembered for a three-volume treatise on transformation groups. Like Klein, he brought group theory to bear on geometry.

life tables Tables that show life expectancies from current age for different groups in the population. While these tables are based on historical observations, they are adjusted according to expert opinion to take account of new conditions such as improving health care, increasing incidence of cancer, etc. They form part of the basis on which actuaries carry out their analysis for insurance companies when setting premiums and annuity rates.

(⊕) SEE WEB LINKS
• Gives access to the latest life tables.

lift *See* AERODYNAMIC DRAG.

light In mechanics, an object such as a string or a rod is said to be light if its weight may be considered negligible compared with that of other objects involved.

light framework A set of light rods that make a rigid framework. As the masses of the rods are negligible the forces acting in the rods can be determined by considering the forces acting at the joints.

Lighthill, Sir Michael James FRS (1924–98) British mathematician who held various academic positions including the Lucasian Professor of Mathematics at Cambridge for ten years between Paul *Dirac and Stephen *Hawking. His main interests lay in fluid motion of all sorts, and his work was important in projects as diverse as the Thames Barrier and Concorde, and in the reduction of noise from jet engines as well as biofluiddynamics.

light year The distance travelled by light in 1 year. Since the speed of light is 299 792 km per sec, and there are $365 \times 24 \times 60 \times 60 = 31\ 536\ 000$ seconds in one year, 1 light year is 9.46×10^{12} km. Because of the vast distances in space, the light year is the basic unit of distance in astronomy.

likelihood function For a sample from a population with an unknown *parameter, the likelihood function is the probability that the sample could have occurred at random; it is a function of the unknown parameter. The method of maximum likelihood selects the value of the parameter that maximizes the likelihood function. For algebraic convenience, it is common to work with the logarithm of the likelihood function. The concept applies equally well when there are two or more unknown parameters.

likelihood ratio test A sophisticated hypothesis test which compares the likelihood of the observed value happening for different possible values of the parameter under investigation. In its simplest form, the comparison is between two possible values, but more generally the alternatives are ranges of the parameter value and the comparison is for maximum likelihoods in the interval.

like terms Terms in an algebraic expression which are the same in relation to all variables and their powers, and which are separated by addition or subtraction signs. Like terms can then be combined into a single term. So in $3x^2y + 6xy - 5xy^2 - 2xy$ there are two like terms, the terms involving xy, but the other two terms are not 'like terms'. This expression could be simplified to $3x^2y + 4xy - 5xy^2$ but no further.

lim inf Abbreviation for *limit inferior.

limit (of $f(x)$) The limit, if it exists, of $f(x)$ as x tends to a is a number l with the property that, as x gets closer and closer to a, $f(x)$ gets closer and closer to l. This is written

$$\lim_{x \to a} f(x) = l.$$

Such an understanding may be adequate at an elementary level. It is important to realize that this limit may not equal $f(a)$; indeed, $f(a)$ may not necessarily be defined.

At a more advanced level, the definition just given needs to be made more precise: a formal statement says that $f(x)$ can be made as close to l as we please by restricting x to a sufficiently small neighbourhood of a (but excluding a itself). Thus, it is said that $f(x)$ tends to l as x tends to a, written $f(x) \to l$, as $x \to a$, if, given any positive number ε (however small), there is a positive number δ (which depends upon ε) such that, for all x, except possibly a itself, lying between $a - \delta$ and $a + \delta$, $f(x)$ lies between $l - \varepsilon$ and $l + \varepsilon$.

Notice that a itself may not be in the domain of f; but there must be a neighbourhood of a, the whole of which (apart possibly from a itself) is in the domain of f. For example, let f be the function defined by

$$f(x) = \frac{\sin x}{x} \quad (x \ne 0).$$

Then 0 is not in the domain of f, but it can be shown that

$$\lim_{x \to 0} \frac{\sin x}{x} = 1.$$

In the above, l is a real number, but it is also possible that $f(x) \to \pm \infty$, where appropriate formal definitions can be given: it is said that $f(x) \to +\infty$ as $x \to a$ if, given any positive number K (however large), there is a positive number δ (which depends upon K) such that, for all x, except possibly a itself, lying between $a - \delta$ and $a + \delta$ $f(x)$ is greater than K. It is clear, for example, that $1/x^2 \to +\infty$ as $x \to 0$. There is a similar definition for $f(x) \to -\infty$ as $x \to a$.

Sometimes the behaviour of $f(x)$ as x gets indefinitely large is of interest and importance: it is said that $f(x) \to l$ as $x \to \infty$ if, given any positive number ε (however small), there is a number X (which depends upon ε)

such that, for all $x > X$, $f(x)$ lies between $l - \varepsilon$ and $l + \varepsilon$. This means that $y = l$ is a horizontal *asymptote for the graph $y = f(x)$. Similar definitions can be given for $f(x) \to l$ as $x \to -\infty$, and for $f(x) \to \pm \infty$ as $x \to \infty$ and $f(x) \to \pm \infty$ as $x \to -\infty$.

limit (of a sequence) The limit, if it exists, of an infinite sequence a_1, a_2, a_3, ... is a number l with the property that a_n gets closer and closer to l as n gets indefinitely large. Such an understanding may be adequate at an elementary level.

At a more advanced level, the statement just given needs to be made more precise: it is said that the sequence a_1, a_2, a_3, ... has the limit l if, given any positive number ε (however small), there is a number N (which depends upon ε) such that, for all $n > N$, a_n lies between $l - \varepsilon$ and $l + \varepsilon$. This can be written $a_n \to l$.

The sequence $0, \frac{1}{2}, \frac{3}{4}, \frac{7}{8}, \frac{15}{16}, \ldots$, for example, has the limit 1. To take another example, the sequence $-1, \frac{1}{2}, -\frac{1}{3}, \frac{1}{4}, -\frac{1}{5}, \ldots$ has the limit 0; since this is the sequence whose n-th term is $(-1)^n/n$, this fact can be stated as $(-1)^n/n \to 0$.

There are, of course, sequences that do not have a limit. These can be classified into different kinds.

(i) A formal definition says that $a_n \to \infty$ if, given any positive number K (however large), there is an integer N (which depends upon K) such that, for all $n > N$, a_n is greater than K. For example, $a_n \to \infty$ for the sequence 1, 4, 9, 16, ..., in which $a_n = n^2$.

(ii) There is a similar definition for $a_n \to -\infty$, and an example is the sequence $-4, -5, -6, \ldots$, in which $a_n = -n - 3$.

(iii) If a sequence does not have a limit but is *bounded, such as the sequence $-\frac{1}{2}, \frac{2}{3}, -\frac{3}{4}, \frac{4}{5}, \ldots$, in which $a_n = (-1)^n \, n/(n+1)$, it may be said to oscillate finitely.

(iv) If a sequence is not bounded, but it is not the case that $a_n \to \infty$ or $a_n \to -\infty$ (the sequence 1, 2, 1, 4, 1, 8, 1, ... is an example), the sequence may be said to oscillate infinitely.

limit from the left and **right** The statement that $f(x)$ tends to l as x tends to a from the left can be written: $f(x) \to l$ as $x \to a-$. Another way of writing this is

$$\lim_{x \to a-} f(x) = l.$$

The formal definition says that this is so if, given $\varepsilon > 0$, there is a number $\delta > 0$ such that, for all x strictly between $a - \delta$ and a, $f(x)$ lies between $l - \varepsilon$ and $l + \varepsilon$. In place of $x \to a-$, some authors use $x \nearrow a$. In the same way, the statement that $f(x)$ tends to l as x tends to a from the right can be written: $f(x) \to l$ as $x \to a+$. Another way of writing this is

$$\lim_{x \to a+} f(x) = l.$$

The formal definition says that this is so if, given $\varepsilon > 0$, there is a number $\delta > 0$ such that, for all x strictly between a and $a + \delta$, $f(x)$ lies between $l - \varepsilon$ and $l + \varepsilon$. In place of $x \to a+$, some authors use $x \searrow a$. For example, if $f(x) = x - [x]$, then

$$\lim_{x \to 1-} f(x) = 1, \quad \lim_{x \to 1+} f(x) = 0.$$

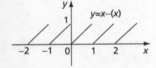

limit inferior = LOWER BOUND. *See* BOUND.

limit of integration In the definite integral

$$\int_a^b f(x)\, dx,$$

the lower limit (of integration) is a and the upper limit (of integration) is b.

limit point *See* ACCUMULATION POINT.

limit superior = UPPER BOUND. *See* BOUND.

lim sup Abbreviation for limit superior.

Lindemann, (Carl Louis) Ferdinand von (1852–1939) German mathematician, professor at Königsberg and then at Munich, who proved in 1882 that π is a *transcendental number.

line Defined in *Euclid as a length without breadth. It is usually taken to refer to a curve, and the term 'straight line' used where appropriate. However, where no ambiguity results, the word straight is omitted. Similarly, a line is usually taken as extending indefinitely in both directions, and the term 'line segment' is used to specify the part of a line joining two points.

line (in two dimensions) Usually implies a straight line, though in certain contexts it can mean a curve. A straight line can be determined by two points on the line or by a single point with the direction of the line. In two dimensions the direction can be specified by the gradient. There are a

number of related standard formats for lines in a plane. The following refer
to the equation of the line through A (x_1, y_1) and B (x_2, y_2):

Slope-intercept form:

$y = mx + c$. The line has a gradient m and intercepts the y-axis at $(0, c)$.
When m is 0 the line is horizontal and the equation is $y = c$, and for a
vertical line the gradient is infinite and the equation is $x = k$ (in this case
there is no intercept on the y-axis). When m is negative the line slopes
down from left to right. To obtain the equation of a line in this form from
two points it is usual to calculate the gradient first and then substitute the
values of (x, y) from one of the known points to find c which is now the
only unknown value.

Two-point form.

The condition that the general point P (x, y) lies on the line through AB is
equivalent to requiring the gradient of $AP =$ the gradient of
$AB = \dfrac{y - y_1}{x - x_1} = \dfrac{y_2 - y_1}{x_2 - x_1}$. This is entirely equivalent to the previous form, in
that the right-hand side of this formula is the gradient m, but it is a single-
stage process that only requires algebraic manipulation to simplify it.

Point-slope form.

If the $\dfrac{y_2 - y_1}{x_2 - x_1}$ in the two-point form is replaced by m, then multiplying
through by $x - x_1$ gives $y - y_1 = m(x - x_1)$.

Two-intercept form.

When the equation is written as $\dfrac{x}{a} + \dfrac{y}{b} = 1$ the line cuts the axes at $(a, 0)$
and $(0, b)$, making it very easy to draw the line.

line (in three dimensions) As in two dimensions there are a number of
standard formats which can be used. With the exception of the two-point
form the 2-dimensional forms are not extensible to three dimensions,
while the parametric and vector forms given below can be reduced to two
dimensions, but are not commonly used. The following refer to the
equation of the line through A (x_1, y_1, z_1) and B (x_2, y_2, z_2).

Two-point form.

The condition that the general point P (x, y, z) lies on the line through
AB is equivalent to requiring the direction ratios of $AP =$ the direction
ratios of AB.

$$\frac{x - x_1}{x_2 - x_1} = \frac{y - y_1}{y_2 - y_1} = \frac{z - z_1}{z_2 - z_1}.$$

Parametric form.

If the vector $\begin{pmatrix} l \\ m \\ n \end{pmatrix}$ is parallel to AB then the point P can be expressed in

the form $(x_1 + \lambda l, y_1 + \lambda m, z_1 + \lambda n)$, in terms of the parameter λ.

Vector form.

If the vector $\begin{pmatrix} l \\ m \\ n \end{pmatrix}$ is parallel to AB then the position vector of the point

P can be expressed in the form $\mathbf{r} = \overrightarrow{OP} = \begin{pmatrix} x_1 \\ y_1 \\ z_1 \end{pmatrix} + \lambda \begin{pmatrix} l \\ m \\ n \end{pmatrix}$, from which

the parametric form can be derived immediately. If $\mathbf{a} = \overrightarrow{OA} = \begin{pmatrix} x_1 \\ y_1 \\ z_1 \end{pmatrix}$ is

the position vector of the point A, and $\mathbf{b}$ is defined similarly, the vector
form can be written as $\mathbf{r} = \mathbf{a} + \lambda(\mathbf{b} - \mathbf{a})$ or alternatively defined by $(\mathbf{r} - \mathbf{a}) \times (\mathbf{b} - \mathbf{a}) = 0$. This last form is relying on the fact that A, B, P are in a straight
line, so the angle between AB and AP is zero, and therefore the *vector
product will be zero.

linear algebra The topics of linear equations, matrices, vectors, and of
the algebraic structure known as a vector space, are intimately linked, and
this area of mathematics is known as linear algebra.

linear complexity *See* ALGORITHMIC COMPLEXITY.

linear congruence equation *See* CONGRUENCE EQUATION.

linear differential equation A *differential equation of the form

$$a_n(x)y^{(n)}(x) + a_{n-1}(x)y^{(n-1)}(x) + \cdots$$
$$+ a_1(x)y'(x) + a_0(x)y(x) = f(x),$$

where $a_0, a_1, \ldots, a_n$ and f are given functions, and $y', \ldots, y^{(n)}$ are
*derivatives of y. *See also* LINEAR DIFFERENTIAL EQUATION WITH CONSTANT
COEFFICIENTS and LINEAR FIRST-ORDER DIFFERENTIAL EQUATION.

linear differential equation with constant coefficients For
simplicity, consider such an equation of second-order,

$$a\frac{d^2y}{dx^2} + b\frac{dy}{dx} + cy = f(x),$$ 1

where a, b and c are given constants and f is a given function. (Higher-order equations can be treated similarly.) Suppose that f is not the zero function. Then the equation

$$a\frac{d^2y}{dx^2} + b\frac{dy}{dx} + cy = 0$$ 2

is the *homogeneous equation that corresponds to the non-homogeneous equation **1**. The two are connected by the following result:

Theorem

If $y = G(x)$ is the general solution of **2** and $y = y_1(x)$ is a particular solution of **1**, then $y = G(x) + y_1(x)$ is the general solution of **1**.

Thus the problem of solving **1** is reduced to the problem of finding the complementary function (C.F.) $G(x)$, which is the general solution of **2**, and a particular solution $y_1(x)$ of **1**, usually known in this context as a particular integral (P.I.).

The complementary function is found by looking for solutions of **2** of the form $y = e^{mx}$ and obtaining the auxiliary equation $am^2 + bm + c = 0$. If this equation has distinct real roots m_1 and m_2, the C.F. is $y = Ae^{m_1x} + Be^{m_2x}$ if it has one (repeated) root m, the C.F. is $y = (A + Bx)e^{mx}$; if it has non-real roots $\alpha \pm \beta$ i, the C.F. is $y = e^{\alpha x}(A\cos\beta x + B\sin\beta x)$.

The most elementary way of obtaining a particular integral is to try something similar in form to $f(x)$. Thus, if $f(x) = e^{kx}$, try as the P.I. $y_1(x) = pe^{kx}$. If $f(x)$ is a polynomial in x, try a polynomial of the same degree. If $f(x) = \cos kx$ or $\sin kx$, try $y_1(x) = p\cos kx + q\sin kx$. In each case, the values of the unknown coefficients are found by substituting the possible P.I. into the equation **1**. If $f(x)$ is the sum of two terms, a P.I. corresponding to each may be found and the two added together.

Using these methods, the general solution of $y'' - 3y' + 2y = 4x + e^{3x}$, for example, is found to be $y = Ae^x + Be^{2x} + 2x + 3 + \frac{1}{2}e^{3x}$.

linear equation Consider, in turn, linear equations in one, two and three variables. A linear equation in one variable x is an equation of the form $ax + b = 0$. If $a \neq 0$, this has the solution $x = -b/a$. A linear equation in two variables x and y is an equation of the form $ax + by + c = 0$. If a and b are not both zero, and x and y are taken as Cartesian coordinates in the plane, the equation is an equation of a *straight line. A linear equation in three variables x, y and z is an equation of the form $ax + by + cz + d = 0$. If a, b and c are not all zero, and x, y and z are taken as Cartesian coordinates in 3-dimensional space, the equation is an equation of a *plane. *See also* SIMULTANEOUS LINEAR EQUATIONS.

linear first-order differential equation A *differential equation of the form

$$\frac{dy}{dx} + P(x)y = Q(x),$$

where P and Q are given functions. One method of solution is to multiply both sides of the equation by an integrating factor $\mu(x)$, given by

$$\mu(x) = \exp\left(\int P(x)\,dx\right).$$

This choice is made because then the left-hand side of the equation becomes

$$\mu(x)\frac{dy}{dx} + \mu(x)P(x)y,$$

which is the exact derivative of $\mu(x)y$, and the solution can be found by integration.

In the case where $Q(x) = 0$, this can also be treated as a *separable-variable first-order differential equation.

linear function In real analysis, a linear function is a function f such that $f(x) = ax + b$ for all x in $\mathbf{R}$, where a and b are real numbers with, normally, $a \neq 0$.

linear interpolation *See* INTERPOLATION.

linearly dependent and **independent** A set of vectors $\mathbf{u}_1, \mathbf{u}_2, \ldots, \mathbf{u}_r$ is linearly independent if $x_1\mathbf{u}_1 + x_2\mathbf{u}_2 + \ldots + x_r\mathbf{u}_r = \mathbf{0}$ holds only if $x_1 = 0, x_2 = 0, \ldots, x_r = 0$. Otherwise, the set is linearly dependent. In 3-dimensional space, any set of four or more vectors must be linearly dependent. A set of three vectors is linearly independent if and only if the three are not coplanar. A set of two vectors is linearly independent if and only if the two are not parallel or, in other words, if and only if neither is a scalar multiple of the other.

linear momentum The linear momentum of a particle is the product of its mass and its velocity. It is a vector quantity, usually denoted by $\mathbf{p}$, and $\mathbf{p} = m\mathbf{v}$. *See also* CONSERVATION OF LINEAR MOMENTUM.

Linear momentum has the dimensions MLT^{-1}, and the SI unit of measurement is the kilogram metre per second, abbreviated to 'kg m s^{-1}'.

linear programming The branch of mathematics concerned with maximizing or minimizing a linear function subject to a number of linear constraints. It has applications in economics, town planning, management in industry, and commerce, for example. In its simplest form, with two variables, the constraints determine a feasible region, which is the interior of a polygon in the plane. The objective function to be maximized or minimized attains its maximum or minimum value at a vertex of the feasible region. For example, consider the problem of maximizing $4x_1 - 3x_2$ subject to $x_1 - 2x_2 \geq -4$, $2x_1 + 3x_2 \leq 13$, $x_1 - x_2 \leq 4$, and $x_1 \geq 0$, $x_2 \geq 0$. The feasible region is the interior of the polygon $OABCD$ shown in the figure, and the objective function $4x_1 - 3x_2$ attains its maximum value of 17 at the point B with coordinates $(5, 1)$.

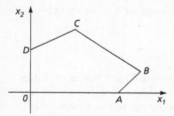

Often integer values are required and in such cases the vertex is not always admissible, and it will be necessary to test all points with integer values which lie close to the vertex. It is also possible, though unusual, that the objective function may be parallel to a constraining condition. In this case, all points on the boundary representing that constraint will be optimal.

linear regression *See* REGRESSION.

linear scale A scale in which the same difference in value is always represented by the same size of interval.

```
                              linear scale
  ┌──┬──┬──┬──┬──┬──┬──┬──┐
 -2 -1  0  1  2  3  4  5  6
```

linear space = VECTOR SPACE.

linear transformation *See* TRANSFORMATION (of the plane).

line of action The line of action of a force is a straight line through the *point of application of the force and parallel to the direction of the force.

line of symmetry *See* SYMMETRICAL ABOUT A LINE.

line segment If A and B are two points on a straight line, the part of the line between and including A and B is a line segment. This may be denoted by AB or BA. (The notation AB may also be used with a different meaning, as a real number, the *measure of $\overrightarrow{AB}$ on a directed line.) Some authors use 'line segment' to mean a *directed line segment.

Liouville, Joseph (1809–82) Prolific and influential French mathematician, who was founder and editor of a notable French mathematical journal and proved important results in the fields of number theory, differential equations, differential geometry and complex analysis. In 1844, he proved the existence of *transcendental numbers and constructed an infinite class of such numbers.

Liouville numbers Irrational number x with the property that for every integer n, there is a rational number $\dfrac{p}{q}$ such that $\left| x - \dfrac{p}{q} \right| < \dfrac{1}{q^n}$. All Liouville numbers are *transcendental numbers.

Lipschitz condition A function f between two metric spaces M and N satisfies the Lipschitz condition if there exists a constant k for which $d(f(x), f(y)) \leq k \times d(x, y)$ for all points x, y in M, i.e. that the distance between the function values is bounded by a constant multiple of the distance between the values.

litre A unit, not an SI unit, used in certain contexts for measuring volume, abbreviated to 'l'. A litre is equal to one cubic decimetre, which equals 1000 cm^3. One millilitre, abbreviated to 1 ml, is equal to 1 cm^3. At one time, a litre was defined as the volume of one kilogram of water at 4°C at standard atmospheric pressure.

Littlewood, John Edensor (1885–1977) British mathematician, based at Cambridge University. Most famous for his collaboration with Godfrey *Hardy, on summability theory, *Fourier series, analytic number theory and the *zeta function.

ln Abbreviation for *logarithm and symbol for logarithm in base e.

load *See* MACHINE.

Lobachevsky, Nikolai Ivanovich (1792–1856) Russian mathematician who in 1829 published his discovery, independently of Bolyai, of *hyperbolic geometry. He continued to publicize his ideas, but his work received wide recognition only after his death.

local maximum (maxima) For a function f, a local maximum is a point c that has a neighbourhood at every point of which, except for c, $f(x) < f(c)$.

If f is *differentiable at a local maximum c, then $f'(c) = 0$; that is, the local maximum is a *stationary point.

local minimum (minima) For a function f, a local minimum is a point c that has a neighbourhood at every point of which, except for c, $f(x) > f(c)$. If f is *differentiable at a local minimum c, then $f'(c) = 0$; that is, the local minimum is a *stationary point.

located vector A vector only requires a magnitude and direction to specify it. A located vector has a specified starting point in addition.

location In statistics, a measure of location is a single figure which gives a typical or, in some sense, central value for a distribution or sample. The most common measures of location are the *mean, the *median and, less usefully, the *mode. For several reasons, the mean is usually the preferred measure of location, but when a distribution is *skew the median may be more appropriate.

locus (loci) A locus (in the plane) is the set of all points in the plane that satisfy some given condition or property. For example, the locus of all points that are a given distance from a fixed point is a circle; the locus of all points equidistant from two given points, A and B, is the perpendicular bisector of AB. If, in a given coordinate system, a locus can be expressed in the form $\{(x, y), | f(x, y) = 0\}$, the equation $f(x, y) = 0$ is called an equation of the locus (in the given coordinate system). The equation may be said to 'represent' the locus.

log Abbreviation and symbol for *logarithm.

logarithm Let a be any positive number, not equal to 1. Then, for any real number x, the meaning of a^x can be defined (see approach 1 to the *exponential function to base a). The logarithmic function to base a, denoted by $\log_a$, is defined as the inverse of this function. So $y = \log_a x$ if and only if $x = a^y$. Thus $\log_a x$ is the index to which a must be raised in order to get x. Since any power a^y of a is positive, x must be positive for $\log_a x$ to be defined, and so the domain of the function $\log_a$ is the set of positive real numbers. (See INVERSE FUNCTION for a more detailed explanation about the domain of an inverse function.) The value $\log_a x$ is called simply the logarithm of x to base a. The notation $\log x$ may be used when the base intended is understood. The following properties hold, where x, y and r are real, with $x > 0$ and $y > 0$:

(i) $\log_a(xy) = \log_a x + \log_a y$.
(ii) $\log_a(1/x) = -\log_a x$.
(iii) $\log_a(x/y) = \log_a x - \log_a y$.
(iv) $\log_a(x^r) = r \log_a x$.

(v) Logarithms to different bases are related by the formula

$$\log_b x = \frac{\log_a x}{\log_a b}.$$

Logarithms to base 10 are called common logarithms. Tables of common logarithms were in the past used for arithmetical calculations. Logarithms to base e are called natural or Napierian logarithms, and the notation ln can be used instead of $\log_e$. However, this presupposes that the value of e has been defined independently. It is preferable to define the *logarithmic function ln in quite a different way. From this, e can be defined and then the equivalence of ln and $\log_e$ proved.

logarithmic derivative The derivative of the logarithm of the function i.e. $\dfrac{d}{dx}(\ln f(x)) = \dfrac{f'(x)}{f(x)}.$

logarithmic differentiation A technique for solving equations or differentiating products, since taking logarithms gives a sum of terms from a product of terms.

logarithmic function The logarithmic function ln must be distinguished from the logarithmic function to base a (*see* LOGARITHM). Here are two approaches:

   1. Suppose that the value of the number e has already been obtained independently. Then the *logarithm of x to base e can be defined and denoted by $\log_e x$, and the logarithmic function ln can be taken to be just this function $\log_e$. The problem with this approach is its reliance on a prior definition of e and the difficulty of subsequently proving some of the important properties of ln.

   2. The following is more satisfactory. Let f be the function defined, for $t > 0$, by $f(t) = 1/t$. Then the logarithmic function ln is defined, for $x > 0$, by

$$\ln x = \int_1^x \frac{1}{t}\, dt.$$

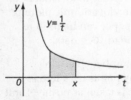

The intention is best appreciated in the case when $x > 1$, for then ln x gives the area under the graph of f in the interval $[1, x]$. This function ln is *continuous and increasing; it is *differentiable and, from the fundamental relationship between differentiation and integration, its *derived function is the function *f*. Thus it has been established that

$$\frac{d}{dx}(\ln x) = \frac{1}{x} \quad (x > 0).$$

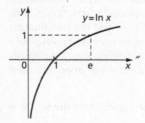

From the definition, the following properties can be obtained, where x, y and r are real, with $x > 0$ and $y > 0$:

   (i) $\ln(xy) = \ln x + \ln y$.
   (ii) $\ln(1/x) = -\ln x$.
   (iii) $\ln(x/y) = \ln x - \ln y$.
   (iv) $\ln(x^r) = r \ln x$.

With this approach, exp can be defined as the inverse function of ln, and the number e defined as exp 1. Finally, it is shown that $\ln x$ and $\log_e x$ are identical.

logarithmic function to base *a* *See* LOGARITHM.

logarithmic plotting Two varieties of logarithmic graph paper are commonly available. One, known as semi-log or single log paper, has a standard scale on the x-axis and a *logarithmic scale on the y-axis. If given data satisfy an equation $y = ba^x$ (where a and b are constants), exhibiting *exponential growth, then on this special graph paper the corresponding points lie on a straight line. The other variety is known as log–log or double log paper and has a logarithmic scale on both axes. In this case, data satisfying $y = bx^m$ (where b and m are constants) produce points that lie on a straight line. Given some experimental data, such logarithmic plotting on one variety or the other of logarithmic graph paper can be used to determine what relationship might have given rise to such data.

logarithmic scale A method of representing numbers (certainly positive and usually greater than or equal to 1) by points on a line as follows. Suppose that one direction along the line is taken as positive, and that the point O is taken as origin. The number x is represented by the point P in such a way that OP is proportional to $\log x$, where logarithms are to base 10. Thus the number 1 is represented by O; and, if the point A represents 10, the point B that represents 100 is such that $OB = 2OA$.

The measurement of sound using decibels, and of the size of earthquakes using the Richter scale are two examples of logarithmic scales.

logarithmic series The *power series

$$x - \frac{x^2}{2} + \frac{x^3}{3} + \cdots + (-1)^{n+1}\left(\frac{x^n}{n}\right) \cdots$$ For $-1 < x \leq 1$ this series converges to $\ln(1+x)$.

logarithmic spiral = EQUIANGULAR SPIRAL.

logic The study of deductive reasoning, by which conclusions are derived from sets of premises. There is a distinction between the form of the argument (which is the logic) and the content, especially if the premises forming the starting point of the argument are not universally accepted. Informally, the term is used to refer to the essential reasoning process in a mathematical proof, so two proofs which differ only in the manipulative detail and not in the essential structure of the argument would be termed 'logically equivalent'.

logically equivalent Two *compound statements involving the same components are logically equivalent if they have the same *truth tables. This means that, for all possible truth values of the components, the resulting truth values of the two statements are the same. For example, the truth table for the statement $(\neg p) \vee q$ can be completed as follows:

p	q	$\neg p$	$(\neg p) \vee q$
T	T	F	T
T	F	F	F
F	T	T	T
F	F	T	T

By comparing the last column here with the truth table for $p \Rightarrow q$ (*see* IMPLICATION), it can be seen that $(\neg p) \vee q$ and $p \Rightarrow q$ are logically equivalent.

logistic function (logistic curve) The curve where $y = \dfrac{k}{1 + e^{(a-bx)}}$, where $b > 0$ and usually $k > 0$. It has horizontal asymptotes at $y = 0$, $y = k$

and intercept at $\left(0, \dfrac{k}{1 + e^a}\right)$ and provides a good model to describe population growth where saturation constraints exist.

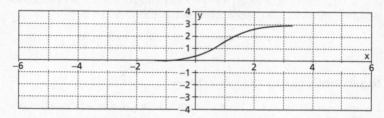

lognormal distribution A positive valued *random variable X is said to have a lognormal distribution if the random variable $Y = \ln X$ has a *normal distribution.

log paper Graph paper where one of the axes uses a *logarithmic scale. If both axes use logarithmic scales it may be known as log–log paper or double log paper.

longitude The longitude of a point P on the Earth's surface is the angle, measured in degrees east or west, between the *meridian through P and the *prime meridian through Greenwich (the meridians being taken here as semicircles). If the meridian through P and the meridian through Greenwich meet the equator at P' and G' respectively, the longitude is the angle $P'OG'$, where O is the centre of the Earth. The longitude of any point may be up to 180° east or up to 180° west. Longitude and *latitude uniquely fix the position of a point on the Earth's surface.

longitudinal study A study of the development of characteristics of the same individuals over a period of time. If different groups are studied simultaneously at different ages, it is difficult to attribute observed differences to any factor, so longitudinal studies are designed to remove the individual differences in an analogous manner to *paired-sample tests. However, they are very expensive to run on a long scale, and require a long-term commitment from both researcher and subjects.

longitudinal wave A form of wave motion is which energy is transmitted along the direction of the wave motion as the wave moves through a medium; sound waves are an example of a longitudinal wave.

loop *See* GRAPH.

Lorentz, Hendrik Antoon FRS (1853–1928) Dutch mathematician and theoretical physicist whose work on the mathematical theory of the electron was awarded the Nobel Prize for Physics in 1902, jointly with his student Pieter Zeeman who had verified the mathematical results experimentally. He is also famous for the *Lorentz–Fitzgerald contraction and *Lorentz transformations.

Lorentz–Fitzgerald contraction To an observer a body moving at high speed appears shorter (in the direction of motion) than it really is. Since the factor is $\sqrt{1 - \dfrac{v^2}{c^2}} : 1$ where c is the speed of light, the speed has to be very high, or the measurement very precise for this to be observable. It was observed 20 years before *Einstein's special theory of relativity provided the explanation for it.

Lorentz force law The force experienced by a charged particle in an *electromagnetic field. The law states that a particle with charge q, moving with a velocity $\mathbf{v}$ through an electric field $\mathbf{E}$ and a magnetic field $\mathbf{B}$, will experience a force $\mathbf{F}$ given by $\mathbf{F} = q(\mathbf{E} + \mathbf{v} \times \mathbf{B})$.

Lorentz transformation The equations for transforming the behaviour of bodies in one frame of reference to another. If x is taken in the direction of the relative velocity of the two frames, and v is the magnitude of the relative velocity then $x' = \beta\,(x - vt)$, $y' = y$, $z' = z$, $\quad t' = \beta\left(t - \dfrac{vx}{c^2}\right)$ where $\beta = \dfrac{1}{\sqrt{1 - \left(\dfrac{v}{c}\right)^2}}$. This formalized the concept that time is not a universal constant and that *space–time needs to be treated as a 4-dimensional construct when high speeds are involved.

Lorenz attractor The Lorenz attractor is one of the most famous of the *fractals and arises as the solution to a set of parameterized *differential equations which describe the flow of fluid in a box which is heated along the bottom, which Lorenz used as model for the behaviour of weather. It became one of landmarks in *chaos theory. The equations producing the Lorenz attractor are:

$$\frac{dx}{dt} = \alpha(y - x)$$

$$\frac{dy}{dt} = x(\beta - z) - y$$

$$\frac{dz}{dt} = xy - \gamma z$$

lower bound *See* BOUND.

lower limit *See* LIMIT OF INTEGRATION.

lower triangular matrix *See* TRIANGULAR MATRIX.

lowest common denominator = LEAST COMMON DENOMINATOR.

Lucas numbers The sequence 2, 1, 3, 4, 7, 11, 18, 29 which is derived from the same relation as the Fibonacci sequence, i.e. that each number is the sum of the previous two numbers in the sequence, but it has different starting values.

lurking variable A potential *confounding variable that has not been measured and not discussed in the interpretation of an *experiment or *observational study.

m Abbreviation and symbol for *metre.

m Abbreviation for milli-used in symbols in the metric system for thousandths of a unit, for example mm.

M Roman *numeral for 1000.

machine A device that enables energy from one source to be modified and transmitted as energy in a different form or for a different purpose. Simple examples, in mechanics, are *pulleys and *levers, which may be used as components in a more complex machine. The force applied to the machine is the effort, and the force exerted by the machine is equal and opposite to the load.

If the effort has magnitude P and the load has magnitude W, the mechanical advantage of the machine is the ratio W/P. The velocity ratio is the ratio of the distance travelled by the effort to the distance travelled by the load. It is a constant for a particular machine, calculable from first principles. For example, in the case of a pulley system, the velocity ratio equals the number of ropes supporting the load or, equivalently, the number of pulleys in the system.

The efficiency is the ratio of the work obtained from the machine to the work put into the machine, often stated as a percentage. In the real world, the efficiency of a machine is less than 1 because of factors such as friction. In an ideal machine for which the efficiency is 1, the mechanical advantage equals the velocity ratio.

Maclaurin, Colin (1698–1746) Scottish mathematician who was the outstanding British mathematician of the generation following *Newton's. He developed and extended the subject of calculus. His textbook on the subject contains important original results, but the *Maclaurin series, which appears in it, is just a special case of the *Taylor series known considerably earlier. He also obtained notable results in geometry and wrote a popular textbook on algebra.

Maclaurin series (expansion) Suppose that f is a *real function, all of whose *derived functions $f^{(r)}(r = 1, 2, \ldots)$ exist in some interval containing 0. It is then possible to write down the power series

$$f(0) + \frac{f'(0)}{1!}x + \frac{f''(0)}{2!}x^2 + \cdots + \frac{f^{(n)}(0)}{n!}x^n + \cdots.$$

This is the Maclaurin series (or expansion) for f. For many important functions, it can be proved that the Maclaurin series is convergent, either for all x or for a certain range of values of x, and that for these values the sum of the series is $f(x)$. For these values it is said that the Maclaurin series is a 'valid' expansion of $f(x)$. The Maclaurin series for some common functions, with the values of x for which they are valid, are given in the Table of series (*Appendix 9). The function f, defined by $f(0) = 0$ and $f(x) = e^{-1/x^2}$ for all $x \neq 0$, is notorious in this context. It can be shown that all of its derived functions exist and that $f^{(r)}(0) = 0$ for all r. Consequently, its Maclaurin series is convergent and has sum 0, for all x. This shows, perhaps contrary to expectation, that, even when the Maclaurin series for a function f is convergent, its sum is not necessarily $f(x)$. *See also* TAYLOR'S THEOREM.

magic square A square array of numbers arranged in rows and columns in such a way that all the rows, all the columns and both diagonals add up to the same total. Often the numbers used are $1, 2, \ldots, n^2$, where n is the number of rows and columns. The example on the left in the figure is said to be of ancient Chinese origin; that on the right is featured in an engraving by Albrecht Dürer.

4	9	2
3	5	7
8	1	6

16	3	2	13
5	10	11	8
9	6	7	12
4	15	14	1

(⊕) SEE WEB LINKS

• An interactive 3 x 3 magic square.

magnetic field A *vector field which exerts a force on moving electric charges.

magnitude (of a vector) *See* VECTOR. Also, if the vector $\mathbf{a}$ is given in terms of its components (with respect to the standard vectors $\mathbf{i}$, $\mathbf{j}$ and $\mathbf{k}$) in the form $\mathbf{a} = a_1\mathbf{i} + a_2\mathbf{j} + a_3\mathbf{k}$, the magnitude of $\mathbf{a}$ is given by the formula

$$|\mathbf{a}| = \sqrt{a_1^2 + a_2^2 + a_3^2}.$$

magnitude of a statistically significant result A statistically significant result is a function of both the sample size and of the magnitude of the difference being measured. With advances in technology, sample sizes now can be extremely large, and in these cases almost any difference will turn out to be significant. It becomes important in such cases to consider whether the difference is large enough to be of practical importance.

main diagonal In the $n \times n$ matrix $[a_{ij}]$, the entries $a_{11}, a_{22}, \ldots, a_{nn}$ form the main diagonal.

major arc *See* ARC.

major axis *See* ELLIPSE.

Mandelbrot, Benoît (1924–2010) Polish mathematician who showed the range of applications of *fractals in mathematics and in nature, and whose work with computer graphics popularized the beautiful fractal images.

Mandelbrot set The set of points c in the complex plane for which the application of the function $f(z) = z^2 + c$ repeatedly to the point $z = 0$ produces a bounded sequence. The set is an extremely complicated object whose boundary is a *fractal. It can also be defined as the set of points c for which the *Julia set of the above function is a connected set. The Mandelbrot set is very frequently used as an illustration of a geometrical fractal.

(((( ))) SEE WEB LINKS

• A Java applet which allows visualizations of Mandelbrot and Julia sets.

Manhattan norm = TAXICAB NORM.

manifold A *topological space X in which every point has a *neighbourhood which is *homeomorphic to Euclidean space. Lines and circles are 1-dimensional manifolds, while a figure of eight is not. 2-dimensional manifolds are called surfaces. The plane, the sphere, and the torus are all manifolds that can be physically manifested in the real world, but the *Klein bottle and the *real projective plane are also manifolds although they cannot be physically constructed.

Note that the requirement for a manifold is that locally the manifold has the properties of Euclidean space, but it does not need to have these properties globally—so the surface of a sphere is a manifold even though it is not itself a Euclidean space.

Mann–Whitney U test *See* WILCOXON RANK-SUM TEST.

mantissa *See* FLOATING-POINT NOTATION.

many-to-one A function or mapping for which more than one element of the domain maps onto an element in the range. So $f(x) = x^2$ is many-to-one if x is defined on the real numbers since $f(a) = f(-a)$. A many-to-one function is not invertible, since there is no way to define a function to return more than one value. However, a many-to-one function can be restricted to a domain within which it is a one-to-one function, and for which the inverse function exists. If $f(x) = x^2$ is restricted to $x \geq 0$ then $f^{-1}(x) = \sqrt{x}$ exists. Electronic calculators do this by default, for example the trigonometric functions will return inverse values over an interval of width 180° or π radians.

mapping A mapping (or *function) f from S to T, where S and T are non-empty sets, is a rule which associates, with each element of S, a unique element of T. The set S is the domain and T is the codomain of f. The phrase 'f from S to T' is written '$f: S \rightarrow T$'. For s in S, the unique element of T that f associates with s is the image of s under f and is denoted by $f(s)$. If $f(s) = t$, it is said that f maps s to t, written $f: s \rightarrow t$. The subset of T consisting of those elements that are images of elements of S under f, that is, the subset $\{t \mid t = f(s)$, for some s in $S\}$, is the image (or range) of f, denoted by $f(S)$. For the mapping $f: S \rightarrow T$, the subset $\{(s, f(s) \mid s \in S\}$ of the *Cartesian product $S \times T$ is the *graph of f. The graph of a mapping has the property that, for each s in S, there is a unique element (s, t) in the graph. Some authors define a mapping from S to T to be a subset of $S \times T$ with this property; then the image of s under this mapping is defined to be the unique element t such that (s, t) is in this subset.

marginal distribution Suppose that X and Y are discrete random variables with *joint probability mass function p. Then the marginal distribution of X is the distribution with probability mass function p_1, where

$$p_1(x) = \sum_y p(x, y).$$

A similar formula gives the marginal distribution of Y.

Suppose that X and Y are continuous random variables with joint probability density function f. Then the marginal distribution of X is the distribution with probability density function f_1, where

$$f_1(x) = \int_{-\infty}^{\infty} f(x, t)\, dt.$$

marginal probability density function The *probability density function of a *marginal distribution.

Markov, Andrei Andreevich (1856–1922) Russian mathematician who, mostly after 1900, proved important results in probability, developing the notion of a *Markov chain and initiating the study of *stochastic processes.

Markov chain Consider a *stochastic process $X_1, X_2, X_3, \ldots$ in which the state space is discrete. This is a Markov chain if the probability that X_{n+1} takes a particular value depends only on the value of X_n and not on the values of $X_1, X_2, \ldots, X_{n-1}$. (This definition can be adapted so as to apply to a stochastic process with a continuous state space, or to a more general stochastic process $\{X(t), t \in T\}$, to give what is called a Markov process.) In most Markov chains, the probability that $X_{n+1} = j$ given that $X_n = i$, denoted by p_{ij}, does not depend on n. In that case, if there are N states, these values p_{ij} are called the transition probabilities and form the transition matrix $[p_{ij}]$, an $N \times N$ row *stochastic matrix.

((●)) SEE WEB LINKS
• An interactive simulation for up to four states with user-defined transition matrix.

martingale A sequence of random variables $\{X_i\}$ such that the expected value of the $(n+1)$-th random variable, once the first n outcomes are known, is just the current value X_n. The simple *random walk, which moves left or right one step with probability 0.5 is an example of a martingale.

mass With any body there are associated two parameters: the *gravitational mass, which occurs in the *inverse square law of gravitation, and the *inertial mass, which occurs in *Newton's second law of motion. By suitable scaling, the two values can be taken to be equal, and their common value is the mass of the particle.
   The SI unit of mass is the *kilogram.

mass–energy equation $E = mc^2$ where E = energy, m = mass and c = the speed of light in a vacuum. This is the equation proposed by Albert * Einstein as part of the special theory of relativity showing the relationship between mass and energy. The implication of this equation is that mass can be considered as potential energy, though it is extremely difficult to make the conversion, and release the stored energy. Two such methods, which form the basis of nuclear power and nuclear weapons, are nuclear fusion and nuclear fission, both of which involve creating a new molecular structure from existing molecules where the new structure contains less mass, with the difference being converted into energy.

matched-pairs design An experimental design structure which matches subjects as closely as possible, and then assigns one of each pair to each of the treatment group and control group. This might involve cutting pieces of cloth in two when testing detergents, or pairing patients in a clinical trial by age, gender, weight etc. as closely as possible. The allocation of treatments to the paired subjects should be randomized. Then the differences in outcome are measured. The strength of this design lies in reducing the amount of variation between subjects so that any actual differences due to the experimental conditions are more easily identified.

matching In a *bipartite graph a matching is a subset of the edges in the graph for which no two edges share a common vertex. A maximal matching is a matching in which the number of edges is as large as possible.

Mathemapedia A wiki (a type of website which allows users to add, remove, edit, and change content quickly and easily) for mathematics education developed by the *National Centre for Excellence in the Teaching of Mathematics (NCETM).

(((⊕))) SEE WEB LINKS
• The home page of Mathemapedia.

mathematical expectation = EXPECTED VALUE.

mathematical induction The method of proof 'by mathematical induction' is based on the following principle:

Principle of mathematical induction
Let there be associated, with each positive integer n, a proposition $P(n)$, which is either true or false. If

   (i) $P(1)$ is true,
   (ii) for all k, $P(k)$ implies $P(k+1)$,

then $P(n)$ is true for all positive integers n.

This essentially describes a property of the positive integers; either it is accepted as a principle that does not require proof or it is proved as a theorem from some agreed set of more fundamental axioms. The following are typical of results that can be proved by induction:

   (i) For all positive integers n, $\sum_{r=1}^{n} r^2 = \frac{1}{6}n(n+1)(2n+1)$.
   (ii) For all positive integers n, the n-th derivative of $\frac{1}{x}$ is $(-1)^n \frac{n!}{x^{n+1}}$.
   (iii) For all positive integers n, $(\cos\theta + i\sin\theta)^n = \cos n\theta + i\sin n\theta$.

In each case, it is clear what the proposition $P(n)$ should be, and that (i) can be verified. The method by which the so-called induction step (ii) is proved depends upon the particular result to be established.

A modified form of the principle is this. Let there be associated, with each integer $n \geq n_0$, a proposition $P(n)$, which is either true or false. If (i) $P(n_0)$ is true, and (ii) for all $k \geq n_0$, $P(k)$ implies $P(k+1)$, then $P(n)$ is true for all integers $n \geq n_0$. This may be used to prove, for example, that $3^n > n^3$ for all integers $n \geq 4$.

mathematical logic *See* LOGIC.

mathematical model A problem of the real world is a problem met by a physicist, an economist, an engineer, or indeed anyone in their normal working conditions or everyday life. Mathematics may help to solve the problem. But to apply the mathematics it is often necessary to develop an abstract mathematical problem, called a mathematical model of the original, that approximately corresponds to the real world problem. Developing such a model may involve making assumptions and simplifications. The mathematical problem can then be investigated and perhaps solved. When interpreted in terms of the real world, this may provide an appropriate solution to the original problem.

mathematics The branch of human enquiry involving the study of numbers, quantities, data, shape and space and their relationships, especially their generalizations and abstractions and their application to situations in the real world. As a broad generalization *pure mathematics is the study of the relationships between abstract quantities according to a well-defined set of rules and *applied mathematics is the application and use of mathematics in the context of the real world. Pure mathematics includes *algebra, *abstract algebra, *calculus, *geometry, *number theory, *topology and *trigonometry. Applied mathematics includes *mechanics, *probability and statistics, *quantum mechanics and *relativity.

Mathieu's equation A second-order differential equation of the form $\frac{d^2y}{dx^2} + (a + b\cos 2x)y = 0$ which occurs in the study of vibrations. The general solution is in the form $Ae^{kx}\phi(x) + Be^{-kx}\phi(-x)$ where k is a constant and ϕ is periodic with period 2π.

matrix (matrices) A rectangular array of entries displayed in rows and columns and enclosed in brackets. The entries are elements of some suitable set, either specified or understood. They are often numbers, perhaps integers, real numbers or complex numbers, but they may be, say, polynomials or other expressions. An $m \times n$ matrix has m rows and n columns and can be written

$$\begin{bmatrix} a_{11} & a_{12} & \cdots & a_{1n} \\ a_{21} & a_{22} & \cdots & a_{2n} \\ \vdots & \vdots & \ddots & \vdots \\ a_{m1} & a_{m2} & \cdots & a_{mn} \end{bmatrix}.$$

Round brackets may be used instead of square brackets. The subscripts are read as though separated by commas: for example, a_{23} is read as 'a, two, three'. The matrix above may be written in abbreviated form as $[a_{ij}]$, where the number of rows and columns is understood and a_{ij} denotes the entry in the i-th row and j-th column. *See also* ADDITION (of matrices), MULTIPLICATION (of matrices) and INVERSE MATRIX.

matrix game A *game in which a *matrix contains numerical data giving information about what happens according to the strategies chosen by two opponents or players. By convention, the matrix $[a_{ij}]$ gives the *payoff to one of the players, R: when R chooses the i-th row and the other player, C, chooses the j-th column, then C pays to R an amount of a_{ij} units. (If a_{ij} is negative, then in fact R pays C a certain amount.) This is an example of a zero-sum game because the total amount received by the two players is zero: R receives a_{ij} units and C receives $-a_{ij}$ units.

If the game is a *strictly determined game, and R and C use *conservative strategies, C will always pay R a certain amount, which is the value of the game. If the game is not strictly determined, then by the *Fundamental Theorem of Game Theory, there is again a value for the game, being the expectation (loosely, the average payoff each time when the game is played many times) when R and C use *mixed strategies that are optimal.

matrix norm If **A** is a square matrix, with real or complex elements, a matrix norm $|\mathbf{A}|$ is a non-negative number with the properties that

$|\mathbf{A}| > 0$ if $\mathbf{A} \neq 0$, and $|\mathbf{A}| = 0$ if $\mathbf{A} = 0$

$|k\mathbf{A}| = k|\mathbf{A}|$ for any positive scalar k,

$|\mathbf{A} + \mathbf{B}| \leq |\mathbf{A}| + |\mathbf{B}|$,

$|\mathbf{AB}| \leq |\mathbf{A}| \times |\mathbf{B}|$.

matrix of coefficients For a set of m linear equations in n unknowns $x_1, x_2, \ldots, x_n$:

$$a_{11}x_1 + a_{12}x_2 + \cdots + a_{1n}x_n = b_1,$$
$$a_{21}x_1 + a_{22}x_2 + \cdots + a_{2n}x_n = b_2,$$
$$\vdots$$
$$a_{m1}x_1 + a_{m2}x_2 + \cdots + a_{mn}x_n = b_m,$$

the matrix of coefficients is the $m \times n$ matrix $[a_{ij}]$.

matrix of cofactors For **A**, a square matrix $[a_{ij}]$, the matrix of cofactors is the matrix, of the same order, that is obtained by replacing each entry a_{ij} by its *cofactor A_{ij}. It is used to find the *adjoint of **A**, and hence the inverse $\mathbf{A}^{-1}$.

Maupertuis, Pierre-Louis Moreau de (1698–1759) French scientist and mathematician known as the first to formulate a proposition called the principle of least action. He was an active supporter in France of *Newton's gravitational theory. He led an expedition to Lapland to measure the length of a degree along a meridian, which confirmed that the Earth is an oblate spheroid.

max-flow/min-cut The value of the total flow in any network is $\leq$ the capacity of any cut, and the maximum flow will equal the minimum cut for any network.

maximal matching *See* MATCHING.

maximin strategy *See* CONSERVATIVE STRATEGY.

maximum For maximum value, *see* GREATEST VALUE. *See also* LOCAL MAXIMUM.

maximum likelihood estimator The *estimator for an unknown parameter given by the method of maximum likelihood. *See* LIKELIHOOD.

Maxwell, James Clerk (1831–79) British mathematical physicist, who to physicists probably ranks second to *Newton in stature. Of considerable importance in applied mathematics are the differential equations known as Maxwell's equations, which are fundamental to the field theory of electromagnetism.

Maxwell's equations A set of differential equations summarizing the relationships between electricity and magnetism.

$$\nabla \cdot \mathbf{D} = \rho$$
$$\nabla \cdot \mathbf{B} = 0$$

$$\nabla \times \mathbf{E} = -\frac{\partial \mathbf{B}}{\partial t}$$

$$\nabla \times \mathbf{H} = \mathbf{J} + \frac{\partial \mathbf{D}}{\partial t}$$

where **B** is the magnetic flux density, **D** is the electric displacement, **E** is the electric field, **H** is the magnetic field strength, **J** is the current density, ρ is the free electric charge density, and t is time.

meagre set A Baire first category set (*see* BAIRE CATEGORY). The name derives from the definition, which requires that a meagre set is small in a precise sense. Further, any subset of a meagre set is meagre and the union of *countably many meagre sets is meagre.

mean The mean of the numbers $a_1, a_2, \ldots, a_n$ is equal to

$$\frac{a_1 + a_2 + \cdots + a_n}{n}.$$

This is the number most commonly used as the average. It may be called the arithmetic mean to distinguish it from other means such as those described below. When each number a_i is to have weight w_b, the weighted mean is equal to

$$\frac{w_1 a_1 + w_2 a_2 + \cdots + w_n a_n}{w_1 + w_2 + \cdots + w_n}.$$

The geometric mean of the positive numbers $a_1, a_2, \ldots, a_n$ is $\sqrt[n]{a_1 a_2 \ldots a_n}$. Given two positive numbers a and b, suppose that $a < b$. The arithmetic mean m is then equal to $\frac{1}{2}(a + b)$, and a, m, b is an arithmetic sequence. The geometric mean g is equal to $\sqrt{ab}$, and a, g, b is a geometric sequence. The arithmetic mean of 3 and 12 is $7\frac{1}{2}$, and the geometric mean is 6. It is a theorem of elementary algebra that, for any positive numbers $a_1, a_2, \ldots, a_n$, the arithmetic mean is greater than or equal to the geometric mean; that is,

$$\frac{1}{n}(a_1 + a_2 + \cdots + a_n) \geq \sqrt[n]{a_1 a_2 \ldots a_n}.$$

The harmonic mean of positive numbers $a_1, a_2, \ldots, a_n$ is the number h such that $1/h$ is the arithmetic mean of $1/a_1, 1/a_2, \ldots, 1/a_n$. Thus

$$h = \frac{n}{(1/a_1) + \cdots + (1/a_n)}.$$

For any set of positive numbers, $a_1, a_2, \ldots, a_n$ the harmonic mean is less than or equal to the geometric mean, and therefore also the arithmetic mean, and equality takes place for all three measures if and only if all the terms in the set are equal.

In statistics, the mean of a set of observations $x_1, x_2, \ldots, x_n$ called a sample mean, is denoted by $\bar{x}$ so that

$$\bar{x} = \frac{x_1 + x_2 + \cdots + x_n}{n}.$$

The mean of a random variable X is equal to the *expected value $E(X)$. This may be called the population mean and denoted by μ. A sample mean $\bar{x}$. may be used to estimate μ.

mean absolute deviation For the sample $x_1, x_2, \ldots, x_n$, with mean $\bar{x}$, the mean absolute deviation is equal to

$$\frac{\sum |x_i - \bar{x}|}{n}.$$

It is a measure of *dispersion, but is not often used.

mean deviation = MEAN ABSOLUTE DEVIATION.

mean squared deviation For the sample $x_1, x_2, \ldots, x_n$, with mean $\bar{x}$, the mean squared deviation (about the mean) is the second moment about the mean and is equal to

$$\frac{\sum (x_i - \bar{x})^2}{n}.$$

It is in fact the *variance, in the form with n as the denominator.

mean squared error For an *estimator X of a parameter θ, the mean squared error is equal to $E(X - \theta)^2)$. For an unbiased estimator this equals $\text{Var}(X)$ (*see* EXPECTED VALUE and VARIANCE).

mean value (of a function) Let f be a function that is continuous on the closed interval $[a, b]$. The mean value of f over the interval $[a, b]$ is equal to

$$\frac{1}{b-a} \int_a^b f(x) \, dx.$$

The mean value $\bar{y}$ has the property that the area under the curve $y = f(x)$ between $x = a$ and $x = b$ is equal to the area under the horizontal line $y = \bar{y}$ between $x = a$ and $x = b$, which is the area of a rectangle with width $b - a$ and height $\bar{y}$.

For example, the function may represent the temperature measured at a certain place over a period of 24 hours. The mean value gives what may be considered an 'average' value for the temperature over that period.

mean value theorem The following theorem, which has very important consequences in differential calculus:

Theorem

Let f be a function that is continuous on $[a, b]$ and differentiable in (a, b). Then there is a number c with $a < c < b$ such that

$$f'(c) = \frac{f(b) - f(a)}{b - a}.$$

The result stated in the theorem can be expressed as a statement about the graph of f: if A, with coordinates $(a, f(a))$, and B, with coordinates $(b, f(b))$, are the points on the graph corresponding to the end-points of the interval, there must be a point C on the graph between A and B at which the tangent is parallel to the chord AB.

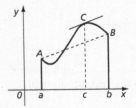

The theorem is normally deduced from *Rolle's Theorem, which is in fact the special case of the mean value theorem in which $f(a) = f(b)$. A rigorous proof of either theorem relies on the non-elementary result that a *continuous function on a closed interval attains its bounds. The mean value theorem has immediate corollaries, such as the following. With the appropriate conditions on f,

(i) if $f'(x) = 0$ for all x, then f is a *constant function,
(ii) if $f'(x) > 0$ for all x, then f is strictly increasing.

The important *Taylor's Theorem can also be seen as an extension of the mean value theorem.

measurable function A function f with a domain D mapping into the set of real numbers is measurable if for every real number k the set $\{x \in D: f(x) > k\}$ is measurable.

measurable set If μ is a *measure defined on a σ-algebra, X, of a set S, then the elements of X are called μ-measurable sets or measurable sets.

measure Let A and B be two points on a *directed line, and let $\overrightarrow{AB}$ be the *directed line segment from A to B. The measure AB is defined by

$$AB = \begin{cases} |AB|, & \text{if } \overrightarrow{AB} \text{ is in the positive direction,} \\ -|AB|, & \text{if } \overrightarrow{AB} \text{ is in the negative direction.} \end{cases}$$

Suppose, for example, that a horizontal line has positive direction to the right (like the usual x-axis). Then the definition of measure gives $AB = 2$ if B is 2 units to the right of A, and $AB = -2$ if B is 2 units to the left of A.
The following properties are easily deduced from the definition:

(i) For points A and B on a directed line, $AB = -BA$.
(ii) For points A, B and C on a directed line, $AB + BC = AC$ (established by considering the different possible relative positions of A, B and C).

measure (S) A function μ defined on a set S which assigns a non-negative value from the *extended real numbers to each subset with the properties that $\mu(\phi) = 0$, $\mu(A \cup B) = \mu(A) + \mu(B)$ if $A \cap B = \phi$. The standard notion of probability is therefore a special case, where the function takes only values in the interval $[0, 1]$. A function with the above properties but taking negative values also is called a signed measure.

measure of dispersion *See* DISPERSION.

measure of location *See* LOCATION.

measure space A set S, with σ-algebra, X, defined on S and a *measure μ defined on X.

measure theory The area of mathematics relating to the study of measurable sets and functions.

mechanical (mechanistic) A procedure which does not require interpretation or judgemental decisions to be made. For example, the solutions of quadratic equations by the quadratic formula. Such procedures should be relatively easy to program into a computer or programmable calculator.

mechanical advantage *See* MACHINE.

mechanics The area of mathematics relating to the study of the behaviour of systems acted on by forces, whether in equilibrium or in motion. Classical or Newtonian mechanics is restricted to systems where *Newton's laws of motion are adequate—broadly speaking any system where large speeds are not encountered, where the general and special theories of relativity need to be considered.

median (in statistics) Suppose that the observations in a set of numerical data are ranked in ascending order. Then the (sample) median is the middle observation if there are an odd number of observations, and is the average of the two middlemost observations if there are an even number. Thus, if there are n observations, the median is the $\frac{1}{2}(n + 1)$-th in ascending order when n is odd, and is the average of the $\frac{1}{2}n$-th and the $(\frac{1}{2}n + 1)$-th when n is even.

    The median of a continuous *distribution with *probability density function f is the number m such that

$$\int_{-\infty}^{m} f(x) \, dx = 0.5.$$

It is the value that divides the distribution into two halves.

median (of a triangle) A line through a vertex of a triangle and the midpoint of the opposite side. The three medians are concurrent at the *centroid.

median–median regression line The *least squares line of regression is relatively easy to compute and has a sound theoretical foundation to justify its use, but *outliers can have a large effect on the line. The median–median regression line is a more *resistant alternative. To obtain the line, divide the data points into three equal groups by size of x (if there is one extra, make the middle group one larger, and if there are two extra make the low and high groups larger). For the low and high groups, obtain the medians of the x and y values separately, giving one summary point for each group. The median–median regression line is the line joining these two points.

median triangle The triangle founded by joining the midpoints of the three sides of a triangle. It is always similar to the original triangle with each side half the length of the original.

   DEF is the median triangle for triangle *ABC*.

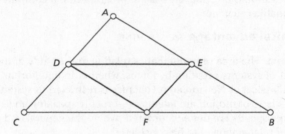

mega- Prefix used with *SI units to denote multiplication by 10^6.

member = ELEMENT.

Menelaus of Alexandria (about AD 100) Greek mathematician whose only surviving work is an important treatise on spherical geometry with applications to astronomy. In this, he was the first to use *great circles extensively and to consider the properties of *spherical triangles. He extended to spherical triangles the theorem about plane triangles that is named after him.

Menelaus' Theorem The following theorem, due to Menelaus of Alexandria:

Theorem

Let L, M and N be points on the sides BC, CA and AB of a triangle (possibly extended). Then L, M and N are collinear if and only if

$$\frac{BL}{LC} \cdot \frac{CM}{MA} \cdot \frac{AN}{NB} = -1.$$

(Note that BC, for example, is considered to be directed from B to C so that LC, for example, is positive if LC is in the same direction as BC and negative if it is in the opposite direction; *see* MEASURE.) *See* CEVA'S THEOREM.

mensuration The measurement or calculation of lengths, angles, areas and volumes associated with geometrical figures.

meridian Strictly, a *great circle on the Earth, assumed to be a sphere, through the north and south poles. However, the meridian through a point P often means not the circle but the semicircle through P with NS as diameter, where N is the north pole and S is the south pole.

meromorphic function A complex-valued function on an *open subset S of the complex plane that is *holomorphic on all of S except for a set of *isolated points which are *poles of the function. Equivalently, a meromorphic function can be defined as the ratio of two *entire functions where the denominator is not the zero function. A meromorphic function can only have a finite order of zeros and isolated poles in its domain, with no *essential singularities.

Mersenne, Marin (1588–1648) French monk, philosopher and mathematician who provided a valuable channel of communication between such contemporaries as *Descartes, *Fermat, *Galileo and *Pascal: 'To inform Mersenne of a discovery is to publish it throughout the whole of Europe.' In an attempt to find a formula for prime numbers, he considered the numbers $2^p - 1$, where p is prime. Not all such numbers are prime: $2^{11} - 1$ is not. Those that are prime are called *Mersenne primes.

Mersenne prime A *prime of the form $2^p - 1$, where p is a prime. The number of known primes of this form is over 30, and keeps increasing as they are discovered by using computers. In 1995, the largest known was $2^{1257787} - 1$. Each Mersenne prime gives rise to an even *perfect number.

(((●))) SEE WEB LINKS

• Home page of the Great Internet Mersenne Prime Search.

mesokurtic *See* KURTOSIS.

method of least squares *See* LEAST SQUARES.

method of maximum likelihood *See* LIKELIHOOD.

method of moments *See* MOMENT (in statistics).

metre In *SI units, the base unit used for measuring length, abbreviated to 'm'. It was once defined as one ten-millionth of the distance from one of the poles to the equator, and later as the length of a platinum–iridium bar kept under specified conditions in Paris. Now it is defined in terms of the wavelength of a certain light wave produced by the krypton-86 atom.

metric (distance function) The function d used to define a *distance between two points in constructing a *metric space, with the following properties:

$$d(x, y) = 0 \text{ if and only if } x = y,$$
$$d(x, y) = d(y, x),$$
$$d(x, y) + d(y, z) \geq d(x, z) \text{ for any points } x, y, z.$$

metric space A set of points together with a *metric (distance function) defined on it.

micro- Prefix used with *SI units to denote multiplication by 10^{-6}.

midpoint Let A and B be points in the plane with Cartesian coordinates (x_1, y_1) and (x_2, y_2). Then, as a special case of the *section formulae, the midpoint of AB has coordinates $\left(\frac{1}{2}(x_1 + x_2) \frac{1}{2}(y_1 + y_2)\right)$. For points A and B in 3-dimensional space, with Cartesian coordinates (x_1, y_1, z_1) and (x_2, y_2, z_2) the midpoint of AB has coordinates

$$\left(\frac{1}{2}(x_1 + x_2), \frac{1}{2}(y_1 + y_2), \frac{1}{2}(z_1 + z_2)\right).$$

If points A and B have position vectors $\mathbf{a}$ and $\mathbf{b}$, the midpoint of AB has position vector $\frac{1}{2}(\mathbf{a} + \mathbf{b})$.

midpoint rule Possibly the simplest structure of numerical integration. If a definite integral $\int_a^b f(x)dx$ is to be calculated using n strips of width $h = \dfrac{b-a}{n}$ then the area under the curve is replaced by a series of rectangles of width h with height equal to the value of the function at its midvalue, i.e. the kth strip has height $f\left(a + \dfrac{(2k-1)}{2} \times h\right)$. Labelling the midpoints $m_1, m_2, \ldots, m_n$ gives the approximation
$\int_a^b f(x)dx = h \times \{f(m_1) + f(m_2) + \cdots + f(m_n)\}.$

midpoint theorem The line joining the midpoints of two sides of a triangle is parallel to the third side, and half its length.

In the figure for the *median triangle *DE*, *EF* and *DF* are parallel to and half the length of *BC*, *AC* and *AB* respectively.

Millennium Prize problems A set of seven classic problems unsolved at the turn of the millennium, which carry a prize of one million US dollars each. They were published by the Clay Mathematics Institute of Cambridge, Massachusetts on 24 May 2000 and the seven problems are:

P versus NP; the Hodge conjecture; the Poincaré conjecture; the *Riemann hypothesis; Yang–Mills existence and mass gap; Navier–Stokes existence and smoothness; the *Birch and Swinnerton-Dyer conjecture.

((⊕)) SEE WEB LINKS

• The Millennium Prize problems home page.

milli- Prefix used with *SI units to denote multiplication by 10^{-3}.

Milnor, John Willard (1931–   ) American mathematician, was awarded the *Fields Medal in 1962 for work in differential topology, and was winner of the *Wolf Prize in 1989. He won the 2011 Abel Prize 'for pioneering discoveries in topology, geometry, and algebra'.

minimax strategy *See* CONSERVATIVE STRATEGY.

Minimax Theorem = FUNDAMENTAL THEOREM OF GAME THEORY.

minimum For minimum value *see* LEAST VALUE. *See also* LOCAL MINIMUM.

minimum capacity (in a network) The minimum flow along an edge. Only of interest when it is more than zero.

minimum connector problem The problem in *graph theory to find the minimum total length of a connected graph. In practical terms this is the situation faced by service companies laying cables or pipes, which is essentially different from constructing any sort of transport links between the points. *Kruskal's algorithm and *Prim's algorithm are two standard approaches to solving this problem.

Minkowski, Hermann (1864–1909) Mathematician, born in Lithuania of German parents, whose concept of space and time as a 4-dimensional entity was a significant contribution to the development of relativity. He also obtained important results in number theory and was a close friend of Hilbert.

minor arc *See* ARC.

minor axis *See* ELLIPSE.

minute (angular measure) *See* DEGREE (angular measure).

mirror-image *See* REFLECTION (of the plane).

mixed fraction *See* FRACTION.

mixed numbers Numbers which have a whole number part and a fractional number part. For example, $3\frac{2}{5}$ is a mixed number.

mixed strategy In a *matrix game, if a player chooses the different rows (or columns) with certain probabilities, the player is using a mixed strategy. If the game is given by an $m \times n$ matrix, a mixed strategy for one player, R, is given by an m-tuple $\mathbf{x}$, where $\mathbf{x} = (x_1, x_2, \ldots, x_m)$, in which $x_i \geq 0$, for $i = 1, 2, \ldots, m$, and $x_1 + x_2 + \ldots + x_m = 1$. Here x_i is the probability that R chooses the i-th row. For example, if $\mathbf{x} = \left(\frac{1}{3}, \frac{1}{3}, \frac{1}{3}\right)$, it means that R chooses each row with equal probability or, loosely, that R chooses each row equally often. If $\mathbf{x} = \left(0, \frac{1}{4}, \frac{3}{4}\right)$, then R never chooses the first row, and chooses the third row three times as often as the second row. A mixed strategy for the other player, C, is defined similarly.

Möbius, August Ferdinand (1790–1868) German astronomer and mathematician, whose work in mathematics was mainly in geometry and topology. One of his contributions was his conception of the *Möbius band.

Möbius band (strip) A continuous flat loop with one twist in it. Between any two points on it, a continuous line can be drawn on the surface without crossing an edge. Thus the band has only one surface and likewise only one edge.

(())) SEE WEB LINKS

• The Möbius band in art.

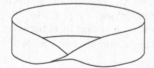

modal interval The interval which has the greatest frequency in a set of grouped data. As with the *mode there is no guarantee that there will be a unique modal interval.

mode The most frequently occurring observation in a sample or, for grouped data, the group with the highest frequency. For a continuous random variable, a value at which the probability density function has a local maximum is called a mode.

model *See* MATHEMATICAL MODEL.

model theory The branch of logic concerned with mathematical structures (for example groups, rings, or graphs) with close links to abstract algebra.

modulo *n*, addition and **multiplication** The word 'modulo' means 'to the modulus'. For any positive integer *n*, let *S* be the *complete set of residues $\{0, 1, 2, \ldots, n-1\}$. Then addition modulo *n* on *S* is defined as follows. For *a* and *b* in *S*, take the usual sum of *a* and *b* as integers, and let *r* be the element of *S* to which the result is *congruent (modulo *n*); the sum $a + b$ (mod *n*) is equal to *r*. Similarly, multiplication modulo *n* is defined by taking *ab* (mod *n*) to be equal to *s*, where *s* is the element of *S* to which the usual product of *a* and *b* is congruent (modulo *n*). For example, addition and multiplication modulo 5 are given by the following tables:

+	0	1	2	3	4
0	0	1	2	3	4
1	1	2	3	4	0
2	2	3	4	0	1
3	3	4	0	1	2
4	4	0	1	2	3

×	0	1	2	3	4
0	0	0	0	0	0
1	0	1	2	3	4
2	0	2	4	1	3
3	0	3	1	4	2
4	0	4	3	2	1

modulus of a complex number (moduli) If *z* is a *complex number and $z = x + yi$, the modulus of *z*, denoted by $|z|$ (read as 'mod *z*'), is equal to $\sqrt{x^2 + y^2}$. (As always, the sign ✓ means the non-negative square root.) If *z* is represented by the point *P* in the *complex plane, the modulus of *z* equals the distance $|OP|$. Thus $|z| = r$, where (r, θ) are the polar coordinates of *P*. If *z* is real, the modulus of *z* equals the *absolute value of the real number, so the two uses of the same notation | | are consistent, and the term 'modulus' may be used for 'absolute value'.

modulus of a congruence *See* CONGRUENCE.

modulus of elasticity A parameter, usually denoted by λ associated with the material of which an elastic string is made. It occurs in the constant of proportionality in *Hooke's law.

modus ponens (in logic) The rule of inference which states that if a *conditional statement and the *antecedent contained in it are both true, then the *consequent statement in it must also be true. Also known as the rule of detachment.

modus tollens The rule of inference which states that if a in *conditional statement and the *negation of the *consequent contained in it are both true, then the *antecedent statement it must be false.

moment (in mechanics) A means of measuring the turning effect of a force about a point. For a system of coplanar forces, the moment of one of the forces **F** about any point A in the plane can be defined as the product of the magnitude of **F** and the distance from A to the line of action of **F**, and is considered to be acting either clockwise or anticlockwise. For example, suppose that forces with magnitudes F_1 and F_2 act at B and C, as shown in the figure. The moment of the first force about A is $F_1 d_1$ clockwise, and the moment of the second force about A is $F_2 d_2$ anticlockwise. The *principle of moments considers when a system of coplanar forces produces a state of equilibrium.

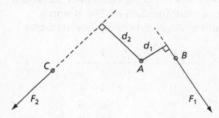

However, a better approach to measuring the turning effect is to define the moment of a force about a point as a vector, as follows. The moment of the force **F**, acting at a point B, about the point A is the vector $(\mathbf{r}_B - \mathbf{r}_A) \times \mathbf{F}$, where this involves a *vector product. The use of vectors not only eliminates the need to distinguish between clockwise and anticlockwise directions, but facilitates the measuring of the turning effects of non-coplanar forces acting on a 3-dimensional body.

Similarly, for a particle P with position vector **r** and *linear momentum **p**, the moment of the linear momentum of P about the point A is the vector $(\mathbf{r} - \mathbf{r}_A) \times \mathbf{p}$. This is the *angular momentum of the particle P about the point A.

Suppose that a *couple consists of a force **F** acting at B and a force $- \mathbf{F}$ acting at C. Let $\mathbf{r}_B$ and $\mathbf{r}_C$ be the position vectors of B and C, and let $\mathbf{r}_A$ be the position vector of the point A. The moment of the couple about A is equal to $(\mathbf{r}_B - \mathbf{r}_A) \times \mathbf{F} + (\mathbf{r}_C - \mathbf{r}_A) \times (-\mathbf{F}) = (\mathbf{r}_B - \mathbf{r}_C) \times \mathbf{F}$, which is independent of the position of A.

moment (in statistics) For a set of observations $x_1, x_2, \ldots, x_n$, the j-th (sample) moment about p is equal to

$$\frac{\sum (x_1 - p)^j}{n}.$$

For a random variable X, the j-th (population) moment about p is equal to $E((X-p)^j)$. (*See* EXPECTED VALUE.) The first moment about 0 is the *mean. For the second moment about the mean, *see* VARIANCE.

   Suppose that a sample is taken from a population with k unknown parameters. In the method of moments, the first k population moments are equated to the first k sample moments to find k equations in k unknowns. These can be solved to find the moment estimates of the parameters.

moment estimate *See* MOMENT (in statistics).

moment estimator The *estimator found by the method of moments. *See* MOMENT (in statistics).

moment generating function $M(t) = E(e^{tX})$ is the moment generating function (mgf) for the random variable X. When $M(t)$ exists then, since $e^{tX} = \sum_{r=0}^{\infty} \dfrac{t^r X^r}{r!}$ and E is an *additive function, it follows that the coefficient of $\dfrac{t^r}{r!}$ will be $E(X^r)$, allowing the moments of X to be recovered from (or generated by) knowledge of the function.

moment of inertia A quantity relating to a rigid body and a given axis, derived from the way in which the mass of the rigid body is distributed relative to the axis. It arises in the calculation of the *kinetic energy and the *angular momentum of the rigid body. With Cartesian coordinates, let I_{xx}, I_{yy} and I_{zz} denote the moments of inertia of the rigid body about the x-axis, the y-axis and the z-axis. Suppose then that a rigid body consists of a system of n particles $P_1, P_2, \ldots, P_n$, where P_i has mass m_i and position vector $\mathbf{r}_i$, and $\mathbf{r}_i = x_i\mathbf{i} + y_i\mathbf{j} + z_i\mathbf{k}$. Then

$$I_{xx} = \sum_{i=1}^{n} m_i(y_i^2 + z_i^2), \qquad I_{yy} = \sum_{i=1}^{n} m_i(z_i^2 + x_i^2),$$

$$I_{zz} = \sum_{i=1}^{n} m_i(x_i^2 + y_i^2).$$

For solid bodies, a moment of inertia is given by an integral. For example,

$$I_{xx} = \int_V \rho(\mathbf{r})(y^2 + z^2)\, dV, \qquad I_{yy} = \int_V \rho(\mathbf{r})(z^2 + x^2)\, dV,$$

$$I_{zz} = \int_V \rho(\mathbf{r})(x^2 + y^2)\, dV,$$

where $\rho(\mathbf{r})$ is the density at the point with position vector $\mathbf{r}$ and $\mathbf{r} = x\mathbf{i} + y\mathbf{j} + z\mathbf{k}$.

For a uniform rod of length l and mass m, the moment of inertia about an axis running along the length of the rod is zero. The moment of inertia about an axis perpendicular to the rod, through the centre of mass, equals $\frac{1}{12}ml^2$. The moment of inertia about an axis perpendicular to the rod, through one end of the rod, equals $\frac{1}{3}ml^2$.

For a uniform circular plate of radius a and mass m, the moment of inertia about an axis perpendicular to the plate, through the centre of the circle, equals $\frac{1}{2}ma^2$.

For a uniform rectangular plate of length a, width b and mass m, the moment of inertia about an axis along a side of length a equals $\frac{1}{3}mb^2$.

Moments of inertia about other axes may be calculated by using the *parallel axis theorem and the *perpendicular axis theorem.

moment of momentum = ANGULAR MOMENTUM.

momentum *See* LINEAR MOMENTUM and ANGULAR MOMENTUM. Often 'momentum' is used to mean linear momentum.

Monge, Gaspard (1746–1818) French mathematician who was a high-ranking minister in Napoleon's government and influential in the founding of the École Polytechnique in Paris. The courses in geometry that he taught there were concerned to a great extent with the subjects of his research: he was the originator of what is called descriptive geometry and the founder of modern analytical solid geometry.

monic polynomial A *polynomial in one variable in which the coefficient of the highest power is 1. The polynomial $a_n x^n + a_{n-1} x^{n-1} + \ldots + a_1 x + a_0$ is monic if $a_n = 1$.

monotonic function A *real function f is monotonic on or over an interval I if it is either increasing on I ($f(x_1) \leq f(x_2)$ whenever $x_1 < x_2$) or decreasing on I ($f(x_1) \geq f(x_2)$ whenever $x_1 < x_2$). Also, f is strictly monotonic if it is either strictly increasing or strictly decreasing.

monotonic sequence A sequence $a_1, a_2, a_3, \ldots$ is *monotonic if it is either increasing ($a_i \leq a_{i+1}$ for all i) or decreasing ($a_i \geq a_{i+1}$ for all i), and strictly monotonic if it is either strictly increasing or strictly decreasing.

Monte Carlo methods The approximate solution of a problem using repeated sampling experiments, and observing the proportion of times some property is satisfied. Estimates of complex probabilities may be computed by running an electronic simulation of the context repeatedly and using the *relative frequency. However, it can also be used in calculating complex or multiple integrals.

Morley's Theorem The following theorem due to Frank Morley (1860–1937):

Theorem

In any triangle, the points of intersection of adjacent trisectors of the angles form the vertices of an equilateral triangle.

It is remarkable that this elegant fact of *Euclidean geometry was not discovered until so long after Euclid's time.

morphism An abstraction of a function or mapping between two spaces.

moving average For a sequence of observations $x_1, x_2, x_3, \ldots$, the moving average of order n is the sequence of arithmetic means

$$\frac{x_1 + \cdots + x_n}{n}, \qquad \frac{x_2 + \cdots + x_{n+1}}{n}, \qquad \frac{x_3 + \cdots + x_{n+2}}{n}, \ldots$$

A weighted moving average such as

$$\frac{x_1 + 2x_2 + x_3}{4}, \qquad \frac{x_2 + 2x_3 + x_4}{4}, \qquad \frac{x_3 + 2x_4 + x_5}{4}, \ldots$$

may also be used. Its main purpose is to offer a view of the underlying behaviour of volatile series, especially those which have periodic variation.

(((■))) SEE WEB LINKS

• Illustrations of the moving average from financial markets.

Müller, Johann *See* REGIOMONTANUS.

multi- Prefix denoting many, for example in multicollinearity, multinomial.

multicollinearity In statistics, when two or more of the explanatory variables in a multiple regression analysis are very strongly correlated.

multinomial The general description for an algebraic expression that is a sum of two or more terms. However, where powers of variables, such as in $3x^4 + x^2 - x + 1$ are involved, 'polynomial' is more common, and for only two terms 'binomial' is the usual description.

multinomial coefficient The number $\begin{pmatrix} n \\ n_1 \ n_2 \ \ldots \ n_k \end{pmatrix} = \dfrac{n!}{n_1! n_2! \cdots n_k!}$

when $\sum_{i=1}^{k} n_i = n$. This is the coefficient of $x_1^{n_1} x_2^{n_2} \ldots x_k^{n_k}$ in the expansion of $(x_1 + x_2 + \ldots x_k)^n$. It also gives the number of ways of choosing $n_i, i = 1, 2, \ldots, k$ of each of k types when choosing a total of n objects, without regard to the order in which they are chosen.

multinomial distribution The distribution of the multinomial random variable which is the generalized form of the binomial distribution to allow more than two outcomes in each of the individual trials. So if $\{Y_i\}, i = 1, 2, \ldots, n$, is a series of independent and identically distributed random variables with $\Pr\{Y_i = x_j\} = p_j$ for $j = 1, 2, \ldots, k$ with

$\sum p_j = 1$ then the probabilities of the possible outcomes are given by the coefficients in the expansion of $(p_1x_1 + p_2x_2 + \ldots + p_kx_k)^n$.

multinomial theorem A generalization of the binomial theorem, giving the expansions of positive integral powers of a *multinomial expression $(x_1 + x_2 + \cdots + x_k)^n = \sum \left(\dfrac{n!}{n_1!n_2! \ldots n_k!} x_1^{n_1} x_2^{n_2} \ldots x_k^{n_k} \right)$ where the sum is over all combinations of $n_1, n_2, \ldots, n_k$ which total n.

multiple *See* DIVIDES.

multiple edges *See* GRAPH.

multiple integral (repeated integral) An integral involving two or more integrations with respect to different variables, where the integrations are carried out sequentially and variables other than the one being integrated with respect to are treated as constants. A double integral is the special case where two integrations are involved. Multiple integrals can be definite or indefinite. In the indefinite case, the first constant of integration will then integrate to be a function of the second variable etc. Where a surface is defined by a function $z = f(x, y)$, the volume under the surface can be obtained by evaluating $\int \int f(x, y) dx\, dy$. Strictly this should be written as $\int \{ \int f(x, y) dx \} dy$ but the brackets are usually omitted. For example, to find the volume under $z = x + 3y + 5$ in the ranges $0 \le x \le 2$, $0 \le y \le 3$ the required integral would be

$$\int_0^3 \int_0^2 (x + 3y + 5)\ dx\ dy = \int_0^3 \left[\frac{x^2}{2} + 3xy + 5 \right]_0^2 dy$$

$$= \int_0^3 (6y + 7) dy = [3y^2 + 7y]_0^3 = 48.$$

multiple precision *See* PRECISION.

multiple regression *See* REGRESSION.

multiple root *See* ROOT.

multiplication Generally, a *binary operation in which the two entities, which can both be numbers, matrices, vectors etc. or can be two entities of different types, are combined by a specified rule to form a product.

multiplication (of complex numbers) For complex numbers z_1 and z_2, given by $z_1 = a + bi$ and $z_2 = c + di$, the product is defined by $z_1z_2 = (ac - bd) + (ad + bc)i$. In some circumstances it may be more convenient to write z_1 and z_2 in *polar form, thus:

$$z_1 = r_1 e^{i\theta_1} = r_1(\cos \theta_1 + i \sin \theta_1),$$
$$z_2 = r_2 e^{i\theta_2} = r_2(\cos \theta_2 + i \sin \theta_2).$$

Then $z_1 z_2 = r_1 r_2 e^{i(\theta_1 + \theta_2)} = r_1 r_2(\cos (\theta_1 + \theta_2) + i \sin (\theta_1 + \theta_2))$. Thus 'you multiply the moduli and add the arguments'.

multiplication (of fractions) Fractions are multiplied by multiplying the numerators and denominators separately, i.e. $\dfrac{a}{b} \times \dfrac{c}{d} = \dfrac{ac}{bd}$. If mixed numbers are involved, they can be converted into improper fractions before multiplication.

multiplication (of numbers) The multiplication of integers is defined by repeated addition, so $3 \times 2 = 2 + 2 = 6$ (and $= 3 + 3$). Multiplication of rationals can then be defined in terms of multiplication of integers and its inverse, division.

multiplication (of matrices) Suppose that matrices **A** and **B** are *conformable for multiplication, so that **A** has order $m \times n$ and **B** has order $n \times p$. Let $\mathbf{A} = [a_{ij}]$ and $\mathbf{B} = [b_{ij}]$. Then (matrix) multiplication is defined by taking the product **AB** to be the $m \times p$ matrix **C**, where $\mathbf{C} = [c_{ij}]$ and

$$c_{ij} = a_{i1}b_{1j} + a_{i2}b_{2j} + \cdots + a_{in}b_{nj} = \sum_{r=1}^{n} a_{ir}b_{rj}.$$

The product **AB** is not defined if **A** and **B** are not conformable for multiplication. Matrix multiplication is not *commutative; for example, if

$$\mathbf{A} = \begin{bmatrix} 0 & 1 \\ 0 & 0 \end{bmatrix} \quad \text{and} \quad \mathbf{B} = \begin{bmatrix} 1 & 0 \\ 0 & 0 \end{bmatrix},$$

then $\mathbf{AB} \neq \mathbf{BA}$. Moreover, it is not true that $\mathbf{AB} = \mathbf{O}$ implies that either $\mathbf{A} = \mathbf{O}$ or $\mathbf{B} = \mathbf{O}$, as the same example of **A** and **B** shows. However, matrix multiplication is associative: $\mathbf{A}(\mathbf{BC}) = (\mathbf{AB})\mathbf{C}$, and the distributive laws hold: $\mathbf{A}(\mathbf{B} + \mathbf{C}) = \mathbf{AB} + \mathbf{AC}$ and $(\mathbf{A} + \mathbf{B})\mathbf{C} = \mathbf{AC} + \mathbf{BC}$. Strictly, in each case, what should be said is that, if **A**, **B** and **C** are such that one side of the equation exists, then so does the other side and the two sides are equal.

multiplication (of polynomials) Polynomials are multiplied out by using the *distributive law of multiplication over addition, so each term in the first polynomial is multiplied by each term in the second. For example, $(3x^3 + 5x^2 - 3x + 2)(x^2 - 2x + 3) = 3x^3(x^2 - 2x + 3) + 5x^2(x^2 - 2x + 3) - 3x(x^2 - 2x + 3) + 3(x^2 - 2x + 3)$ and then each of these brackets can be expanded before collecting like terms together.

multiplication modulo *n* *See* MODULO *N*, ADDITION AND
MULTIPLICATION.

multiplication sign The symbol × used to denote multiplication of
number or any other binary operation with similar properties. Where
characters and/or brackets are involved rather than just numbers, it
is common to use a dot, or to use no symbol at all. So 3×5, $5(3+6)$,
$4x^2y^3$, $x \cdot y$ all contain instructions to multiply quantities together.

multiplicative group A *group in which the operation is called
multiplication, usually denoted by '.' but with the product $a. b$ normally
written as ab.

multiplicative identity An identity under a multiplication operation,
so for real and complex numbers 1 is the multiplicative identity and for
3×3 matrices it is $\begin{pmatrix} 1 & 0 & 0 \\ 0 & 1 & 0 \\ 0 & 0 & 1 \end{pmatrix}$.

multiplicative inverse *See* INVERSE ELEMENT.

multiplicity *See* ROOT.

multiplying factor (in differential equations) = INTEGRATING FACTOR.
See LINEAR FIRST-ORDER DIFFERENTIAL EQUATION.

multiply out (expressions) To expand an expression which is in a
contracted form. For example, $(3x+1)(2x-3) = 6x^2 - 7x - 3$.

multivariate Relating to more than two random variables. *See* JOINT
CUMULATIVE DISTRIBUTION FUNCTION, JOINT DISTRIBUTION, JOINT
PROBABILITY DENSITY FUNCTION and JOINT PROBABILITY MASS FUNCTION.

mutually disjoint = PAIRWISE DISJOINT.

mutually exclusive events Events A and B are mutually exclusive if
A and B cannot both occur; that is, if $A \cap B = \emptyset$. For mutually exclusive
events, the probability that either one event or the other occurs is given by
the addition law, $\Pr(A \cap B) = \Pr(A) + \Pr(B)$.

For example, when a die is thrown, the probability of obtaining a 'one' is
$\frac{1}{6}$, and the probability of obtaining a 'two' is $\frac{1}{6}$. These events are mutually
exclusive. So the probability of obtaining a 'one' or a 'two' is equal to
$\frac{1}{6} + \frac{1}{6} = \frac{1}{3}$.

mutually exclusive sets Two sets A, B are mutually exclusive if their
intersection is the empty set i.e. $A \cap B = \emptyset$.

mutually prime = RELATIVELY PRIME.

n Abbreviation for *nano.

N Symbol for *newton.

n- Prefix when denoting a finite but unspecified number in a description. For example, an n-gon is a polygon with n sides.

naïve set theory The area of mathematics which attempts to formalize the nature of a set using a minimal number of independent *axioms. Bertrand *Russell worked in this area, but discovered that paradoxes appear, such as *Russell's paradox, and the less ambitious *axiomatic set theory must be used.

nand the logic gate in combinatorial circuits used to represent the connector *not and.

nano- Prefix used with *SI units to denote multiplication by 10^{-9}.

Napier, John (1550–1617) Scottish mathematician responsible for the publication in 1614 of the first tables of logarithms. His mathematical interests, pursued in spare time from church and state affairs, were in spherical trigonometry and, notably, in computation. His logarithms were defined geometrically rather than in terms of a base. In effect, he took the logarithm of x to be y, with

$$x = 10^7 \times \left(1 - \frac{1}{10^7}\right)^y.$$

Napierian logarithm *See* LOGARITHM.

Napier's bones An invention by Napier for carrying out multiplication. It consisted of a number of rods made of bone or ivory, which were marked with numbers from the multiplication tables. Multiplication using the rods involved some addition of digits and was very similar to the familiar method of long multiplication. Logarithms were not involved at all.

Nash, John Forbes (1928–2015) American mathematician who won the Nobel Prize for Economic Science in 1994 for his work in non-cooperative game theory which was published in 1949. Despite the importance of this work in economics, and in diplomatic and military strategy to which Nash applied it for the RAND Corporation during the Cold War, Nash viewed himself as a pure mathematician. He made important contributions in the theory of real algebraic varieties, Riemann geometry and manifolds, and particularly on parabolic and elliptical equations. He was thought to be in contention for the *Fields Medal in 1958, but had not yet published his work on parabolic and elliptical equations and was not successful. By the time the 1962 awards were being made, Nash's long battle with schizophrenia was dominating his life. The book and film *A Beautiful Mind* catalogued the story of his brilliance and his struggle with mental illness. He was awarded the Abel Prize for 2015 along with Louis Nirenberg 'for striking and seminal contributions to the theory of nonlinear partial differential equations and its applications to geometric analysis.'

National Centre for Excellence in the Teaching of Mathematics (NCETM) Set up following the publication of Adrian *Smith's inquiry into the state of post-14 mathematics education in the UK to provide effective strategic leadership for mathematics-specific professional development for teachers.

((⊕)) SEE WEB LINKS
• The NCETM home page.

natural logarithm *See* LOGARITHM.

natural number One of the numbers 1, 2, 3, ... Some authors also include 0. The set of natural numbers is often denoted by **N**.

nautical mile A unit of length used in navigation. Now standardized as 1852 metres, or 1.15 miles it was originally defined as the average length of 1 minute of *latitude.

n-cube A *simple graph, denoted by Q_n, whose vertices and edges correspond to the vertices and edges of an n-dimensional *hypercube.

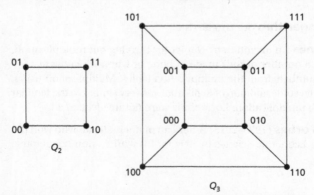

Q_2

Q_3

Thus, there are 2^n vertices that can be labelled with the *binary words of length n, and there is an edge between two vertices if they are labelled with words that differ in exactly one digit. The graphs Q_2 and Q_3 are shown in the figure.

n-dimensional space Points in the plane can be specified by means of Cartesian coordinates (x, y), and points in 3-dimensional space may be assigned Cartesian coordinates (x, y, z). It can be useful, similarly, to consider space of n dimensions, for general values of n, by defining a point to be given by n coordinates. Many familiar ideas from geometry of two and three dimensions can be generalized to space of higher dimensions. For example, if the point P has coordinates $(x_1, x_2, \ldots, x_n)$, and the point Q has coordinates $(y_1, y_2, \ldots, y_n)$, then the distance PQ can be defined to be equal to

$$\sqrt{(x_1 - y_1)^2 + (x_2 - y_2)^2 + \cdots + (x_n - y_n)^2}.$$

Straight lines can be defined, and so can the notion of angle between them. The set of points whose coordinates satisfy a linear equation $a_1x_1 + a_2x_2 + \cdots + a_nx_n = b$ is called a hyperplane, which divides the space into two *half-spaces, as a plane does in three dimensions. There are other so-called subspaces of different dimensions, a straight line being a subspace of dimension 1 and a hyperplane being a subspace of dimension $n-1$. Other curves and surfaces can also be considered. In n dimensions, the generalization of a square in two dimensions and a cube in three dimensions is a *hypercube.

nearest point A point which is not in a subset A of a *metric space that has the least distance to any point in A.

necessary and sufficient condition If statement B is true whenever A is true, and false when A is false then each is a necessary and sufficient condition for the other. For example, for a number to be divisible by 10 it is necessary and sufficient that the number be both even and divisible by 5.

necessary condition A condition which is required to be true for a statement to be true. For example, for a number to be divisible by 10 it is necessary that the number be even. Note that if A is a necessary condition for B, B is a sufficient condition for A, and vice versa, so in the above example knowing that a number is divisible by 10 is sufficient to know that the number is even.

needle problem *See* BUFFON'S NEEDLE.

negation If p is a statement, then the statement 'not p', denoted by $\neg p$, is the negation of p. It states, in some suitable wording, the opposite of p. For example, if p is 'It is raining', then $\neg p$ is 'It is not raining'; if p is '2 is not an integer', then $\neg p$ is 'It is not the case that 2 is not an integer' or, in other words, '2 is an integer'. If p is true then $\neg p$ is false, and vice versa. So the *truth table of $\neg p$ is:

p	$\neg p$
T	F
F	T

negative *See* INVERSE ELEMENT.

negative (of a matrix) The negative of an $m \times n$ matrix **A**, where $\mathbf{A} = [a_{ij}]$, is the $m \times n$ matrix **C**, where $\mathbf{C} = [c_{ij}]$ and $c_{ij} = -a_{ij}$. It is denoted by $-\mathbf{A}$.

negative (of a vector) Given a vector **a**, let $\overrightarrow{AB}$ be a *directed line segment representing **a**. The negative of **a**, denoted by $-\mathbf{a}$, is the vector represented by $\overrightarrow{BA}$ The following properties hold, for all **a**:

(i) $\mathbf{a} + (-\mathbf{a}) = (-\mathbf{a}) + \mathbf{a} = 0$, where **0** is the zero vector,

(ii) $-(-\mathbf{a}) = \mathbf{a}$.

negative binomial distribution The generalization of the *geometric distribution to the waiting time for the k-th success in a sequence of independent experiments, all with the same probabilities of success. The probability mass function is given by

$\Pr\{X = r\} = \binom{r-1}{k-1} p^k (1-p)^{r-k}$ and can be thought of as conditioning

on observing $k-1$ successes in the first $r-1$ trials. By the binomial

distribution this has probability $\binom{r-1}{k-1} p^{k-1}(1-p)^{(r-1)-(k-1)}$ or

$\binom{r-1}{k-1} p^{k-1}(1-p)^{(r-k)}$. To get the required event, the r-th trial has to be

a success, which occurs with probability p.

negative correlation *See* CORRELATION.

negative direction *See* DIRECTED LINE.

negative number A real number which is less than zero.

neighbourhood On the *real line, a neighbourhood of the real number a is an open interval $(a-\delta, a+\delta)$, where $\delta > 0$, with its centre at a.

More generally, in a *Euclidean or *metric space it is the *open set of all points whose *distance d from a fixed point a is strictly less than a specified value, i.e. the set of points x for which $d(x, a) < \delta$.

nested multiplication A polynomial $f(x)$, such as $2x^3 - 7x^2 + 5x + 11$, can be evaluated for $x = h$ by calculating h^2 and h^3, multiplying by appropriate coefficients and summing the terms. But fewer operations are required if the polynomial is rewritten as $((2x-7)x+5)x+11$ and then evaluated. This method, known as nested multiplication, is therefore more efficient and is recommended when evaluation is to be carried out by hand or by computer. A polynomial $a_5x^5 + a_4x^4 + a_3x^3 + a_2x^2 + a_1x + a_0$ of degree 5, for example, would be rewritten for this purpose as $((((a_5x + a_4)x + a_3)x + a_2)x + a_1)x + a_0$. The steps involved in this evaluation correspond exactly to the calculations that are made in the process of *synthetic division, in which the remainder gives $f(h)$.

nested sets An ordered collection of sets (or intervals) for which each set (or interval) is contained in the preceding set (or interval) and the *diameter of the set (or length of the interval) tends to zero as the number of sets (or intervals) tends to infinity.

net Remaining after all deductions have been made. For example, the net income from the sale of a house will be the sale price less fees paid to lawyers, estate agents, etc. involved in the sale.

net (of a solid) A plane figure which can be folded (without cutting) to construct a polygon. There is likely to be more than one net for any polyhedron, but some plane figures which look like nets do not construct the required polyhedron without cutting.

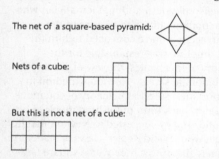

The net of a square-based pyramid:

Nets of a cube:

But this is not a net of a cube:

network A *digraph in which every arc is assigned a *weight (normally some non-negative number). In some applications, something may be thought of as flowing or being transported between the vertices of a network, with the weight of each arc giving its 'capacity'. In some other cases, the vertices of a network may represent steps in a process and the weight of the arc joining u and v may give the time that must elapse between step u and step v.

network flow A set of non-negative values assigned to each arc of a *network which does not exceed the *capacity of that arc, and for which the total amount entering and leaving each vertex is the same. Many optimization problems can be characterized by networks in which it is required to maximize the flow.

Neumann, John Von *See* VON NEUMANN.

Neumann condition A *boundary condition for a partial differential equation.

neutral element An element e is a neutral element for a *binary operation $\circ$ on a set S if, for all a in S, $a \circ e = e \circ a = a$. If the operation is called multiplication, a neutral element is normally called an identity element and may be denoted by 1. If the operation is called addition, such an element is normally denoted by 0, and is often called a zero element. However, there is a case for preferring the term 'neutral element', as there is an alternative definition for the term 'zero element' (*see* ZERO ELEMENT).

neutral equilibrium *See* EQUILIBRIUM.

newton The SI unit of *force, abbreviated to 'N'. One newton is the force required to give a mass of one kilogram an acceleration of one metre per second per second.

Newton, Isaac (1642–1727) English physicist and mathematician who dominated and revolutionized the mathematics and physics of the 17th century. *Gauss and *Einstein seem to have conceded to him the place of top mathematician of all time. He was responsible for the essentials of the calculus, the theory of mechanics, the law of gravity, the theory of planetary motion, the *binomial series, *Newton's method in numerical analysis, and many important results in the theory of equations. A morbid dislike of criticism held him back from publishing much of his work. For example, in 1684, Edmond *Halley suggested to Newton that he investigate the law of attraction that would yield Kepler's laws of planetary motion. Newton replied immediately that it was the inverse square law— he had worked that out many years earlier. Rather startled by this, Halley set to work to persuade Newton to publish his results. This Newton eventually did in the form of his *Principia*, published in 1687, perhaps the single most powerful work in the long history of mathematics.

Newton quotient The quotient $(f(a+h)-f(a))/h$, which is used to determine whether f is *differentiable at a, and if so to find the *derivative.

Newton–Raphson method = NEWTON'S METHOD.

Newton's interpolating polynomial The polynomial of degree n which interpolates $n+1$ evenly spaced data points (x_0, y_0), (x_1, y_1), ..., (x_n, y_n) where $x_k = x_0 + kh$ is given by

$$f(x) = y_0 + \frac{(x-x_0)}{h}\Delta y_0 + \frac{(x-x_0)(x-x_1)}{2!h^2}$$

$$\Delta^2 y_0 + \cdots + \frac{(x-x_0)(x-x_1)\cdots(x-x_n)}{n!h^n}\Delta^n y_0 \text{ where } \Delta y_0 = y_1 - y_0,$$

$\Delta^2 y_0 = \Delta y_1 - \Delta y_0$ etc. which are obtained by constructing a forward difference table.

Newton's interpolation formula *See* GREGORY–NEWTON FORWARD DIFFERENCE.

Newton's inverse square law of gravitation *See* INVERSE SQUARE LAW OF GRAVITATION.

Newton's law of restitution *See* COEFFICIENT OF RESTITUTION.

Newton's laws of motion Three laws of motion, applicable to particles of constant mass, formulated by *Newton in his *Principia*. They can be stated as follows:

(i) A particle continues to be at rest or moving with constant velocity unless the total force acting on the particle is non-zero.

(ii) The rate of change of linear momentum of a particle is proportional to the total force acting on the particle.

(iii) Whenever a particle exerts a force on a second particle, the second particle exerts an equal and opposite force on the first.

The first law can be interpreted as saying that a force is required to alter the velocity of a particle. The linear momentum **p** of a particle, mentioned in the second law, is given by $\mathbf{p} = m\mathbf{v}$, where m is the mass and **v** is the velocity. So, when m is constant, $d\mathbf{p}/dt = m(d\mathbf{v}/dt) = m\mathbf{a}$, where **a** is the acceleration. The second law is therefore often stated as $m\mathbf{a} = \mathbf{F}$, where **F** is the total force acting on the particle. The third law states that 'to every action there is an equal and opposite reaction'.

Newton's method The following method of finding successive approximations to a root of an equation $f(x) = 0$. Suppose that x_0 is a first approximation, known to be quite close to a root. If the root is in fact $x_0 + h$, where h is 'small', taking the first two terms of the Taylor series gives $f(x_0 + h) \approx f(x_0) + hf'(x_0)$. Since $f(x_0 + h) = 0$, it follows that $h \approx -f(x_0)/f'(x_0)$. Thus x_1, given by

$$x_1 = x_0 - \frac{f(x_0)}{f'(x_0)},$$

is likely to be a better approximation to the root. To see the geometrical significance of the method, suppose that P_0 is the point $(x_0, f(x_0))$ on the curve $y = f(x)$, as shown in the left-hand figure. The value x_1 is given by the point at which the tangent to the curve at P_0 meets the x-axis. It may be possible to repeat the process to obtain successive approximations $x_0, x_1, x_2, \ldots$, where

$$x_{n+1} = x_n - \frac{f(x_n)}{f'(x_n)}.$$

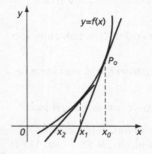
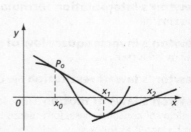

These may be successively better approximations to the root as required, but in the right-hand figure is shown the graph of a function, with a value x_0 close to a root, for which x_1 and x_2 do not give successively better approximations.

Neyman, Jerzy (1894–1981) Statistician known for his work in the theories of *hypothesis testing and *estimation. He introduced the notion of the *confidence interval. Born in Romania, he spent some years at University College, London before going to the University of California at Berkeley in 1938.

nine-point circle Consider a triangle with vertices A, B and C. Let A', B' and C' be the midpoints of the three sides. Let D, E and F be the feet of the perpendiculars from A, B and C, respectively, to the opposite sides. Then AD, BE and CF are concurrent at the *orthocentre H. Let P, Q and R be the midpoints of AH, BH and CH. The points A', B', C', D, E, F, P, Q and R lie on a circle called the nine-point circle. The centre of this circle lies on the *Euler line of the triangle at the midpoint of OH, where O is the *circumcentre. Feuerbach's Theorem states that the nine-point circle touches the *incircle and the three *excircles of the triangle. This was proved in 1822 by Karl Wilhelm Feuerbach (1800–34).

node *See* GRAPH and TREE.

Noether, Amalie ('Emmy') (1882–1935) German mathematician, known for her highly creative work in the theory of *rings, non-commutative algebras and other areas of *abstract algebra. She was responsible for the growth of an active group of algebraists at Göttingen in the period up to 1933.

noise A descriptive name for *random error or variation in observations which is not explained by the model.

nominal *See* DATA.

nominal scale A set of categories of data which cannot be ordered, but by which data can be classified, for example by colour or sex. *Compare with* ORDINAL.

nonagon = ENNEAGON.

non-basic variables *See* BASIC SOLUTION.

non-denumerable An infinite set which cannot be put in one-to-one correspondence with the set of *natural numbers. For example, the set of real numbers between zero and one is non-denumerable, and contains more numbers than all the integers, or even all the rational numbers, both of which are *denumerable.

non-Euclidean geometry After unsuccessful attempts had been made at proving that the *parallel postulate could be deduced from the other postulates of Euclid's, the matter was settled by the discovery of non-Euclidean geometries by *Gauss, *Lobachevsky and *Bolyai. In these, all Euclid's axioms hold except the parallel postulate. In the so-called hyperbolic geometry, given a point not on a given line, there are at least two lines through the point parallel to the line. In elliptic geometry, given a point not on a given line, there are no lines through the point parallel to the line.

non-homogeneous linear differential equation *See* LINEAR DIFFERENTIAL EQUATION WITH CONSTANT COEFFICIENTS.

non-homogeneous set of linear equations A set of m linear equations in n unknowns $x_1, x_2, \ldots, x_n$ that has the form

$$a_{11}x_1 + a_{12}x_2 + \cdots + a_{1n}x_n = b_1,$$
$$a_{21}x_1 + a_{22}x_2 + \cdots + a_{2n}x_n = b_2,$$
$$\vdots$$
$$a_{m1}x_1 + a_{m2}x_2 + \cdots + a_{mn}x_n = b_m,$$

where $b_1, b_2, \ldots, b_m$ are not all zero. (*Compare with* HOMOGENEOUS SET OF LINEAR EQUATIONS.) In matrix notation, this set of equations may be written as $\mathbf{Ax} = \mathbf{b}$, where $\mathbf{A}$ is the $m \times n$ matrix $[a_{ij}]$, and $\mathbf{b}(\neq \mathbf{0})$ and $\mathbf{x}$ are column matrices:

$$\mathbf{b} = \begin{bmatrix} b_1 \\ b_2 \\ \vdots \\ b_m \end{bmatrix}, \qquad \mathbf{x} = \begin{bmatrix} x_1 \\ x_2 \\ \vdots \\ x_m \end{bmatrix}.$$

Such a set of equations may be inconsistent, have a unique solution, or have infinitely many solutions (*see* SIMULTANEOUS LINEAR EQUATIONS). For a set consisting of the same number of linear equations as unknowns, the matrix of coefficients $\mathbf{A}$ is a square matrix, and the set of equations has a unique solution, namely $\mathbf{x} = \mathbf{A}^{-1}\mathbf{b}$, if and only if $\mathbf{A}$ is invertible.

non-negative Not negative, so it can be either zero or positive.

non-parametric methods Methods of *inference that make no assumptions about the underlying population *distribution. Non-parametric tests are often concerned with hypotheses about the median of a population and use the ranks of the observations. Examples of non-parametric tests are the *Wilcoxon rank-sum test, the *Kolmogorov-Smirnov test and the *sign test.

non-response bias Unless a survey carries some element of compulsion in it, even if a sample is well constructed, there is a likelihood that some people will choose not to respond to it, or that some people may not be able to be contacted. Strong opinions are likely to be over-represented in any survey with a volunteer response, and the response rate is important. The lower the response rate, the less reasonable it is to try to extrapolate from the survey results to say something about the population.

non-significant result interpretation If a statistical test does not show a significant result, it is important not to interpret this as positive support for the null hypothesis, but as an absence of strong evidence against it.

non-singular A square matrix $\mathbf{A}$ is non-singular if it is not *singular; that is, if det $\mathbf{A} \neq 0$, where det, $\mathbf{A}$ is the determinant of $\mathbf{A}$. *See also* INVERSE MATRIX.

non-symmetric (of a relation) Not *symmetric, or *asymmetric, or *antisymmetric. The relation has to hold for some pairs in both orders, and hold for only one order for some other pairs, i.e. there exist elements a, b, c, d for which $a \sim b, b \sim a$, whereas $c \sim d$, but $d \sim c$ does not hold.

non-transitive (of a relation) Neither *transitive nor *intransitive. The transitive relationship has to hold for some triples, and not for others, i.e. there exist elements a, b, c, d, e, f for which $a \sim b, b \sim c, a \sim c$, whereas $d \sim e, e \sim f$, but $d \sim f$ does not hold.

norm The norm of a mathematical object is a measure which describes some sense of the length or size of the object. So the *absolute value of real numbers, *modulus of a complex number, *matrix norms and *vector norms are all examples of norms.

See also PARTITION (of an interval).

normal (to a curve) Suppose that P is a point on a curve in the plane. Then the normal at P is the line through P perpendicular to the tangent at P.

normal (to a plane) A line perpendicular to the plane. A normal is perpendicular to any line that lies in the plane.

normal (to a surface) *See* TANGENT PLANE.

normal space A *topological space X in which every two disjoint *closed sets of X have disjoint *open neighbourhoods.

normal distribution The continuous probability *distribution with *probability density function f given by

$$f(x) = \frac{1}{\sqrt{2\pi\sigma^2}} \exp\left(-\frac{(x-\mu)^2}{2\sigma^2}\right),$$

denoted by N (μ, σ^2). It has mean μ and variance σ^2. The distribution is widely used in statistics because many experiments produce data that are approximately normally distributed, the sum of random variables from non-normal distributions is approximately normally distributed (*see* CENTRAL LIMIT THEOREM), and it is the limiting distribution of distributions such as the *binomial, *Poisson and *chi-squared distributions. It is called the standard normal distribution when $\mu = 0$ and $\sigma^2 = 1$.

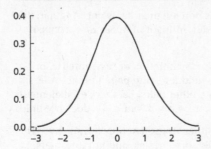

If X has the distribution $N(\mu, \sigma^2)$ and $Z = (X - \mu)/\sigma$, then Z has the distribution $N(0, 1)$. The diagram shows the graph of the probability density function of $N(0, 1)$.

The following table gives, for each value z, the percentage of observations which exceed z, for the standard normal distribution $N(0, 1)$. Thus the values are to be used for one-tailed tests. Interpolation may be used for values of z not included.

z	0.0	0.5	1.0	1.28	1.5	1.64	1.96	2.33	2.57	3.0	3.5
%	50	30.9	15.9	10.0	6.7	5.0	2.5	1.0	0.5	0.14	0.02

normal reaction *See* CONTACT FORCE.

normal subgroup If H is a *subgroup of a *group G, and for any element, x of G, the left and right cosets of H are equal, then H is a normal subgroup.

normal vector (to a plane) A vector whose direction is perpendicular to the plane. A normal vector is perpendicular to any vector whose direction lies in the plane.

not *See* NEGATION.

not and If p and q are statements then 'p *not and* q', denoted by $p \uparrow q$, is true unless both p and q are true. Since 'not and' is cumbersome language, this is often referred to as *nand in logic. The *truth table is as follows:

p	q	$p \uparrow q$
T	T	F
T	F	T
F	T	T
F	F	T

nought = ZERO.

nowhere dense Describing a set in a *topological space whose *closure has empty *interior.

nowhere-differentiable function A *continuous function on the real line that is not differentiable at a single point.

***n*-th derivative** *See* HIGHER DERIVATIVE.

***n*-th-order partial derivative** *See* HIGHER-ORDER PARTIAL DERIVATIVE.

***n*-th root** For any real number a, an n-th root of a is a number x such that $x^n = a$. (When $n = 2$, it is called a *square root, and when $n = 3$ a cube root.)

First consider n even. If a is negative, there is no real number x such that $x^n = a$. If a is positive, there are two such numbers, one positive and one negative. For $a \geq 0$, the notation $\sqrt[n]{a}$ is used to denote quite specifically the non-negative n-th root of a. For example, $\sqrt[4]{16} = 2$, and 16 has two fourth roots, namely 2 and −2.

Next consider n odd. For all values of a, there is a unique number x such that $x^n = a$, and it is denoted by $\sqrt[n]{a}$ For example, $\sqrt[3]{-8} = -2$.

***n*-th root of unity** A complex number z such that $z^n = 1$. The n distinct n-th roots of unity are $e^{i2k\pi/n}(k = 0, 1, \ldots, n-1)$, or

$$\cos \frac{2k\pi}{n} + i \sin \frac{2k\pi}{n} \quad (k = 0, 1, \ldots, n-1).$$

They are represented in the *complex plane by points that lie on the unit circle and are vertices of a regular n-sided polygon. The n-th roots of unity for $n = 5$ and $n = 6$ are shown in the figure. The n-th roots of unity always include the real number 1, and also include the real number −1 if n is even. The non-real n-th roots of unity form pairs of *conjugates. *See also* CUBE ROOT OF UNITY and FOURTH ROOT OF UNITY.

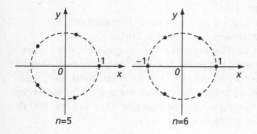

$n=5$ $n=6$

n-tuple An n-tuple consists of n objects normally taken in a particular order, denoted, for example, by $(x_1, x_2, \ldots, x_n)$. The term is a generalization of a *triple when $n = 3$, and a *quadruple when $n = 4$. When $n = 2$ it is an *ordered pair.

null hypothesis *See* HYPOTHESIS TESTING.

nullity The nullity of a matrix $\mathbf{A}$ is a number relating to the set of solutions of $\mathbf{Ax} = \mathbf{0}$, the corresponding *homogeneous set of linear equations: it can be defined as the number of *parameters required when the solutions of the equations are expressed in terms of parameters. In other words, it is the number of unknowns that are free to take arbitrary values. A way of obtaining the solutions in terms of parameters is by transforming the matrix to one in *reduced echelon form. There is an important connection between the nullity and the *rank of a matrix: for an $m \times n$ matrix $\mathbf{A}$, the nullity of $\mathbf{A}$ equals n minus the rank of $\mathbf{A}$.

null matrix = ZERO MATRIX.

null measure = ZERO MEASURE.

null sequence A sequence whose limit is 0.

null set = EMPTY SET.

null vector A vector that has zero magnitude. Since vectors normally have a magnitude and a direction associated with them, the null vector is unusual in that it does not have a direction or sense.

number line = REAL LINE.

number systems The early Egyptian number system used different symbols for 1, 10, 100, and so on, with each symbol repeated the required number of times. Later, the Babylonians had symbols for 1 and 10 repeated similarly, but for larger numbers they used a positional notation with base 60, so that groups of symbols were positioned to indicate the number of different powers of 60.

The Greek number system used letters to stand for numbers. For example, α, β, γ and δ represented 1, 2, 3 and 4, and ι, κ, λ and μ represented 10, 20, 30 and 40. The Roman number system is still known today and used for some special purposes. Roughly speaking, each Roman numeral is repeated as often as necessary to give the required total, with the larger numerals appearing before the smaller, except that if a smaller precedes a larger its value is subtracted. For example, IX, XXVI and CXLIV represent 9, 26 and 144. *See* APPENDIX 16.

The Hindu–Arabic number system, in which numbers are generally written today, uses the Arabic numerals and a positional notation with base 10. It originated in India, where records of its use go back to the 6th century. It was introduced to Europe in the 12th century, promoted by *Fibonacci and others.

number theory (higher arithmetic) The area of mathematics concerning the study of the arithmetic properties of integers and related number systems such as prime numbers. Representations of numbers as sums of squares etc. appear very abstract, but number theory has provided the basis for secure encryption in electronic communications where the messages can easily be intercepted, and subjected to analysis by very powerful computers.

numeral A symbol used to denote a number. The Roman numerals I, V, X, L, C, D and M represent 1, 5, 10, 50, 100, 500 and 1000 in the Roman number system. The Arabic numerals 0, 1, 2, 3, 4, 5, 6, 7, 8 and 9 are used in the Hindu–Arabic number system to give numbers in the form generally familiar today. *See* NUMBER SYSTEMS.

numerator *See* FRACTION.

numerical analysis Many mathematical problems to which answers are required in practical situations cannot be solved in all generality. The only way may be to use a numerical method; that is, to consider the problem in such a way that, from given data, a solution—or an approximation to one—is obtained essentially by numerical calculations. The subject dealing with the derivation and analysis of such methods is called numerical analysis.

The advent of very powerful computers means that such computational methods are increasingly important in modelling complex systems. For example, the design of aircraft and high-performance cars can now be extensively tested by simulations prior to physically building and testing models.

numerical differentiation The use of formulae, usually expressed in terms of *finite differences, for estimating derivatives of a function $f(x)$ from values of the function. For example $f'(x) = \dfrac{f(x+h) - f(x)}{h}$ and $f''(x) = \dfrac{f(x+h) - 2f(x) + f(x-h)}{h^2}$ where h is a small positive increment.

numerical integration The methods of numerical integration are used to find approximate values for *definite integrals and are useful for the following reasons. It may be that there is no analytical method of finding an *antiderivative of the *integrand. Possibly, such a method may exist but be very complicated. In another case, the integrand may not be known explicitly; it might be that only its values at certain points within the interval of integration are known. Among the elementary methods of numerical integration are the *trapezium rule and *Simpson's rule.

numerical value = ABSOLUTE VALUE.

objective function *See* LINEAR PROGRAMMING.

objective row *See* SIMPLEX METHOD.

oblique A pair of intersecting lines, to be taken, for example, as coordinate axes, are oblique if they are not (or at least not necessarily) perpendicular. Three concurrent lines, similarly, are oblique if they are not mutually perpendicular.

oblong = RECTANGLE or RECTANGULAR.

observation A particular value taken by a *random variable. Normally, *n* observations of the random variable *X* are denoted by $x_1, x_2, \ldots, x_n$.

observational study A statistical study in which the researchers observe or question participants but without designing an experiment in which interventions are used. The lack of randomization in allocating participants into groups means that the issue of *confounding variables is present. *See also* CASE CONTROL STUDIES.

observer With a *frame of reference it is convenient to associate an observer who is thought of as viewing the motion with respect to this frame of reference and having the capability to measure positions in space and the duration of time intervals.

obtuse angle An angle that is greater than a *right angle and less than two right angles. An obtuse-angled triangle is one in which one angle is obtuse.

oct- Prefix denoting eight.

octagon An eight-sided polygon.

octahedron (octahedra) A *polyhedron with 8 faces, often assumed to be regular. The regular octahedron is one of the *Platonic solids and its faces are equilateral triangles. It has 6 vertices and 12 edges.

octal A number system using the base of eight, often used in computing.

octant In a Cartesian coordinate system in 3-dimensional space, the axial planes divide the rest of the space into eight regions called octants. The set of points $\{(x, y, z) \mid x > 0, y > 0, z > 0\}$ may be called the *positive (or possibly the first) octant.

odd function The *real function f is an odd function if $f(-x) = -f(x)$ for all x (in the domain of f). Thus the graph $y = f(x)$ of an odd function is symmetrical about the origin; that is, it has a half-turn symmetry about the origin, because whenever (x, y) lies on the graph then so does $(-x, -y)$. For example, f is an odd function when $f(x)$ is defined as any of the following: $2x$, x^3, $x^7 - 8x^3 + 5x$, $1/(x^3 - x)$, $\sin x$, $\tan x$.

odd integer Not exactly divisible by 2. General form of an odd number is $2n + 1$ where n is an integer.

odd part The largest odd factor of a given integer, so the odd part of 14, which can be written as $\mathrm{odd}(14)$ is 7; $\mathrm{odd}(9) = 9$ and $\mathrm{odd}(8) = 1$ as is odd (2^n) for any positive integer n.

odds Betting odds are expressed in the form $r{:}s$ corresponding in theory to a probability of $\dfrac{r}{r + s}$ of winning. In reverse, if p is the probability of an event happening, the odds on it will be $p{:}(1{-}p)$. In practice, odds offered by a bookmaker will only be an approximation to this probability, as the bookmaker wishes to make money, and will change the odds to reflect the amount of money being bet on particular events even if they have no reason to change their view of the probabilities of those events.

odds ratio Defined as the ratio of the odds of an event happening in two distinct groups. If the probabilities of the event in the two groups are p and q then the odds are $p{:}(1{-}p)$ and $q{:}(1{-}q)$ and the odds ratio is
$$\frac{p/1 - p}{q/1 - q} = \frac{p(1 - q)}{q(1 - p)}.$$

off diagonal In a square matrix or array, the *diagonal from the bottom-left to the top-right corner.

ogive A graph of the *cumulative frequency distribution of a sample or of the *cumulative distribution function of a random variable.

omega Greek letter o, symbol ω, often used to represent *angular velocity, and the smallest infinite ordinal number, i.e. the order of the natural numbers.

one The smallest natural number, symbol 1. The multiplicative identity for real and complex numbers.

one-sided surface A surface where any two points can be joined without leaving the surface or crossing an edge, for example the *Möbius band.

one-sided test *See* HYPOTHESIS TESTING.

one-tailed test *See* HYPOTHESIS TESTING.

one-to-many A rule which associates a single member of the domain of more than one number of the range. Where the rule is acting on numbers, a one-to-many mapping is not a function. For example, $\tan^{-1} x$ takes not only the principal value, α, returned by a calculator, lying between $-90°$ and $+90°$, but also $\alpha \pm 180n$ for any integer n.

one-to-one correspondence A *one-to-one mapping between two sets that is also an *onto mapping. Thus each element of the first set is made to correspond with exactly one element of the second set, and vice versa. Such a correspondence is also called a *bijection.

one-to-one mapping A *mapping $f: S \rightarrow T$ is one-to-one if, whenever s_1 and s_2 are distinct elements of S, their images $f(s_1)$ and $f(s_2)$ are distinct elements of T. So f is one-to-one if $f(s_1) = f(s_2)$ implies that $s_1 = s_2$.

only if *See* CONDITION, NECESSARY AND SUFFICIENT.

onto mapping A *mapping $f: S \rightarrow T$ is onto if every element of the codomain T is the image under f of at least one element of the domain S. So f is onto if the image (or range) $f(S)$ is the whole of T.

open ball In a *metric space, it is the set of all points whose distance d from a fixed point a is strictly less than a specified value, i.e. the set of points x for which $d(x, a) < \delta$. *See* METRIC.

open cover An open cover of a set X is a collection of open sets whose union contains X as a subset.

open disc *See* DISC.

open half-plane *See* HALF-PLANE.

open half-space *See* HALF-SPACE.

open interval The open interval (a, b) is the set

$$\{x \mid x \in R \text{ and } a < x < b\}$$

For open interval determined by I, see INTERVAL.

open set A set of points around each of which there is an *open ball lying within the set.

operation An operation on a set S is a rule that associates with some number of elements of S a resulting element. If this resulting element is always also in S, then S is said to be closed under the operation. An operation that associates with one element of S a resulting element is called a *unary operation; one that associates with two elements of S a resulting element is a *binary operation.

operator A symbol used to indicate that a mathematical operation is to be performed on one or more quantities. So $\sqrt{\ }$ is an operator acting on one quantity, and $\cap$ is an operator which requires two.

opposite side In a right-angled triangle, where one of the other angles has been specified, the opposite side is the side which is not an *arm of the specified angle.

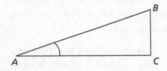

BC is opposite to the angle at A.

optimal A value or solution is optimal if it is, in a specified sense, the best possible. For example, in a *linear programming problem the maximum or minimum value of the objective function, whichever is required, is the optimal value and a point at which it is attained is an optimal solution.

optimality condition *See* SIMPLEX METHOD.

optimal strategy FUNDAMENTAL THEOREM OF GAME THEORY.

optimization The process of finding the best possible solution to a problem. In mathematics, this often consists of maximizing or minimizing the value of a certain function, perhaps subject to given constraints.

or *See* DISJUNCTION.

orbit The path traced out by a body experiencing a *central force, such as the gravitational force associated with a second body. When the force is given by the *inverse square law of gravitation, the orbit is either elliptic, parabolic or hyperbolic, with the second body at a focus of the conic.

order (of a differential equation) *See* DIFFERENTIAL EQUATION.

order (of a group) The order of a *group G is the number of elements in G.

order (of a matrix) An $m \times n$ matrix is said to have order $m \times n$ (read as 'm by n'). An $n \times n$ matrix may be called a square matrix of order n.

order (of a partial derivative) *See* HIGHER-ORDER PARTIAL DERIVATIVE.

order (of a root) *See* ROOT.

ordered pair An ordered pair consists of two objects considered in a particular order. Thus, if $a \neq b$, the ordered pair (a, b) is not the same as the ordered pair (b, a). *See* CARTESIAN PRODUCT.

ordered set A sequence of elements where both the nature and the order of the elements is important so that p, q, r is not the same as p, r, q. For example, in badminton singles, a point is awarded to the player serving if they win the passage of play, but if they lose it then the service changes to the opponent. So if player A is serving and the passages of play are won by B, A, B then no points have been scored, but the sequence A, B, B would register scores of $1 - 0$, $1 - 0$ (change server), $1 - 1$.

order statistics Statistics which depend on their value for the place they occupy in an ordering of the data. So *minimum, *maximum, *median, *quartiles and *percentiles are all order statistics.

ordinal data Data that can be ordered. For example, measurements on a linear scale, preferences in some contexts, but where multiple complex criteria are concerned it is not always possible to construct an ordered set of preferences.

ordinal number A number denoting the position in a sequence, so 'first', 'second', 'third' is the start of the ordinal numbers.

ordinal scale A scale in which data are ranked but without units of measurement to indicate the size of differences. So the values used in calculating *Spearman's or another *rank correlation coefficient are on an ordinal scale.

ordinary differential equation *See* DIFFERENTIAL EQUATION.

ordinary point A point at which the function is uniquely defined and is *continuously differentiable.

ordinate The y-coordinate in a Cartesian coordinate system in the plane.

orientable Describing a *surface for which it is possible to make a consistent choice of a positive direction for a vector which is *normal to the surface. If you start at a point P with normal vector $\mathbf{v}$ on a *Möbius strip and travel along the length of the strip until you return to P, you will have passed a point P' on the other side of the Möbius strip at which the normal vector is $-\mathbf{v}$. So the Möbius strip is not orientable. For physical solids, it is often easiest to identify orientability by the solid having an inside and an outside, but the Klein bottle is the generalization of the Möbius strip for which there is no inside and outside, in the same way that the Möbius strip does not have a top and a bottom.

origin *See* COORDINATES (on a line), COORDINATES (in the plane) and COORDINATES (in 3-dimensional space).

orthocentre The point at which the *altitudes of a triangle are concurrent. The orthocentre lies on the *Euler line of the triangle.

orthogonal basis A *basis for a *vector space in which the components of the basis are mutually orthogonal. *See also* ORTHONORMAL.

orthogonal curves Two curves, or straight lines, are orthogonal if they intersect at right angles.

orthogonal matrix A square matrix $\mathbf{A}$ is orthogonal if $\mathbf{A}^T\mathbf{A} = \mathbf{I}$, where $\mathbf{A}^T$ is the *transpose of $\mathbf{A}$. The following properties hold:
  (i) If $\mathbf{A}$ is orthogonal, $\mathbf{A}^{-1} = \mathbf{A}^T$ and so $\mathbf{A}\mathbf{A}^T = \mathbf{I}$.
  (ii) If $\mathbf{A}$ is orthogonal, det $\mathbf{A} = \pm 1$.
  (iii) If $\mathbf{A}$ and $\mathbf{B}$ are orthogonal matrices of the same order, then $\mathbf{A}\mathbf{B}$ is orthogonal.

orthogonal polynomials A set of *polynomials $\{p_k(x)\}$, $k = 1, 2, 3, \ldots$, are said to be orthogonal over an interval $[a, b]$ if for some weighting function $w(x)$, $\int_a^b w(x)p_i(x)p_j(x)dx = 0$ when $i \neq j$. If $\int_a^b w(x)p_i(x)p_i(x)dx = 1$ it is a set of orthonormal polynomials.

orthogonal projection A projection of a figure onto a line or plane so that each element of the figure is mapped onto the closest point on the line or plane. If X is the point and X' its image, this means that XX' will be orthogonal (or at right angles) to the line or plane.

orthogonal sequence A sequence of pairwise *orthogonal vectors, which often functions like the *Legendre polynomials, or the Laguerre, Jacobi, or Hermite polynomials. So the vectors in each pair of vectors in the sequence are orthogonal to one another.

orthogonal vectors Two non-zero vectors are orthogonal if their directions are perpendicular. Thus non-zero vectors **a** and **b** are orthogonal if and only if $\mathbf{a} \cdot \mathbf{b} = 0$.

orthonormal An orthonormal set of vectors is a set of mutually orthogonal unit vectors. In 3-dimensional space, three mutually orthogonal unit vectors form an orthonormal basis. The standard orthonormal basis in a Cartesian coordinate system consists of the unit vectors **i**, **j** and **k** along the three coordinate axes.

orthonormal polynomials *See* ORTHOGONAL POLYNOMIALS.

orthonormal sequence An *orthogonal sequence of vectors which are normalized, i.e. each is of unit length.

Osborne's rule The rule which summarizes the correspondence between trigonometric and hyperbolic function identities. It states that the identities are the same except where a product of two sins or sinhs is involved, where an extra negative is required. For example,

$$\sin(A + B) = \sin A \cos B + \sin B \cos A \text{ gives}$$
$$\sinh(A + B) = \sinh A \cosh B + \sinh B \cosh A, \text{ but}$$

$$\cos(A + B) = \cos A \cos B - \sin A \sin B \text{ gives}$$
$$\cosh(A + B) = \cosh A \cosh B - \sinh B \sinh A.$$

oscillate finitely and **infinitely** *See* LIMIT (of a sequence).

oscillations A particle or rigid body performs oscillations if it travels to and fro in some way about a central position, usually a point of stable *equilibrium. Examples include the swings of a simple or compound pendulum, the bobbing up and down of a particle suspended by a spring, and the vibrations of a violin string. The oscillations are *damped when there is a resistive force and *forced when there is an applied force.

osculate Two curves osculate when they meet at a point where they share a common *tangent.

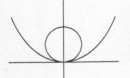

The parabola $y=x^2$ and the circle $x^2+(y-1)^2=1$ share a common tangent at the origin.

osculation *See* CUSP.

osculinflection *See* CUSP.

outlier An observation that is deemed to be unusual and possibly erroneous because it does not follow the general pattern of the data in the sample. However, in some contexts it is the outliers which are the most important observations. For example, the differences between low tide and the following high tide vary according to weather conditions, the season, and the stage of the lunar cycle. Unusually large tidal range occurrences can cause considerable damage.

p Abbreviation for *pico-.

pair A set with two elements, often with an order implicit in it, such as the coordinate pair (x, y).

paired-sample tests (in statistics) A situation in which there is a specific link between the two observations in each pair, for example if pieces of cloth are divided in two and the two halves of each piece make a pair. Paired-sample tests are generally much more powerful than the analogous two-sample test because they remove a major source of variation. By looking only at the difference in outcome between the two members of each pair, the variability between subjects has been removed, making it easier to identify any genuine difference in the experimental treatments.

pairwise Applying to all possible pairs which can be constructed from the elements of a set. See *pairwise disjoint for an example.

pairwise disjoint Sets $A_1, A_2, \ldots, A_n$ are said to be pairwise disjoint if $A_i \cap A_j = \emptyset$ for all i and j, with $i \neq j$. The term can also be applied to an infinite collection of sets.

Pappus of Alexandria (about AD 320) Considered to be the last great Greek geometer. He wrote commentaries on *Euclid and *Ptolemy, but most valuable is his *Synagoge* ('Collection'), which contains detailed accounts of much Greek mathematics, some of which would otherwise be unknown.

Pappus' Theorems The following two theorems are about surfaces and solids of revolutions:

Theorem

Suppose that an arc of a plane curve is rotated through one revolution about a line in the plane that does not cut the arc. Then the area of the surface of revolution obtained is equal to the length of the arc times the distance travelled by the centroid of the curve.

Theorem

Suppose that a plane region is rotated through one revolution about a line in the plane that does not cut the region. Then the volume of the solid of revolution obtained is equal to the area of the region times the distance travelled by the centroid of the region.

The theorems can be used to find surface areas and volumes, such as those of a *torus. They can also be used to find the positions of centroids. For example, using the second theorem and the known volume of a sphere, the position of the centroid of the region bounded by a semicircle and a diameter can be found.

parabola A *conic with eccentricity equal to 1. Thus a parabola is the locus of all points P such that the distance from P to a fixed point F (the focus) is equal to the distance from P to a fixed line l (the directrix). It is obtained as a plane section of a cone in the case when the plane is parallel to a generator of the cone (*see* CONIC). A line through the focus perpendicular to the directrix is the axis of the parabola, and the point where the axis cuts the parabola is the vertex. It is possible to take the vertex as origin, the axis of the parabola as the x-axis, the tangent at the vertex as the y-axis and the focus as the point $(a, 0)$. In this coordinate system, the directrix has equation $x = -a$ and the parabola has equation $y^2 = 4ax$. When investigating the properties of a parabola, it is normal to choose this convenient coordinate system.

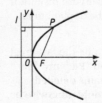

Different values of a give parabolas of different sizes, but all parabolas are the same shape. For any value of t, the point $(at^2, 2at)$ satisfies $y^2 = 4ax$, and conversely any point of the parabola has coordinates $(at^2, 2at)$ for some value of t. Thus $x = at^2$, $y = 2at$ may be taken as *parametric equations for the parabola $y^2 = 4ax$.

One important property of a parabola is this. For a point P on the parabola, let α be the angle between the tangent at P and a line through P parallel to the axis of the parabola, and let β be the angle between the tangent at P and a line through P and the focus, as shown in the figure;

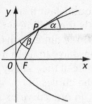

then $\alpha = \beta$. This is the basis of the parabolic reflector: if a source of light is placed at the focus of a parabolic reflector, each ray of light is reflected parallel to the axis, so producing a parallel beam of light.

(⊕) SEE WEB LINKS
• Properties of the parabola with links to animated illustrations.

parabolic cylinder A *cylinder in which the fixed curve is a *parabola and the fixed line to which the generators are parallel is perpendicular to the plane of the parabola. It is a *quadric, and in a suitable coordinate system has equation

$$\frac{x^2}{a^2} = \frac{2y}{b}.$$

parabolic spiral A spiral in which the length of the radius vector r is defined by the *polar equation $r^2 = k\theta$. In the diagram $k = 1$.

(⊕) SEE WEB LINKS
• An animation of a parabolic spiral.

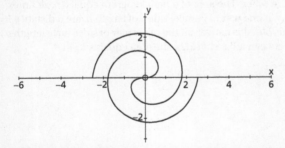

paraboloid See ELLIPTIC PARABOLOID and HYPERBOLIC PARABOLOID.

paradox A situation in which an apparently reasonable assumption leads to an unreasonable conclusion. It may seem that, if a certain statement is true, then it follows that it is false; and also that, if it is false, then it is true. Examples of this are *Grelling's paradox, the *liar paradox and *Russell's paradox. See also ZENO OF ELEA and SIMPSON'S PARADOX.

parallel Two or more lines or planes that are always equidistant from one another, however far they are extended. Consequently they will never meet, though this is not a sufficient condition for lines in three dimensions to be parallel, as *skew lines also never meet. Parallel lines have to lie in a plane. Sometimes curves or surfaces are described as parallel if they satisfy the equidistant condition, where the distance between the curves or surfaces is defined as the shortest distance from a point on one curve or surface to any point lying on the other.

parallel axis theorem The following theorem about *moments of inertia of a rigid body:

Theorem

Let I_C be the moment of inertia of a rigid body about an axis through C, the centre of mass of the rigid body. Then the moment of inertia of the body about some other axis parallel to the first axis is equal to $I_C + md^2$, where m is the mass of the rigid body and d is the distance between the two parallel axes.

parallel computation The simultaneous execution of parts of the same task on multiple processors to obtain faster results. *Compare* SERIAL COMPUTATION.

parallelepiped A *polyhedron with six faces, each of which is a parallelogram. (The word is commonly misspelt.)

parallelogram A quadrilateral in which (i) both pairs of opposite sides are parallel, and (ii) the lengths of opposite sides are equal. Either property, (i) or (ii), in fact implies the other. The area of a parallelogram equals 'base times height'. That is to say, if one pair of parallel sides, of length b, are a distance h apart, the area equals bh. Alternatively, if the other pair of sides have length a and θ is the angle between adjacent sides, the area equals $ab \sin \theta$.

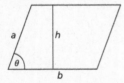

parallelogram law *See* ADDITION (of vectors).

parallel postulate The axiom of Euclidean geometry which says that, if two straight lines are cut by a *transversal and the interior angles on one side add up to less than two right angles, then the two lines meet on that side. It is equivalent to Playfair's axiom, which says that, given a point not on a given line, there is precisely one line through the point parallel to the line. The parallel postulate was shown to be independent of the other axioms of Euclidean geometry in the 19th century, when *non-Euclidean geometries were discovered in which the other axioms hold but the parallel postulate does not.

parameter (in pure mathematics) A variable that is to take different values, thereby giving different values to certain other variables. For example, a parameter t could be used to write the solutions of the equation $5x_1 + 4x_2 = 7$ as $x_1 = 3 - 4t$, $x_2 = -2 + 5t(t \in \mathbf{R})$. *See also* PARAMETRIC EQUATIONS (of a line in space) and PARAMETRIZATION.

parameter (in statistics) A parameter for a *population is some quantity that relates to the population, such as its mean or median. A parameter for a population may be estimated from a sample by using an appropriate *statistic as an *estimator. For a *distribution, a constant that appears in the *probability mass function or *probability density function of the distribution is called a parameter. In this sense, the *normal distribution has two parameters and the *Poisson distribution has one parameter, for example.

parametric equations (of a curve) *See* PARAMETRIZATION.

parametric equations (of a line in space) Given a line in 3-dimensional space, let (x_1, y_1, z_1) be coordinates of a point on the line, and l, m, n be *direction ratios of a direction along the line. Then the line consists of all points P whose coordinates (x, y, z) are given by

$$x = x_1 + tl, \qquad y = y_1 + tm, \qquad z = z_1 + tn,$$

for some value of the *parameter t. These are parametric equations for the line. They are most easily established by using the *vector equation of the line and taking components. If none of l, m, n is zero, the equations can be written

$$\frac{x - x_1}{l} = \frac{y - y_1}{m} = \frac{z - z_1}{n} \quad (= t),$$

which can be considered to be another form of the parametric equations, or called the equations of the line in 'symmetric form'. If, say, $n = 0$ and l and m are both non-zero, the equations are written

$$\frac{x - x_1}{l} = \frac{y - y_1}{m}, \qquad z = z_1.$$

if, say, $m = n = 0$, they become $y = y_1$, $z = z_1$.

parametric statistics The branch of statistics dealing with inference about parameters of a population on the basis of observations and measurements taken from a sample.

parametrization (of a curve) A method of associating, with every value of a *parameter t in some interval I (or some other subset of $\mathbf{R}$), a point $P(t)$ on the curve such that every point of the curve corresponds to some value of t. Often this is done by giving the x- and y-coordinates of P as functions of t, so that the coordinates of P may be written $(x(t), y(t))$. The equations that give x and y as functions of t are parametric equations for the curve. For example, $x = at^2$, $y = 2at$ $(t \in \mathbf{R})$ are parametric equations for the parabola $y^2 = 4ax$; and $x = a \cos \theta$, $y = b \sin \theta$ $(\theta \in [0, 2\pi))$ are parametric equations for the ellipse

$$\frac{x^2}{a^2} + \frac{y^2}{b^2} = 1.$$

The gradient dy/dx of the curve at any point can be found, if $x'(t) \neq 0$, from $dy/dx = y'(t)/x'(t)$.

parentheses Brackets, usually ( ), which can be used in algebraic expressions to change the order in which arithmetic operations are to be carried out.

Pareto chart A bar chart in which the categories are arranged in the order of their frequencies, starting with the most frequent. This reveals what are the most important factors in any given situation, and enables a realistic cost–benefit analysis of what measures might be undertaken to improve performance.

parity The attribute of an integer of being even or odd. Thus, it can be said that 6 and 14 have the same parity (both are even), whereas 7 and 12 have opposite parity.

parity check *See* CHECK DIGIT.

partial derivative Suppose that f is a real function of n variables x_1, $x_2, \ldots, x_n$, so that its value at a typical point is $f(x_1, x_2, \ldots, x_n)$. If

$$\frac{f(x_1 + h, x_2, \ldots, x_n) - f(x_1, x_2, \ldots, x_n)}{h}$$

tends to a limit as $h \to 0$, this limit is the partial derivative of f, at the point $(x_1, x_2, \ldots, x_n)$, with respect to x_1; it is denoted by $f_1(x_1, x_2, \ldots, x_n)$ or $\partial f/\partial x_1$ (read as 'partial df by dx_1'). For a particular function, this partial derivative may be found using the normal rules of differentiation, by differentiating as though the function were a function of x_1 only and treating $x_2, \ldots, x_n$ as constants. The other partial derivatives,

$$\frac{\partial f}{\partial x_2}, \ldots, \frac{\partial f}{\partial x_n},$$

are defined similarly. The partial derivatives may also be denoted by $f_{x_1}, f_{x_2}, \ldots, f_{x_n}$. For example, if $f(x, y) = xy^3$, then the partial derivatives are $\partial f/\partial x$, or f_x and $\partial f/\partial y$, or f_y; and $f_x = y^3, f_y = 3xy^2$. With the value of f at (x, y) denoted by $f(x, y)$, the notation f_1 and f_2 is also used for the partial derivatives f_x and f_y, and this can be extended to functions of more variables. *See also* HIGHER-ORDER PARTIAL DERIVATIVE.

partial differential equation *See* DIFFERENTIAL EQUATION.

partial differentiation The process of obtaining one of the *partial derivatives of a function of more than one variable. The partial derivative $\partial f/\partial x_i$ is said to be obtained from f by 'differentiating partially with respect to x_i'.

partial fractions Suppose that $f(x)/g(x)$ defines a *rational function, so that $f(x)$ and $g(x)$ are polynomials, and suppose that the degree of $f(x)$ is less than the degree of $g(x)$. In general, $g(x)$ can be factorized into a product of some different linear factors, each to some index, and some different irreducible quadratic factors, each to some index. Then the original expression $f(x)/g(x)$ can be written as a sum of terms: corresponding to each $(x - \alpha)^n$ in $g(x)$, there are terms

$$\frac{A_1}{x - \alpha} + \frac{A_2}{(x - \alpha)^2} + \cdots + \frac{A_n}{(x - \alpha)^n},$$

and corresponding to each $(ax^2 + bx + c)^n$ in $g(x)$, there are terms

$$\frac{B_1 x + C_1}{ax^2 + bx + c} + \frac{B_2 x + C_2}{(ax^2 + bx + c)^2} + \cdots + \frac{B_n x + C_n}{(ax^2 + bx + c)^n},$$

where the real numbers denoted here by capital letters are uniquely determined. The expression $f(x)/g(x)$ is said to have been written in partial fractions. The method, which sounds complicated when stated in general, as above, is more easily understood from examples:

$$\frac{3}{(x - 1)(x + 2)} = \frac{A}{x - 1} + \frac{B}{x + 2},$$

$$\frac{3x^2 + 2x + 1}{(x - 1)^3} = \frac{A}{x - 1} + \frac{B}{(x - 1)^2} + \frac{C}{(x - 1)^3},$$

$$\frac{3x + 2}{(x - 1)(x^2 + x + 1)^2} = \frac{A}{x - 1} + \frac{Bx + C}{x^2 + x + 1} + \frac{Dx + E}{(x^2 + x + 1)^2},$$

$$\frac{3x + 2}{(x - 1)^2(x^2 + x + 1)} = \frac{A}{x - 1} + \frac{B}{(x - 1)^2} + \frac{Cx + D}{x^2 + x + 1}.$$

The values for the numbers $A, B, C, \ldots$ are found by first multiplying both sides of the equation by the denominator $g(x)$. In the last example, this gives

$$3x + 2 = A(x - 1)(x^2 + x + 1) + B(x^2 + x + 1) + (Cx + D)(x - 1)^2.$$

This has to hold for all values of x, so the coefficients of corresponding powers of x on the two sides can be equated, and this determines the unknowns. In some cases, setting x equal to particular values (in this example, $x = 1$) may determine some of the unknowns more quickly. The method of partial fractions is used in the integration of rational functions.

partial product When an infinite product

$$\prod_{r=1}^{\infty} a_r$$

is formed from a sequence $a_1, a_2, a_3, \ldots$, the product $a_1 a_2 \cdots a_n$ of the first n terms is called the n-th partial product.

partial sum The n-th partial sum s_n of a series $a_1 + a_2 + \cdots$ is the sum of the first n terms; thus $s_n = a_1 + a_2 + \ldots + a_n$.

particle An object considered as having no size, but having mass, position, velocity, acceleration, and so on. It is used in *mathematical models to represent an object in the real world of negligible size. The subject of particle dynamics is concerned with the study of the motion of one or more particles experiencing a system of forces.

particular integral *See* LINEAR DIFFERENTIAL EQUATION WITH CONSTANT COEFFICIENTS.

particular solution *See* DIFFERENTIAL EQUATION.

partition (of an interval) Let $[a, b]$ be a closed interval. A set of $n + 1$ points $x_0, x_1, \ldots, x_n$ such that

$$a = x_0 < x_1 < x_2 < \cdots < x_{n-1} < x_n = b$$

is a partition of the interval $[a, b]$. A partition divides the interval into n subintervals $[x_i, x_{i+1}]$. The *norm of the partition P is equal to the length of the largest subinterval and is denoted by $\|P\|$. Such partitions are used in defining the *Riemann integral of a function over $[a, b]$ (*see* INTEGRAL).

partition (of a positive integer) A partition of the positive integer n is obtained by writing $n = n_1 + n_2 + \ldots + n_k$, where $n_1, n_2, \ldots, n_k$ are positive integers, and the order in which $n_1, n_2, \ldots, n_k$ appear is unimportant. The number of partitions of n is denoted by $p(n)$. For example, the partitions of 5 are

$$5, \ 4+1, \ 3+2, \ 3+1+1, \ 2+2+1, \ 2+1+1+1, \ 1+1+1+1+1,$$

and hence $p(5) = 7$. The values of $p(n)$ for small values of n are as follows:

n	1	2	3	4	5	6	7	8	9	10
$p(n)$	1	2	3	5	7	11	15	22	30	42

partition (of a set) A partition of a set S is a collection of non-empty subsets of S such that every element of S belongs to exactly one of the subsets in the collection. Thus S is the union of these subsets, and any two distinct subsets are disjoint. Given a partition of a set S, an *equivalence relation $\sim$ on S can be obtained by defining $a \sim b$ if a and b belong to the same subset in the partition. It is an important fact that, conversely, from any equivalence relation on S a partition of S can be obtained.

pascal The SI unit of *pressure, abbreviated to 'Pa'. One pascal is equal to one *newton per square metre.

Pascal, Blaise (1623–62) French mathematician and religious philosopher who in mathematics is noted for his work in geometry, hydrostatics and probability. *Pascal's triangle, the arrangement of the binomial coefficients, was not invented by him but he did use it in his studies in probability, on which he corresponded with *Fermat. He also joined in the work on finding the areas of curved figures, work which was soon to lead to the calculus. Here, his main contribution was to find the area of an arch in the shape of a *cycloid.

Pascal's triangle The figure shows the first seven rows of the arrangement of numbers known as Pascal's triangle. In general, the n-th row consists of the *binomial coefficients $\binom{n}{r}$, or ${}^nC_r$, with $r = 0, 1, \ldots, n$. With the numbers set out in this fashion, it can be seen how the number $\binom{n+1}{r}$ is equal to the sum of the two numbers $\binom{n}{r-1}$ and $\binom{n}{r}$, which are situated above it to the left and right. For example, $\binom{7}{3}$ equals 35, and is the sum of $\binom{6}{2}$, which equals 15, and $\binom{6}{3}$, which equals 20.

```
                    1       1
                1       2       1
            1       3       3       1
        1       4       6       4       1
    1       5      10      10       5       1
  1     6      15      20      15       6       1
1     7     21      35      35      21       7       1
```

(((⊕))) SEE WEB LINKS
• An illustration of some of the patterns in Pascal's triangle.

path (in a graph) Let u and v be vertices of a *graph. A path from u to v is traced out by travelling from u to v along edges. In a precise definition, it is normal to insist that, in a path, no edge is used more than once and no vertex is visited more than once. Thus a path may be defined as a sequence $v_0, e_1, v_1, \ldots, e_k, v_k$ of alternately vertices and edges (where e_i is an edge joining v_{i-1} and v_i), with all the edges different and all the vertices different; this is a path from v_0 to v_k.

path connected Describing a *topological space in which there is a *path between any two points.

Pauli, Wolfgang Ernst FRS (1900–58) Austrian mathematician and theoretical physicist whose work revolutionized the understanding of atomic behaviour. He first proposed the existence of neutral particles with zero mass, which he termed 'neutrons' but have subsequently become known as 'neutrinos', and he proposed a quantum spin number for electrons. He was awarded the Nobel Prize for Physics in 1945 for his discovery of the Pauli exclusion principle which states that no two electrons can have the same quantum numbers.

payoff (in game theory) The positive or negative amount each player is credited with once all players have chosen their strategy.

p.d.f. = PROBABILITY DENSITY FUNCTION.

Peano, Giuseppe (1858–1932) Italian mathematician whose work towards the axiomatization of mathematics was very influential. He was responsible for the development of symbolic logic, for which he introduced important notation. His axioms for the integers were an important development in the formal analysis of arithmetic. He was also the first person to produce examples of so-called space-filling curves.

Peano axioms In 1889 Giuseppe *Peano published a set of five axioms which attempted to provide a rigorous basis for the infinite set of natural numbers to be generated by a finite set of symbols and rules. In 1888 Richard *Dedekind had published a set of axioms on the same topic, but Peano's were more precisely stated. However, they are sometimes referred to as the Peano–Dedekind axioms in recognition of the foundation laid by Dedekind.
The axioms are:
- (i) Zero is a natural number.
- (ii) Every natural number has a successor in the natural numbers.
- (iii) Zero is not the successor of any natural number.
- (iv) If the successors of two natural numbers n, m are the same number, then $n = m$.

(v) If a set S contains zero and the successor of every number in S is also in S, then the set S contains the natural numbers.

The last of these is known as the principle of induction. The reasoning is important in proof in that it allows a property to be established as true for an infinite number of cases by establishing that if a statement is true for some starting point, and knowledge that it is true for any particular case is sufficient to know that it must also be true for the next case, this is sufficient to establish that it must always be true.

Peano curve Take the diagonal of a square, as shown on the left in the diagram. Replace this by the nine diagonals of smaller squares (drawn here in a way that indicates the order in which the diagonals are to be traced out). Now replace each straight section of this by the nine diagonals of even smaller squares, to obtain the result shown on the right. The Peano curve is the curve obtained when this process is continued indefinitely. It has the remarkable property that it passes through every point of the square, and it is therefore described as space-filling. Any similar space-filling curve constructed in the same kind of way may also be called a Peano curve.

 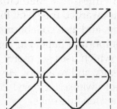 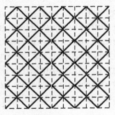

Pearson, Karl (1857–1936) British statistician influential in the development of statistics for application to the biological and social sciences. Stimulated by problems in evolution and heredity, he developed fundamental concepts such as the *standard deviation, the *coefficient of variation and, in 1900, the *chi-squared test. This work was contained in a series of important papers written while he was professor of applied mathematics, and then eugenics, at University College, London.

Pearson's product moment correlation coefficient *See* CORRELATION.

pedal curve The curve generated by taking the foot of the perpendiculars from a fixed point to all the tangents to a given curve.

(⊕) SEE WEB LINKS
• An animation of a pedal curve.

pedal triangle In the triangle with vertices A, B, and C, let D, E and F be the feet of the perpendiculars from A, B and C, respectively, to the opposite sides, so that AD, BE and CF are the *altitudes. The triangle DEF is called the pedal triangle. The altitudes of the original triangle bisect the angles of the pedal triangle.

Pell, John (1610–85) English mathematician remembered in the name of *Pell's equation, which was wrongly attributed to him.

Pell's equation The *Diophantine equation $x^2 = ny^2 + 1$, where n is a positive integer that is not a perfect square. Methods of solving such an equation have been sought from as long ago as the time of *Archimedes. *Bhāskara solved particular cases. *Fermat apparently understood that infinitely many solutions always exist, a fact that was proved by *Lagrange. * Pell's name was incorrectly given to the equation by *Euler.

pendulum See SIMPLE PENDULUM, CONICAL PENDULUM, COMPOUND PENDULUM and FOUCAULT PENDULUM.

Penrose, Roger FRS (1931–) British mathematician and theoretical physicist who was Rouse Ball Professor of Mathematics at Oxford University for 25 years. He was awarded the *Wolf Prize for Physics in 1988 jointly with Stephen *Hawking, and also collaborated with other pure mathematicians, producing important papers in cosmology, topology, manifolds and twistor theory which uses geometry and algebra in an attempt to unite quantum theory and relativity. He is well known as the author of popular science books whose lucid style makes cutting-edge scientific ideas accessible to a wide audience.

penta- Prefix denoting five.

pentagon A five-sided polygon.

pentagonal number A number which can be arranged in the shape of nested pentagons as shown in the figure. The sequence is 1, 5, 12, 22, 35, ... which can be generated by $\dfrac{n}{2}(3n-1)$ for $n = 1, 2, 3, ...$

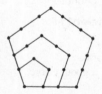

pentagram The plane figure shown, formed by joining alternate vertices of a regular *pentagon.

per cent (%) Literally from the Latin 'in every hundred', so it is a fraction with the denominator of 100 replaced by this phrase, or by the symbol %. So 20 per cent = 20% = 20/100 which is 1/5 or 0.2.

percentage A proportion, rate or ratio expressed with a denominator of 100.

percentage error The *relative error expressed as a percentage. When 1.9 is used as an approximation for 1.875, the relative error equals 0.025/1.875 (or 0.025/1.9) = 0.013, to 2 significant figures. So the percentage error is 1.3%, to 2 significant figures.

percentile *See* QUANTILE.

perfect number An integer that is equal to the sum of its positive divisors (not including itself). Thus, 6 is a perfect number, since its positive divisors (not including itself) are 1, 2 and 3, and $1 + 2 + 3 = 6$; so too are 28 and 496, for example. At present there are over 30 known perfect numbers, all even. If $2^p - 1$ is prime (so that it is a *Mersenne prime), then $2^{p-1}(2^p - 1)$ is perfect; moreover, these are the only even perfect numbers. It is not known if there are any odd perfect numbers; none has been found, but it has not been proved that one cannot exist.

((()) SEE WEB LINKS
• A brief history of perfect numbers.

perfect square An integer of the form n^2, where n is a positive integer. The perfect squares are 1, 4, 9, 16, 25, ...

perigee *See* APSE.

perihelion *See* APSE.

perimeter The perimeter of a plane figure is the length of its boundary. Thus, the perimeter of a rectangle of length L and width W is $2L + 2W$. The perimeter of a circle is the length of its circumference.

period, periodic If, for some value p, $f(x + p) = f(x)$ for all x, the real function f is periodic and has period p. For example, $\cos x$ is periodic with period 2π, since $\cos(x + 2\pi) = \cos x$ for all x; or, using degrees, $\cos x°$ is periodic with period 360, since $\cos(x + 360)° = \cos x°$ for all x. Some authors restrict the use of the term 'period' to the smallest positive value of p with this property.

In mechanics, any phenomenon that repeats regularly may be called periodic, and the time taken before the phenomenon repeats itself is then called the period. The motion may be said to consist of repeated cycles, the period being the time taken for the execution of one cycle. Suppose that $x = A \sin(\omega t + \alpha)$, where A (> 0), ω and α are constants. This may, for example, give the displacement x of a particle, moving in a straight line, at time t. The particle is thus oscillating about the origin. The period is the time taken for one complete oscillation and is equal to $2\pi/\omega$.

peripheral vertex (in a graph) Any vertex in a graph for which the distance between it and a *central vertex is equal to the *radius of the graph. In other words, any point which is as far away from the centre as it is possible to get.

permutation At an elementary level, a permutation of n objects can be thought of as an arrangement or a rearrangement of the n objects. The number of permutations of n objects is equal to $n!$. The number of 'permutations of n objects taken r at a time' is denoted by $^{n}P_{r}$, and equals n $(n-1)\ldots(n-r+1)$, which equals $n!/(n-r)!$. For example, there are 12 permutations of A, B, C, D taken two at a time: AB, AC, AD, BA, BC, BD, CA, CB, CD, DA, DB, DC.

Suppose that n objects are of k different kinds, with r_1 alike of one kind, r_2 alike of a second kind, and so on. Then the number of distinct permutations of the n objects equals

$$\frac{n!}{r_1!r_2!\ldots r_k!},$$

where $r_1 + r_2 + \ldots + r_k = n$. For example, the number of different anagrams of the word '*CHEESES*' is $7!/3!\,2!$, which equals 420. At a more advanced level, a permutation of a set X is defined as a *one-to-one onto mapping from X to X.

permutation group (substitution group) A set of permutations which constitutes a group when multiplication is defined as applying the permutations successively. One important special case is the set of all permutations of any set of n objects. This is a permutation group for any positive integer n, and there will be $n!$ elements in the full permutation group, known as the *symmetric group. The *subgroup consisting of all the *even permutations is known as the alternating group.

permutation matrix An n-by-n matrix with a single 1 in each row and each column and 0s everywhere else which can be used to represent a permutation of the n objects in a set. For example, to map 1, 2, 3 to 1, 3, 2 the matrix $\begin{pmatrix} 1 & 0 & 0 \\ 0 & 0 & 1 \\ 0 & 1 & 0 \end{pmatrix}$ would be used because

$$\begin{pmatrix} 1 & 0 & 0 \\ 0 & 0 & 1 \\ 0 & 1 & 0 \end{pmatrix} \begin{pmatrix} 1 \\ 2 \\ 3 \end{pmatrix} = \begin{pmatrix} 1 \\ 3 \\ 2 \end{pmatrix}.$$

permutations and combinations Counting the number of ways in which some particular arrangements of objects can be achieved.

perpendicular At right angles to one another. Can be of two lines, two planes, a line and a plane, or a line and a surface.

perpendicular axis theorem The following theorem about *moments of inertia of a *lamina:

Theorem
Suppose that a rigid body is in the form of a lamina lying in the plane $z = 0$. Let I_{Ox}, I_{Oy} and I_{Oz} be the moments of inertia of the body about the three coordinate axes. Then $I_{Oz} = I_{Ox} + I_{Oy}$.

perpendicular bisector The perpendicular bisector of a line segment AB is the straight line perpendicular to AB through the midpoint of AB.

perpendicular distance The distance from a point to a line or plane measured along the line perpendicular to the line or plane which passes through the given point. It will be the shortest distance between the point and any given point on the line or plane.

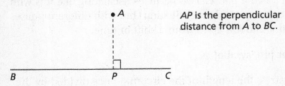

AP is the perpendicular distance from A to BC.

perpendicular lines In coordinate geometry of the plane, a useful necessary and sufficient condition that two lines, with gradients m_1 and m_2, are perpendicular is that $m_1 m_2 = -1$. (This is taken to include the cases when $m_1 = 0$ and m_2 is infinite, and vice versa.)

perpendicular planes Two planes in 3-dimensional space are perpendicular if *normals to the two planes are perpendicular. If $\mathbf{n}_1$ and $\mathbf{n}_2$ are vectors normal to the planes, this is so if $\mathbf{n}_1 . \mathbf{n}_2 = 0$.

perspective The art and mathematics of realistically representing three dimensions in a 2-dimensional diagram. Straight lines in space appear as straight in the image. Lines which are parallel to the image plane appear as parallel in the image. Other parallel lines in space meet at a vanishing point, and each new direction will require a separate vanishing point. See diagram for an example with two vanishing points. *See also* PROJECTIVE GEOMETRY.

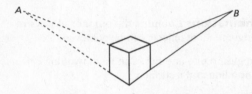

perturbation A small change in the parameter values in an equation or optimization problem, usually done to investigate the stability of a situation or to help identify the required solution.

perturbation (in mechanics) A small change in the path of a particle, usually while in an orbit.

peta- Prefix used with *SI units to denote multiplication by 10^{15}.

p.g.f. *See* PROBABILITY GENERATING FUNCTION.

phase Suppose that $x = A \sin(\omega t + \alpha)$, where $A(> 0)$, ω and α are constants. This may, for example, give the displacement x of a particle, moving in a straight line, at time t. The particle is thus oscillating about the origin. The constant α is the phase. Two particles oscillating like this with the same *amplitude and *period (of oscillation) but with different phases are executing the same motion apart from a shift in time.

phi The Greek letter phi, symbol ϕ.

pi For circles of all sizes, the length of the circumference divided by the length of the diameter is the same, and this number is the value of π. It is equal to 3.141 592 65 to 8 decimal places. Sometimes π is taken to be equal to $\frac{22}{7}$, but it must be emphasized that this is only an approximation. The decimal expansion of π is neither finite nor recurring, for it was shown in 1761 by Lambert, using continued fractions, that π is an *irrational number. In 1882, Lindemann proved that π is also a *transcendental number. The number appears in some contexts that seem to have no connection with the definition relating to a circle. For example:

$$\frac{\pi}{4} = 1 - \frac{1}{3} + \frac{1}{5} - \frac{1}{7} + \cdots, \qquad \frac{\pi^2}{6} = 1 + \frac{1}{2^2} + \frac{1}{3^2} + \frac{1}{4^2} + \cdots,$$

$$\frac{\pi}{2} = \frac{2}{1} \times \frac{2}{3} \times \frac{4}{3} \times \frac{4}{5} \times \frac{6}{5} \times \frac{6}{7} \times \frac{8}{7} \times \frac{8}{9} \times \cdots$$

$$= \prod_{n=1}^{\infty} \frac{4n^2}{4n^2 - 1} \text{ (\textbf{Wallis's Product})}.$$

Picard iteration An iterative method for solving ordinary *differential equations.

pico- Prefix used with *SI units to denote multiplication by 10^{-12}.

pie chart Suppose that some finite set is partitioned into subsets, and that the proportion of elements in each subset is expressed as a percentage. A pie chart is a diagram consisting of a circle divided into sectors whose areas are proportional to these percentages. In a similar way, when nominal data have been collected the data can be presented as a pie chart in which the circle is divided into sectors whose areas are proportional to the frequencies. The figure shows the kinds of vehicles recorded in a small traffic survey.

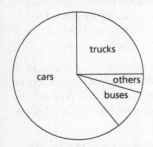

piecewise continuous A function is piecewise continuous on the interval (a, b) if the interval can be subdivided into a finite number of subintervals in which the function is continuous with finite endpoints.

Pigeonhole Principle The observation that, if m objects are distributed into n boxes, with $m > n$, then at least one box receives at least two objects. This can be applied in some obvious ways; for example, if you take any 13 people, then at least two of them have birthdays that fall in the same month. It also has less trivial applications.

pilot survey A small scale study conducted before a complete survey to test the effectiveness of the questionnaire.

pivot (in the simplex method) *See* SIMPLEX METHOD.

pivot, pivot operation Each step in *Gaussian elimination or *Gauss–Jordan elimination consists of making one of the entries a_{ij} of the matrix equal to 1 and using this to produce zeros elsewhere in the j-th column. This process is known as a pivot operation, with a_{ij} as pivot. The pivot, of course, must be non-zero. This corresponds to solving one of the equations for one of the unknowns x_j and substituting this into the other equations.

pivotal column (in the simplex method) *See* SIMPLEX METHOD.

pivotal row (in the simplex method) *See* SIMPLEX METHOD.

pivot table An interactive table in an electronic spreadsheet that computes summary statistics for a table of data. It is of particular use in removing the tedium of repeated calculations of the same statistic in a 2-way table, or n-way table.

place value The power of the base of a number representation system that uses the place value notation.

place value notation A system of representing numbers by an ordered sequence of digits where both the digit and its place value have to be known to determine the value. For example, the 5 in 53 indicates fifty because the place value is ten, but the 5 in 35 is just 5 units. The binary and decimal place-value systems are the two most commonly used.

placebo A placebo appears the same as the treatment to a participant in a study, often a clinical trial of a drug or other treatment, but has no active ingredients. They are used in *control groups because of the positive effects patients can experience simply from taking something they believe will make them better.

planar graph A *graph that can be drawn in such a way that no two edges cross. The *complete graph K_4, for example, is planar as either drawing of it in the figure shows. Neither the complete graph K_5 nor the complete bipartite graph $K_{3,3}$, for example, is planar.

Planck, Max Karl Ernst Ludwig FRS (1858–1947) German mathematician and theoretical physicist whose work in thermodynamics concerning the relationship between energy and wavelength won him the Nobel Prize for Physics in 1918, and provided some of the foundations on which *quantum mechanics was later developed.

plane (in Cartesian coordinates) A plane is represented by a linear equation, in other words by an equation of the form $ax + by + cz + d = 0$, where the constants a, b and c are not all zero. Here a, b and c are *direction ratios of a direction *normal to the plane. *See also* VECTOR EQUATION (of a plane).

plane geometry The area of mathematics relating to the properties of figures and lines drawn in a plane, and the relations between them.

plane of symmetry A plane with respect to which a 3-dimensional figure is symmetric. So for a cylinder, any plane containing parallel diameters in the two circular faces of the cylinder will be a plane of symmetry, as will the plane parallel to the two circular ends and equidistant from them. *See figure.*

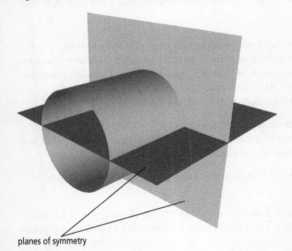

planes of symmetry

Platonic solid A convex *polyhedron is regular if all its faces are alike and all its vertices are alike. More precisely, this means that

    (i) all the faces are regular polygons having the same number p of edges, and

    (ii) the same number q of edges meet at each vertex.

Notice that the polyhedron shown here, with 6 triangular faces, satisfies (i), but is not regular because it does not satisfy (ii).

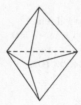

There are five regular convex polyhedra, known as the Platonic solids:

  (i) the regular *tetrahedron, with 4 triangular faces ($p = 3$, $q = 3$),
 (ii) the *cube, with 6 square faces ($p = 4$, $q = 3$),
(iii) the regular *octahedron, with 8 triangular faces ($p = 3$, $q = 4$),
 (iv) the regular *dodecahedron, with 12 pentagonal faces ($p = 5$, $q = 3$),
  (v) the regular *icosahedron, with 20 triangular faces ($p = 3$, $q = 5$).

platykurtic *See* KURTOSIS.

Playfair's axiom The axiom which says that, given a point not on a given line, there is precisely one line through the point parallel to the line. It is equivalent to the axiom of *Euclid's known as the *parallel postulate. The name arises from its occurrence in a book by John Playfair (1748–1819), a British geologist and mathematician.

plot To mark points on a coordinate graph. Also used in relating to constructing the curve of a function—by plotting a number of points and inferring the behaviour of the curve between the plotted points.

plus Addition. So $3 + 5$ means add 3 and 5, but is normally read as '3 plus 5'.

plus sign The symbol $+$ used to denote addition of numbers or any other binary operation with similar properties. Can be used to denote a positive quantity, though the absence of $+$ or $-$ is taken to imply the positive.

p.m.f. $=$ PROBABILITY MASS FUNCTION.

p-norm In a *metric space X, the norm defined by

$$||X||_p = \left(\sum_{i=1}^{n} |x_i|^p \right)^{\frac{1}{p}}.$$

Note that when $p = 2$ this is just the standard *Euclidean norm, or the length of the vector, and for $p = 1$ it is the Manhattan or *taxicab norm.

Poincaré, (Jules) Henri (1854–1912) French mathematician, who can be regarded as the last mathematician to be active over the entire field of mathematics. He moved from one area of mathematics to another, making major contributions to most of them—'a conqueror rather than a colonist', it was said. Within pure mathematics, he is seen as one of the main founders of topology and the discoverer of what are called automorphic functions. In applied mathematics, he is remembered for his theoretical work on the qualitative aspects of celestial mechanics, which was probably the most important work in that area since *Laplace and *Lagrange. Connecting and motivating both was his qualitative theory of differential equations. Not the least of his achievements was a stream of books, popular in both senses, still strongly recommended to any aspiring young mathematician.

point A geometrical construct which has position but no size. The position is often specified by coordinates.

point at infinity The point added to the set of complex numbers to obtain the *extended complex plane.

point estimate *See* ESTIMATE.

point mass (in mechanics) The simplifying assumption that all the mass of a body is concentrated at a single point so that considerations such as rotation can be ignored.

point of application In the real world, a force acting on a body is usually applied over a region, such as a part of the surface of the body. When someone is pushing a trolley by hand, there is a region of the handle of the trolley where the hand pushes. When a tugboat is towing a ship, the rope that is being used is attached to a bollard on the deck of the ship. Some forces, such as gravitational forces, act throughout a body. In contrast, in a *mathematical model, it is assumed that every force acts at a particular point, the point of application of the force. In some mathematical models, each point of the surface of a body may experience a force (when the surface is experiencing a *pressure), or every point throughout the volume occupied by a body may experience a force (when the body is experiencing a *field of force).

point of inflexion A point on a graph $y = f(x)$ at which the *concavity changes. Thus, there is a point of inflexion at a if, for some $\delta > 0$, $f''(x)$

exists and is positive in the open interval $(a - \delta, a)$ and $f''(x)$ exists and is negative in the open interval $(a, a + \delta)$, or vice versa.

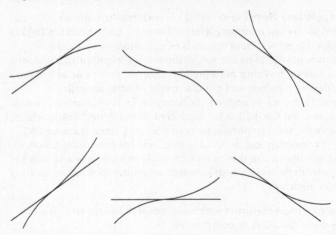

If the change (as x increases) is from concave up to concave down, the graph and its tangent at the point of inflexion look like one of the cases shown in the first row of the figure. If the change is from concave down to concave up, the graph and its tangent look like one of the cases shown in the second row. The middle diagram in each row shows a point of inflexion that is also a *stationary point, since the tangent is horizontal.

If f'' is continuous at a, then for $y = f(x)$ to have a point of inflexion at a it is necessary that $f''(a) = 0$, and so this is the usual method of finding possible points of inflexion. However, the condition $f''(a) = 0$ is not sufficient to ensure that there is a point of inflexion at a: it must be shown that $f''(x)$ is positive to one side of a and negative to the other side. Thus, if $f(x) = x^4$ then $f''(0) = 0$; but $y = x^4$ does not have a point of inflexion at 0 since $f''(x)$ is positive to both sides of 0. Finally, there may be a point of inflexion at a point where $f''(x)$ does not exist, for example at the origin on the curve $y = x^{1/3}$, as shown in the figure.

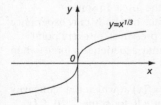

pointwise convergence A sequence of functions on a given set D such that for any point x in D, for every $\varepsilon > 0$, there is an integer N for which $|f_m(x) - f_n(x)| < \varepsilon$ for all m, $n > N$, is said to have pointwise convergence. Note that N appears only after both x and ε are defined, and it may depend on both of them. This is in contrast to *uniform convergence, where a single N has to exist for all values of x, for any given ε.

Poisson, Siméon-Denis (1781–1840) French mathematician who was a pupil and friend of *Laplace and *Lagrange. He extended their work in celestial mechanics and made outstanding contributions in the fields of electricity and magnetism. In 1837, in an important paper on probability, he introduced the distribution that is named after him and formulated the law of large numbers.

Poisson distribution The discrete probability *distribution with *probability mass function given by $\Pr(X = r) = \exp(-\lambda)\lambda^r / r!$, for $r = 0, 1, 2, \ldots$, where λ is a positive parameter. The mean and variance are both equal to λ. The Poisson distribution gives the number of occurrences in a certain time period of an event which occurs randomly but at a constant average rate. It can be used as an approximation to the *binomial distribution, where n is large and p is small, by taking $\lambda = np$.

polar coordinates Suppose that a point O in the plane is chosen as origin, and let Ox be a directed line through O, with a given unit of length. For any point P in the plane, let $r = |OP|$ and let θ be the angle (in radians) that OP makes with Ox, the angle being given a positive sense anticlockwise from Ox. The angle θ satisfies $0 \leq \theta < 2\pi$. Then (r, θ) are the polar coordinates of P. (The point O gives no corresponding value for θ, but is simply said to correspond to $r = 0$.) If P has polar coordinates (r, θ) then $(r, \theta + 2k\pi)$, where k is an integer, may also be permitted as polar coordinates of P.

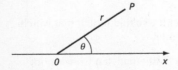

Suppose that Cartesian coordinates are taken with the same origin and the same unit of length, with positive x-axis along the directed line Ox. Then the Cartesian coordinates (x, y) of a point P can be found from (r, θ), by $x = r \cos \theta$, $y = r \sin \theta$. Conversely, the polar coordinates can be found from the Cartesian coordinates as follows: $r = \sqrt{x^2 + y^2}$, and θ is such that

$$\cos \theta = \frac{x}{\sqrt{x^2 + y^2}}, \quad \sin \theta = \frac{y}{\sqrt{x^2 + y^2}}.$$

(It is true that, when $x \neq 0$, $\theta = \tan^{-1}(y/x)$, but this is not sufficient to determine θ.) In certain circumstances, authors may allow r to be negative, in which case the polar coordinates $(-r, \theta)$ give the same point as $(r, \theta + \pi)$.

In mechanics, it is useful, when a point P has polar coordinates (r, θ), to define vectors $\mathbf{e}_r$ and $\mathbf{e}_\theta$ by $\mathbf{e}_r = \mathbf{i} \cos \theta + \mathbf{j} \sin \theta$ and $\mathbf{e}_\theta = -\mathbf{i} \sin \theta + \mathbf{j} \cos \theta$, where $\mathbf{i}$ and $\mathbf{j}$ are unit vectors in the directions of the positive x- and y-axes. Then $\mathbf{e}_r$ is a unit vector along OP in the direction of increasing r, and $\mathbf{e}_\theta$ is a unit vector perpendicular to this in the direction of increasing θ. These vectors satisfy $\mathbf{e}_r \cdot \mathbf{e}_\theta = 0$ and $\mathbf{e}_r \times \mathbf{e}_\theta = \mathbf{k}$, where $\mathbf{k} = \mathbf{i} \times \mathbf{j}$.

polar equation An equation of a curve in *polar coordinates is usually written in the form $r = f(\theta)$. Polar equations of some common curves are given in the table.

Curve		Polar equation	
Circle	$x^2 + y^2 = 1$	$r = 1$	
Half-line	$y = x \ (x > 0)$	$\theta = \pi / 4$	
Line	$x = 1$	$r = \sec \theta$	$(-\tfrac{1}{2}\pi < \theta < \tfrac{1}{2}\pi)$
Line	$y = 1$	$r = \operatorname{cosec} \theta$	$(0 < \theta < \pi)$
Circle	$x^2 + y^2 - 2ax = 0 \ (a > 0)$	$r = 2a \cos \theta$	$(-\tfrac{1}{2}\pi < \theta \leq \tfrac{1}{2}\pi)$
Circle	$x^2 + y^2 - 2by = 0 \ (b > 0)$	$r = 2b \sin \theta$	$(0 \leq \theta < \pi)$
Cardioid	See CARDIOID	$r = 2a(1 + \cos \theta)$	$(-\pi < \theta \leq \pi)$
Conic	See CONIC	$l/r = 1 + e \cos \theta$	$(\cos \theta \neq -1/e)$

polar form of a complex number For a complex number z, let $r = |z|$ and $\theta = \arg z$. Then $z = r(\cos \theta + i \sin \theta)$, and this is the polar form of z. It may also be written $z = re^{i\theta}$.

pole (in analysis) A *singularity in a complex-valued function at which the function goes to infinity like a polynomial.

Polya, George (1887–1985) Hungarian mathematician whose work included collaboration with *Hardy and *Littlewood, with papers on mathematical physics, geometry, complex analysis, combinatorics and probability theory. He is perhaps best known for his contribution to mathematical education through his book *How To Solve It* published in its second edition in 1957, and still widely regarded half a century later as one of the best expositions of the art of doing mathematics. In a remarkably easy prose style for a book about mathematics, Polya argues that problem solving requires the study of *heuristics, and summarizes the problem-solving cycle in four stages: understanding the problem, devising a plan, carrying out the plan, looking back, or 'see, plan, do, check'.

polygon A plane figure bounded by some number of straight sides. This definition would include polygons like the one shown below, but often

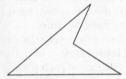

figures of this sort are intended to be excluded and it is implicitly assumed that a polygon is to be *convex. Thus a (convex) polygon is a plane figure consisting of a finite region bounded by some finite number of straight lines, in the sense that the region lies entirely on one side of each line. The exterior angles of the polygon shown below are those marked,

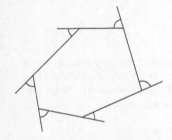

and the sum of the exterior angles is always equal to $360°$. The sum of the interior angles of an n-sided (convex) polygon equals $2n - 4$ right angles. In a regular (convex) polygon, all the sides have equal length and the interior angles have equal size; and the vertices lie on a circle.

   Polygons with the following numbers of edges have the names given:

5	pentagon	9	enneagon (or nonagon)
6	hexagon	10	decagon
7	heptagon	11	hendecagon
8	octagon	12	dodecagon

polygon of forces Suppose that the forces $\mathbf{F}_1, \mathbf{F}_2, \ldots, \mathbf{F}_n$ act on a particle. Make a construction of directed line segments as follows. Let

$\overrightarrow{A_1A_2}$ represent $\mathbf{F}_1$, $\overrightarrow{A_2A_3}$ represent $\mathbf{F}_2$, and so on, with $\overrightarrow{A_nA_{n+1}}$ representing $\mathbf{F}_n$. (These line segments are not necessarily coplanar.) Then the total force $\mathbf{F}$ acting on the particle, given by $\mathbf{F} = \sum \mathbf{F}_b$ is represented by the directed line segment $\overrightarrow{A_1A_{n+1}}$ In particular, if the total force is zero, the point A_{n+1} coincides with A_1 and the directed line segments form the n sides of a polygon, called a **polygon of forces**.

polyhedron (polyhedra) A solid figure bounded by some number of plane polygonal faces. This definition would include the one shown here, and it may well be that figures such as this are to be excluded from consideration. So it is often assumed that a polyhedron is *convex. Thus a (convex) polyhedron is a finite region bounded by some number of planes, in the sense that the region lies entirely on one side of each plane. Each edge of the polyhedron joins two vertices, and each edge is the common edge of two faces.

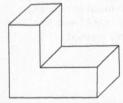

The numbers of vertices, edges and faces of a convex polyhedron are connected by an equation given in *Euler's Theorem (for polyhedra).

Certain polyhedra are called regular, and these are the five *Platonic solids; the thirteen *Archimedean solids are semi-regular.

polynomial Let $a_0, a_1, \ldots, a_n$ be real numbers. then

$$a_n x^n + a_{n-1}x^{n-1} + \ldots + a_1 x + a_0$$

is a **polynomial** in x (with real coefficients). When $a_0, a_1, \ldots, a_n$ are not all zero, it can be assumed that $a_n \neq 0$ and the polynomial has DEGREE n. For example, $x^2 - \sqrt{2}x + 5$ and $3x^4 + \frac{7}{2}x^2 - x$ are polynomials of degrees 2 and 4 respectively. The number a_r is the coefficient of x^r (for $r = 1, 2, \ldots, n$) and a_0 is the constant term. A polynomial can be denoted by $f(x)$ (so that f is a *polynomial function), and then $f(-1)$, for example, denotes the value of the polynomial when x is replaced by -1. In the same way, it is possible to consider polynomials in x, or more commonly in z, with complex coefficients, such as $z^2 + 2(1 - i)z + (15 + 6i)$, and the same terminology is used.

polynomial equation An equation $f(x) = 0$, where $f(x)$ is a *polynomial.

polynomial function In real analysis, a polynomial function is a function f defined by a formula $f(x) = a_n x^n + a_{n-1}x^{n-1} + \ldots + a_1 x + a_0$ for all x in **R**, where $a_0, a_1, \ldots, a_n$ are real numbers. *See also* POLYNOMIAL; terminology applying to a polynomial is also used to apply to the corresponding polynomial function.

Poncelet, Jean-Victor (1788–1867) French military engineer and mathematician, recognized in mathematics as the founder of projective geometry.

pons asinorum The Latin name, meaning 'bridge of asses', familiarly given to the fifth proposition in the first book of Euclid: that the base angles of an isosceles triangle are equal. It is said to be the first proof found difficult by some readers.

pooled estimate of variance (standard deviation) If it is reasonable to assume that the variances of two populations that you wish to compare are equal, then the most discriminatory test available uses a single estimate of the common variance and uses all of the available data from the two samples. The pooled estimate is given by

$$s_p^2 = \frac{\sum_{i=1}^{n_X} (x_i - \bar{x})^2 + \sum_{j=n}^{n_Y} (y_j - \bar{y})^2}{n_X + n_Y - 2}$$ where $\{x_i\}$, $\{y_j\}$ are the two sets of sample observations with sample sizes n_X, n_Y respectively and sample means $\bar{x}, \bar{y}$.

population A set about which *inferences are to be drawn based on a sample taken from the set. The sample may be used to accept or reject a hypothesis about the population.

population mean *See* MEAN.

position ratio If A and B are two given points, and P is a point on the line through A and B, then the ratio $AP : PB$ may be called the position ratio of P.

position vector Suppose that a point O is chosen as origin (in the plane, or in 3-dimensional space). Given any point P, the position vector $\mathbf{p}$ of the point P is the *vector represented by the directed line segment $\overrightarrow{OP}$. Authors who refer to $\overrightarrow{OP}$ as a vector call $\overrightarrow{OP}$ the position vector of P.

In mechanics, the position vector of a particle is often denoted by $\mathbf{r}$, which is a function of the time t, giving the position of the particle as it moves. If Cartesian coordinates are used, then $\mathbf{r} = x\mathbf{i} + y\mathbf{j} + z\mathbf{k}$, where x, y and z are functions of t.

positive A value greater than zero.

positive angle Measured in an anticlockwise direction.

positive correlation *See* CORRELATION.

positive direction *See* DIRECTED LINE.

possible Capable of being true, not containing any inherent contradiction. For example, it is possible to solve $x^2 + 1 = 0$ within the set of complex numbers, but not within the set of reals. It is not possible for x to be both even and odd.

posterior distribution *See* PRIOR DISTRIBUTION.

posterior probability *See* PRIOR PROBABILITY.

post-multiplication When the product **AB** of matrices **A** and **B** is formed (*see* MULTIPLICATION (of matrices)), **A** is said to be post-multiplied by **B**.

post-optimal analysis When a computer program is used to solve a *linear programming problem, the solution can provide further information, e.g. about any *slack there is in the variables, or, the value attached to relaxing a constraint. Since these pieces of information are available only after the algorithm has produced the optimum solution, the general term post-optimal analysis is used to describe procedures to obtain them.

postulate = AXIOM.

potential energy With every *conservative force there is associated a potential energy, denoted by E_p or V, defined up to an additive constant as follows. The change in potential energy during a time interval is defined to be the negative of the *work done by the force during that interval. In detail, the change in E_p during the time interval from $t = t_1$ to $t = t_2$ is equal to

$$-\int_{t_1}^{t_2} \mathbf{F}.\mathbf{v}\, dt,$$

where **v** is the velocity of the point of application of the conservative force **F**. For example, if $\mathbf{F} = -k(x - l)\mathbf{i}$, as in *Hooke's law, then $E_p = \frac{1}{2}k(x - l)^2$, taking the potential energy to be zero when $x = l$. In problems involving *gravity near the Earth's surface it may be assumed that the gravitational force is given by $\mathbf{F} = -mg\mathbf{k}$, and then, taking the potential energy to be zero when $z = 0$, $E_p = mgz$. *See also* GRAVITATIONAL POTENTIAL ENERGY.

Potential energy has the dimensions ML^2T^{-2}, and the SI unit of measurement is the *joule.

pound Symbol lb. A unit of the British system of mass, equal to 0.454 *kilogram. 1 kg = 2.2 lbs.

power (of a matrix) If **A** is a square matrix, then **AA**, **AAA**, **AAAA**, ... are defined, and these powers of **A** are denoted by $\mathbf{A}^2$, $\mathbf{A}^3$, $\mathbf{A}^4$, ... For all positive integers p and q, (i) $\mathbf{A}^p \mathbf{A}^q = \mathbf{A}^{p+q} = \mathbf{A}^q \mathbf{A}^p$, and (ii) $(\mathbf{A}^p)^q = \mathbf{A}^{pq}$. By definition, $\mathbf{A}^0 = \mathbf{I}$. Moreover, if **A** is *invertible, then $\mathbf{A}^2$, $\mathbf{A}^3$, ... are invertible, and it can be shown that $(\mathbf{A}^{-1})^p$ is the inverse of $\mathbf{A}^p$, that is, that $(\mathbf{A}^{-1})^p = (\mathbf{A}^p)^{-1}$. So either of these is denoted by $\mathbf{A}^{-p}$. Thus, when **A** is invertible, $\mathbf{A}^p$ has been defined for all integers (positive, zero and negative) and properties (i) and (ii) above hold for *all* integers p and q.

power (in mechanics) The power associated with a force is the rate at which the force does work. The power P is given by $P = \mathbf{F} \cdot \mathbf{v}$, where **v** is the velocity of the point of application of the force **F**. The *work done by the force during a certain time interval is equal to the definite integral, with respect to t, of the power, between the appropriate limits.

In everyday usage, the power of an engine can be thought of as the rate at which the engine can do work, or as the rate at which it can produce energy.

Power has the dimensions ML^2T^{-3}, and the SI unit of measurement is the *watt.

power (of a real number) When a real number a is raised to the *index p to give a^p, the result is a power of a. The same index notation is used in other contexts, for example to give a power $\mathbf{A}^p$ of a square matrix (*see* POWER (of a matrix) or a power g^p of an element g in a multiplicative group. When a^p is formed, p is sometimes called the power, but it is more correctly called the index to which a is raised.

power (of a test) In *hypothesis testing, the probability that a test rejects the null hypothesis when it is indeed false. It is equal to $1 - \beta$, where β is the probability of a *Type II error.

power series A series $a_0 + a_1x + a_2x^2 + a_3x^3 + \ldots + a_nx^n + \ldots$, in ascending powers of x, with coefficients a_0, a_1, a_2, ..., is a power series in x. For example, the *geometric series $1 + x + x^2 + \ldots + x^n + \ldots$ is a power series; this has a sum to infinity (*see* SERIES) only if $-1 < x < 1$. Further examples of power series are given in Appendix 6. Notice that it is necessary to say for what values of x each series has the given sum; this is the case with any power series.

power set The set of all subsets of a set S is the power set of S, denoted by $\mathscr{P}(S)$. Suppose that S has n elements a_1, a_2, ..., a_n, and let A be a subset of S. For each element a_i of S, there are two possibilities: either $a_i \in A$ or not. Considering all n elements leads to 2^n possibilities in all. Hence, S has 2^n subsets; that is, $\mathscr{P}(S)$ has 2^n elements. If $S = \{a, b, c\}$, the $8(=2^3)$ elements of $\mathscr{P}(S)$ are $\varnothing$, $\{a\}$, $\{b\}$, $\{c\}$, $\{a, b\}$, $\{a, c\}$, $\{b, c\}$ and $\{a, b, c\}$.

precision (in statistics) A measurement of the quality of a statistic as an *estimator of a *parameter. The precision is measured by the *standard error of the statistic, so in general the precision may be improved by increasing the sample size. The precision is used in comparing the *efficiency of two or more estimators of a parameter.

precision (numerical analysis) The accuracy to which a given calculation is performed. Single precision is usually up to 16 digits, double precision uses two connected single-precision blocks, and more than two connected blocks is variously described as high, multiple or extended precision.

predator–prey equations To model the behaviour in a predator–prey situation the normal dependencies on birth and death rates for each population are modified by terms reflecting their interdependency: since the rate of encounter between predator and prey is approximately proportional to the product of the population sizes this would give rise to linked differential equations of the form $\dfrac{dX}{dt} = (b_X - d_X)X - kXY$, $\dfrac{dX}{dt} = (b_Y - d_Y)Y + lXY$, The model is simpler than in real life but solutions of these linked differential equations give insights into the conditions in which predator–prey populations can coexist.

((())) SEE WEB LINKS
• Simulations of fox and rabbit populations.

predicate (in logic) An expression which ascribes a property to one or more subjects. 'Louise is a girl' is a predicate with a single subject. A predicate with more than one subject will be a *relation. So, 'Louise, Joanne and Laura are sisters' is a predicate with multiple binary relations between pairs of subjects.

predicted variable = DEPENDENT VARIABLE.

predictor variable = EXPLANATORY VARIABLE.

pre-multiplication When the product **AB** of matrices **A** and **B** is formed (*see* MULTIPLICATION (of matrices)), **B** is said to be pre-multiplied by **A**.

pressure Force per unit area. The simplest situation to consider is when a force is uniformly distributed over a plane surface, the direction of the force being perpendicular to the surface. Suppose that a rectangular box, of length a and width b, has weight mg. Then the area of the base is ab, and the gravitational force creates a pressure of mg/ab acting at each point of the ground on which the box rests.

Now consider a liquid of density ρ at rest in an open container. The weight of liquid supported by a horizontal surface of area ΔA situated at a depth h is $\rho(\Delta A)hg$. The resulting pressure (in excess of *atmospheric pressure) at each point of this horizontal surface at a depth h is ρgh. Further investigations involving elemental regions show that pressure can be defined at a point in a liquid, and that pressure at a point is independent of direction. In general, pressure in a fluid (a liquid or a gas) is a scalar function of position and time.

Pressure has dimensions $ML^{-1}T^{-2}$, and the SI unit of measurement is the *pascal.

prime A positive integer p is a prime, or a prime number, if $p \neq 1$ and its only positive divisors are 1 and itself.

It is known that there are infinitely many primes. Euclid's proof argues by contradiction as follows. Suppose that there are finitely many primes $p_1, p_2, \ldots, p_n$. Consider the number $p_1 p_2, \ldots, p_n + 1$. This is not divisible by any of $p_1, p_2, \ldots, p_n$, so it is either another prime itself or is divisible by primes not so far considered. It follows that the number of primes is not finite.

Large prime numbers can be discovered by using computers. At any time, the largest known prime is usually the largest known *Mersenne prime.

prime decomposition *See* UNIQUE FACTORIZATION THEOREM.

prime factorization, prime representation = PRIME DECOMPOSITION.

prime meridian The *meridian through Greenwich from which *longitude is measured.

prime number theorem For a positive real number x, let $\pi(x)$ be the number of *primes less than or equal to x. The prime number theorem says that, as $x \to \infty$,

$$\frac{\pi(x)}{x/\ln x} \to 1.$$

In other words, for large values of x, $\pi(x)$ is approximately equal to $x/\ln x$. This gives, in a sense, an idea of what proportion of integers are prime. Proved first in 1896 by Jacques *Hadamard and Charles De La

*Vallée-Poussin independently, all proofs are either extremely complicated or based on advanced mathematical ideas.

prime to each other = RELATIVELY PRIME.

primitive (n-th root of unity) An *n-th root of unity z is primitive if every n-th root of unity is a power of z. For example, i is a primitive fourth root of unity, but -1 is a fourth root of unity that is not primitive.

primitive polynomial A polynomial over an *integral domain with the property that the greatest common divisor of its coefficients is 1.

Prim's algorithm (to solve the minimum connector problem) This method is more effective than *Kruskal's algorithm when a large number of vertices and/or when the distances are listed in tabular form rather than shown on a graph. Since all vertices will be on the minimum connected route, it does not matter which vertex you choose as a starting point, so make an arbitrary choice, call it P_1 say. Now choose the point with the shortest edge connecting it to P_1 (and any time there is a tie for the shortest edge, choose either one at random) and call it P_2. At each stage add the edge and new point P_i which adds the shortest distance to the total and does not create any loops. Once P_n has been reached, the minimum connected path has been identified.

principal axes (in mechanics) *See* PRODUCTS OF INERTIA.

principal axes (of a quadric) A set of principal axes of a *quadric is a set of axes of a coordinate system in which the quadric has an equation in *canonical form.

principal moments of inertia *See* PRODUCTS OF INERTIA.

principal value A unique member of a set of values a variable can take, especially when taking the inverse of a periodic function, or taking square roots. The principal value is the one returned by a scientific calculator when evaluating these inverse functions. If $x^2 = 9$, x could be 3 or -3, but a calculator will give $\sqrt{9} = 3$. Similarly $\sin 30° = \sin 150° = \sin 390° = \sin(-210°) = \sin(-330°) = 0.5$, and the calculator will give $\sin^{-1} 0.5 = 30°$. *See also* ARGUMENT.

principle of conservation of energy *See* CONSERVATION OF ENERGY.

principle of conservation of linear momentum *See* CONSERVATION OF LINEAR MOMENTUM.

principle of mathematical induction *See* MATHEMATICAL INDUCTION.

principle of moments The principle that, when a system of coplanar forces acts on a body and produces a state of equilibrium, the sum of the *moments of the forces about any point in the plane is zero.

Suppose, for example, that a light rod is supported by a pivot and that a particle of mass m_1 is suspended from the rod at a distance d_1 to the right of the pivot, and another of mass m_2 is suspended from the rod at a distance d_2 to the left of the pivot. The force due to gravity on the first particle has magnitude m_1g, and its moment about the pivot is equal to $(m_1g)d_1$ clockwise. Corresponding to the second particle, there is a moment $(m_2g)d_2$ anticlockwise. By the principle of moments, there is equilibrium if $(m_1g)d_1 = (m_2g)d_2$.

To use the vector definition of moment in this example, take the pivot as the origin O, the x-axis along the rod with the positive direction to the right and the y-axis vertical with the positive direction upwards. The force due to gravity on the first particle is $-m_1\,\mathbf{gj}$, and the moment of the force about O equals $(d_1\mathbf{i}) \times (-m_1\mathbf{gj}) = -(m_1g)d_1\mathbf{k}$. Corresponding to the second particle, the moment equals $(-d_2\mathbf{i}) \times (-m_2\mathbf{gj}) = (m_2g)d_2\mathbf{k}$. The principle of moments gives the same condition for equilibrium as before.

principle of the excluded middle The statement that every proposition is either true or false.

prior distribution The *distribution attached to a *parameter before certain data are obtained. The posterior distribution is the distribution attached to it after the data are obtained. The posterior distribution may then be considered as the prior distribution before further data are obtained.

prior probability The probability attached to an *event before certain data are obtained. The posterior probability is the probability attached to it after the data are obtained. The posterior probability may be calculated by using *Bayes' Theorem. It may then be considered as the prior probability before further data are obtained.

prism A convex *polyhedron with two 'end' faces that are congruent convex polygons lying in parallel planes in such a way that, with edges joining corresponding vertices, the remaining faces are parallelograms. A right-regular prism is one in which the two end faces are regular polygons and the remaining faces are rectangular. A right-regular prism in which the rectangular faces are square is semi-regular (*see* ARCHIMEDEAN SOLID).

prisoner's dilemma (in game theory) Any generalized situation which is essentially similar to the following: two prisoners are interviewed separately about a crime. If both deny the charge there is no evidence to convict them and both will be released. If one confessed and implicates the other, who denies it, then the first will be released and the other will be heavily punished. If both confess then both receive a lesser punishment than the prisoner in the case where one denied it and the other confessed. The 'dilemma' for the prisoner is that denying the charge can result in both the best possible outcome and the worst possible outcome depending on the action of the other prisoner. An example of this is the arms race, where both sides disarming is the best outcome but unilateral disarmament could result in annihilation.

(⊕) SEE WEB LINKS
• Play the prisoner's dilemma against the computer.

probability The probability of an *event *A*, denoted by $\Pr(A)$, is a measure of the possibility of the event occurring as the result of an experiment. For any event *A*, $0 \le \Pr(A) \le 1$. If *A* never occurs, then $\Pr(A) = 0$; if *A* always occurs, then $\Pr(A) = 1$. If an experiment could be repeated *n* times and the event *A* occurs *m* times, then the limit of m/n as $n \to \infty$ is equal to $\Pr(A)$.

If the *sample space *S* is finite and the possible outcomes are all equally likely, then the probability of the event *A* is equal to $n(A)/n(S)$, where $n(A)$ and $n(S)$ denote the number of elements in *A* and *S*. The probability that a randomly selected element from a finite population belongs to a certain category is equal to the proportion of the population belonging to that category.

The probability that a discrete *random variable *X* takes the value x_i is denoted by $\Pr(X = x_i)$. The probability that a continuous random variable *X* takes a value less than or equal to *x* is denoted by $\Pr(X \le x)$. This notation may be extended in a natural way.

See also CONDITIONAL PROBABILITY and PRIOR PROBABILITY.

probability density function For a continuous *random variable X, the probability density function (or p.d.f.) of X is the function f such that

$$\Pr(a \leq X \leq b) = \int_a^b f(x)dx.$$

Suppose that, for the random variable X, a sample is taken and a corresponding *histogram is drawn with class intervals of a certain width. Then, loosely speaking, as the number of observations increases and the width of the class intervals decreases, the histogram assumes more closely the shape of the graph of the probability density function.

probability generating function If X is a random variable where X takes only non-negative integer values then $P(t) = E(t^X)$ is the probability generating function (pgf). $P(t) = \sum_{i=0}^{\infty} p_i t^i$. If this series is differentiated n times, the terms in t^i where $i < n$ have become identically zero, the n-th differential of $p_n t^n$ is $n!p_n$ and all higher terms still include a power of t and therefore evaluating the n-th differential of $P(t)$ at zero will generate the probability X takes the value n since $P^{(n)}(0) = n! \times \Pr\{X = n\}$. The mean and variance of X can also be computed using the pgf since $E(X) = P'(1)$ and $E(X^2) = P''(1) + P'(1)$ so Var $(X) = P''(1) + P'(1) - \{P'(1)\}^2$. One of the reasons pgfs are powerful tools is the *convolution theorem which states that if X and Y are two independent random variables defined on the set of non-negative integers with pgfs $P_{X(t)}$ and $P_{Y(t)}$ then $P_{X+Y}(t) = P_X(t) \times P_Y(t)$, i.e. the pgf of the sum is the product of the pgfs, allowing elegant derivations of expectation algebra theorems and results such as the sum of two independent Poisson distributions being another Poisson.

probability mass function For a discrete *random variable X, the probability mass function (or p.m.f.) of X is the function p such that $p(x_i) = \Pr(X = x_i)$, for all i.

probability measure See PROBABILITY SPACE.

probability paper Graph paper scaled so that the *cumulative distribution function for a specified probability distribution will be a straight line. This then provides an informal test as to whether data might reasonably be supposed to come from that family of distributions, by considering how closely the observed data fit a straight line. If it is a satisfactory fit, it can also provide estimates of parameter values. The most commonly tested distribution is the *normal, so this is the most common probability paper, often generated now within a statistical computer package while testing assumptions for sophisticated procedures.

probability space A finite *measure space with associated probability measure that assigns unit measure to the complete space.

probable error (in statistics) The half-width of a 50% confidence interval. The observed value from a normal distribution satisfies $|X - \mu| < 0.67\sigma$ with probability 0.5, so the probable error for a normal distribution is 0.67σ.

product (of complex numbers) *See* MULTIPLICATION (of complex numbers).

product (of matrices) *See* MULTIPLICATION (of matrices).

product moment correlation coefficient *See* CORRELATION.

product notation For a finite sequence $a_1, a_2, \ldots, a_n$, the product $a_1 a_2 \ldots a_n$ may be denoted, using the capital Greek letter pi, by

$$\prod_{r=1}^{n} a_r.$$

(The letter r used here could equally well be replaced by any other letter.) For example,

$$\prod_{r=1}^{9} \left(1 - \frac{1}{r+1}\right) = \left(1 - \frac{1}{2}\right)\left(1 - \frac{1}{3}\right)\cdots\left(1 - \frac{1}{10}\right).$$

Similarly, from an infinite sequence $a_1, a_2, a_3, \ldots$, an *infinite product $a_1 a_2 a_3 \ldots$ can be formed, and it is denoted by

$$\prod_{r=1}^{\infty} a_r.$$

This notation is also used for the value of the infinite product, if it exists. For example

$$\prod_{r=2}^{\infty} \left(1 - \frac{1}{r^2}\right) = \frac{1}{2}.$$

product of inertia A quantity similar to a *moment of inertia, but relating to a rigid body and a pair of given perpendicular axes. The products of inertia occur in the inertia matrix which arises in connection with the *angular momentum and the *kinetic energy of the rigid body. With Cartesian coordinates, let I_{yz}, I_{zx} and I_{xy} denote the products of inertia of the rigid body associated with the y- and z-axes, the z- and x-axes, and the x- and y-axes. Suppose that a rigid body consists of a system of n particles $P_1, P_2, \ldots, P_n$, where P_i has mass m_i and position vector $\mathbf{r}_i$, with $\mathbf{r}_i = x_i\mathbf{i} + y_i\mathbf{j} + z_i\mathbf{k}$. Then

$$I_{yz} = \sum_{i=1}^{n} m_i y_i z_i, \qquad I_{zx} = \sum_{i=1}^{n} m_i z_i x_i, \qquad I_{xy} = \sum_{i=1}^{n} m_i x_i y_i.$$

For solid bodies, the products of inertia are defined by corresponding integrals.

When the coordinate planes are planes of symmetry of the rigid body, the products of inertia are zero. Furthermore, it can be shown that, for any point fixed in a rigid body, such as the centre of mass, there is a set of three perpendicular axes, with origin at the point, such that the corresponding products of inertia are all zero. These are called the principal axes, and the corresponding moments of inertia are the principal moments of inertia. The use of the principal axes produces a considerable simplification in the expressions for the angular momentum and the kinetic energy of the rigid body.

product rule (for differentiation) *See* DIFFERENTIATION.

product set = CARTESIAN PRODUCT.

progression A sequence in which each term is obtained from the previous one by some rule. The most common progressions are *arithmetic sequences, *geometric sequences and *harmonic sequences.

projectile When a body is travelling near the Earth's surface, subject to no forces except the uniform gravitational force and possibly air resistance, it may be called a projectile. The standard *mathematical model uses a particle to represent the body and a horizontal plane to represent the Earth's surface.

When there is no air resistance, the trajectory, the path traced out by the projectile, is a parabola whose vertex corresponds to the point at which the projectile attains its maximum height. Take the origin at the point of projection, the x-axis horizontal and the y-axis vertical with the positive direction upwards. Then the equation of motion is $m\ddot{\mathbf{r}} = -mg\mathbf{j}$, subject to $\mathbf{r} = \mathbf{0}$ and $\dot{\mathbf{r}} = (v \cos \theta)\mathbf{i} + (v \sin \theta)\mathbf{j}$ at $t = 0$, where v is the speed of projection and θ is the *angle of projection. This gives $x = (v \cos \theta)t$ and $y = (v \sin \theta)t - \frac{1}{2}gt^2$, for $t \geq 0$. From these, it is found that $y = 0$ when $t = \dfrac{2v \sin \theta}{g}$ and then $x = \dfrac{v^2}{g} \sin 2\theta$. Hence the *range on the horizontal plane through the point of projection is equal to $\dfrac{v^2}{g} \sin 2\theta$, and the time of flight is $\dfrac{2v \sin \theta}{g}$. The maximum height is attained at $t = \dfrac{v \sin \theta}{g}$, halfway through the flight. The maximum range, for any given value of v, is obtained when $\theta = \frac{1}{4}\pi$. Similar investigations can be carried out for a projectile projected from a point a given height above a horizontal plane, or projected up or down an inclined plane.

For a body projected vertically upwards with initial speed v, the equation $y = vt - \frac{1}{2}gt^2$ is obtained. The time of flight is $2v/g$, and the maximum height attained is $v^2/2g$.

A more sophisticated model, based on a sphere, is necessary if the motion of a long-range projectile such as an intercontinental missile is to be investigated. The rotation of the sphere needs to be considered if the effect of the Coriolis force is to be included.

projection (of a line on a plane) Given a line l and a plane p, the locus of all points N in the plane p such that N is the projection on p of some point on l is a straight line, the projection of l on p.

projection (of a point on a line) Given a line l and a point P not on l, the projection of P on l is the point N on l such that PN is perpendicular to l. The length $|PN|$ is the distance from P to l. The point N is called the foot of the perpendicular from P to l.

projection (of a point on a plane) Given a plane p and a point P not in p, the projection of P on p is the point N in p such that PN is perpendicular to p. The length $|PN|$ is the distance from P to p. The point N is called the foot of the perpendicular from P to p.

projection (of a vector on a vector) *See* VECTOR PROJECTION (of a vector on a vector).

projective geometry The area of geometry concerned with the properties, including invariance, of geometric figures under projection.

proof A chain of reasoning, starting from axioms, usually also with assumptions on which the conclusion then depends, that leads to a conclusion and which satisfies the logical rules of inference.

proof by contradiction A *direct proof of a statement q is a logically correct argument establishing the truth of q. A proof by contradiction assumes that q is false and derives the truth of some statement r and of its negation $\neg r$. This contradiction shows that the initial assumption cannot hold, hence establishing the truth of q. A more complicated example is a proof that 'if p then q'. A proof by contradiction assumes that p is true and that q is false, and derives the truth of some statement r and of its negation $\neg r$. This contradiction shows that the initial assumptions cannot both hold, and so a valid proof has been given that, if p is true, then q is true.

proper divisor = PROPER FACTOR.

proper factor A factor other than the number itself, so 1, 2, 3, 6 are all factors of 6 but only 1, 2 and 3 are proper factors.

proper fraction *See* FRACTION.

properly included *See* PROPER SUBSET.

proper subset Let A be a subset of B. Then A is a proper subset of B if A is not equal to B itself. Thus there is some element of B not in A. The subset A is then said to be properly (or strictly) included in B, and this is written $A \subset B$. Some authors use $A \subset B$ to mean $A \subseteq B$ (*see* SUBSET), but they then have no easy means of indicating proper inclusion.

proportion If two quantities x and y are related by an equation $y = kx$, where k is a constant, then y is said to be (directly) proportional to x, which may be written $y \propto x$. The constant k is the constant of proportionality. It is also said that y varies directly as x. When y is plotted against x, the graph is a straight line through the origin.

If $y = k/x$, then y is inversely proportional to x. This is written $y \propto 1/x$, and it is said that y varies inversely as x.

proposition A mathematical statement for which a proof is either required or provided.

proposition (in mathematical logic) = STATEMENT.

pseudo-metric One of the conditions for a distance measure d to be a *metric is that $d(x, y) = 0$ implies that $x = y$. If this condition is relaxed so that the distance between two different points can be zero, then d is said to be a pseudo-metric, and the space is said to be a pseudo-metric space.

pseudo-prime The positive integer n is a pseudo-prime if $a^n \equiv a$ (mod n), for all integers a. According to *Fermat's Little Theorem, all primes are pseudo-primes. There are comparatively few pseudo-primes that are not primes; the first is 561. To determine whether an integer is prime or composite, it may be useful to test first whether or not it is a pseudo-prime. For most composite numbers, this will establish that they are composite.

pseudo-random numbers *See* RANDOM NUMBERS.

Ptolemy (Claudius Ptolemaeus) (2nd century AD) Greek astronomer and mathematician, responsible for the most significant work of trigonometry of ancient times. Usually known by its Arabic name *Almagest* ('The Greatest'), it contains, amongst other things, tables of chords, equivalent to a modern table of sines, and an account of how they were obtained. Use is made of *Ptolemy's Theorem, from which the familiar addition formulae of trigonometry can be shown to follow.

Ptolemy's Theorem The following theorem of *Euclidean geometry:

Theorem

Suppose that a convex quadrilateral has vertices A, B, C and D, in that order. Then the quadrilateral is cyclic if and only if

$$AB.CD + AD.BC = AC.BD.$$

pulley A grooved wheel around which a rope can pass. When supported on an axle in some way, the device can be used to change the direction of a force: for example, pulling down on a rope may enable a load to be lifted. If the contact between the pulley and the axle is smooth, the magnitude of the force is not changed. A system of pulleys can be constructed that enables a large load to be raised a small distance by a small effort moving through a large distance. In such a *machine, the velocity ratio is equal to the number of ropes supporting the load or, equivalently, the number of pulleys in the system.

pure imaginary A *complex number is pure imaginary if its real part is zero.

pure mathematics The area of mathematics concerning the relationships between abstract systems and structures and the rules governing their behaviours, motivated by its intrinsic interest or elegance rather than its application to solving problems in the real world. Much modern applied mathematics is based on what was viewed as very esoteric pure mathematics when it was devised. For example, matrix algebra is now the cornerstone of video-game technologies, computer-aided design, etc.

pure strategy In a *matrix game, if a player always chooses one particular row (or column) this is a pure strategy. Compare this with *mixed strategy.

p-value (statistics) The probability that a given test statistic would take the observed value, or one which was less likely, if the null hypothesis had been true. While most text books construct hypothesis tests by finding critical values or constructing critical regions for a pre-determined level of significance for the test, almost all published statistics now report test outcomes in terms of p-values.

pyramid A convex *polyhedron with one face (the *base) a convex polygon and all the vertices of the base joined by edges to one other vertex (the apex); thus the remaining faces are all triangular. A right-regular pyramid is one in which the base is a regular polygon and the remaining faces are isosceles triangles.

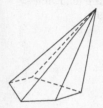

Pythagoras (died about 500 BC) Greek philosopher and mystic who, with his followers, seems to have been the first to take mathematics seriously as a study in its own right as opposed to being a collection of formulae for practical calculation. The Pythagoreans are credited with the discovery of the well-known *Pythagoras' Theorem on right-angled triangles. They were also much concerned with *figurate numbers, for semi-philosophical reasons. It is said that they regarded whole numbers as the fundamental constituents of reality, a view that was shattered by the discovery of *irrational numbers.

Pythagoras' Theorem Probably the best-known theorem in geometry, which gives the relationship between the lengths of the sides of a right-angled triangle:

Theorem

In a right-angled triangle, the square on the hypotenuse is equal to the sum of the squares on the other two sides.

Thus, if the hypotenuse, the side opposite the right angle, has length c and the other two sides have lengths a and b, then $a^2 + b^2 = c^2$. One elegant proof is obtained by dividing up a square of side $a + b$ in two different ways as shown in the figure, and equating areas.

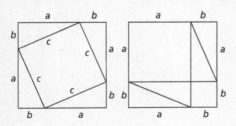

Pythagorean triple A set of three positive integers a, b and c such that $a^2 + b^2 = c^2$ (*see* PYTHAGORAS' THEOREM). If $\{a, b, c\}$ is a Pythagorean triple, then so is $\{ka, kb, kc\}$ for any positive integer k. Pythagorean triples that have *greatest common divisor equal to 1 include the following: $\{3, 4, 5\}$, $\{5, 12, 13\}$, $\{8, 15, 17\}$, $\{7, 24, 25\}$ and $\{20, 21, 29\}$.

Q *See* RATIONAL NUMBER.

QED Abbreviation for quod erat demonstrandum. Latin for 'which was to be proved'. Often written at the end of a proof.

QEF Abbreviation for quod erat faciendum. Latin for 'which was to be done'. Often written at the end of a construction.

quadrangle *See* COMPLETE QUADRANGLE.

quadrant In a Cartesian coordinate system in the plane, the axes divide the rest of the plane into four regions called quadrants. By convention, they are usually numbered as follows: the first quadrant is $\{(x, y)|\, x > 0, y > 0\}$, the second is $\{(x, y)|\, x < 0, y > 0\}$, the third is $\{(x, y)|\, x < 0, y < 0\}$, the fourth is $\{(x, y)|\, x > 0, y < 0\}$.

quadratic complexity *See* ALGORITHMIC COMPLEXITY.

quadratic equation A quadratic equation in the unknown x is an equation of the form $ax^2 + bx + c = 0$, where a, b and c are given real numbers, with $a \neq 0$. This may be solved by *completing the square or by using the formula

$$x = \frac{-b \pm \sqrt{b^2 - 4ac}}{2a},$$

which is established by completing the square. If $b^2 > 4ac$, there are two distinct real roots; if $b^2 = 4ac$, there is a single real root (which it may be convenient to treat as two equal or coincident roots); and, if $b^2 < 4ac$, the equation has no real roots, but there are two complex roots:

$$x = -\frac{b}{2a} \pm i\frac{\sqrt{4ac - b^2}}{2a}.$$

If α and β are the roots of the quadratic equation $ax^2 + bx + c = 0$, then $\alpha + \beta = -b/a$ and $\alpha\beta = c/a$. Thus a quadratic equation with given numbers α and β as its roots is $x^2 - (\alpha + \beta)x + \alpha\beta = 0$.

quadratic form In one, two, and three variables, functions in quadratic form are:

$$F(x) = ax^2$$
$$F(x, y) = ax^2 + by^2 + cxy$$
$$F(x, y, z) = ax^2 + by^2 + cz^2 + dxy + exz + fyz$$

Generally, if **x** is a column of n variables and **A** is an $n \times n$ matrix then $\mathbf{x}^T\mathbf{A}\mathbf{x}$ will be in quadratic form. **A** is normally *symmetric.

quadratic function In real analysis, a quadratic function is a *real function f such that $f(x) = ax^2 + bx + c$ for all x in **R**, where a, b and c are real numbers, with $a \neq 0$. (In some situations, $a = 0$ may be permitted.) The graph $y = f(x)$ of such a function is a *parabola with its axis parallel to the y-axis, and with its vertex downwards if $a > 0$ and upwards if $a < 0$. The graph cuts the x-axis where $ax^2 + bx + c = 0$, so the points (if any) are given by the roots (if real) of this *quadratic equation. The position of the vertex can be determined by completing the square or by finding the *stationary point of the function using differentiation. If the graph cuts the x-axis in two points, the x-coordinate of the vertex is midway between these two points. In this way the graph of the quadratic function can be sketched, and information can be deduced.

quadratic polynomial A polynomial of degree two.

quadrature A method of quadrature is a numerical method that finds an approximation to the area of a region with a curved boundary; the area of the region may then be found by some kind of limiting process.

quadric (quadric surface) A locus in 3-dimensional space that can be represented in a Cartesian coordinate system by a polynomial equation in x, y and z of the second degree; that is, an equation of the form

$$ax^2 + by^2 + cz^2 + 2fyz + 2gzx + 2hxy + 2ux + 2vy + 2wz + d = 0,$$

where the constants a, b, c, f, g and h are not all zero. When the equation represents a non-empty locus, it can be reduced by translation and rotation of axes to one of the following *canonical forms, and hence classified:

(i) Ellipsoid: $\dfrac{x^2}{a^2}+\dfrac{y^2}{b^2}+\dfrac{z^2}{c^2}=1$.

(ii) Hyperboloid of one sheet: $\dfrac{x^2}{a^2}+\dfrac{y^2}{b^2}-\dfrac{z^2}{c^2}=1$.

(iii) Hyperboloid of two sheets: $\dfrac{x^2}{a^2}+\dfrac{y^2}{b^2}-\dfrac{z^2}{c^2}=-1$.

(iv) Elliptic paraboloid: $\dfrac{x^2}{a^2}+\dfrac{y^2}{b^2}=\dfrac{2z}{c}$.

(v) Hyperbolic paraboloid: $\dfrac{x^2}{a^2}-\dfrac{y^2}{b^2}=\dfrac{2z}{c}$.

(vi) Quadric cone: $\dfrac{x^2}{a^2}+\dfrac{y^2}{b^2}=\dfrac{z^2}{c^2}$.

(vii) Elliptic cylinder: $\dfrac{x^2}{a^2}+\dfrac{y^2}{b^2}=1$.

(viii) Hyperbolic cylinder: $\dfrac{x^2}{a^2}-\dfrac{y^2}{b^2}=1$.

(ix) Parabolic cylinder: $\dfrac{x^2}{a^2}=\dfrac{2y}{b}$.

(x) Pair of non-parallel planes: $\dfrac{x^2}{a^2}=\dfrac{y^2}{b^2}$ $\left(\text{that is, } y=\pm\dfrac{b}{a}x\right)$.

(xi) Pair of parallel planes: $\dfrac{x^2}{a^2}=1$ (that is, $x=\pm a$).

(xii) Plane: $\dfrac{x^2}{a^2}=0$ (that is, $x=0$).

(xiii) Line: $\dfrac{x^2}{a^2}+\dfrac{y^2}{b^2}=0$ (that is, $x=y=0$).

(xiv) Point: $\dfrac{x^2}{a^2}+\dfrac{y^2}{b^2}+\dfrac{z^2}{c^2}=0$ (that is, $x=y=z=0$).

Forms (i), (ii), (iii), (iv) and (v) are the non-*degenerate quadrics.

quadric cone A *quadric whose equation in a suitable coordinate system is

$$\frac{x^2}{a^2}+\frac{y^2}{b^2}=\frac{z^2}{c^2}.$$

Sections by planes parallel to the xy-plane are ellipses (circles if $a = b$), and sections by planes parallel to the other axial planes are hyperbolas. These sections are commonly known as *conic sections.

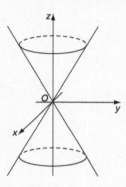

quadrilateral A polygon with four sides. *See also* COMPLETE QUADRILATERAL.

quadrillion A thousand million million (10^{15}). In Britain it used to mean a million to the power of 4 (10^{24}) but this is no longer used on the rare occasions the word is used.

quadruple Four objects normally taken in a particular order, denoted, for example, by (x_1, x_2, x_3, x_4).

quality control The application of statistical methods to the maintenance of quality standards in a production or other process. Methods include *acceptance sampling to make decisions on whether to take delivery of products, and *control charts to identify when modification to a process is required.

quantifier The two expressions 'for all ...' and 'there exists ...' are called quantifiers. A phrase such as 'for all x' or 'there exists x' may stand in front of a sentence involving a symbol x and thereby create a statement that makes sense and is either true or false. There are different ways in English of expressing the same sense as 'for all x', but it is sometimes useful to standardize the language to this particular form. This is known as a universal quantifier and is written in symbols as '$\forall x$'. Similarly, 'there exists x' may be used as the standard form to replace any phrase with this meaning, and is an existential quantifier, written in symbols as '$\exists x$'.

For example, the statements, 'if x is any number greater than 3 then x is positive' and 'there is a real number satisfying $x^2 = 2$', can be written in more standard form: 'for all x, if x is greater than 3 then x is positive', and 'there exists x such that x is real and $x^2 = 2$'. These can be written, using the symbols of mathematical logic, as: $(\forall x)(x > 3 \Rightarrow x > 0)$, and $(\exists x)(x \in \mathbf{R} \wedge x^2 = 2)$.

quantile Let X be a continuous *random variable. For $0 < p < 1$, the p-th quantile is the value x_p such that $\Pr(X \leq x_p) = p$. In other words, the fraction of the population less than or equal to x_p is p. For example, $x_{0.5}$ is the (population) median.

Often percentages are used. The n-th percentile is the value $x_{n/100}$ such that n per cent of the population is less than or equal to $x_{n/100}$. For example, 30% of the population is less than or equal to the 30th percentile. The 25th, 50th and 75th percentiles are the *quartiles.

Alternatively, the population may be divided into tenths. The n-th decile is the value $x_{n/10}$ such that n-tenths of the population is less than or equal to $x_{n/10}$. For example, three-tenths of the population is less than or equal to the 3rd decile.

The terms can be modified, though not always very satisfactorily, to be applicable to a discrete random variable or to a large sample ranked in ascending order.

quantity An entity that has a numerical value or magnitude.

quantum mechanics The area of mechanics concerned with the behaviour of particles at small scales where the discrete nature of matter becomes important. The results of experiments early in the 20th century could only be explained by quantities such as momentum and energy only being able to take certain discrete values, yet the behaviour of particles taken together was that of a wave, giving rise to the wave-particle duality. *See also* HEISENBERG'S UNCERTAINTY PRINCIPLE.

quartic equation A polynomial equation of degree four.

quartic polynomial A polynomial of degree four.

quartile deviation = SEMI-INTERQUARTILE RANGE.

quartiles For numerical data ranked in ascending order, the quartiles are values derived from the data which divide the data into four equal parts. If there are n observations, the first quartile (or lower quartile) Q_1 is the $\frac{1}{4}(n + 1)$-th, the second quartile (which is the *median) Q_2 is the $\frac{1}{2}(n + 1)$-th and the third quartile (or upper quartile) Q_3 is the $\frac{3}{4}(n + 1)$-th in ascending order. When $\frac{1}{4}(n + 1)$ is not a whole number, it is sometimes thought necessary to take the (weighted) average of two observations, as is done for the median. However, unless n is very small, an observation that is nearest will normally suffice. For example, for the sample 15, 37, 43, 47, 54, 55, 57, 64, 76, 98, we may take $Q_1 = 43$, $Q_2 = 54.5$ and $Q_3 = 64$.

For a random variable, the quartiles are the *quantiles $x_{0.25}$, $x_{0.5}$ and $x_{0.75}$; that is, the 25th, 50th and 75th percentiles.

quasi-metric One of the conditions for a distance measure d to be a *metric is that $d(x, y) = d(y, x)$. If this condition is relaxed so that the distance between two points can be different depending on the direction of travel, then d is said to be a quasi-metric, and the space is said to be a quasi-metric space. Quasi-metrics are common in real life—the times taken to walk between two points can be quite different if there is a steep gradient between them, for example, and one-way systems in towns mean there is often a different route required to go from B to A than will take you from A to B.

quaternion The complex number system can be obtained by taking a complex number to be an expression $a + bi$, where a and b are real numbers, and defining addition and multiplication in the natural way with the understanding that $i^2 = -1$. In an extension of this idea, Hamilton introduced the following notion, originally for use in mechanics. Define a quaternion to be an expression $a + bi + cj + dk$, where a, b, c and d are real numbers, and define addition and multiplication in the natural way, with

$$i^2 = j^2 = k^2 = -1, \quad ij = -ji = k, \quad jk = -kj = 1, \quad ki = -ik = j.$$

All the normal laws of algebra hold except that multiplication is not *commutative. That is to say, the quaternions form a *ring which is not commutative but in which every non-zero element has an *inverse.

q

queuing theory A study of the processes relating customer wait and service time patterns where there is a random element involved in one or other, or in most cases both parts. In various contexts, the probability model for each part can be quite different, and the randomness of each stage makes analysis appropriate for simulations where the behaviour of the queue can be observed repeatedly for a variety of parameter values in order to investigate the effect of a variety of management strategies.

quick sort algorithm As the name suggests this method is quicker than a standard *bubble sort algorithm, because it sorts smaller groups only. The method involves choosing a pivot value for any list or sublist, usually by taking the middle (or one of the two middle items in an even-numbered list) and placing all items smaller than the pivot on the left and the larger items on the right, then the two groups created being treated as lists in the next stage. The process ends when all sublists are of length 1.

Quillen, Daniel Grey (1940–2011) American mathematician whose work on the higher algebraic K-theory, which combined geometric and topological methods in dealing with algebraic problems in ring theory and module theory, was awarded the *Fields Medal in 1978.

quintic equation A polynomial equation of degree five.

quintic polynomial A polynomial of degree five.

quota sample *See* SAMPLE.

quotient *See* DIVISION ALGORITHM.

quotient rule (for differentiation) *See* DIFFERENTIATION.

quotient space If X is a *topological space and $\sim$ is an *equivalence relation defined on X, then the quotient space $Y = X/\sim$ is defined to be the set of equivalence classes of elements of X. So $Y = \{\{u \in X : u \sim x\} : x \in X\}$, where the open sets are defined to be those sets of equivalence classes whose unions are open sets in X. Quotient spaces are also known as identification spaces and factor spaces.

q

R *See* REAL NUMBER.

rad Abbreviation and symbol for *radian.

radial and **transverse components** When a point P has polar coordinates (r, θ), the vectors $\mathbf{e}_r$ and $\mathbf{e}_\theta$ are defined by $\mathbf{e}_r = \mathbf{i} \cos \theta + \mathbf{j} \sin \theta$ and $\mathbf{e}_\theta = -\mathbf{i} \sin \theta + \mathbf{j} \cos \theta$, where $\mathbf{i}$ and $\mathbf{j}$ are unit vectors in the directions of the positive x- and y-axes. Then $\mathbf{e}_r$ is a unit vector along OP in the direction of increasing r, and $\mathbf{e}_\theta$ is a unit vector perpendicular to this in the direction of increasing θ. Any vector $\mathbf{v}$ can be written in terms of its components in the directions of $\mathbf{e}_r$ and $\mathbf{e}_\theta$. Thus $\mathbf{v} = v_1 \mathbf{e}_r + v_2 \mathbf{e}_\theta$, where $v_1 = \mathbf{v} \cdot \mathbf{e}_r$ and $v_2 = \mathbf{v} \cdot \mathbf{e}_\theta$. The component v_1 is the radial component, and the component v_2 is the transverse component.

radial set *See* ABSORBING SET.

radian In elementary work, angles are measured in degrees, where one revolution measures 360°. In more advanced work, it is essential that angles are measured differently. Suppose that a circle centre O meets two lines through O at A and B. Take the length of arc AB and divide it by the length of OA. This value is independent of the radius of the circle, and depends only upon the size of $\angle AOB$. So the value is called the size of $\angle AOB$, measured in radians.

The angle measures 1 radian when the length of the arc AB equals the length of OA. This happens when $\angle AOB$ is about 57°. More accurately, 1 radian = 57.296° = 57° 17′45″, approximately. Since the length of the circumference of a circle of radius r is $2\pi r$, one revolution measures 2π radians. Consequently, $x° = \pi x/180$ radians. In much theoretical work, particularly involving calculus, radian measure is essential. When *trigonometric functions are evaluated with a calculator, it is essential to be sure that the correct measure is being used.

The radian is the SI unit for measuring angle, and is abbreviated to 'rad'.

radical axis The radical axis of two circles is the straight line containing all points P such that the lengths of the tangents from P to the two circles are equal. Each figure shows a point P on the radical axis and tangents touching the two circles at T_1 and T_2, with $PT_1 = PT_2$. If the circles intersect in two points, as in the figure on the right, the radical axis is the straight line through the two points of intersection. In this case, there are some points on the radical axis inside the circles, from which tangents to the two circles cannot be drawn. For the circles with equations $x^2 + y^2 + 2g_1x + 2f_1y + c_1 = 0$ and $x^2 + y^2 + 2g_2x + 2f_2y + c_2 = 0$, the radical axis has equation $2(g_1 - g_2)x + 2(f_1 - f_2)y + (c_1 - c_2) = 0$.

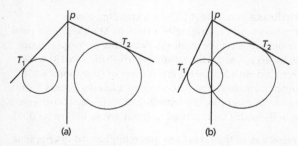

(a)   (b)

radical sign The sign $\sqrt{}$ used in connection with square roots, cube roots and n-th roots, for larger values of n. The notation $\sqrt{a}$ indicates square root, $\sqrt[3]{a}$ indicates cube root and $\sqrt[n]{a}$ indicates *n-th root of a. See *square root and *n-th root for detailed explanation of the correct usage of the notation.

radius (radii) A radius of a circle is a line segment joining the centre of the circle to a point on the circle. All such line segments have the same length, and this length is also called the radius of the circle. The term also applies in both senses to a sphere. *See also* CIRCLE and SPHERE.

radius (of a graph) The minimum *eccentricity of any vertex in a *graph. *See also* CENTRAL VERTEX (GRAPH).

radius of convergence If R is the largest real number for which the *power series $\sum_{n=0}^{\infty} a_n z^n$ converges for all z satisfying $|z| < R$, then R is the radius of convergence for that power series. If the power series is of a real variable x, then this condition generates an interval on the real line, but the term 'radius of convergence' is still used.

radius of curvature *See* CURVATURE.

radius of gyration The square root of the ratio of a *moment of inertia to the mass of a rigid body. Thus if I is the moment of inertia of a rigid body of mass m about a specified axis, and k is the radius of gyration, then $I = mk^2$. It means that, when rotating about this axis, the body has the same moment of inertia as a ring of the same mass and of radius k.

radius vector Suppose that a point O is taken as origin in the plane. If a point P in the plane has *position vector $\mathbf{p}$, then $\mathbf{p}$ may also be called the radius vector, particularly when P is a typical point on a certain curve or when P is thought of as a point or particle moving in the plane.

raise (to a power) To multiply by itself to the number of times specified by the power so $5^3 = 5 \times 5 \times 5 = 125$ and $(a + b)^2 = a^2 + 2ab + b^2$.

Ramanujan, Srinivasa (1887–1920) The outstanding Indian mathematician of modern times. Originally a clerk in Madras, he studied and worked on mathematics totally unaided. Following correspondence with G. H. *Hardy, he accepted an invitation to visit Britain in 1914. He studied and collaborated with Hardy on the subject of *partitions and other topics, mainly in number theory. He was considered a genius for his inexplicable ability in, for example, the handling of series and *continued fractions. Because of ill-health he returned to India a year before he died.

random The general use of the term is to mean haphazard or irregular, but in statistics it carries a more precise meaning indicating that the outcome can only be described in probability terms, and specifically in many contexts that outcomes are equally likely. Therefore, a random number generator should generate each of the n numbers in its outcome space with probability $\frac{1}{n}$, and a random sample size from the population means that all possible combinations of that sample size from the population are equally likely.

random error (noise) The difference between predicted values and observed values treated as a random variable. In effect it is the variation which is unexplained by the model, and hence the term 'error' is rather unhelpful. In regression modelling it is also termed the *residual.

randomization in design of experiments It is important that researchers do not have the flexibility to choose which subjects are assigned to different experimental groups if potential sources of bias are to be minimized, as they may (even subconsciously) assign subjects to the groups which they would expect to favour the outcome they expect or hope for. Depending on the nature of the trial, this may mean randomizing the nature of the treatment received or the order in which treatments are applied etc.

randomize To order data, or select data, in a deliberately random manner.

randomized blocks Where groups within a population are likely to respond differently to an experiment, the variation between subjects may be so great that simple randomized experiments will not reveal differences in treatments which actually exist. The randomized blocks design is analogous to stratified sampling for estimation purposes in the same situation, where the randomization is done within groupings of similar subjects—these groupings are the 'blocks' of the name. Each treatment is then applied to equal numbers of subjects in each block, but the choice of which subject gets which treatment is randomized within each block. For example, if two methods of improving memory are to be compared it might be considered likely that age and sex would be factors in memory performance, in which case the blocks might consist of males or females of different age-groupings.

random numbers Tables of random numbers give lists of the digits 0, 1, 2, . . . , 9 in which each digit is equally likely to occur at any stage. There is no way of predicting the next digit. Such tables can be used to select items at random from a *population. Numbers generated by a deterministic algorithm that appear to pass statistical tests for randomness are called pseudo-random numbers. Such an algorithm may be called a random number generator.

random sample *See* SAMPLE.

random variable A quantity that takes different numerical values according to the result of a particular experiment. A random variable is *discrete if the set of possible values is finite or countably infinite. For a discrete random variable, the probability of its taking any particular value is given by the *probability mass function. A random variable is continuous if the set of possible values forms an interval, finite or infinite. For a continuous random variable, the probability of its taking a value in a particular sub-interval may be calculated from the *probability density function.

random vector An ordered set of n random variables, often representing the outcomes of a repeated experiment. For example, if I throw a die four times the outcome of the experiment could be described by the vector (x_1, x_2, x_3, x_4) where each x_i is a uniform distribution on the set $(1, 2, \ldots, 6)$. If I score 5, then 2, then 5, then 6, the observed outcome would be recorded as $(5, 2, 5, 6)$.

random walk Consider a *Markov chain $X_1, X_2, X_3, \ldots$ in which the state space is $\{\ldots, -2, -1, 0, 1, 2, 3, \ldots\}$. With the integers positioned as they occur on the real line, imagine an object moving from integer to integer a step at a time. That is to say, from position i the object either moves to $i - 1$ or $i + 1$, or possibly stays where it is. Such a Markov chain is called a one-dimensional random walk: if $X_n = i$, then $X_{n+1} = i - 1$, i or $i + 1$. The state i is an absorbing state if, whenever $X_n = i$, then $X_{n+1} = i$; in other words, when the object reaches this position it stays there. When a gambler, playing a sequence of games, either wins or loses a fixed amount in each game, his winnings give an example of a random walk.

Random walks in two or more dimensions can be defined similarly.

range (of a function or mapping) *See* FUNCTION and MAPPING.

range (in mechanics) The range of a *projectile on a horizontal or inclined plane passing through the point of projection is the distance from the point of projection to the point at which the projectile lands on the plane.

range (in statistics) The difference between the maximum and minimum observations in a set of numerical data. It is a possible measure of *dispersion of a sample.

rank (of a matrix) Let $\mathbf{A}$ be an $m \times n$ matrix. The column rank of $\mathbf{A}$ is the largest number of elements in a *linearly independent set of columns of $\mathbf{A}$. The row rank of $\mathbf{A}$ is the largest number of elements in a linearly independent set of rows of $\mathbf{A}$. It can be shown that elementary row operations on a matrix do not change the column rank or the row rank. Consequently, the column rank and row rank of $\mathbf{A}$ are equal, both being equal to the number of non-zero rows in the matrix in *reduced echelon form to which $\mathbf{A}$ can be transformed. This common value is the rank of $\mathbf{A}$. It can also be shown that the rank of $\mathbf{A}$ is equal to the number of rows and columns in the largest square submatrix of $\mathbf{A}$ that has non-zero determinant. An $n \times n$ matrix is *invertible if and only if it has rank n.

rank (in statistics) The observations in a sample are said to be ranked when they are arranged in order according to some criterion. For example, numbers can be ranked in ascending or descending order, people can be ranked according to height or age, and products can be ranked according to their sales. The rank of an observation is its position in the list when the sample has been ranked. *Non-parametric methods frequently make use of ranking rather than the exact values of the observations in the sample.

rank correlation coefficient A lot of bivariate data does not meet the requirements necessary to use the product moment correlation coefficient (pmcc). This is often because at least one of the variables is measured on a scale which is *ordinal but not *interval. Where the measurement is reported on a non-numerical scale it is transparently not interval but often gradings, based on subjective evaluations, are reported on a numerical scale.

Spearman's rank correlation coefficient is the product moment correlation applied to the ranks. In the case of where there are no tied ranks, this is algebraically equivalent to computing the value of

$1 - \dfrac{6\sum d_i^2}{n(n^2 - 1)}$ where d_i = difference in ranks of the i-th pair and n is the

number of data pairs. When n is reasonably small, this computation is very quick. When there are tied ranks the above formula is not algebraically equivalent to the pmcc applied to the ranks, but is often used as a reasonable approximation.

Kendall's rank correlation coefficient: Sir Maurice Kendall proposed a measure of rank correlation based on the number of neighbour-swaps needed to move from one rank ordering to another. If Q is the minimum number of neighbour swaps needed in a set of data with n values then

Kendall's coefficient is given by $\tau = 1 - \dfrac{4Q}{n(n - 1)}$. Significance tests can

then be carried out using critical values from books of statistical tables.

rate of change Suppose that the quantity y is a function of the quantity x, so that $y = f(x)$. If f is *differentiable, the rate of change of y with respect to x is the derivative dy/dx or $f'(x)$. The rate of change is often with respect to time t. Suppose, now, that x denotes the *displacement of a particle, at time t, on a directed line with origin O. Then the velocity is dx/dt or $\dot{x}$, the rate of change of x with respect to t, and the acceleration is d^2x/dt^2 or $\ddot{x}$, the rate of change of the velocity with respect to t.

In the reference to velocity and acceleration in the preceding paragraph, a common convention has been followed, in which the unit vector $\mathbf{i}$ in the positive direction along the line has been suppressed. Velocity and acceleration are in fact vector quantities, and in the 1-dimensional case above are equal to $\dot{x}\mathbf{i}$ and $\ddot{x}\mathbf{i}$. When the motion is in two or three dimensions, vectors are used explicitly. If $\mathbf{r}$ is the *position vector of a particle, the velocity of the particle is the vector $\dot{\mathbf{r}}$ and the acceleration is $\ddot{\mathbf{r}}$.

ratio The quotient of two numbers or quantities giving their relative size. The ratio of x to y is written as $x : y$ and is unchanged if both quantities are multiplied or divided by the same quantity. So $2 : 3$ is the same as $6 : 9$ and $1 : \frac{3}{2}$. Where a ratio is expressed in the form $1 : a$ it is called a unitary ratio, and this form makes comparison of ratios much easier.

rational function In real analysis, a rational function is a *real function f such that, for x in the domain, $f(x) = g(x)/h(x)$, where $g(x)$ and $h(x)$ are polynomials, which may be assumed to have no common factor of degree greater than or equal to 1. The domain is usually taken to be the whole of **R**, with any zeros of the denominator $h(x)$ omitted.

rationalize To remove radicals from an expression or part of it without changing the value of the whole expression. For example, in the expression $1/(2 - \sqrt{x})$ the denominator can be rationalized by multiplying the numerator and the denominator by $2 + \sqrt{x}$ to give

$$\frac{2 + \sqrt{x}}{4 - x}.$$

rational number A number that can be written in the form a/b, where a and b are integers, with $b \neq 0$. The set of all rational numbers is usually denoted by **Q**. A real number is rational if and only if, when expressed as a decimal, it has a finite or recurring expansion (*See* DECIMAL REPRESENTATION). For example,

$$\frac{5}{4} = 1.25, \qquad \frac{2}{3} = 0.\dot{6}, \qquad \frac{20}{7} = 2.\dot{8}5714\dot{2}.$$

A famous proof, attributed to Pythagoras, shows that $\sqrt{2}$ is not rational, and e and π are also known to be irrational.

The same rational number can be expressed as a/b in different ways; for example, $\frac{2}{3} = \frac{6}{9} = \frac{-4}{-6}$. In fact, $a/b = c/d$ if and only if $ad = bc$. But a rational number can be expressed uniquely as a/b if it is insisted that a and b have *greatest common divisor 1 and that $b > 0$. Accepting the different forms for the same rational number, the explicit rules for addition and multiplication are that

$$\frac{a}{b} + \frac{c}{d} = \frac{ad + bc}{bd} \quad \text{and} \quad \frac{a}{b} \cdot \frac{c}{d} = \frac{ac}{bd}.$$

The set of rational numbers is closed under addition, subtraction, multiplication and division (not allowing division by zero). Indeed, all the axioms for a *field can be seen to hold.

A more rigorous approach sets up the field **Q** of rational numbers as follows. Consider the set of all *ordered pairs (a, b), where a and b are integers, with $b \neq 0$. Introduce an *equivalence relation $\sim$ on this set, by defining $(a, b) \sim (c, d)$ if $ad = bc$, and let $[(a, b)]$ be the *equivalence class containing (a, b). The intuitive approach above suggests that addition and multiplication between equivalence classes should be defined by

$$[(a, b)] + [(c, d)] = [(ad + bc, bd)] \quad \text{and} \quad [(a, b)][(c, d)] = [(ac, bd)],$$

where it is necessary, in each case, to verify that the class on the right-hand side is independent of the choice of elements (a, b) and (c, d) taken as *representatives of the equivalence classes on the left-hand side. It can then be shown that the set of equivalence classes, with this addition and multiplication, form a field **Q**, whose elements, according to this approach, are called rational numbers.

ratio scale (statistics) A scale measurement which has a fixed zero and allows comparison in ratio terms. So the length of objects is a ratio scale, and a page may be twice as long as it is broad, but temperature is not a ratio scale—the choice of zero is arbitrary and 'twice as hot' is not a meaningful statement. While time is not a ratio scale, the length of time taken on a task is.

ratio test A test for convergence or divergence of an infinite series $\{a_n\}$. If the ratio $\left| \dfrac{a_{n+1}}{a_n} \right| \to k$ as $n \to \infty$ then when k is < 1 the series converges absolutely, and for $k > 1$ the series diverges. If $k = 1$, the ratio test is not sufficient to determine whether it converges.

raw data Data which have not been analysed. The values of the variables as they were recorded.

ray = HALF-LINE.

reachable set = ATTAINABLE SET.

real Involving real numbers only, i.e. having no *imaginary part.

real axis In the *complex plane, the x-axis is called the real axis; its points represent the real numbers.

real function A function from the set **R** of real numbers (or a subset of **R**) to **R**. Thus, if f is a real function, then, for every real number x in the domain, a corresponding real number $f(x)$ is defined. In analysis, a function $f: S \to \mathbf{R}$ is often defined by giving a formula for $f(x)$, without specifying the domain S (See FUNCTION). In that case, it is usual to assume that the domain is the largest possible subset S of **R**. For example, if $f(x) = 1/(x - 2)$, the domain would be taken to be $\mathbf{R} \setminus \{2\}$; that is, the set of all real numbers not equal to 2. If $f(x) = \sqrt{9 - x^2}$, the domain would be the closed interval $[-3, 3]$.

real line On a horizontal straight line, choose a point O as origin, and a point A, to the right of O, such that $|OA|$ is equal to 1 unit. Each positive real number x can be represented by a point on the line to the right of O, whose distance from O equals x units, and each negative number by a point on the line to the left of O. The origin represents zero. The line is called the real line when its points are taken in this way to represent the real numbers.

real number The numbers generally used in mathematics, in scientific work and in everyday life are the real numbers. They can be pictured as points of a line, with the integers equally spaced along the line and a real number b to the right of the real number a if $a < b$. The set of real numbers is usually denoted by **R**. It contains such numbers as 0, $\frac{1}{2}$, -2, 4.75, $\sqrt{2}$ and π. Indeed, **R** contains all the rational numbers but also numbers such as $\sqrt{2}$ and π that are irrational. Every real number has an expression as an infinite *decimal fraction.

The set of real numbers, with the familiar addition and multiplication, form a *field and, since there is a notion of 'less than' that satisfies certain basic axioms, **R** is called an 'ordered' field. However, a statement of a complete set of axioms that characterize **R** will not be attempted here. There are too a number of rigorous approaches that, assuming the existence of the field **Q** of *rational numbers, construct a system of real numbers with the required properties.

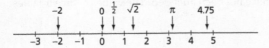

real part A *complex number z may be written $x + yi$, where x and y are real, and then x is the real part of z. It is denoted by Re z or $\Re z$.

real projective plane An abstract 2-dimensional, non-*orientable (one-sided) surface which has no boundary. Since the *Möbius strip has an edge (boundary), it can be constructed physically without self-intersection, but like the *Klein bottle there is no way for the surface of the real projective plane to go round itself when it needs to close up, so 3-dimensional models intersect or pass through themselves. However, 4-dimensional models can describe it without self-intersection.

The definition of the real projective plane is the space of all lines through the origin in Euclidean 3-dimensional space. Topologically, this can be defined simply as the direction vectors for the unit sphere. Then, choosing a hemisphere, since each vector in the other hemisphere defines the same line as one in the chosen hemisphere, the identification of opposite points provides an equivalence, whose *quotient space is the real projective plane. That hemisphere surface can be represented by its projection onto a plane (giving a disc), and the only 'problem' revolves around the 'equator' of the hemisphere, which is the unit circle surrounding the disc. To satisfy the topological properties required, this needs to be represented by a Möbius strip attached to the edge of the unit disc—which is not physically achievable without self-intersection.

real world *See* MATHEMATICAL MODEL.

rearrangement A positioning of a set of objects where together they occupy the same places, but each individual object is not necessarily in the same position.

reciprocal The multiplicative inverse of a quantity may, when the operation of multiplication is *commutative, be called its reciprocal. Thus the reciprocal of 2 is $\frac{1}{2}$, the reciprocal of $3x + 4$ is $1/(3x + 4)$ and the reciprocal of $\sin x$ is $1/\sin x$.

reciprocal rule (for differentiation) *See* DIFFERENTIATION.

rectangle A four-sided plane figure containing four right angles. Opposite sides will necessarily be parallel and of equal length. In the special case where all four sides are equal, the figure will be a *square.

rectangular (oblong) Shaped like a rectangle.

rectangular coordinate system *See* COORDINATES (in the plane).

rectangular distribution = UNIFORM DISTRIBUTION.

rectangular hyperbola A *hyperbola whose *asymptotes are perpendicular. With the origin at the centre and the coordinate x-axis along the transverse axis, it has equation $x^2 - y^2 = a^2$. Instead, the coordinate axes can be taken along the asymptotes in such a way that the two branches of the hyperbola are in the first and third quadrants. This coordinate system can be obtained from the other by a rotation of axes. The rectangular hyperbola then has equation of the form $xy = c^2$. For example, $y = 1/x$ is a rectangular hyperbola. For $xy = c^2$, it is customary to take $c > 0$ and to use, as parametric equations, $x = ct$, $y = c/t$ $(t \neq 0)$.

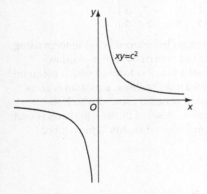

rectangular number Any number which can be expressed non-trivially as a rectangular array of dots, i.e. excluding the trivial case of a single row. Thus rectangular numbers are non-prime in that they can be expressed as $a \times b$ where neither a nor b is 1. For many rectangular numbers, the representation is not unique, so 12 could be drawn as a 4×3 or a 6×2 rectangle. In the case where there is a representation with equal factors it is a *square number, so 16 is a square number although it can also be represented as an 8×2 rectangle.

rectilinear motion Motion in a straight line, allowing movement in both directions.

recurrence relation = DIFFERENCE EQUATION.

recurring decimal *See* DECIMAL REPRESENTATION.

recursion theory A branch of mathematical logic and computer science concerned with the study of *computable functions.

reduce Modify or simplify a form expression or a number; for example, $\dfrac{6}{9}$ can be reduced to the equivalent form $\dfrac{2}{3}$ and $\dfrac{x^2 - 1}{x - 1}$ can be reduced to $x + 1$ provided $x \neq 1$.

reduced echelon form Suppose that a row of a matrix is called zero if all its entries are zero. Then a matrix is in reduced echelon form if

 (i) all the zero rows come below the non-zero rows,
 (ii) the first non-zero entry in each non-zero row is 1 and occurs in a column to the right of the leading 1 in the row above,
 (iii) the leading 1 in each non-zero row is the only non-zero entry in the column that it is in.

(If (i) and (ii) hold, the matrix is in *echelon form.) For example, these two matrices are in reduced echelon form:

$$\begin{bmatrix} 1 & 6 & 0 & 0 & 2 \\ 0 & 0 & 1 & 0 & -3 \\ 0 & 0 & 0 & 1 & 5 \end{bmatrix}, \quad \begin{bmatrix} 0 & 0 & -1 & 4 & 2 \\ 0 & 1 & 2 & -3 & 5 \\ 0 & 0 & 0 & 0 & 0 \end{bmatrix}.$$

Any matrix can be transformed to a matrix in reduced echelon form using elementary row operations, by a method known as *Gauss–Jordan elimination. For any matrix, the reduced echelon form to which it can be transformed is unique. The solutions of a set of linear equations can be immediately obtained from the reduced echelon form to which the augmented matrix has been transformed. A set of linear equations is said to be in reduced echelon form if its augmented matrix is in reduced echelon form.

reduced set of residues For a positive integer n, the number of positive integers, less than n, *relatively prime to n, is denoted by $\phi(n)$ (*See* EULER'S FUNCTION). A reduced set of residues modulo n is a set of $\phi(n)$ integers, one *congruent (modulo n) to each of the positive integers less than n, relatively prime to n. Thus $\{1, 5, 7, 11\}$ is a reduced set of residues modulo 12, and so is $\{1, -1, 5, -5\}$.

reductio ad absurdum The Latin phrase meaning 'reduction to absurdity' used to describe the method of *proof by contradiction.

reduction formula Let I_n be some quantity that is dependent upon the integer $n (\geq 0)$. It may be possible to establish some general formula, expressing I_n in terms of some of the quantities $I_{n-1}, I_{n-2}, \ldots$. Such a formula is a reduction formula and can be used to evaluate I_n for a particular value of n. The method is useful in integration. For example, if

$$I_n = \int_0^{\pi/2} \sin^n x \, dx,$$

it can be shown, by integration by parts, that $I_n = ((n-1)/n)I_{n-2}$ $(n \geq 2)$. It is easy to see that $I_0 = \pi/2$, and then the reduction formula can be used, for example, to find that

$$I_6 = \frac{5}{6}I_4 = \frac{5}{6} \times \frac{3}{4}I_2 = \frac{5}{6} \times \frac{3}{4} \times \frac{1}{2}I_0 = \frac{5}{6} \times \frac{3}{4} \times \frac{1}{2} \times \frac{\pi}{2} = \frac{5\pi}{32}.$$

redundant If an equation or inequality does not make any difference by its existence, it is said to be redundant. For example, if $2x + y = 7$, $3x - y = 3$ and $5x + y = 13$, any one of these equations can be termed redundant because the other two are sufficient to identify $x = 2$, $y = 3$ as the only solution. If $3x + 2y > 4$ and $6x + 4y > 9$ then the first inequality is redundant because if $6x + 4y > 9$ it follows that $3x + 2y$ must be > 4.5 and the first inequality is automatically satisfied.

re-entrant An interior angle of a polygon is re-entrant if it is greater than two right angles.

reflection (of the plane) Let l be a line in the plane. Then the mirror-image of a point P is the point P' such that PP' is perpendicular to l and l cuts PP' at its midpoint. The reflection of the plane in the line l is the transformation of the plane that maps each point P to its mirror-image P'. Suppose that the line l passes through the origin O and makes an angle α with the x-axis. If P has polar coordinates (r, θ), its mirror-image P' has polar coordinates $(r, 2\alpha - \theta)$. In terms of Cartesian coordinates, reflection in the line *l* maps *P* with coordinates (x, y) to P' with coordinates (x', y'), where

$$x' = x \cos 2\alpha + y \sin 2\alpha,$$
$$y' = x \sin 2\alpha - y \cos 2\alpha.$$

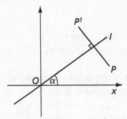

reflex angle An angle that is greater than 2 *right angles and less than 4 right angles.

reflexive relation A *binary relation $\sim$ on a set S is reflexive if $a \sim a$ for all a in S.

Regiomontanus (1436–76) A central figure in mathematics in the 15th century. Born Johann Müller, he took as his name the Latin form of Königsberg, his birthplace. His *De triangulis omnimodis* ('On All Classes of Triangles') was the first modern account of trigonometry and, even though it did not appear in print until 1533, it was influential in the revival of the subject in the West.

region A *connected subset of 2-dimensional space. For example, the set of points (x, y) satisfying $(x - 3)^2 + (y + 1)^2 < 4$ is the interior of a circle, radius 2, centred at $(3, -1)$ and is an open region. If the inequality was $\leq$ then the circle itself would be included and it would be a closed region.

regression A statistical procedure to determine the relationship between a *dependent variable and one or more *explanatory variables. The purpose is normally to enable the value of the dependent variable to be predicted from given values of the explanatory variables. It is multiple regression if there are two or more explanatory variables. Usually the model supposes that $E(Y)$, where Y is the dependent variable, is given by some formula involving certain unknown parameters. In simple linear regression, $E(Y) = b_0 + b_1 X$. In multiple linear regression, with k explanatory variables $X_1, X_2, \ldots, X_k$, $E(Y) = b_0 + b_1 X_1 + b_2 X_2 + \ldots + b_k X_k$. Here $b_0, b_1, \ldots, b_k$ are the regression coefficients. *See also* LEAST SQUARES.

regression to the mean = REVERSION TO THE MEAN.

regular graph A *graph in which all the vertices have the same *degree. It is r-regular or regular of degree r if every vertex has degree r.

regular polygon *See* POLYGON.

regular polyhedron *See* PLATONIC SOLID.

regular space A *topological space X in which every non-empty *closed subset S of X and a point p of X which is not in S have non-overlapping neighbourhoods.

regular tessellation *See* TESSELLATION.

relation A relation on a set S is usually a *binary relation on S, though the notion can be extended to involve more than two elements. An example of a ternary relation, involving three elements, is 'a lies between b and c', where a, b and c are real numbers.

relative address In spreadsheets, a formula will often use the contents of another cell or cells. If these are always in the same position relative to where the formula appears, then relative addressing is used in the formula which can then be copied and pasted into other locations to perform the required calculations. A formula can contain a mixture of relative and *absolute addresses.

relative complement If the set A is included in the set B, the *difference set $B \setminus A$ is the (relative) complement of A in B, or the complement of A relative to B.

relative efficiency *See* ESTIMATOR.

relative error Let x be an approximation to a value X and let $X = x + e$. The relative error is $|e/X|$. When 1.9 is used as an approximation for 1.875, the relative error equals $0.025/1.875 = 0.013$, to 3 decimal places. (This may be expressed as a *percentage error of 1.3%.) Notice, in this example, that $0.025/1.9 = 0.013$, to 3 decimal places, too. In general, when e is small, it does not make much difference if the relative error is taken as $|e/x|$, instead of $|e/X|$; this has to be done if the exact value is not known but only the approximation. The relative error may be a more helpful figure than the absolute *error. An absolute error of 0.025 in a value of 1.9, as above, may be acceptable. But the same absolute error in a value of 0.2, say, would give a relative error of $0.025/0.2 = 0.125$ (a percentage error of $12\frac{1}{2}\%$), which would probably be considered quite serious.

relative frequency The ratio $\dfrac{n}{N}$ where n is the observed number of a particular event, and N is the number of trials. As N gets large, the *weak law of large numbers says that $\frac{n}{N}$ will tend to the probability of that event with a probability of 1. In cases where no way of calculating a probability exists (for example, by using an equally likely argument in throwing a fair die) then the value of $\frac{n}{N}$ can be used to estimate this probability. The larger the value of N, the better this statistic is as an estimator.

relatively prime Integers a and b are relatively prime if their *greatest common divisor (g.c.d.) is 1. Similarly, any number of integers $a_1, a_2, \ldots, a_n$ are relatively prime if their g.c.d. is 1.

relative measure of dispersion A measure that indicates the magnitude of a measure of *dispersion relative to the magnitude of the *mean, such as the *coefficient of variation.

relative position, relative velocity and relative acceleration Let $\mathbf{r}_P$ and $\mathbf{r}_Q$ be the position vectors of particles P and Q with respect to some *frame of reference with origin O, as shown in the diagram. The position vector of P relative to Q is $\mathbf{r}_P - \mathbf{r}_Q$. If $\mathbf{v}_P = \dot{\mathbf{r}}_P$ and $\mathbf{v}_Q = \dot{\mathbf{r}}_Q$, then $\mathbf{v}_P$ and $\mathbf{v}_Q$ are the velocities of P and Q relative to the frame of reference with origin O and $\mathbf{v}_P - \mathbf{v}_Q$ is the velocity of P relative to Q. If $\mathbf{a}_P = \dot{\mathbf{v}}_P$ and $\mathbf{a}_Q = \dot{\mathbf{v}}_Q$, then $\mathbf{a}_P$ and $\mathbf{a}_Q$ are the accelerations of P and Q relative to the frame of reference with origin O, and $\mathbf{a}_P - \mathbf{a}_Q$ is the acceleration of P relative to Q. These quantities may be called the relative position, the relative velocity and the relative acceleration of P with respect to Q.

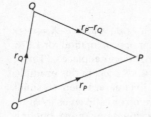

These notions are important when there are two or more frames of reference, each with an associated observer. For example, in a problem involving a ship and an aeroplane, the ship's captain and an observer on the land may both be viewing the aeroplane. The velocity of the aeroplane, for example, depends on which of them is making the measurement. The distinction must be made between the velocity of the aeroplane relative to the ship and the velocity of the aeroplane relative to the land.

relative risk The ratio of the *risks of the same outcome for two different groups, or of a different outcome for the same group. So the relative risk of a car accident when the driver has consumed the legal limit of alcohol for driving, compared with having consumed no alcohol, may be 2.5. Similarly, you could compute the relative risk of travelling by plane compared to travelling by car.

relativity theory The theory in physics put forward by Albert *Einstein in the early 20th century which fundamentally changed commonly held views of the universe, and concepts of space and time which had previously been regarded as independent of one another. The special theory of relativity deals with the behaviour of particles in different frames of reference which are moving at constant relative speeds, the behaviour of fast-travelling particles for which time passes more slowly, and includes the *mass–energy equation. The general theory of relativity describes gravitational forces in terms of the curvature of space, caused by the presence of mass, and deals with the behaviour of particles in different frames of reference which are accelerating relative to one another.

((⊕)) SEE WEB LINKS

• An article about relativity with a related video.

reliability The sampling variance of a statistic. For example, the reliability of $\bar{X}_n = \frac{1}{n}\sum_{i=1}^{n} X_i$ is $\frac{\sigma^2}{n}$ where $\sigma^2 = \text{Var}(X)$ so, as the sample size increases, $\bar{X}_n$ becomes a better estimator in the sense that the variability decreases.

remainder *See* DIVISION ALGORITHM and TAYLOR'S THEOREM.

remainder theorem The following result about polynomials:

Theorem

If a polynomial $f(x)$ is divided by $x - h$, then the remainder is equal to $f(h)$.

   It is proved as follows. Divide the polynomial $f(x)$ by $x - h$ to get a quotient $q(x)$ and a remainder which will be a constant, r. This means that $f(x) = (x - h)q(x) + r$. Replacing x by h in this equation gives $r = f(h)$, thus proving the theorem. An important consequence of the remainder theorem is the *Factor Theorem.

removable singularity If there is a value which $f(z)$ could be assigned at a *singular point which would make the function analytic at that point, then that point is termed a removable singularity. For example, if $f(z) = 1/z$, then $|f(z)| \to \infty$ as $|z| \to 0$, so this is not a removable singularity, but for $f(z) = \dfrac{\sin z}{z}$, which also has a singularity at $z = 0$, if we define $f(0) = 1$ the function is analytic at 0, and this is an example of a removable singularity.

repeated measures designs An *experimental design where the same participants are measured repeatedly under different conditions. By using the same participants for each condition a major source of variability is removed.

repeated root *See* ROOT.

repeating decimal *See* DECIMAL REPRESENTATION.

repetition codes One of the simplest possible error-correcting procedures in data transmission or storage. The principle is that the sender repeats the message several times, with the hope that the noisy channel interferes with only a small proportion of these messages, and not in a consistent way. The receiver can check if they have the same message each time: if they have, then all is well, but if not, they can try to recover the original message by taking the various sections of the data stream which occurred most often. Generally, other error-correcting mechanisms such as *checksums offer better performance but the ease of implementation of repetition codes makes them attractive.

replicable Able to be repeated. For example, an experiment which measures how far an elastic spring is extended when light masses are suspended.

replication In a designed experiment the same treatment is often applied to a number of subjects to provide more information. An experiment using three replications of each treatment will be more powerful than single observations on each treatment.

representation (of a vector) When the directed line segment $\overrightarrow{AB}$ represents the *vector **a**, then $\overrightarrow{AB}$ is a representation of **a**.

representation theory The branch of mathematics that studies abstract algebraic structures in terms of linear transformations of vector spaces, in particular relating to the symmetries of different structures or groups. One of its attractions is that it can reduce problems in abstract algebra to problems in linear algebra, which is an area of mathematics that is well developed.

representative Given an *equivalence relation on a set, any one of the *equivalence classes can be specified by giving one of the elements in it. The particular element a used can be called a representative of the class, and the class can be denoted by $[a]$.

representative sample A *sample which shares certain characteristics of the population, usually the proportions of members of

particular groups who might be expected to behave differently in the
variable(s) of interest. Quota sampling and stratified sampling both give
representative samples.

residual The difference between an observed value and the value
predicted by some statistical model. The residuals may be checked to
assess how well the model fits the data, perhaps by using a *chi-squared
test. A large residual may indicate an *outlier.

residual variation The variation unaccounted for by a model which
has been fitted to a set of data.

residue class (modulo n) An *equivalence class for the *equivalence
relation of *congruence modulo n. So, two integers are in the same class if
they have the same remainder upon division by n. If $[a]$ denotes the
residue class modulo n containing a, the residue classes modulo n can be
taken as $[0], [1], [2], \ldots, [n-1]$. The sum and product of residue classes
can be defined by

$$[a] + [b] = [a + b], \quad [a][b] = [ab],$$

where it is necessary to show that the definitions here do not depend upon
which *representatives a and b are chosen for the two classes. With this
addition and multiplication, the set, denoted by $\mathbf{Z}_n$, of residue classes
modulo n forms a *ring (in fact, a commutative ring with identity). If n is
composite, the ring $\mathbf{Z}_n$ has *divisors of zero, but when p is prime $\mathbf{Z}_p$ is a
*field.

resistant statistic A statistic which is relatively unaffected by unusual
observations. The median and inter-quartile range are examples of
resistant statistics, while the mean, standard deviation, and range are not.

resistive force A force that opposes the motion of a body. It acts on the
body in a direction opposite to the direction of the velocity of the body. For
example, a ball-bearing falling in a cylindrical glass jar containing oil
experiences the downward pull of gravity and a resistive force upwards
caused by the oil acting to oppose the motion of the falling ball-bearing.
Other examples are *friction and *aerodynamic drag. When there is a
resistive force, the principle of *conservation of energy does not hold, but
the *work–energy principle does hold.

resolution The process of writing a vector in terms of its *components in
two or three mutually perpendicular directions. The process is used, for
example, when a particle is acted on by two or more forces. It may then be
useful to resolve all the forces in two or three mutually perpendicular
directions. In one problem, it may be appropriate to resolve the forces

horizontally and vertically; in another, along and perpendicular to an inclined plane. The equation of motion in vector form may then be replaced by two or three scalar equations of motion.

resolve *See* RESOLUTION.

resonance Suppose that a body capable of performing *oscillations is subject to an applied force which is itself oscillatory, and not subject to any *resistive force. In certain circumstances the body may oscillate with an amplitude that, in theory, increases indefinitely. This happens when the applied force has the same period as the period of the natural oscillations of the body, and then resonance is said to occur. For example, if the equation $m\ddot{x} + kx = F \cos \Omega t$ holds, there is resonance when $\Omega = \sqrt{k/m}$, in which case the solution of the equation involves the particular integral

$$\frac{Ft}{2\sqrt{km}} \sin \sqrt{\frac{k}{m}} t,$$

which gives oscillations of increasing amplitude.

When a resistive force is included, the amplitude of the forced oscillations may take a maximum value for a certain value of the *angular frequency of the applied force. This example of resonance is important in the design of seismographs, instruments for measuring the strength of earthquakes.

(⊕) **SEE WEB LINKS**
• Animations and videos of forced oscillation and resonance.

response bias Occurs when participants in a survey respond differently from the way they actually feel. This may be because the questions have been asked in a loaded manner, or because the information sought is sensitive and the participant is not prepared to be honest. *See also* ANONYMITY OF SURVEY DATA.

response variable = DEPENDENT VARIABLE.

restriction (of a mapping) A mapping $g : S_1 \rightarrow T_1$ is a restriction of the mapping $f : S \rightarrow T$ if $S_1 \subseteq S$, $T_1 \subseteq T$ and $g(s_1) = f(s_1)$ for all s_1 in S_1. Thus a restriction is obtained by taking, perhaps, a subset of the domain or codomain or both, but otherwise following the same rule for defining the mapping.

resultant The sum of two or more vectors, particularly if they represent forces, may be called their resultant.

retail price index (RPI) A weighted average of the general level of prices of goods and services used by most households in the UK, expressed as an *index. The RPI was introduced in its current form in 1947 and is the best known measure of inflation in the UK. One of the inherent difficulties in such an index is that the range of goods and services in the calculation of the index are revised in January of each year. This is necessary in order for the index to reflect consumer spending trends, but it does mean that it is not a direct comparison year on year.

(((⊕))) SEE WEB LINKS

• Gives access to the latest indices on price inflation, including the RPI values since June 1947.

retardation = DECELERATION.

reversion to the mean In a sequence of independent observations of a *random variable, the greater the deviation from its *mean of an observation, the greater the probability that the next observation will be closer to the mean. This appears to be counter-intuitive to the principle of independence, but it is actually a direct consequence of it, since probability measure is strictly increasing over an increasing interval. So, as you move away from the mean you necessarily increase the proportion of the distribution lying closer to the mean.

revolution One complete turn or cycle of a periodic motion. For example a car engine may operate most efficiently at 3500 revs/min (revolutions per minute).

revolve Rotate about an axis or point.

rhombohedron A *polyhedron with six faces, each of which is a rhombus. It is thus a *parallelepiped whose edges are all the same length.

rhombus (rhombi) A quadrilateral all of whose sides have the same length. A rhombus is both a *kite and a *parallelogram.

Riemann, (Georg Friedrich) Bernhard (1826–66) German mathematician who was a major figure in 19th-century mathematics. In many ways, he was the intellectual successor of Gauss. In geometry, he started the development of those tools which Einstein would eventually use to describe the universe and which in the 20th century would be turned into the theory of manifolds. His basic geometrical ideas were presented in his famous inaugural lecture at Göttingen, to an audience including Gauss. He did much significant work in analysis, in which his name is preserved in the *Riemann integral, the *Cauchy–Riemann equations and Riemann surfaces. He also made connections between prime number theory and analysis: he formulated the *Riemann hypothesis, a conjecture concerning the so-called zeta function, which if proved would give information about the distribution of prime numbers.

Riemann hypothesis (Riemann zeta hypothesis) The hypothesis that the *zeta function has no non-trivial zeros unless $\operatorname{Re}(z) = \frac{1}{2}$.

Riemann integral, Riemann sum *See* INTEGRAL.

Riemann–Roch Theorem Gives the number of independent analytic functions that can be defined on a given *Riemann surface in terms of various topological invariants that are easily calculated. The theorem has applications linking the study of topology with algebraic geometry.

Riemann sphere The representation of the *extended complex plane by the *stereographic projection.

Riemann surface A surface which is a 1-dimensional complex manifold. The *Riemann sphere is the simplest example of a Riemann surface which is not a subset of the complex plane. Riemann surfaces can be thought of as deformations of the complex plane in the sense that the local topology can be that of the complex plane but the global topology may be quite different—like a sphere, a torus, or sheets glued together. They are particularly useful in relation to the study of *analytic functions.

right angle A quarter of a complete revolution. It is thus equal to 90° or $\pi/2$ radians.

right-angled triangle A triangle containing a right angle. *Pythagoras' Theorem and the *trigonometric functions for acute angles can be used for calculations in right-angled triangles.

right-circular *See* CONE and CYLINDER.

right derivative *See* LEFT AND RIGHT DERIVATIVE.

right-hand limit (at a) The real number line is usually represented with values increasing from left to right, so the right-hand limit is the limit of the function from above, i.e. where $x \to a$ but takes only values greater than a.

right-handed system Let Ox, Oy and Oz be three mutually perpendicular directed lines, intersecting at the point O. In the order Ox, Oy, Oz, they form a right-handed system if a person standing with their head in the positive z-direction and facing the positive y-direction would have the positive x-direction to their right. Putting it another way, when seen from a position facing the positive z-direction, a rotation from the positive x-direction to the positive y-direction passes through a right angle clockwise. Following the normal practice, the figures in this book that use a Cartesian coordinate system for 3-dimensional space have Ox, Oy and Oz forming a right-handed system.

The three directed lines Ox, Oy and Oz (in that order) form a left-handed system if, taken in the order Oy, Ox, Oz, they form a right-handed system. If the direction of any one of three lines of a right-handed system is reversed, the three directed lines form a left-handed system.

Similarly, an ordered set of three oblique directed lines may be described as forming a right- or left-handed system. Three vectors, in a given order, form a right- or left-handed system if directed line segments representing them define directed lines that do so.

right-regular *See* PRISM *and* PYRAMID.

rigid body An object with the property that it does not change shape whatever forces are applied to it. It is used in a *mathematical model to represent an object in the real world. It may be a system of particles held in a rigid formation, or it may be a distribution of mass in the form of a rod, a lamina or some 3-dimensional shape. Problems on the motion of a rigid body are concerned with such matters as the position in space of the rigid body; the angular velocity, angular momentum and kinetic energy of the rigid body; and the position vector, velocity and acceleration of its centre of mass. In general, a rigid body has six *degrees of freedom. One equation of motion in vector form governs the motion of the centre of mass, and another relates the rate of change of angular momentum to the moment of the forces acting on the rigid body.

ring Sets of entities with two operations, often called addition and multiplication, occur in different situations in mathematics and sometimes share many of the same properties. It is useful to recognize these similarities by identifying certain of the common characteristics. One such set of properties is specified in the definition of a ring: a ring is a set R, closed under two operations called addition and multiplication, denoted in the usual way, such that

(i) for all a, b and c in R, $a + (b + c) = (a + b) + c$,

(ii) for all a and b in R, $a + b = b + a$,

(iii) there is an element 0 in R such that $a + 0 = a$ for all a in R,

(iv) for each a in R, there is an element $-a$ in R such that $a + (-a) = 0$,

(v) for all a, b and c in R, $a(bc) = (ab)c$,

(vi) for all a, b and c in R, $a(b + c) = ab + ac$ and $(a + b)c = ac + bc$.

The element guaranteed by **3** is a *neutral element for addition. It can be shown that in a ring this element is unique and has the extra property that

$a0 = 0$ for all a in R, so it is usually called the *zero element. Also, for each a, the element $-a$ guaranteed by **4** is unique and is the *negative of a. The ring is a commutative ring if it is true that

7. for all a and b in R, $ab = ba$,

and it is a commutative ring with identity if also

8. there is an element $1(\neq 0)$ such that $a1 = a$ for all a in R.

If certain further properties are added, the definitions of an *integral domain and a *field are obtained. So any integral domain and any field is a ring. Further examples of rings (which are not integral domains or fields) are the set of 2×2 real matrices and the set of all even integers, each with the appropriate addition and multiplication. Another example of a ring is $\mathbf{Z}_n$, the set $\{0, 1, 2, \ldots, n-1\}$ with addition and multiplication modulo n.

A ring may be denoted by $\langle R, +, \times \rangle$ and another ring by, say, $\langle R', \oplus, \otimes \rangle$ when it is necessary to distinguish the operations in one ring from the operations in the other. But it is sufficient to refer simply to the ring R when the operations intended are clear.

rise The difference between the *ordinates (y values) of a pair of points in a 2-dimensional coordinate system. Used with the *run in calculating the *gradient of the line joining the points.

risk Describes the proportion or rate of incidence of a particular outcome amongst a group. So the risk of lung cancer amongst those smoking more than 20 cigarettes a day can be expressed as a percentage which equates to the statistical probability that an individual in that category will contract lung cancer.

robust A test or *estimator is robust if it is not sensitive to small discrepancies in assumptions, such as the assumption of the normality of the underlying distribution. A test is said to be robust to *outliers if it is not unduly affected by their presence.

rod An object considered as being 1-dimensional, having length and density but having no width or thickness. It is used in a *mathematical model to represent a thin straight object in the real world. It may be rigid or flexible, depending on the circumstances.

Rodrigues's formula *See* LEGENDRE POLYNOMIALS.

Rolle, Michel (1652–1719) French mathematician, remembered primarily for the theorem that bears his name, which appears in a book of his published in 1691.

Rolle's Theorem The following result concerning the existence of *stationary points of a function f:

Theorem

Let f be a function which is continuous on $[a, b]$ and differentiable in (a, b), such that $f(a) = f(b)$. (Some authors require that $f(a) = f(b) = 0$.) Then there is a number c with $a < c < b$ such that $f'(c) = 0$.

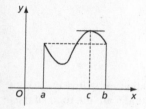

The result stated in the theorem can be expressed as a statement about the graph of f: with appropriate conditions on f, between any two points on the graph $y = f(x)$ that are level with each other, there must be a stationary point; that is, a point at which the tangent is horizontal. The theorem is, in fact, a special case of the *mean value theorem; however, it is normal to establish Rolle's Theorem first and deduce the mean value theorem from it. A rigorous proof relies on the non-elementary result that a *continuous function on a closed interval attains its bounds.

rolling condition The relationship between the linear and angular speeds of a cylinder or sphere when it is rolling on a plane surface without slipping. If v is the speed of the centre of mass of the cylinder or sphere, with radius r, and ω is the angular speed of the rotation, then $v = r\omega$.

Roman numeral *See* NUMERAL.

root (of an equation) Let $f(x) = 0$ be an equation that involves the indeterminate x. A root of the equation is a value h such that $f(h) = 0$. Such a value is also called a zero of the function f. Some authors use 'root' and 'zero' interchangeably.

If $f(x)$ is a polynomial, then $f(x) = 0$ is a polynomial equation. By the *Factor Theorem, h is a root of this equation if and only if $x - h$ is a factor of $f(x)$. The value h is a simple root if $x - h$ is a factor but $(x - h)^2$ is not a factor of $f(x)$; and h is a root of order (or multiplicity) n if $(x - h)^n$ is a factor but $(x - h)^{n+1}$ is not. A root of order 2 is a double root; a root of order 3 is a triple root. A root of order n, where $n \geq 2$, is a multiple (or repeated) root.

If h is a double root of the polynomial equation $f(x) = 0$, then close to $x = h$ the graph $y = f(x)$ looks something like one of the diagrams in the first row of the figure. If h is a triple root, then the graph looks like one of the diagrams in the second row of the figure. The value h is a root of order at least n if and only if $f(h) = 0, f'(h) = 0, \ldots, f^{(n-1)}(h) = 0$.

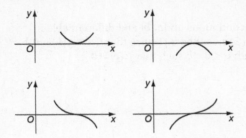

If α and β are the roots of the quadratic equation $ax^2 + bx + c = 0$, with $a \neq 0$, then $\alpha + \beta = -b/a$ and $\alpha\beta = c/a$. If α, β and γ are the roots of the cubic equation $ax^3 + bx^2 + cx + d = 0$, with $a \neq 0$, then $\alpha + \beta + \gamma = -b/a$, $\beta\gamma + \gamma\alpha + \alpha\beta = c/a$ and $\alpha\beta\gamma = -d/a$. Similar results hold for polynomial equations of higher degree.

root (of a tree) *See* TREE.

root mean squared deviation The positive square root of the *mean squared deviation.

root of unity *See* N-TH ROOT OF UNITY.

rose A curve consisting of a number of loops meeting at the origin, which resemble the petals of a flower like the rose. In polar coordinates the equation is of the form $r = a \cos(n\theta)$ or $r = a \sin(n\theta)$ where a determines the distance from the origin to the tip of each loop. The polar axis is always an axis of symmetry for curves of this form with cos, while for curves with sin the perpendicular to the polar axis is an axis of symmetry. If n is even, then there will be $2n$ loops, while if n is odd there are only n loops with gaps between the loops. The reason for this discrepancy is that when n is odd, and r returns a negative value, it traces over a loop created 180° out of phase, where r returned a positive value. When n is even, the value of r returned for any two angles which are 180° different is necessarily the same sign, and a new loop is created.

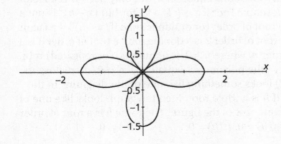

The rose created by $r = 1.5 \cos(2\theta)$.

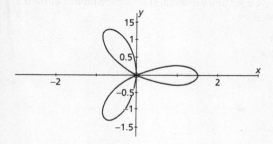

The rose created by $r = 1.5 \cos(3\theta)$.

rotation (of the plane) A rotation of the plane about the origin O through an angle α is the *transformation of the plane in which O is mapped to itself, and a point P with polar coordinates (r, θ) is mapped to the point P' with polar coordinates $(r, \theta + \alpha)$. In terms of Cartesian coordinates, P with coordinates (x, y) is mapped to P' with coordinates (x', y'), where

$$x' = x \cos \alpha - y \sin \alpha,$$
$$y' = x \sin \alpha + y \cos \alpha.$$

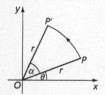

rotational symmetry A plane figure has rotational symmetry about a point O if the figure appears the same when it is rotated about O through some positive angle less than one complete revolution. For example, an equilateral triangle (and, indeed, any regular polygon) has rotational symmetry about its centre.

rotation of axes (in the plane) Suppose that a Cartesian coordinate system has a given x-axis and y-axis with origin O and given unit length, so that a typical point P has coordinates (x, y). Consider taking a new coordinate system with the same origin O and the same unit length, with X-axis and Y-axis, such that a rotation through an angle α (with the positive direction taken anticlockwise) carries the x-axis to the X-axis and the y-axis to the Y-axis. With respect to the new coordinate system, the point P has coordinates (X, Y). Then the old and new coordinates in such a rotation of axes are related by

$$x = X \cos \alpha - Y \sin \alpha,$$
$$y = X \sin \alpha + Y \cos \alpha.$$

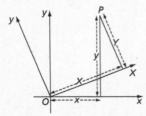

In matrix notation, these equations become

$$\begin{bmatrix} x \\ y \end{bmatrix} = \begin{bmatrix} \cos \alpha & -\sin \alpha \\ \sin \alpha & \cos \alpha \end{bmatrix} \begin{bmatrix} X \\ Y \end{bmatrix}$$

and, conversely,

$$\begin{bmatrix} X \\ Y \end{bmatrix} = \begin{bmatrix} \cos \alpha & \sin \alpha \\ -\sin \alpha & \cos \alpha \end{bmatrix} \begin{bmatrix} x \\ y \end{bmatrix}.$$

For example, in a rotation of axes through an angle of $-\pi/4$ radians, the coordinates are related by

$$x = \frac{1}{\sqrt{2}}(X + Y), \qquad y = \frac{1}{\sqrt{2}}(-X + Y),$$

and the curve with equation $x^2 - y^2 = 2$ has equation $XY = 1$ in the new coordinate system.

rough surface *See* CONTACT FORCE.

round Shaped like a *circle or *sphere.

rounding Suppose that a number has more digits than can be conveniently handled or stored. In rounding (as opposed to *truncation), the original number is replaced by the number with the required number of digits that is closest to it. Thus, when rounding to 1 decimal place, the number 1.875 becomes 1.9 and the number 1.845 becomes 1.8. It is said that the number is accordingly rounded up or rounded down. When the original is precisely at a halfway point (for example, if 1.85 is to be rounded to 1 decimal place), it may be rounded up (to 1.9) or rounded down (to 1.8). Some authors like to recommend a particular way of deciding which to do. *See also* DECIMAL PLACES and SIGNIFICANT FIGURES.

rounding error = ROUND-OFF ERROR.

round-off error When a number X is rounded to a certain number of digits to obtain an approximation x, the *error is called the round-off error. For some authors this is $X - x$, and for others it is $x - X$. For example, if 1.875 is rounded to 1 decimal place, or to 2 significant figures, to give 1.9, the round-off error is either 0.025 or -0.025. When a number is rounded to k decimal places, the round-off error lies between $\pm 5 \times 10^{-(k+1)}$; for example, when rounding to 3 decimal places, the round-off error lies between ± 0.0005. For some authors, the error is $|X - x|$ (*See* ABSOLUTE VALUE) and so, for them, it is always greater than or equal to zero.

route inspection problem (Chinese postman problem) (in graph theory) A postman needs to travel along each road, or edge (arc) in graph theory, while travelling the shortest possible route. If a closed Eulerian trail can be found, it must constitute the minimum possible since it requires each edge to be travelled exactly once, but most graphs do not have such a route available. The best approach is as follows. If all the vertices (nodes) are of even degree, then there will be a Eulerian trail. If exactly two of the vertices are odd, then use Dijkstra's method of solving the *shortest path problem to identify the shortest path between those two vertices, and the postman will have to walk the edges on that route twice, and all other edges once. However, when more than two vertices are odd, there are various ways they could be paired off, and what is required is the minimum sum of shortest paths between the pairs. The edges on each of these shortest paths will have to be repeated.

row A horizontal line of elements in an *array, especially in a matrix.

row equivalence (of matrices) Two matrices when one can be transformed into the other by a finite set of *elementary matrix operations on its rows.

row matrix A *matrix with exactly one row, that is, a $1 \times n$ matrix of the form $[a_1 \, a_2 \ldots a_n]$. Given an $m \times n$ matrix, it may be useful to treat its rows as individual row matrices.

row operation *See* ELEMENTARY ROW OPERATION.

row rank *See* RANK.

row stochastic matrix *See* STOCHASTIC MATRIX.

row vector = ROW MATRIX.

RSA (public-key cryptography) A cryptographic algorithm which depends on the difficulty of factorizing large *semiprimes. The creator publishes a semiprime together with an auxiliary value, which form a key that anyone can use to encrypt a message (hence the term 'public-key'), but only someone who knows the two prime factors of the semiprime can decode the message.

ruled surface A surface that can be traced out by a moving straight line; in other words, every point of the surface lies on a straight line lying wholly in the surface. Examples are a *cone, a *cylinder, a *hyperboloid of one sheet and a *hyperbolic paraboloid.

run (in a coordinate system) The difference between the abscissae (x-values) of a pair of points in a 2-dimensional coordinate system. Used with the *rise in calculating the *gradient of the line joining the points.

run (in statistics) A sequence of consecutive observations from a sample that share a specified property. Often the observations have either one or other of two properties A and B. For example, A may be 'above the median' and B 'below the median'. If the sequence of observations then gives the sequence $AABAAABBAB$, for example, the number of runs equals 6. This statistic is used in some *non-parametric methods.

Runge–Kutta methods A numerical method of solving *differential equations in the form $\frac{dy}{dx} = f(x, y)$ which uses the midpoint of interval(s) to improve accuracy. So if the value of the function is known at (x_n, y_n), and the estimate y_{n+1} of y is required at $x_{n+1} = x_n + h$. The second-order Runge–Kutta formula is

$$k_1 = h \times f(x_n, y_n)$$
$$k_2 = h \times f(x_n + \tfrac{1}{2}h, y_n + \tfrac{1}{2}k_1)$$
$$y_{n+1} = y_n + k_2 + O(h^3)$$

and the fourth-order Runge–Kutta formula is

$$k_1 = h \times f(x_n, y_n)$$
$$k_2 = h \times f(x_n + \tfrac{1}{2}h, y_n + \tfrac{1}{2}k_1)$$
$$k_3 = h \times f(x_n + \tfrac{1}{2}h, y_n + \tfrac{1}{2}k_2)$$
$$k_4 = h \times f(x_n + h, y_n + k_3)$$
$$y_{n+1} = y_n + \tfrac{1}{6}k_1 + \tfrac{1}{3}k_2 + \tfrac{1}{3}k_3 + \tfrac{1}{6}k_4 + O(h^5)$$

Russell, Bertrand Arthur William (1872–1970) British philosopher, logician and writer on many subjects. He is remembered in mathematics as the author, with A. N. *Whitehead, of *Principia mathematica*, published between 1910 and 1913 in three volumes, which set out to show that pure mathematics could all be derived from certain fundamental logical axioms. Although the attempt was not completely successful, the work was highly influential. He was also responsible for the discovery of *Russell's paradox.

Russell's paradox By using the notation of set theory, a set can be defined as the set of all x that satisfy some property. Now it is clearly possible for a set not to belong to itself: any set of numbers, say, does not belong to itself because to belong to itself it would have to be a number. But it is also possible to have a set that does belong to itself: for example, the set of all sets belongs to itself. In 1901, Bertrand Russell drew attention to what has become known as Russell's paradox, by considering the set R, defined by $R = \{x \mid x \notin x\}$. If $R \in R$ then $R \notin R$; and if $R \notin R$ then $R \in R$. The paradox points out the danger of the unrestricted use of abstraction, and various solutions have been proposed to avoid the paradox.

Rutherford, Lord FRS (1871–1937) Born in New Zealand to a Scottish father and English mother, Ernest Rutherford went to Cambridge University in 1894 following an undergraduate degree in mathematics and physics from Canterbury College. He then spent ten years at McGill University in Montreal before returning to England, eventually as Cavendish Professor of Physics at Cambridge. He was awarded the 1908 Nobel Prize in Chemistry 'for his investigations into the disintegration of the elements, and the chemistry of radioactive substances', and named many of the basic components of atomic physics such as alpha, beta and gamma rays, the proton, the neutron and half-life. He was the first to realize that almost all of the mass of an atom, and all its positively charged components, were concentrated in a tiny proportion of the atom's size which came to be known as the 'nucleus'. Since he regarded himself primarily as a physicist, Lord Rutherford was reportedly somewhat disgruntled that his Nobel Prize was in Chemistry.

rv Abbreviation for *random variable.

saddle-point Suppose that a surface has equation $z = f(x, y)$, with, as usual, the z-axis vertically upwards. A point P on the surface is a saddle-point if the *tangent plane at P is horizontal and if P is a local minimum on the curve obtained by one vertical cross-section and a local maximum on the curve obtained by another vertical cross-section. It is so called because the central point on the seat of a horse's saddle has this property. The *hyperbolic paraboloid, for example, has a saddle-point at the origin. *See also* STATIONARY POINT (in two variables).

sample A subset of a *population selected in order to make *inferences about the population. (Strictly speaking, it is not a subset because elements may be repeated.) It is a random sample if it is chosen in such a way that every sample of the same size has an equal chance of being selected. If the sample is chosen in such a way that no member of the population can be selected more than once, this is sampling without replacement. In sampling with replacement, an element has a chance of being selected more than once. A quota sample is a sample in which a predetermined number of elements have to be selected from each category in a specified list. *See also* STRATIFIED SAMPLE.

sample space The set of all possible outcomes of an experiment. For example, suppose that the aim of the experiment is to toss a coin three times and record the results. Then the sample space could be expressed as {HHH, HHT, HTH, HTT, THH, THT, TTH, TTT} where, for example, 'HTT' indicates that the coin came up 'heads' on the first toss and 'tails' on the second and third tosses. If the aim of the experiment is to toss a coin three times and count the numbers of heads obtained, the sample space could be taken as {0, 1, 2, 3}.

sampling distribution Every *statistic is a random variable because its value varies from one sample to another. The *distribution of this random variable is a sampling distribution.

sampling frame A full listing of a population for the purposes of constructing a sample. For example, if a sample was to be taken of pupils in a school, then an alphabetical list of all pupils or alphabetical lists by year group would be two obvious sampling frames which could be used.

saturated (in networks) An edge where the flow is up to its *capacity.

scalar (for matrices) In work with *matrices, the entries must belong to some particular set S. An element of S may be called a scalar to emphasize that it is not a matrix. For example, the occasion may arise when a matrix $\mathbf{A}$ is to be multiplied by a scalar k to form $k\mathbf{A}$, or when a certain row of a matrix is to be multiplied by a scalar.

scalar (for vectors) In work with *vectors, a quantity that is a real number, in other words, *not* a vector, is called a scalar.

scalar matrix A diagonal matrix where all the entries in the leading diagonal are equal, to k say, and all other entries are zero. Multiplication by this matrix is equivalent to multiplication by the scalar k.

scalar multiple (of a matrix) Let $\mathbf{A}$ be an $m \times n$ matrix, with $\mathbf{A} = [a_{ij}]$, and k a scalar. The scalar multiple $k\mathbf{A}$ is the $m \times n$ matrix $\mathbf{C}$, where $\mathbf{C} = [c_{ij}]$ and $c_{ij} = ka_{ij}$. Multiplication by scalars has the following properties:

  (i) $(h + k)\mathbf{A} = h\mathbf{A} + k\mathbf{A}$.
  (ii) $k(\mathbf{A} + \mathbf{B}) = k\mathbf{A} + k\mathbf{B}$.
  (iii) $h(k\mathbf{A}) = (hk)\mathbf{A}$.
  (iv) $0\mathbf{A} = \mathbf{O}$, the zero matrix.
  (v) $(-1)\mathbf{A} = -\mathbf{A}$, the negative of $\mathbf{A}$.

scalar multiple (of a vector) Let $\mathbf{a}$ be a non-zero vector and k a non-zero scalar. The scalar multiple of $\mathbf{a}$ by k, denoted by $k\mathbf{a}$, is the vector whose magnitude is $|k|\|\mathbf{a}\|$ and whose direction is that of $\mathbf{a}$, if $k > 0$, and that of $-\mathbf{a}$, if $k < 0$. Also, $k\mathbf{0}$ and $0\mathbf{a}$ are defined to be $\mathbf{0}$, for all k and $\mathbf{a}$. Multiplication by scalars has the following properties:

  (i) $(h + k)\mathbf{a} = h\mathbf{a} + k\mathbf{a}$.
  (ii) $k(\mathbf{a} + \mathbf{b}) = k\mathbf{a} + k\mathbf{b}$.
  (iii) $h(k\mathbf{a}) = (hk)\mathbf{a}$.
  (iv) $1\mathbf{a} = \mathbf{a}$.
  (v) $(-1)\mathbf{a} = -\mathbf{a}$, the negative of $\mathbf{a}$.

scalar product For vectors $\mathbf{a}$ and $\mathbf{b}$, the scalar product $\mathbf{a} \cdot \mathbf{b}$ is defined by $\mathbf{a} \cdot \mathbf{b} = |\mathbf{a}||\mathbf{b}| \cos \theta$, where θ is the angle, in radians with $0 \leq \theta \leq \pi$, between $\mathbf{a}$ and $\mathbf{b}$. This is a scalar quantity; that is, a real number, not a vector. The scalar product has the following properties:

  (i) $\mathbf{a} \cdot \mathbf{b} = \mathbf{b} \cdot \mathbf{a}$.
  (ii) For non-zero vectors $\mathbf{a}$ and $\mathbf{b}$, $\mathbf{a} \cdot \mathbf{b} = 0$ if and only if $\mathbf{a}$ is perpendicular to $\mathbf{b}$.

(iii) $\mathbf{a} \cdot \mathbf{a} = |\mathbf{a}|^2$; the scalar product $\mathbf{a} \cdot \mathbf{a}$ may be written $\mathbf{a}^2$.

(iv) $\mathbf{a} \cdot (\mathbf{b} + \mathbf{c}) = \mathbf{a} \cdot \mathbf{b} + \mathbf{a} \cdot \mathbf{c}$, the distributive law.

(v) $\mathbf{a} \cdot (k\mathbf{b}) = (k\mathbf{a}) \cdot \mathbf{b} = k(\mathbf{a} \cdot \mathbf{b})$.

(vi) If vectors $\mathbf{a}$ and $\mathbf{b}$ are given in terms of their components (with respect to the standard vectors $\mathbf{i}$, $\mathbf{j}$ and $\mathbf{k}$) as $\mathbf{a} = a_1\mathbf{i} + a_2\mathbf{j} + a_3\mathbf{k}$, $\mathbf{b} = b_1\mathbf{i} + b_2\mathbf{j} + b_3\mathbf{k}$, then $\mathbf{a} \cdot \mathbf{b} = a_1b_1 + a_2b_2 + a_3b_3$.

scalar projection (of a vector on a vector) *See* VECTOR PROJECTION (of a vector on a vector).

scalar triple product For vectors $\mathbf{a}$, $\mathbf{b}$ and $\mathbf{c}$, the *scalar product, $\mathbf{a} \cdot (\mathbf{b} \times \mathbf{c})$, of $\mathbf{a}$ with the vector $\mathbf{b} \times \mathbf{c}$ (*See* VECTOR PRODUCT), is called a scalar triple product. It is a scalar quantity and is denoted by $[\mathbf{a}, \mathbf{b}, \mathbf{c}]$. It has the following properties:

(i) $[\mathbf{a}, \mathbf{b}, \mathbf{c}] = -[\mathbf{a}, \mathbf{c}, \mathbf{b}]$.

(ii) $[\mathbf{a}, \mathbf{b}, \mathbf{c}] = [\mathbf{b}, \mathbf{c}, \mathbf{a}] = [\mathbf{c}, \mathbf{a}, \mathbf{b}]$.

(iii) The vectors $\mathbf{a}$, $\mathbf{b}$ and $\mathbf{c}$ are coplanar if and only if $[\mathbf{a}, \mathbf{b}, \mathbf{c}] = 0$

(iv) If the vectors are given in terms of their components (with respect to the standard vectors $\mathbf{i}$, $\mathbf{j}$ and $\mathbf{k}$) as $\mathbf{a} = a_1\mathbf{i} + a_2\mathbf{j} + a_3\mathbf{k}$, $\mathbf{b} = b_1\mathbf{i} + b_2\mathbf{j} + b_3\mathbf{k}$, and $\mathbf{c} = c_1\mathbf{i} + c_2\mathbf{j} + c_3\mathbf{k}$, then
$$[\mathbf{a}, \mathbf{b}, \mathbf{c}] = a_1(b_2c_3 - b_3c_2) + a_2(b_3c_1 - b_1c_3) + a_3(b_1c_2 - b_2c_1)$$
$$= \begin{bmatrix} a_1 & a_2 & a_3 \\ b_1 & b_2 & b_3 \\ c_1 & c_2 & c_3 \end{bmatrix}.$$

(v) Let $\overrightarrow{OA}$, $\overrightarrow{OB}$ and $\overrightarrow{OC}$ represent $\mathbf{a}$, $\mathbf{b}$ and $\mathbf{c}$ and let P be the *parallelepiped with OA, OB and OC as three of its edges. The *absolute value of $[\mathbf{a}, \mathbf{b}, \mathbf{c}]$ then gives the volume of the parallelepiped P. (If $\mathbf{a}$, $\mathbf{b}$ and $\mathbf{c}$ form a right-handed system, then $[\mathbf{a}, \mathbf{b}, \mathbf{c}]$ is positive; and if $\mathbf{a}$, $\mathbf{b}$ and $\mathbf{c}$ form a left-handed system, then $[\mathbf{a}, \mathbf{b}, \mathbf{c}]$ is negative.)

scale A marking of values along a line to use in making measurements or in representing measurements. A linear scale is one in which the same measurement interval is always represented by the same interval on the scale. Other scales are the logarithmic and probability.

scalene triangle A triangle in which all three sides have different lengths.

scatter diagram A 2-dimensional diagram showing the points corresponding to n paired-sample observations (x_1, y_1), (x_2, y_2), ..., (x_n, y_n), where $x_1, x_2, \ldots, x_n$ are the observed values of the *explanatory variable and $y_1, y_2, \ldots, y_n$ are the observed values of the *dependent variable.

scatter plot = SCATTER DIAGRAM.

scheduling (in critical path analysis) In its simplest form, critical path analysis assumes each activity requires one worker, but in reality this is often not true. It can be further complicated if certain activities can only be carried out by certain workers, as is the case in building projects involving skilled tradespeople. Scheduling is the task of assigning workers to the activities with the aim of completing the project with as few workers as possible, or if the number of workers is limited, completing it in the shortest possible time with the available workers.

Schrödinger, Erwin Rudolf Alexander (1887–1961) Austrian mathematician and theoretical physicist who developed *Heisenberg's initial work in *quantum mechanics specifically relating to wave mechanics and the general theory of relativity. He shared the Nobel Prize for Physics in 1933 with Paul *Dirac for this work.

scientific notation A number is said to be in scientific notation when it is written as a number between 1 and 10, times a power of 10; that is to say, as $a \times 10^n$, where $1 \leq a < 10$ and n is an integer. Thus 634.8 and 0.002 34 are written in scientific notation as 6.348×10^2 and 2.34×10^{-3}. The notation is particularly useful for very large and very small numbers.

sd Abbreviation for *standard deviation.

se Abbreviation for *standard error.

seasonal variation *See* TIME SERIES.

secant *See* TRIGONOMETRIC FUNCTION.

secant (of a curve) A line that cuts a given curve, usually in more than one point.

secant method Given two successive approximations to a, the secant method calculates where the secant through these two points

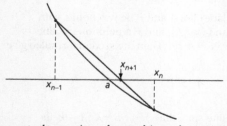

meets the x-axis and uses this as the next approximation. So

$$x_{n+1} = x_n - \frac{(x_n - x_{n-1})}{f(x_n) - f(x_{n-1})} \times f(x_n) \text{ for } n = 1, 2, 3, \ldots.$$

sech *See* HYPERBOLIC FUNCTION.

second (angular measure) *See* DEGREE (angular measure).

second (time) In *SI units, the base unit used for measuring time, abbreviated to 's'. It was once defined as 1/86 400 of the mean solar day, which is the average time it takes for the Earth to rotate relative to the Sun. Now it is defined in terms of the radiation emitted by the caesium-133 atom.

second derivative *See* HIGHER DERIVATIVE.

second derivative test *See* DERIVATIVE TEST.

second-order partial derivative *See* HIGHER-ORDER PARTIAL DERIVATIVE.

section The plane figure obtained when a surface or solid is intersected by a plane. If the figure has an axis of symmetry and the plane is at right angles to the axis then it is a cross-section. For example, all the cross-sections of a right circular cone are circles, and the radius of the cross-section is proportional to the distance of the vertex from the plane.

section formula (in vectors) Let A and B be two points with *position vectors $\mathbf{a}$ and $\mathbf{b}$, and P a point on the line through A and B, such that $AP : PB = m : n$. Then P has position vector $\mathbf{p}$, given by the section formula

$$\mathbf{p} = \frac{1}{m+n}(m\mathbf{b} + n\mathbf{a}).$$

It is possible to choose m and n such that $m + n = 1$ and thus to suppose, changing notation, that $AP : PB = k : 1 - k$. Then $\mathbf{p} = (1-k)\mathbf{a} + k\mathbf{b}$.

section formulae (in the plane) Let A and B be two points with Cartesian coordinates (x_1, y_1) and (x_2, y_2), and P a point on the line through A and B such that $AP : PB = m : n$. Then the section formulae give the coordinates of P as

$$\left(\frac{mx_2 + nx_1}{m+n}, \frac{mx_2 + ny_1}{m+n}\right).$$

If, instead, P is a point on the line such that $AP : PB = k : 1 - k$, its coordinates are $((1 - k)x_1 + kx_2, (1 - k)y_1 + ky_2)$.

section formulae (in 3-dimensional space) Let A and B be two points with Cartesian coordinates (x_1, y_1, z_1) and (x_2, y_2, z_2), and P a point on the line through A and B. If $AP : PB = m : n$, the section formulae give the coordinates of P as

$$\left(\frac{mx_2 + nx_1}{m + n}, \frac{my_2 + ny_1}{m + n}, \frac{mz_2 + nz_1}{m + n} \right),$$

and, if P is a point on the line such that $AP : PB = k : 1 - k$, its coordinates are $((1 - k)x_1 + kx_2, (1 - k)y_1 + ky_2, (1 - k)z_1 + kz_2)$.

sector A sector of a circle, with centre O, is the region bounded by an arc AB of the circle and the two radii OA and OB. The area of a sector is equal to $\frac{1}{2}r^2\theta$, where r is the radius and θ is the angle in radians.

segment A segment of a circle is the region bounded by an arc AB of the circle and the chord AB.

selection The number of selections of n objects taken r at a time (that is, the number of ways of selecting r objects out of n) is denoted by ^nC_r and is equal to

$$\frac{n!}{r!(n - r)!}.$$

(*See* BINOMIAL COEFFICIENT, where the alternative notation $\binom{n}{r}$ is defined.) For example, from four objects A, B, C and D, there are six ways of selecting two: AB, AC, AD, BC, BD, CD. The property that $^{n+1}C_r = {}^nC_{r-1} + {}^nC_r$ can be seen displayed in *Pascal's triangle.

selection bias Bias which occurs because of the method of selection of a sample. For example, if people are interviewed on a Monday morning in town, various groups such as teachers and rural dwellers are likely to be under-represented.

self-inverse An element of a group, ring, etc. which is its own inverse, i.e. an element a for which $a^2 = e$ where e is the identity element. So the identity is always a self-inverse element; in transformations any reflection is a self-inverse, and so is a rotation through $180°$.

self-reference A statement which refers to itself. This can give rise to paradoxes. For example 'this statement is false' makes no sense because if it is true then it must be false, and if it is false then it must be true!

self-selected samples These are almost the worst sort of samples. Where a newspaper, radio or television station asks those interested to respond the results are likely only to reflect the most strongly held opinions, and the results are virtually worthless.

semi- Prefix denoting half.

semicircle One half of a circle cut off by a diameter.

semi-interquartile range A measure of *dispersion equal to half the difference between the first and third *quartiles in a set of numerical data.

semi-metric One of the conditions for a distance measure d to be a *metric is the triangle inequality. That means that $d(x, z) \leq d(x, y) + d(y, z)$ for any points x, y, and z. If this condition is relaxed then d is said to be a semi-metric, and the space is said to be a semi-metric space.

semi-norm A function $\|x\|$ on a *vector space X for which the following conditions hold for all x and y in X:

   (i) $\|x\| \geq 0$;
   (ii) $\|kx\| = |k| \|x\|$;
   (iii) $\|x + y\| \leq \|x\| + \|y\|$.

This is a generalization of the *norm which removes the restriction that $\|x\| = 0$ only if $x = 0$.

semiprime The product of two primes, so $6 = 2 \times 3$ is a semiprime but 12 is not. Very large semiprimes form the basis of many cryptographic protocols.

semi-regular polyhedron See ARCHIMEDEAN SOLID.

semi-regular tessellation See TESSELLATION.

semi-vertical angle See CONE.

sense One of the two possible directions on a line, or of rotation, i.e. clockwise or anticlockwise. So the vectors **a** and $-$**a** have the same magnitude, and their directions are parallel, but in opposite senses.

sensitivity analysis The varying of *parameters in a *simulation to discover which parameters have the greatest influence on the features of interest.

separable (of a function) A function which can be written so that the variables are separated additively or multiplicatively, for example $2x^2y + 4y = 2y(x^2 + 2)$ or $x^2 + y^2 - 4x + 6y + 13 = (x - 2)^2 + (y + 3)^2$. Where a function is separable, optimization procedures can be applied to each of the variables individually, which is much simpler.

separable first-order differential equation A first-order differential equation $dy/dx = f(x, y)$ in which the function f can be expressed as the product of a function of x and a function of y. The differential equation then has the form $dy/dx = g(x)h(y)$, and its solution is given by the equation

$$\int \frac{1}{h(y)} dy = \int g(x) \ dx + c,$$

where c is an arbitrary constant.

separated points A special case of *separated sets where $\{x\}$ and $\{y\}$ are *singleton sets within a *topological space X.

separated sets In a *topological space X, two sets for which each set is *disjoint from the *closure of the other set. Any two separated sets must be disjoint, but the closures do not have to be disjoint. For example, for the sets [0, 1) and (1, 2], 1 belongs to the closure of both sets but they are separated because 1 itself is in neither set.

separation axioms Topology is all about the essential properties of shapes and space and how they behave under continuous distortions. Central to this is that we need points not just to be distinct but also to be distinguishable topologically; and we need subsets of a topological space to be not just disjoint but also separated. The separation axioms all say, essentially, that points and subsets which are distinguishable in some weak sense must also be separated in another, stronger sense.

sept- Prefix denoting seven.

sequence A finite sequence consists of n terms $a_1, a_2, \ldots, a_n$, one corresponding to each of the integers $1, 2, \ldots, n$, where n, some positive integer, is the LENGTH of the sequence. An infinite sequence consists of terms $a_1, a_2, a_3, \ldots$, one corresponding to each positive integer. Sometimes, it is more convenient to denote the terms of a sequence by $a_0, a_1, a_2, \ldots$. *See also* LIMIT (of a sequence).

sequence of functions A list of functions $\{f_n, n = 1, 2, \ldots\}$ where each f_n maps a given subset D of $\mathbf{R}$ onto $\mathbf{R}$.

sequential sampling A statistical process to decide which of two hypotheses to accept as true. Observations are made one at a time, and a test is carried out to decide whether one of the hypotheses is to be accepted or whether more observations should be made. The sampling stops when it has been decided which hypothesis to accept.

serial (of a relation) A *binary relation which is *connected, *asymmetric and *transitive, imposing an ordering on the whole set. For example, 'is greater than' imposes an order on the set of real numbers.

serial computation The sequential execution of all parts of the same task on a single processor. *Compare* PARALLEL COMPUTATION.

serial correlation If $\{X_i\}$ is a sequence of *random variables and correlation $(X_i, X_{i+1}) \neq 0$ then $\{X_i\}$ is said to have serial correlation.

series A finite series is written as $a_1 + a_2 + \ldots + a_n$, where $a_1, a_2, \ldots, a_n$ are n numbers called the terms in the series, and n, some positive integer, is the length of the series. The sum of the series is simply the sum of the n terms. For certain finite series, such as *arithmetic series and *geometric series, the sum of the series is given by a known formula. The following can also be established:

$$\sum_{r=1}^{n} r^2 = 1^2 + 2^2 + \cdots + n^2 = \frac{1}{6}n(n+1)(2n+1),$$
$$\sum_{r=1}^{n} r^3 = 1^3 + 2^3 + \cdots + n^3 = \frac{1}{4}n^2(n+1)^2.$$

An infinite series is written as $a_1 + a_2 + a_3 + \ldots$, with terms $a_1, a_2, a_3, \ldots$, one corresponding to each positive integer. Let s_n be the sum of the first n terms of such a series. If the sequence $s_1, s_2, s_3, \ldots$ has a limit s, then the value s is called the SUM (or sum to infinity) of the infinite series. Otherwise, the infinite series has no sum. *See also* ARITHMETIC SERIES, GEOMETRIC SERIES, BINOMIAL SERIES, TAYLOR SERIES and MACLAURIN SERIES.

set A well-defined collection of objects. It may be possible to define a set by listing the elements: $\{a, e, i, o, u\}$ is the set consisting of the vowels of the alphabet, $\{1, 2, \ldots, 100\}$ is the set of the first 100 positive integers. The meaning of $\{1, 2, 3, \ldots\}$ is also clear: it is the set of all positive integers. It may be possible to define a set as consisting of all elements, from some universal set, that satisfy some property. Thus the set of all real numbers that are greater than 1 can be written as either $\{x \mid x \in \mathbf{R} \text{ and } x > 1\}$ or $\{x : x \in \mathbf{R} \text{ and } x > 1\}$, both of which are read as 'the set of x such that x belongs to $\mathbf{R}$ and x is greater than 1'. The same set is sometimes written $\{x \in \mathbf{R} \mid x > 1\}$.

set theory The study of the properties of sets and their relations, originally developed by Georg *Cantor.

Sevre, Jean-Pierre (1926–) French mathematician who made important contributions in topology, complex analysis, number theory and algebraic geometry. Winner of the *Fields Medal in 1954, the *Wolf Prize in 2000 and the first Abel Prize in 2003.

sex- Prefix denoting six.

sexagesimal Based on the number 60, so measurement of time in hours, minutes and seconds and of angles in degrees, minutes and seconds are sexagesimal measurements.

sgn *See* SIGNUM FUNCTION.

shear A transformation in which points move parallel to a fixed line or plane and move a distance proportional to their distance from the fixed line or plane. Areas and volumes are preserved under a shear transformation in two and three dimensions respectively, which transform rectangles into parallelograms and cuboids into parallelepipeds when the fixed line or plane is parallel to one of the sides.

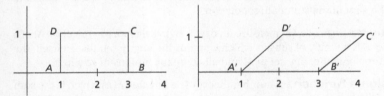

shearing force (mechanics) The internal force perpendicular to the axis of a thin rod or beam.

sheet *See* HYPERBOLOID OF ONE SHEET and HYPERBOLOID OF TWO SHEETS.

SHM = SIMPLE HARMONIC MOTION.

shooting method A method commonly used to solve a *boundary value problem, by reducing it to an *initial value problem.

shortest path algorithm (Dijkstra's method) An algorithm to solve the *shortest path problem. In essence it is a procedure which labels vertices with a minimum distance from the starting position, in order of ascending minimum distances. At each stage therefore you do not have to consider the multiple possible routes to get to any of the vertices already labelled. The procedure stops once the finish point has been labelled in this manner, even though there may be other points not yet labelled—for which the shortest path would be longer than the one of interest.

shortest path problem The problem in *graph theory to find the shortest *connected route between two points or vertices, i.e. the minimum sum of the edges on all possible connected routes between the two vertices.

side One of the lines joining two adjacent vertices in a *polygon. One of the faces in a *polyhedron.

sieve of Eratosthenes The following method of finding all the *primes up to some given number N. List all the positive integers from 2 up to N. Leave the first number, 2, but delete all its multiples; leave the next remaining number, 3, but delete all its multiples; leave the next remaining number, 5, but delete all its multiples, and so on. The integers not deleted when the process ends are the primes.

(⊕) SEE WEB LINKS

• An interactive animation of the sieve of Eratosthenes used to obtain prime numbers.

sigma Greek letter s, written σ, commonly used to denote the standard deviation of a distribution or a set of data. The capital $\sum$ is used to denote the sum in the summation notation.

sigma algebra (in measure theory) A sigma algebra or σ-algebra, X, of a set S is a family of subsets which contains the empty set, the set itself, the complement of any set in X, and all countable unions of sets in X.

sigma function (in number theory) The function $\sigma(n)$ which is the sum of the proper factors of a positive integer n. For a perfect number $\sigma(n) = n$.

sign A symbol denoting an operation, such as $+$, $-$, $\times$ and $\div$ in arithmetic or $\circ$ to denote convolution of functions etc. The positive or negative nature of a quantity is called its sign.

signed minor $=$ COFACTOR.

signed rank test $=$ WILCOXON SIGNED RANK TEST.

significance level *See* HYPOTHESIS TESTING.

significance test *See* HYPOTHESIS TESTING.

significant figures To count the number of significant figures in a given number, start with the first non-zero digit from the left and, moving to the right, count all the digits thereafter, counting final zeros if they are to the right of the decimal point. For example, 1.2048, 1.2040, 0.012 048, 0.001 2040 and 1204.0 all have 5 significant figures. In *rounding or *truncation of a number to n significant figures, the original is replaced by a number with n significant figures.

Note that final zeros to the left of the decimal point may or may not be significant: the number 1 204 000 has at least 4 significant figures, but without more information there is no way of knowing whether or not any more figures are significant. When 1 203 960 is rounded to 5 significant figures to give 1 204 000, an explanation that this has 5 significant figures is required. This could be made clear by writing it in *scientific notation: 1.2040×10^6.

To say that $a = 1.2048$ to 5 significant figures means that the exact value of a becomes 1.2048 after rounding to 5 significant figures; that is to say, $1.204\ 75 \leq a < 1.204\ 85$.

sign test A non-parametric test (*See* NON-PARAMETRIC METHODS) to test the *null hypothesis that a sample is selected from a population with median m. If the null hypothesis is true, the sample of size n is expected to have an equal number of observations above and below m, and the probability that r of them are greater than m has the *binomial distribution $B(n, 0.5)$. The null hypothesis is rejected if the value lies in the critical region determined by the chosen significance level.

The test may also be used to compare the medians of two populations from paired data.

signum function The real function, denoted by sgn, defined by

$$\text{sgn}\,x = \begin{cases} -1, & \text{if } x < 0, \\ 0, & \text{if } x = 0, \\ 1, & \text{if } x > 0. \end{cases}$$

The name and notation come from the fact that the value of the function depends on the sign of x.

similar (of figures) Two geometrical figures are similar if they are of the same shape but not necessarily of the same size. This includes the case when one is a mirror-image of the other, so the three triangles shown in the figure are all similar. For two similar triangles, there is a correspondence between their vertices such that corresponding angles are equal and the ratios of corresponding sides are equal. In the figure, $\angle A = \angle P$, $\angle B = \angle Q$, $\angle C = \angle R$ and $QR/BC = RP/CA = PQ/AB = 3/2$.

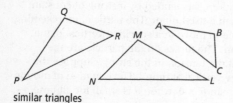

similar triangles

simple curve A continuous plane curve that does not intersect itself.

simple fraction A *fraction in which the numerator and denominator are positive integers, as opposed to a *compound fraction.

simple graph A *graph with no loops or multiple edges.

simple group A *group which has no normal subgroups other than itself and the subgroup consisting of the identity element.

simple harmonic approximation The approximation of the motion of a particle by simpler motion, for example in a single pendulum.

simple harmonic motion Suppose that a particle is moving in a straight line so that its displacement x at time t is given by $x = A \sin(\omega t + \alpha)$, where $A(>0)$, ω and α are constants. Then the particle is executing simple harmonic motion, with *amplitude A, *period $2\pi/\omega$ and *phase α. This equation gives the general solution of the differential equation $\ddot{x} + \omega^2 x = 0$.

An example of simple harmonic motion is the motion of a particle suspended from a fixed support by a spring (*See* HOOKE'S LAW). Also, the motion of a *simple pendulum or a *compound pendulum performing oscillations of small amplitude is approximately simple harmonic motion.

simple interest Suppose that a sum of money P is invested, attracting interest at i per cent a year. When simple interest is given, the interest due each year is $(i/100)P$ and so, after n years, the amount becomes

$$P\left(1 + \frac{ni}{100}\right).$$

When points are plotted on graph paper to show how the amount increases, they lie on a straight line. Most banks and building societies in fact do not operate in this way but use the method of *compound interest.

simple linear regression *See* REGRESSION.

simple pendulum In a *mathematical model, a simple pendulum is represented by a particle of mass m, suspended by a string of constant length l and negligible mass from a fixed point. The particle, representing the bob of the pendulum, is free to move in a specified vertical plane through the point of suspension. The forces on the particle are the uniform gravitational force and the tension in the string. Suppose the string makes an angle θ with the vertical at time t. The equation of motion can be shown to give $\ddot{\theta} + (g/l)\sin\theta = 0$. when θ is small for all time, $\sin\theta \approx \theta$ and the equation becomes $\ddot{\theta} + \omega^2\theta = 0$, where $\omega^2 = g/l$.

It follows that the pendulum performs approximately *simple harmonic motion with period $2\pi\sqrt{l/g}$.

simple root *See* ROOT.

simplex The simplest geometrical figure in a given dimension, so the line, triangle, and tetrahedron are the simplices in one, two, and three dimensions. *See also* κ-SIMPLEX.

simplex method An algebraic method of solving the *standard form of a linear programming problem which allows the solution of multivariable problems, which could not be tackled graphically. The process is as follows, using a two-variable, two-constraint problem to illustrate. So, to maximize $P = 2x + 3y$, subject to $x + 2y \leq 20$, $2x + y \leq 15$, $x \geq 0$, $y \geq 0$:

(1) introduce *slack variables to rewrite each inequality as an equation, giving $x + 2y + s - 20 = 0$ and $2x + y + t - 15 = 0$;

(ii) introduce all other slack variables into each equation with a coefficient of zero, giving $x + 2y + s + 0t - 20 = 0$ and $2x + y + 0s + t - 15 = 0$;

(iii) rewrite the objective function in standard form, including all slack variables with a coefficient of zero, giving $P - 2x - 3y + 0s + 0t = 0$.

The simplex tableau is a table which shows the current values for each variable in each constraint, with the last row, known as the objective row, showing the values in the equation derived from the objective function. The simplex algorithm is the process by which that table is manipulated in order to find the optimal solution. In the solution, the algorithm will produce a series of tableaux.

The initial tableau for this problem will be:

basic variable	x	y	s	t	value
s	1	2	1	0	20
t	2	1	0	1	15
P	-2	-3	0	0	0

The optimality condition is that when the objective row shows zero in any columns representing basic variables and no negative entries, then it represents an optimal solution.

The standard procedure to achieve this condition is to repeat the following set of steps until all the basic variables have been set to zero.

(i) Choose the variable with the largest negative entry in the objective row. The column it is in is called the pivotal column and it is known as the entering variable. In the example y is the entering variable, and the pivotal column is the third from the left.

(ii) For each row of the tableau, divide the value (in the last column) by the entering variable. In the example this gives $20/2 = 10$ and $15/1 = 15$. The smallest of these determines the pivotal row of the tableau at this stage and the leaving variable, which is s in the example. The pivot is the cell in both the pivotal column and the pivotal row, i.e. 2 in this case.

(iii) Divide all entries in the pivotal row by the pivot, so the row reads s, 0.5, 1, 0.5, 0, 10 in the example.

(iv) Add or subtract multiples of this row to (or from) each other to make the entry in the pivotal column zero. So it has to be subtracted once from the t row, and added three times to the P row, leaving the tableau looking like this:

basic variable	x	y	s	t	value
s	0.5	1	0.5	0	10
t	1.5	0	-0.5	1	5
P	-0.5	0	1.5	0	30

The process would now be repeated, and x will now be the entering variable, and the values at step 2 are 20 for s, and 10/3 for t, so t is the leaving variable, and at step 3 the pivotal row will read t, 1, 0, $-1/3$, 2/3, 10/3.

Repeating step 4 gives:

basic variable	x	y	s	t	value
s	0	1	2/3	1/3	25/3
t	1	0	$-1/3$	2/3	10/3
P	0	0	4/3	1/3	95/3

For a many-variable problem this would have to be repeated for each variable in turn, but in this simple example the optimality condition is now satisfied and the maximum value of P is 95/3 occurring when $s = 4/3$ and $t = 1/3$, corresponding to $x = 10/3$ and $y = 25/3$. In this simple problem, the solution could have been found much more quickly by the *vertex method, but the simplex method is an important tool because it can be applied to much more complex problems.

simplex tableau *See* SIMPLEX METHOD.

simplify Reduce an expression by algebraic manipulation. For example, $8x - 2x$ can be simplified to $6x$ and $3(2x + 5y) + 2(x - 3y)$ can be simplified to $8x + 9y$ by multiplying out the brackets and collecting like terms.

simply connected Describing a *topological space which is *path connected if any simple closed curve—that is, any path—can be shrunk continuously to a point in the set. Intuitively, this means that there are no holes in the space, and the complement of the space is also connected. A disc in two dimensions and a sphere in three dimensions are simply connected, while the annulus and torus are examples which are not simply connected.

Simpson's paradox The paradox in which considering the rates of occurrence in a two-way table of the characteristics in two groups separately can lead to the opposite conclusion from when the two groups are combined. For this to occur, it is necessary that there be a substantial difference in the proportions of one category within two groups. For example, if 80% of students applying for science are accepted and 40% of students applying for arts are accepted, irrespective of sex in both cases, then there is no discrimination. However, if 75% of boys apply for science courses and only 30% of girls apply for science, the overall success rate of applications for boys and girls would look very discriminatory. With the above proportions, and 1000 boys and 1000 girls altogether the two-way table would be:

	Accept	Fail
Boys	700	300
Girls	520	420

That is, overall 70% of boys would be accepted compared with only 52% of girls, but the rates of acceptance from each group by type of course were identical.

(((·))) SEE WEB LINKS

• Examples of Simpson's paradox.

Simpson's rule An approximate value can be found for the definite integral

$$\int_a^b f(x)\ dx,$$

using the values of $f(x)$ at equally spaced values of x between a and b, as follows. Divide $[a, b]$ into n equal subintervals of length h by the *partition

$$a = x_0 < x_1 < x_2 < \cdots < x_{n-1} < x_n = b,$$

where $x_{i+1} - x_i = h = (b-a)/n$. Denote $f(x_i)$ by f_i, and let P_i be the point (x_i, f_i). The *trapezium rule uses the line segment $P_i P_{i+1}$ as an approximation to the curve. Instead, take an arc of a certain parabola (in fact, the graph of a polynomial function of degree two) through the points P_0, P_1 and P_2, an arc of a parabola through P_2, P_3 and P_4, similarly, and so on. Thus n must be even. The resulting Simpson's rule gives

$$\tfrac{1}{3}h(f_0 + 4f_1 + 2f_2 + 4f_3 + 2f_4 + \cdots + 2f_{n-2} + 4f_{n-1} + f_n)$$

as an approximation to the value of the integral. This approximation has an *error that is roughly proportional to $1/n^4$. In general, Simpson's rule can be expected to be much more accurate than the trapezium rule, for a given value of n. It was devised by the British mathematician Thomas Simpson (1710–61).

Simson line Given a triangle and a point, P, on its *circumcircle, the feet of the perpendiculars from P to each of the sides of the triangle (or their extensions) are *collinear and define the Simson line. The converse of this is true as well, that if the feet of the perpendiculars from a point P to each of the sides of the triangle (or their extensions) are collinear then P lies on the circumcircle.

simulation An attempt to replicate a physical procedure mathematically, where the system being studied is too complicated for explicit analytic methods to be used. Because of the complexity, simulation is often carried out by computer, perhaps using *Monte Carlo methods. As simulation is often only an approximation to the physical procedure, the use of *sensitivity analysis is crucial.

simultaneous linear differential equations Where $\dfrac{dx}{dt} = f(x, y)$ and $\dfrac{dy}{dt} = g(x, y)$ where f and g are linear functions of x and y. The solution is found by differentiating one of the equations and then substituting to remove one of the variables and leave a second-order linear differential equation. For example, if $\dfrac{dx}{dt} = x + y$ (A) and $\dfrac{dy}{dt} = 4x - 2y$ (B) with $x = 1$, $y = 4$ when $t = 0$. Then writing (A) as $y = \dfrac{dx}{dt} - x$ (C) and differentiating gives $\dfrac{dy}{dt} = \dfrac{d^2x}{dt^2} - \dfrac{dx}{dt}$ (D).

So substituting C and D into B gives $\dfrac{d^2x}{dt^2} - \dfrac{dx}{dt} = 4x - 2\left(\dfrac{dx}{dt} - x\right) \Rightarrow \dfrac{d^2x}{dt^2} + \dfrac{dx}{dt} - 6x = 0$.

The solution to this is $x = Ae^{-3t} + Be^{2t}$, $y = -4Ae^{-3t} + Be^{2t}$ and substituting for the boundary conditions gives $A = 1$ and $B = 0$ and the solutions to the original differential equations are $x = e^{-3t}$, $y = -4e^{-3t}$.

simultaneous linear equations The solution of a set of m linear equations in n unknowns can be investigated by the method of *Gaussian elimination that transforms the augmented matrix to *echelon form. The number of non-zero rows in the echelon form cannot be greater than the number of unknowns, and three cases can be distinguished:

(i) If the echelon form has a row with all its entries zero except for a non-zero entry in the last place, then the set of equations is inconsistent.

(ii) If case (i) does not occur and, in the echelon form, the number of non-zero rows is equal to the number of unknowns, then the set of equations has a unique solution.

(iii) If case (i) does not occur and, in the echelon form, the number of non-zero rows is less than the number of unknowns, then the set of equations has infinitely many solutions.

When the set of equations is consistent, that is, in cases (ii) and (iii), the solution or solutions can be found either from the echelon form using *back-substitution or by using *Gauss–Jordan elimination to find the *reduced echelon form. When there are infinitely many solutions, they can be expressed in terms of *parameters that replace those unknowns free to take arbitrary values.

sine See TRIGONOMETRIC FUNCTION.

sine rule See TRIANGLE.

singleton A set containing just one element.

singular A square matrix $\mathbf{A}$ is singular if det $\mathbf{A} = 0$, where det $\mathbf{A}$ is the determinant of $\mathbf{A}$. A singular matrix is not invertible (See INVERSE MATRIX).

singular point (singularity) A point on a curve where there is not a unique tangent which is itself differentiable. It may be an isolated point, or a point where the curve cuts itself such as a *cusp.

A point at which a complex function is not *analytic. For example, $f(z) = \frac{1}{z}$ has a singularity at $z = 0$, and is said to be an isolated singularity because there is no other singular point in the neighbourhood of $z = 0$ (in fact it is the only singular point of that function).

Singularities occur in three categories: *removable singularities, *poles, and *essential singularities.

sinh *See* HYPERBOLIC FUNCTION.

sink The vertex in a *network towards which all flows are directed.

SI units The units used for measuring physical quantities in the internationally agreed system Système International d'Unités are known as SI units. There are seven base units of which the *metre, for measuring length, the *kilogram, for measuring mass, and the *second, for measuring time, are the commonest in mathematics. The so-called supplementary units are the *radian, for measuring angle, and the steradian, for measuring solid angle. From these base and supplementary units, further derived units are defined, such as the 'square metre' for measuring area and the 'metre per second' for measuring velocity. Some derived units have special names, such as the *newton, *joule, *watt, *pascal and *hertz.

Each base unit has an agreed abbreviation, and in the abbreviations of derived units positive and negative indices are used. There should be a small space between different units involved. For example, the abbreviations for 'square metre' and 'metre per second' are 'm^2' and 'm s^{-1}'. Derived units with special names have their own abbreviations.

Prefixes are used to define multiples of units by powers of 10. Powers in which the index is a multiple of 3 are preferred. In the list below, the abbreviation for the prefix is given in brackets. For example, 1 megawatt equals 10^6 watts, and 1 milligram equals 10^{-3} grams; the abbreviation for megawatt is 'MW' and the abbreviation for milligram is 'mg'.

10^3	kilo-	(k)	10^{-3}	milli-	(m)
10^6	mega-	(M)	10^{-6}	micro-	(μ)
10^9	giga-	(G)	10^{-9}	nano-	(n)
10^{12}	tera-	(T)	10^{-12}	pico-	(p)
10^{15}	peta-	(P)	10^{-15}	femto-	(f)
10^{18}	exa-	(E)	10^{-18}	atto-	(a)

The following additional prefixes sometimes occur:

10	deka-	(da)	10^{-1}	deci-	(d)
10^2	hecto-	(h)	10^{-2}	centi-	(c)

Six-Sigma The name given to a total quality control philosophy—the name deriving from the likelihood, in a normal distribution, that reducing variation to the extent that the tolerance in a production process is at the sixth sigma represents a failure rate of less than 1 in 250 000.

skew lines Two straight lines in 3-dimensional space that do not intersect and are not parallel.

skewness The amount of asymmetry of a *distribution. One measure of skewness is the coefficient of skewness, which is defined to be equal to $\mu_3/(\mu_2)^{3/2}$, where μ_2 and μ_3 are the second and third *moments about the mean. This measure is zero if the distribution is symmetrical about the mean. If the distribution has a long tail to the left, as in the left-hand figure, it is said to be skewed to the left and to have negative skewness, because the coefficient of skewness is negative. If the distribution has a long tail to the right, as in the right-hand figure, it is said to be skewed to the right and to have positive skewness, because the coefficient of skewness is positive. A skew distribution is one that is skewed to the left or skewed to the right.

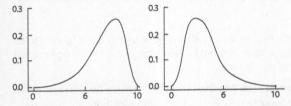

skew-symmetric matrix Let **A** be the square matrix $[a_{ij}]$. Then the matrix **A** is skew-symmetric if $\mathbf{A}^T = -\mathbf{A}$ (*See* TRANSPOSE); that is to say, if $a_{ij} = -a_{ji}$ for all i and j. It follows that in a skew-symmetric matrix the entries in the main diagonal are all zero: $a_{ii} = 0$ for all i.

slack (in critical path analysis) The difference between the latest time and the earliest time at a vertex (node) in an activity network. It represents the allowable delay at that point in the process which would not delay the overall completion. The slack will be zero for *critical events.

slack variables Variables which are introduced into *linear programming problems so that the inequalities representing the constraints can be replaced by equations. Each equation will introduce one slack variable, for example if $2x + 3y \leq 30$ then we introduce a slack variable s satisfying $2x + 3y + s - 30 = 0$. This allows the algebraic method of solution known as the *simplex method to be used.

slant asymptote *See* ASYMPTOTE.

slant height *See* CONE.

slide rule A mechanical device used for mathematical calculations such as multiplication and division. In the simplest form, one piece slides alongside another piece, each piece being marked with a *logarithmic scale. To multiply x and y, the reading 'x' on one scale is placed opposite the '1' on the other scale, and the required product 'xy' then appears opposite the 'y'. More elaborate slide rules have log–log scales or scales giving trigonometric functions. Slide rules have now been superseded by electronic calculators.

(⊕) SEE WEB LINKS
• Descriptions of various slide rules and how they work.

sliding–toppling condition Where a rigid body is at rest on a plane, it can be determined whether it will slide or topple first by calculating what the minimum force is for it to topple, assuming it does not slide, and then calculate what the minimum force is for it to slide, assuming it does not topple first. The lower of these two minima will tell which happens first.

slope $=$ GRADIENT.

small circle A circle on the surface of a sphere that is not a *great circle. It is the curve of intersection obtained when a sphere is cut by a plane not through the centre of the sphere.

Smith, Adrian Frederick Melhuish FRS (1946–) British statistician who conducted the Inquiry into Post-14 Mathematics Education for the UK Secretary of State for Education and Skills in 2003–04. He has been President of the Royal Statistical Society, and in 2008 was appointed as Director General for Science and Research in the Department for Innovation, Universities and Skills.

smooth A curve or function which is differentiable everywhere, and for which the differential is continuous, so $f(x) = x^2$ is smooth but $g(x) = |x|$ is not.

smoothly hinged Implies that the force at the hinge can be treated as a single force at an unknown angle, rather than requiring the consideration of a couple as well.

smooth surface *See* CONTACT FORCE.

snowflake curve The *Koch curve, or any similar curve constructed in a similar way.

software *See* COMPUTER.

solid A closed 3-dimensional geometric figure such as a *cube or *cylinder.

solid angle The 3-dimensional analogue of the 2-dimensional concept of angle. Just as an angle is bounded by two lines, a solid angle is bounded by the generators of a cone.

A solid angle is measured in steradians: this is defined to be the area of the intersection of the solid angle with a sphere of unit radius. Thus the 'complete' solid angle at a point measures 4π steradians (comparable with one complete revolution measuring 2π radians). The steradian is the SI unit for measuring solid angle and is abbreviated to 'sr'.

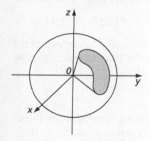

solid of revolution Suppose that a plane region is rotated through one revolution about a line in the plane that does not cut the region. The 3-dimensional region thus obtained is a solid of revolution. *See also* VOLUME OF A SOLID OF REVOLUTION.

solution A solution of a set of equations is an element, belonging to some appropriate *universal set, specified or understood, that satisfies the equations. For a set of equations in n unknowns, a solution may be considered to be an n-tuple $(x_1, x_2, \ldots, x_n)$, or a column matrix

$$\begin{bmatrix} x_1 \\ x_2 \\ \vdots \\ x_n \end{bmatrix},$$

such that $x_1, x_2, \ldots, x_n$ satisfy the equations. *See also* INEQUALITY.

solution set The solution set of a set of equations is the set consisting of all the solutions. It may be considered as a subset of some appropriate *universal set, specified or understood. *See also* INEQUALITY.

sorting algorithms An algorithm which puts the elements of a list in a certain order, usually alphabetical or numerical. Commonly used examples include the *bubble sort, bi-directional bubble sort, selection sort, shell sort, merge sort, insertion sort, heap sort, quick sort, and combinations of these. How efficient these are depends on the size of the list to be sorted, the *algorithmic complexity of the procedure, and on the available memory space in the computer doing the sort.

(⊕) SEE WEB LINKS
• Demonstrations of a variety of sorting algorithms in operation.

source The vertex in a *network away from which all flows are directed.

source (in transportation problems) *See* TRANSPORTATION PROBLEMS.

space A set of points with a structure which defines the behaviour of the space and the relationship between the points. *See* EUCLIDEAN, MEASURE SPACE, VECTOR SPACE.

space-filling curve *See* PEANO CURVE.

space–time A 4-dimensional construct combining time with the three dimensions of space, replacing the conceptual framework of classical ordinary mechanics in which space and time exist separately. The general and special theories of *relativity showed that this is not true when high speeds are involved.

spanning set *See* BASIS.

spanning tree A *subgraph which is a *tree and which passes through all the vertices of the original graph.

sparse matrix A matrix which has a high proportion of zero entries. *Compare* DENSE.

Spearman's rank correlation coefficient *See* RANK CORRELATION.

special relativity *See* RELATIVITY.

speed In mathematics, it is useful to distinguish between *velocity and speed. First, when considering motion of a particle in a straight line, specify a positive direction so that it is a *directed line. Then the velocity of the particle is positive if it is moving in the positive direction and negative if it is moving in the negative direction. The speed of the particle is the *absolute value of its velocity. In more advanced work, when the velocity is a vector **v**, the speed is the magnitude |**v**| of the velocity.

speed of light In a vacuum, speed travels at $299\,792\,458$ ms^{-1}. This is the same for other electromagnetic waves.

sphere The sphere with centre C and radius r is the locus of all points (in 3-dimensional space) whose distance from C is equal to r. If C has Cartesian coordinates (a, b, c), this sphere has equation

$$(x - a)^2 + (y - b)^2 + (z - c)^2 = r^2.$$

The equation $x^2 + y^2 + z^2 + 2ux + 2vy + 2wz + d = 0$ represents a sphere, provided that $u^2 + v^2 + w^2 - d > 0$, and it is then an equation of the sphere with centre $(-u, -v, -w)$ and radius $\sqrt{u^2 + v^2 + w^2 - d}$.

The volume of a sphere of radius r is equal to $\frac{4}{3}\pi r^3$, and the surface area equals $4\pi r^2$.

spherical angle An angle formed by two *great circles meeting on the surface of a sphere, measured as the angle between their tangents at the point of intersection.

spherical cap See ZONE.

spherical n-sided polygon A closed geometrical figure on a sphere with n vertices and n sides that are arcs of great circles.

spherical polar coordinates Suppose that three mutually perpendicular directed lines Ox, Oy and Oz, intersecting at the point O, and forming a right-handed system, are taken as coordinate axes (See COORDINATES (in 3-dimensional space)). For any point P, let M be the projection of P on the xy-plane. Let $r = |OP|$, let θ be the angle $\angle zOP$ in radians ($0 \leq \theta \leq \pi$) and let ϕ be the angle $\angle xOM$ in radians ($0 \leq \phi < 2\pi$). Then (r, θ, π) are the spherical polar coordinates of P. (The point O gives no value for θ or ϕ, but is simply said to correspond to $r = 0$.) The value $\phi + 2k\pi$, where k is an integer, may be allowed in place of ϕ. The Cartesian coordinates (x, y, z) of P can be found from r, θ and ϕ by $x = r \sin\theta \cos\phi$, $y = r \sin\theta \sin\phi$, $z = r \cos\theta$. Spherical polar coordinates may be useful in treating problems involving spheres, for a sphere with centre at the origin then has equation $r = $ constant. See diagram.

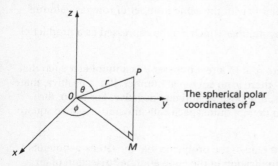

The spherical polar coordinates of P

spherical triangle A triangle on a sphere, with three vertices and three sides that are arcs of *great circles. The angles of a spherical triangle do not add up to 180°. In fact, the sum of the angles can be anything between 180° and 540°. Consider, for example, a spherical triangle with one vertex at the North Pole and the other two vertices on the equator of the Earth.

spherical trigonometry The application of the methods of trigonometry to the study of such matters as the angles, sides and areas of *spherical triangles and other figures on a sphere.

spiral *See* ARCHIMEDEAN and EQUIANGULAR SPIRAL.

spread = DISPERSION.

spring A device, usually made out of wire in the form of a *helix, that can be extended and compressed. It is elastic and so resumes its natural length when the forces applied to extend or compress it are removed. How the *tension in the spring varies with the *extension may be complicated. In the simplest mathematical model, it is assumed that *Hooke's law holds, so that the tension is proportional to the extension.

spurious correlation Where two variables possess a correlation but the link between them is not direct, but through another variable, often time or size. For example, the recorded number of burglaries and the number of teachers in towns in the UK show a strong positive correlation, but this is because both of them have a positive correlation with the size of the town.

square The regular quadrilateral, i.e. with four equal sides and four right angles.

square (as a power) A number or expression raised to the power of 2, i.e. multiplied by itself, so $3^2 = 3 \times 3 = 9$ and $(a+b)^2 = a^2 + 2ab + b^2$.

square matrix A *matrix with the same number of rows as columns.

square number Any number which can be expressed as a product of two equal integers.

square root A square root of a real number a is a number x such that $x^2 = a$. If a is negative, there is no such real number. If a is positive, there are two such numbers, one positive and one negative. For $a \geq 0$, the notation $\sqrt{a}$ is used to denote quite specifically the non-negative square root of a.

squaring the circle One of the problems that the Greek geometers attempted (like the *duplication of the cube and the *trisection of an angle) was to find a construction, with ruler and pair of compasses, to obtain a square whose area was equal to that of a given circle. This is equivalent to a geometrical construction to obtain a length of $\sqrt{\pi}$ from a given unit length. Now constructions of the kind envisaged can give only lengths that are algebraic numbers (and not even all algebraic numbers at that: for instance, $\sqrt[3]{2}$ cannot be obtained). So the proof by Lindemann in 1882 that π is *transcendental established the impossibility of squaring the circle.

stability The nature of the *equilibrium of a particle or rigid body in an equilibrium position, whether stable, unstable or neutral.

stable equilibrium *See* EQUILIBRIUM.

stable numerical analysis In *numerical analysis, the stability of an algorithm is the extent to which an error in the calculation does not grow to be much bigger during the calculation. Since numerical analysis is concerned with finding approximate solutions to problems for which no analytical solution exists, it is an important consideration as to whether rounding or truncation errors in the algorithm are damped or magnified. Algorithms which can be shown not to magnify approximation errors are called numerically stable.

stable system of differential equations Systems of differential equations are used to describe dynamical deterministic relationships involving quantities and their rates of change. The stability of a system is the extent to which small perturbations of the initial conditions cause a change in the solution. A system is called stable if any solution which starts close enough to a *stationary point will revert to that stationary point over time.

stadium paradox One of *Zeno's paradoxes which relate to the issue of whether time and space are made up of minute indivisible parts. It considers two rows of objects of equal size—of the smallest possible size—moving with equal velocity in opposite directions, with a speed equal to the smallest unit of length per smallest unit of time. The argument at the centre of the paradox is that relative to one another, they move at one unit of space in half of the unit of time, contradicting the proposition that there is a smallest unit of time.

standard deviation The positive square root of the *variance, a commonly used measure of the *dispersion of observations in a sample. For a *normal distribution $N(\mu, \sigma^2)$, with mean μ and standard deviation σ, approximately 95% of the distribution lies in the interval between $\mu - 2\sigma$ and $\mu + 2\sigma$.

The standard deviation of an *estimator of a population parameter is the *standard error.

standard error The standard deviation of an *estimator of a population parameter. The standard error of the sample mean, from a sample of size n, is $\sigma/\sqrt{n}$, where σ^2 is the population variance.

standard form (of a linear programming problem) Where the objective function $\sum a_i x_i$ is to be maximized, and all constraints other than the non-negativity conditions can be rewritten in the form $\sum b_j x_j \le c_j$. If the objective function is to be minimized, then the problem can be converted to standard form by making the objective function $-\sum a_i x_i$ and maximizing this, with the proviso that the optimal solution must be multiplied by -1 at the end. *Linear programming problems in standard form can be solved by the *simplex method.

standard form (of a number) = SCIENTIFIC NOTATION.

standardize (in statistics) To transform a random variable so that it has a mean of zero and variance of 1. If $E(X) = \mu$ and Var $(X) = \sigma^2$ the standardized variable Z will have $E(X) = 0$ and Var $(X) = 1$. The most common application of this is in the normal distribution where probabilities for $X \sim N(\mu, \sigma^2)$ are evaluated using the $N(0, 1)$ tables, applying the transformation $Z = \dfrac{X - \mu}{\sigma}$ (alternative form $X = \mu + Z\sigma$).

standard normal distribution *See* NORMAL DISTRIBUTION.

state, state space *See* STOCHASTIC PROCESS.

statement In mathematical logic, the fundamental property of a statement is that it makes sense and is either true or false. For example, 'there is a real number x such that $x^2 = 2$' makes sense and is true; the statement 'if x and y are positive integers then $x - y$ is a positive integer' makes sense and is false. In contrast, '$x = 2$' is not a statement.

static friction *See* FRICTION.

statics A part of *mechanics concerned with the forces acting on a rigid body or system of particles at rest with forces acting in equilibrium.

stationary point (in one variable) A point on the graph $y = f(x)$ at which f is *differentiable and $f'(x) = 0$. The term is also used for the number c such that $f'(c) = 0$. The corresponding value $f(c)$ is a stationary value. A stationary point c can be classified as one of the following, depending upon the behaviour of f in the neighbourhood of c:

    (i) a *local maximum, if $f'(x) > 0$ to the left of c and $f'(x) < 0$ to the right of c,

(ii) a *local minimum, if $f'(x) < 0$ to the left of c and $f'(x) > 0$ to the right of c,

(iii) neither local maximum nor local minimum.

The case (iii) can be subdivided to distinguish between the following two cases. It may be that $f'(x)$ has the same sign to the left and to the right of c, in which case c is a horizontal *point of inflexion; or it may be that there is an interval at every point of which $f'(x)$ equals zero and c is an end-point or interior point of this interval.

stationary point (in two variables) A point P on the surface $z = f(x, y)$ is a stationary point if the *tangent plane at P is horizontal. This is so if $\partial f/\partial x = 0$ and $\partial f/\partial y = 0$. Now let

$$r = \frac{\partial^2 f}{\partial x^2}, \quad s = \frac{\partial^2 f}{\partial x \partial y}, \quad t = \frac{\partial^2 f}{\partial y^2}.$$

If $rt > s^2$ and $r < 0$, the stationary point P is a local maximum (all the vertical cross-sections through P have a local maximum at P). If $rt > s^2$ and $r > 0$, the stationary point is a local minimum (all the vertical cross-sections through P have a local minimum at P). If $rt < s^2$, the stationary point is a *saddle-point.

stationary value *See* STATIONARY POINT.

statistic A function of a sample; in other words, a quantity calculated from a set of observations. Often a statistic is an *estimator for a population parameter. For example, the sample mean, sample variance and sample median are each a statistic. The sum of the observations in a sample is also a statistic, but this is not an estimator.

statistical equilibrium A state in which the probability distribution of states remains constant over time. This is an important idea in statistical physics, and in *Markov chains, where the existence of an equilibrium distribution is not guaranteed. For example, in the simple two-state case with transition matrix $\mathbf{T} = \begin{pmatrix} 0.25 & 0.5 \\ 0.75 & 0.5 \end{pmatrix}$, i.e. state 1 stays in state 1 with probability 0.25 and moves to state 2 with probability 0.75 while if it is in state 2 it will move to state 1 or stay with probability 0.5. Then statistical equilibrium exists with the probabilities of being in state 1 and state 2 respectively being 0.4 and 0.6.

statistically dependent (of two random variables) Not independent. However, the nature of the dependence can be extremely different in different contexts, from one being completely determined by the other, to an extremely small effect. For example, if a bag has 50% red and 50% black balls in it, and event A is that the first ball drawn from the bag is red, and B is the second red, then consider the following two cases. First, if the bag has only two balls in it to start with, then if A happens B cannot, i.e. the likelihood of B is completely determined by knowledge of A. However, if the bag contains 1000 balls altogether, the conditional probability $P(B|A) = \frac{500}{999}$, which is almost the same as $P(B)$ when no information is available about the colour of the first ball.

statistical model A statistical description of an underlying system, intended to match a real situation as nearly as possible. The model for a population is fitted to a sample by estimating the *parameters in the model. It is then possible to perform *hypothesis testing, construct *confidence intervals and draw *inferences about the population.

statistical tables Sets of tables showing values of probabilities or often cumulative probabilities of common distributions for certain values of their parameters. Alternatively, the tables may give the values of the random variable for specific percentages of the distribution, giving only the values commonly used as significance levels in hypothesis tests.

statistics Quantitative data on any subject, usually collected with a purpose such as recording performance of bodies for comparison, or for public record and which can then be analysed by a variety of statistical methods.

stem-and-leaf plot A method of displaying *grouped data, by listing the observations in each group, resulting in something like a histogram on its side. For the observations 45, 25, 67, 49, 12, 9, 45, 34, 37, 61, 23, grouped using class intervals 0–9, 10–19, 20–29, 30–39, 40–49, 50–59 and 60–69, a stem-and-leaf plot is shown on the left. If, as in this case, the class intervals are defined by the digit occurring in the 'tens' position, the diagram is more commonly written as shown on the right. *See figure.*

0–9	9				0	9		
10–19	12				1	2		
20–29	23	25			2	3	5	
30–39	34	37			3	4	7	
40–49	45	45	49		4	5	5	9
50–59					5			
60–69	61	67			6	1	7	

stem-and-leaf plot

step function A real function is a step function if its domain can be partitioned into a number of intervals on each of which the function takes a constant value. An example is the function f defined by $f(x) = [x]$, where this denotes the *integer part of x.

steradian *See* SOLID ANGLE.

stereographic projection Suppose that a sphere, centre O, touches a plane at the point S, and let N be the opposite end of the diameter through S. If P is any point (except N) on the sphere, the line NP meets the plane in a corresponding point P'. Conversely, each point in the plane determines a point on the sphere, so there is a *one-to-one correspondence between the points of the sphere (except N) and the points of the plane. This means of mapping sphere to plane is called stereographic projection. Circles(great or small) not through N on the sphere's surface are mapped to circles in the plane; circles (great or small) through N are mapped to straight lines. The angles at which curves intersect are preserved by the projection.

(⊕) SEE WEB LINKS
• A full description of stereographic projection with two videos to illustrate how it looks.

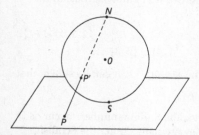

Stevin, Simon (1548–1620) Flemish engineer and mathematician , the author of a popular pamphlet that explained decimal fractions simply and was influential in bringing about their everyday use, replacing sexagesimal fractions. He made discoveries in statics and hydrostatics and was also responsible for an experiment in 1586, in which two lead spheres, one ten times the weight of the other, took the same time to fall 30 feet. This probably preceded any similar experiment by *Galileo.

stiffness A parameter that describes the ability of a spring to be extended or compressed. It occurs as a constant of proportionality in *Hooke's law. The stiffness can be determined as follows. Plot the magnitude of an applied force against the extension that it produces. If Hooke's law holds, the graph produced should be a straight line through the origin and its gradient is the stiffness.

Stirling number of the first kind The number $s(n, r)$ of ways of partitioning a set of n elements into r *cycles. For example, the set $\{1, 2, 3, 4\}$ can be partitioned into two cycles in the following ways:

$[1, 2, 3][4]$, $[1, 3, 2][4]$, $[1, 2, 4][3]$, $[1, 4, 2][3]$,
$[1, 3, 4][2]$, $[1, 4, 3][2]$, $[2, 3, 4][1]$, $[2, 4, 3][1]$,
$[1, 2][3, 4]$, $[1, 3][2, 4]$, $[1, 4][2, 3]$.

So $s(4, 2) = 11$. Clearly $s(n, 1) = (n - 1)!$ and $s(n, n) = 1$. It can be shown that

$$s(n + 1, r) = s(n, r - 1) + ns(n, r)$$

Some authors define these numbers differently so that they satisfy $s(n + 1, r) = s(n, r - 1) - ns(n, r)$. The result is that the values are the same except that some of them occur with a negative sign.

Rather like the *binomial coefficients, the Stirling numbers occur as coefficients in certain identities. They are named after the Scottish mathematician James Stirling (1692–1770).

Stirling number of the second kind The number $S(n, r)$ of ways of partitioning a set of n elements into r non-empty subsets. For example, the set $\{1, 2, 3, 4\}$ can be partitioned into two non-empty subsets in the following ways:

$\{1, 2, 3\} \cup \{4\}$, $\{1, 2, 4\} \cup \{3\}$, $\{1, 3, 4\} \cup \{2\}$, $\{2, 3, 4\} \cup \{1\}$,
$\{1, 2\} \cup \{3, 4\}$, $\{1, 3\} \cup \{2, 4\}$, $\{1, 4\} \cup \{2, 3\}$.

So $S(4, 2) = 7$. Clearly, $S(n, 1) = 1$ and $S(n, n) = 1$. It can be shown that

$$S(n + 1, r) = S(n, r - 1) + r\, S(n, r)$$

Rather like the *binomial coefficients, the Stirling numbers occur as coefficients in certain identities. They are named after the Scottish mathematician James Stirling (1692–1770).

Stirling's formula The formula

$$\frac{\sqrt{2\pi n}(n/e)^n}{n!} \to 1.$$

Named after the Scottish mathematician James Stirling (1692–1770), it was known earlier by *De Moivre. It gives the approximation $n! \approx \sqrt{2\pi\, n}(n/e)^n$ for large values of n.

stochastic matrix A square matrix is row stochastic if all the entries are non-negative and the entries in each row add up to 1. It is column stochastic if all the entries are non-negative and the entries in each column add up to 1. A square matrix is stochastic if it is either row stochastic or column stochastic. It is doubly stochastic if it is both row stochastic and column stochastic.

stochastic process A family $\{X(t), t \in T\}$ of *random variables, where T is some *index set. Often the index set is the set $\mathbf{N}$ of natural numbers, and the stochastic process is a sequence $X_1, X_2, X_3, \ldots$ of random variables. For example, X_n may be the outcome of the n-th trial of some experiment, or the n-th in a set of observations. The possible values taken by the random variables are often called states, and these form the state space. The state space is said to be discrete if it is finite or countably infinite, and continuous if it is an interval, finite or infinite. If there are countably many states, they may be denoted by $0, 1, 2, \ldots$, and if $X_n = i$ then X_n is said to be in state i.

stochastic variable = RANDOM VARIABLE.

Stokes, Sir George Gabriel FRS (1819–1903) Irish mathematician who was Lucasian Professor of Mathematics at Cambridge University. His prolific mathematical research output arose mainly from study of physical processes such as hydrodynamics and hydrostatics, wind pressure systems and spectrum analysis. As well as being one of Britain's foremost scientists Stokes was unusually active in public life, as a member of Parliament for Cambridge University 1881–92, secretary of the Royal Society 1854–84 and then its President 1885–90.

Stokes's Theorem $\int_S \nabla \times \mathbf{F} d\mathbf{S} = \int_C \mathbf{F} \, d\mathbf{r}$ where $\mathbf{F}$ is a vector function defined on a surface S which has a boundary curve C, i.e. that the integral of the *curl of $\mathbf{F}$ over the surface is equal to the integral of $\mathbf{F}$ round the boundary of the surface.

straight edge Tool used for drawing straight lines, but without any measurement of distance, in geometrical constructions.

straight line (in the plane) A straight line in the plane is represented in Cartesian coordinates by a linear equation; that is, an equation of the form $ax + by + c = 0$, where the constants a and b are not both zero. A number of different forms are useful for obtaining an equation of a given line.

(i) The equation $y = mx + c$ represents the line with gradient m that cuts the y-axis at the point $(0, c)$. The value c may be called the intercept.

(ii) The line through a given point (x_1, y_1) with gradient m has equation $y - y_1 = m(x - x_1)$.

(iii) The line through the two points (x_1, y_1) and (x_2, y_2) has, if $x_2 \neq x_1$, equation

$$y - y_1 = \frac{y_2 - y_1}{x_2 - x_1}(x - x_1),$$

and has equation $x = x_1$, if $x_2 = x_1$.

(iv) The line that meets the coordinate axes at the points $(p, 0)$ and $(0, q)$, where $p \neq 0$ and $q \neq 0$, has equation $x/p + y/q = 1$.

straight line (in 3-dimensional space) A straight line in 3-dimensional space can be specified as the intersection of two planes. Thus a straight line is given, in general, by two *linear equations, $a_1x + b_1y + c_1z + d_1 = 0$ and $a_2x + b_2y + c_2z + d_2 = 0$. (If these equations represent identical or parallel planes, they do not define a straight line.) Often it is more convenient to obtain *parametric equations for the line, which can also be written in 'symmetric form'. *See also* VECTOR EQUATION (of a line).

strategy (in game theory) A combination of moves for a player in a game. In some games, a strategy can be decided by the current state of the game—such as in noughts and crosses. In other games, like poker, a variation in approach over a number of games is required so the other players cannot identify your hand, simply by your moves. In game theory this is modelled by a probability distribution on the available choices.

stratified sample Where a population is not homogeneous, but contains groups of individuals (the strata) for which the characteristic of interest is likely to be noticeably different, it is likely that a stratified sample will provide a better estimator of the characteristic than a simple random sample, in the sense that the estimates will be closer to the true value more often. The procedure is to take a random sample from within each group, proportional to the size of the group. This removes one major source of variation in the simple random sample, namely how many of each group is taken. For example, if a junior tennis club had 8, 15 and 12 players in three age groups, then a sample of 5 players which took 1, 2 and 2 players from age groups is likely to be more representative of the population than a simple random sample which could end up with all the sample coming from one of the groups.

stress The force per unit area across a surface of a body. Different types of stress are shear, tensile and compressive, and all bodies undergo it. As with elastic strings and springs most bodies revert to their original shapes once the stress is removed, but if the stress is too great they may remain in the deformed state or even break.

strict Adjective implying that it is more restrictive than the relation itself. So a strictly monotonically increasing function requires $f(x + \delta) > f(x)$ for any $\delta > 0$ and any x.

strictly decreasing *See* DECREASING FUNCTION and DECREASING SEQUENCE.

strictly determined game A *matrix game is strictly determined if there is an entry in the matrix that is the smallest in its row and the largest in its column. If the game is strictly determined then, when the two players

R and C use *conservative strategies, the pay-off is always the same and is the value of the game.

The game given by the matrix on the left in the figure is strictly determined. The conservative strategy for R is to choose Row 3, and the conservative strategy for C is to choose Column 1. The value of the game is 5. The game on the right is not strictly determined.

$$\begin{bmatrix} 2 & 8 & 4 \\ 3 & 1 & 4 \\ 5 & 6 & 7 \end{bmatrix}, \quad \begin{bmatrix} 2 & 8 & 4 \\ 6 & 1 & 3 \\ 5 & 4 & 7 \end{bmatrix}.$$

strictly included *See* PROPER SUBSET.

strictly increasing *See* INCREASING FUNCTION and INCREASING SEQUENCE.

strictly monotonic *See* MONOTONIC FUNCTION and MONOTONIC SEQUENCE.

string *See* ELASTIC STRING and INEXTENSIBLE STRING.

strut A rod in *compression.

Student *See* GOSSET, WILLIAM SEALY.

Student's *t*-distribution = *T*-DISTRIBUTION.

Sturm–Liouville equation A second-order *differential equation in the form $\dfrac{d}{dx}\left[p(x)\dfrac{dy}{dx}\right] + [\lambda\omega(x) - q(x)]y = 0$ where $\omega(x)$ is called the 'weighting function'. The values of λ for which there is a solution are the *characteristic values and the corresponding solutions are the *characteristic functions.

subdivision (of a graph) An edge (arc) joins two vertices (nodes) of a graph. If another vertex is introduced onto that edge, it creates two edges, so in the figures, the insertion of the vertex C has subdivided the edge AB into two edges, namely AC and CB. The extra vertex C is necessarily of degree 2. A subdivision of a graph is a graph modified by the addition of one or more vertices of degree 2 onto an existing edge or edges.

$A \underset{}{\circ}\!\!\!-\!\!\!-\!\!\!-\!\!\!-\!\!\!-\!\!\!-\!\!\!-\!\!\!-\!\!\!\underset{}{\circ} B \qquad A \underset{}{\circ}\!\!\!-\!\!\!-\!\!\!-\!\!\!\underset{}{\circ} C \underset{}{-\!\!\!-\!\!\!-\!\!\!-\!\!\!-\!\!\!}\underset{}{\circ} B$

subdivision (of an interval) = PARTITION (of an interval).

subfield Let F be a *field with operations of addition and multiplication. If S is a subset of F that forms a field with the same operations, then S is a subfield of F. For example, the set **Q** of rational numbers forms a subfield of the field **R** of real numbers.

subgraph A subset of the vertices and edges of a graph which themselves form a graph.

subgroup Let G be a *group with a given operation. If H is a subset of G that forms a group with the same operation, then H is a subgroup of G. For example, $\{1, i, -1, -i\}$ forms a subgroup of the group of all non-zero complex numbers with multiplication.

subject (in logic) The category which a *categorical proposition is about. For example, in the statement 'some trees are evergreen', the subject is 'trees'.

submatrix A submatrix of a *matrix **A** is obtained from **A** by deleting from **A** some number of rows and some number of columns. For example, suppose that **A** is a 4×4 matrix and that $\mathbf{A} = [a_{ij}]$. Deleting the first and third rows and the second column gives the submatrix

$$\begin{bmatrix} a_{21} & a_{23} & a_{24} \\ a_{41} & a_{43} & a_{44} \end{bmatrix}.$$

subring Let R be a *ring with operations of addition and multiplication. If S is a subset of R that forms a ring with the same operations, then S is a subring of R. For example, the set **Z** of all integers forms a subring of the ring **R** of all real numbers, and the set of all even integers forms a subring of **Z**.

subset The set A is a subset of the set B if every element of A is an element of B. When this is so, A is included in B, written $A \subseteq B$, and B includes A, written $B \supseteq A$. The following properties hold:

  (i) For all sets A, $\varnothing \subseteq A$ and $A \subseteq A$.
  (ii) For all sets A and B, $A = B$ if and only if $A \subseteq B$ and $B \subseteq A$.
  (iii) For all sets A, B and C, if $A \subseteq B$ and $B \subseteq C$, then $A \subseteq C$.

See also PROPER SUBSET.

subspace *See* N-DIMENSIONAL SPACE.

substitution The replacement of a term in an expression or equation with another which is known to have the same value. This includes replacing variables with their numerical value to evaluate a formula. *See also* INTEGRATION.

substitution group = PERMUTATION GROUP.

subtend The angle that a line segment or arc with end-points A and B subtends at a point P is the angle APB. For example, one of the circle theorems may be stated as follows: the angle subtended by a diameter of a circle at any point on the circumference is a right angle.

subtraction The mathematical operation which is the inverse operation to *addition which calculates the difference between two numbers or quantities. So $7 - 2 = 5$, and $(3x + 5y) - (x + 2y) = 2x + 3y$.

subtraction (of matrices) For matrices **A** and **B** of the same order, the operation of subtraction is defined by taking $\mathbf{A} - \mathbf{B}$ to mean $\mathbf{A} + (-\mathbf{B})$, where $-\mathbf{B}$ is the *negative of **B**. Thus if $\mathbf{A} = [a_{ij}]$ and $\mathbf{B} = [b_{ij}]$, then $\mathbf{A} - \mathbf{B} = \mathbf{C}$, where $\mathbf{C} = [c_{ij}]$ and $c_{ij} = a_{ij} - b_{ij}$.

sufficient condition A condition which is enough for a statement to be true. For example, for a number to be even, it is sufficient that the number be divisible by 10. Note that if A is a necessary condition for B, B is a sufficient condition for A, and vice versa, so in the above example for a number to be divisible by 10 it is necessary that the number is even.

sufficient estimator An *estimator of a parameter θ that gives as much information about θ as is possible from the sample. The sample mean is a sufficient estimator of the population mean of a normal distribution.

sufficient statistic (for a parameter) A statistic that contains all of the information in a sample which is relevant to a point estimate of the parameter. When a sufficient statistic exists the *maximum likelihood estimator will be a function of the sufficient statistic.

sum The outcome of an *addition.

sum (of matrices) *See* ADDITION (of matrices).

sum (of a series) *See* SERIES.

sum (of sets) = UNION OF SETS.

summability theory The study of quantities which can be added or integrated, and particularly how to assign values to *divergent series and integrals.

summation notation The finite series $a_1 + a_2 + \ldots + a_n$ can be written, using the capital Greek letter sigma, as

$$\sum_{r=1}^{\infty} a_r.$$

Similarly, for example,

$$1^2 + 2^2 + \cdots + 10^2 = \sum_{r=1}^{10} r^2, \quad 1 + x + x^2 + \cdots + x^{n-1} = \sum_{r=0}^{n-1} x^r.$$

(The letter 'r' used here could equally well be replaced by any other letter.)
In a similar way, the infinite series $a_1 + a_2 + \ldots$ can be written as

$$\sum_{r=1}^{\infty} a_r.$$

The infinite series may not have a sum to infinity (*See* SERIES), but if it does
the same abbreviation is also used to denote the sum to infinity. For
example, the harmonic series

$$\sum_{r=1}^{\infty} \frac{1}{r}$$

is the infinite series $1 + \frac{1}{2} + \frac{1}{3} + \frac{1}{4} + \cdots$ and has no sum to infinity. But,
for $-1 < x < 1$, the geometric series $1 + x + x^2 + x^3 + \cdots$ has sum
$1/(1-x)$, and this can be written

$$\sum_{r=0}^{\infty} x^r = \frac{1}{1-x} \quad (-1 < x < 1).$$

sum to infinity *See* SERIES and GEOMETRIC SERIES.

sup Abbreviation for supremum.

supersink (in a network) If a network has two or more *sinks then a
supersink can be introduced connected to all the sinks, with the flow on
each edge equal to the total inflow to the source to which it connects.

supersource (in a network) If a network has two or more *sources then
a supersource can be introduced connected to all the sources, with the
flow on each edge equal to the total outflow from the source to which it
connects.

supplement *See* SUPPLEMENTARY ANGLES.

supplementary angles Two angles that add up to two right angles.
Each angle is the supplement of the other.

supplementary unit *See* SI UNITS.

supremum (suprema) *See* BOUND.

surd An *irrational number in the form of a root of some number (such as $\sqrt{5}$, $\sqrt[3]{2}$ or $4^{2/3}$) or a numerical expression involving such numbers. The term is rarely used nowadays.

surface A set of points (x, y, z) in a Cartesian space whose coordinates satisfy an equation expressed as $z = f(x, y)$, or implicitly as $g(x, y, z) = 0$, or expressed in a parameterized form.

surface (of a solid) The boundary of a geometric shape in three dimensions. For example, the surface of a cylinder has two circular sections and a curved surface.

surface In a *topological space X, a surface is a 2-dimensional *manifold. The most common are the surfaces of solid objects in the world around us, with an inside and an outside, but there are topological surfaces, such as the *Klein bottle, which do not have orientation in the same way—just as the *Möbius strip has only one side.

surface (in topology) A 2-dimensional *manifold. The most familiar examples are the *surfaces of solids, but examples like the *Klein bottle and the *real projective plane are also surfaces.

surface of revolution Suppose that an arc of a plane curve is rotated through one revolution about a line in the plane that does not cut the arc. The surface of the 3-dimensional figure obtained is a surface of revolution. *See also* AREA OF A SURFACE OF REVOLUTION.

surjection = ONTO MAPPING.

surjective mapping = ONTO MAPPING.

syllogism An argument consisting of three *categorical propositions where the truth of the first two requires the third (the conclusion) also to be true. For example, 'All men are mortal; Socrates is a man; therefore Socrates is mortal.'
　　Any syllogism must relate to three entities (which can be a group, an individual entity, or a concept—here 'men', Socrates', and 'mortality'), each of which appears exactly twice.

symbol A letter or sign which is used to represent something else, which could be an operation or relation, a function, a number or a quantity.

symbolic logic The area of mathematics relating to deductive argument and to the structure and relationships of statements and symbols.

symmetrical about a line A plane figure is symmetrical about the line l if, whenever P is a point of the figure, so too is P', where P' is the mirror-image of P in the line l. The line l is called a line of symmetry; and the figure is said to have bilateral symmetry or to be symmetrical by reflection in the line l. The letter A, for example, is symmetrical about the vertical line down the middle.

symmetrical about a point A plane figure is symmetrical about the point O if, whenever P is a point of the figure, so too is P', where O is the midpoint of $P'P$. The point O is called a centre of symmetry; and the figure is said to have half-turn symmetry about O because the figure appears the same when rotated through half a revolution about O. The letter S, for example, is symmetrical about the point at its centre.

symmetric design A block design in which the number of blocks and the numbers of points in each block are the same.

symmetric difference For sets A and B (subsets of some *universal set), the symmetric difference, denoted by $A + B$, is the set $(A \setminus B) \cup (B \setminus A)$. The notation $A \triangle B$ is also used. The set is represented by the shaded regions of the *Venn diagram shown. The following properties hold, for all A, B and C (subsets of some universal set E):

(i) $A + A = \emptyset$, $A + \emptyset = A$, $A + A' = E$, $A + E = A'$.
(ii) $A + B = (A \cup B) \setminus (A \cap B) = (A \cup B) \cap (A' \cup B')$.
(iii) $A + B = B + A$, the commutative law.
(iv) $(A + B) + C = A + (B + C)$, the associative law.
(v) $A \cap (B + C) = (A \cap B) + (A \cap C)$, the operation $\cap$ is distributive over the operation $+$.

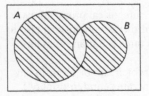

symmetric group For any set X, a permutation of X is a *one-to-one onto mapping from X to X. If X has n elements, there are $n!$ permutations of X and the set of all of these, with *composition of mappings as the operation, forms a *group called the symmetric group of degree n, denoted by S_n.

symmetric matrix Let $\mathbf{A}$ be the square matrix $[a_{ij}]$. Then $\mathbf{A}$ is symmetric if $\mathbf{A}^T = \mathbf{A}$ (*See* TRANSPOSE); that is, if $a_{ij} = a_{ji}$ for all i and j.

symmetric relation A *binary relation $\sim$ on a set S is symmetric if, for all a and b in S, whenever $a \sim b$ then $b \sim a$.

symmetry (of an expression) An expression where the roles of variables can be interchanged without changing the expressions. For example, $x^2 + xy + y^2$ is symmetric in x and y.

symmetry (of a geometrical figure) A figure may have reflective symmetry in a line (axis), or plane, or may have *rotational symmetry.

symmetry (of a graph) Two particular symmetries that a given graph $y = f(x)$ may have are symmetry about the y-axis, if f is an *even function, and symmetry about the origin, if f is an *odd function.

synthetic division A method of finding the quotient and remainder when a polynomial $f(x)$ is divided by a factor $x - h$, in which the numbers are laid out in the form of a table. The rule to remember is that, at each step, 'you multiply and then add'. To divide $ax^3 + bx^2 + cx + d$ by $x - h$, set up a table as follows:

h	a	b	c	d
	0	...	...	...
	...	...	...	...

Working from the left, each total, written below the line, is multiplied by h and entered above the line in the next column. The two numbers in that column are added to form the next total:

h	a	b	c	d
	0	ah	...	...
	a	$ah + b$	...	...

The last total is equal to the remainder, and the other numbers below the line give the coefficients of the quotient. By the *remainder theorem, the remainder equals $f(h)$.

For example, suppose that $2x^3 - 7x^2 + 5x + 11$ is to be divided by $x - 2$. The resulting table is:

2	2	-7	5	11
	0	4	-6	-2
	2	-3	-1	9

So the remainder equals 9, and the quotient is $2x^2 - 3x - 1$. The calculations here correspond exactly to those that are made when the polynomial is evaluated for $x = 2$ by *nested multiplication, to obtain $f(2) = 9$.

synthetic proof Deduction from *axioms. Compare with *analytic proof.

systematic error An error which is not random, and is therefore liable to introduce bias into an estimator. For example, $\sum_{r=1}^{n} \frac{(x_i - \bar{x})^2}{n}$ has a tendency to underestimate the population variance. In a production process if one of ten prongs on a rotating drum is broken, then every tenth item on the production line will be faulty in addition to any random faults which may occur.

systematic sampling Choose a number at random from 1 to n, and then take every n-th value thereafter. This means that all units have an equal chance of being selected, but it does not constitute a *random sample, as not all combinations are equally likely. There is a particular danger with this method, especially for 'production line sampling' as any faults with a particular machine in the system will affect items regularly, and the systematic method of choosing the items to be sampled means that you are very likely either to miss them all or hit them all (or some fraction of them).

Système International d'Unités *See* SI UNITS.

system of forces The collection of forces acting on a particle, a system of particles or a rigid body.

system of particles A collection of particles whose motion is under investigation. In a given mathematical model, the particles may be able to move freely relative to each other, or they may be connected by light rods, springs or strings.

systems analysis The area of mathematics concerned with the analysis of large complex systems, particularly where the systems involve interactions.

T Symbol for transpose of a matrix, written as a superscript, so $\mathbf{M}^\mathrm{T}$ is the transpose of $\mathbf{M}$.

T Symbol for 'true' in *logic and in *truth tables.

tableau *See* SIMPLEX METHOD.

tables A set of values of functions, set out in a tabular form for a range of arguments such as log tables, trigonometric tables and statistical tables. With the advent of hand-held calculators which can compute the values of all the mathematical functions previously tabulated, only statistical tables are in common use now.

tangent *See* TRIGONOMETRIC FUNCTION.

tangent (to a curve) Let P be a point on a (plane) curve. Then the tangent to the curve at P is the line through P that touches the curve at P. The gradient of the tangent at P is equal to the gradient of the curve at P. *See also* GRADIENT (of a curve).

tangent field A *differential equation gives information about the rate of change of the dependent variable but not the actual value of the variable. A graph showing the direction indicators of the tangents from which families of approximate solution curves can be drawn are called tangent fields, or direction fields. Curves along which the direction indicators have the same gradient are called isoclines, so in the example below, all the direction indicators at the same value of y are parallel and any horizontal line is an isocline.

$\dfrac{dy}{dx} = -y$ has the direction field shown in the figure.

The general solution to this differential equation is $y = Ae^{-x}$ and solution curves are shown for the cases where $A = -2$, -0.5 and 1. In the case where an analytical or algebraic solution is available the tangent fields are not especially useful, but they are powerful tools when no exact solution can be obtained.

(⊕) SEE WEB LINKS
• An interactive demonstration of tangent fields.

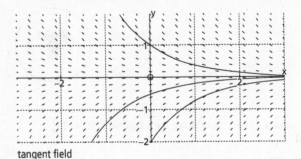

tangent field

tangent line *See* TANGENT PLANE.

tangent plane Let P be a point on a smooth surface. The tangent at P to any curve through P on the surface is called a tangent line at P, and the tangent lines are all perpendicular to a line through P called the normal to the surface at P. The tangent lines all lie in the plane through P perpendicular to this normal, and this plane is called the tangent plane at P. (A precise definition of 'smooth' cannot be given here. A tangent plane does *not* exist, however, at a point on the edge of a cube or at the vertex of a (double) cone, for example.)

At a point P on the surface, it may be that all the points near P (apart from P itself) lie on one side of the tangent plane at P. On the other hand, this may not be so—there may be some points close to P on one side of the tangent plane and some on the other. In this case, the tangent plane cuts the surface in two curves that intersect at P. This is what happens at a *saddle-point.

tangent rule *See* TRIANGLE.

tanh *See* HYPERBOLIC FUNCTION.

Tartaglia, Niccolò (1499–1557) Italian mathematician remembered as the discoverer of a method of solving cubic equations, though he may not have been the first to find a solution. His method was published by Cardano, to whom he had communicated it in confidence. Born Niccolò Fontana, his adopted nickname means 'Stammerer'.

tautochrone As well as being the solution to the *brachistochrone problem, the *cycloid has another property. Suppose that a cycloid is positioned as in the diagram for the brachistochrone problem. If a particle starts from rest at any point of the cycloid and travels along the curve under the force of gravity, the time it takes to reach the lowest point is independent of its starting point. So the cycloid is also called the tautochrone (from the Greek for 'same time').

tautology A *compound statement that is true for all possible truth values of its components. For example, $p \Rightarrow (q \Rightarrow p)$ is a tautology, as can be seen by calculating its *truth table.

taxicab norm The sum of the absolute values of the components of a vector. The name derives from the distance a taxi has to drive in a rectangular street grid to get from the origin to a particular point. It is also known as the Manhattan norm because Manhattan has perhaps the most famous rectangular street grid. The taxicab norm is $p = 1$ in the family of *p-norms.

Taylor, Brook (1685–1731) English mathematician who contributed to the development of calculus. His text of 1715 contains what has become known as the Taylor series. Its importance was not appreciated until much later, but it had in fact been discovered earlier by James *Gregory and others.

Taylor, Richard (1962–) British mathematician who worked with Andrew *Wiles to complete the proof of *Fermat's Last Theorem.

Taylor polynomial, Taylor series (expansion) *See* TAYLOR'S THEOREM.

Taylor's Theorem Applied to a suitable function f, Taylor's Theorem gives a polynomial, called a Taylor polynomial, of any required degree, that is an approximation to $f(x)$.

Theorem
Let f be a function such that, in an interval I, the derived functions $f^{(r)}(r = 1, \ldots, n)$ are continuous, and suppose that $a \in I$. Then, for all x in I,

$$f(x) = f(a) + \frac{f'(a)}{1!}(x - a) + \frac{f''(a)}{2!}(x - a)^2 + \cdots$$
$$+ \frac{f^{(n-1)}(a)}{(n-1)!}(x - a)^{n-1} + R_n,$$

where various forms for the remainder R_n are available.

Two possible forms for R_n are

$$R_n = \frac{1}{(n-1)!} \int_a^x f^{(n)}(t)(x-t)^{n-1} dt \quad \text{and}$$

$$R_n = \frac{f^{(n)}(c)}{n!}(x-a)^n,$$

where c lies between a and x. By taking $x = a + h$, where h is small (positive or negative), the formula

$$f(a+h) = f(a) + \frac{f'(a)}{1!}h + \frac{f''(a)}{2!}h^2 + \cdots + \frac{f^{(n-1)}(a)}{(n-1)!}h^{n-1} + R_n$$

is obtained, where the second form of the remainder now becomes

$$R_n = \frac{f^{(n)}(a+k)}{n!}h^n,$$

and k lies between 0 and h. This enables $f(a+h)$ to be determined up to a certain degree of accuracy, the remainder R_n giving the *error. Suppose now that, for the function f, Taylor's Theorem holds for all values of n, and that $R_n \to 0$ as $n \to \infty$; then an infinite series can be obtained whose sum is $f(x)$. In such a case, it is customary to write

$$f(x) = f(a) + \frac{f'(a)}{1!}(x-a) + \frac{f''(a)}{2!}(x-a)^2 + \cdots.$$

This is the Taylor series (or expansion) for f at (or about) a. The special case with $a = 0$ is the *Maclaurin series for f.

Tchebyshev = CHEBYSHEV.

t-distribution The continuous probability *distribution of a random variable formed from the ratio of a random variable with a standard *normal distribution and the square root of a random variable with a *chi-squared distribution divided by its degrees of freedom. Alternatively, it is formed from the square root of a random variable with an *F-distribution in which the numerator has one degree of freedom. The shape, which depends on the degrees of freedom, is similar in shape to a standard normal distribution, but is less peaked and has fatter tails. The distribution is used to test for significant differences between a sample mean and the population mean, and can be used to test for differences between two sample means. The table gives, for different degrees of freedom, the values corresponding to certain values of α, to be used in a *one-tailed or *two-tailed *t-test, as explained below.
Corresponding to the first column value α, the table gives the one-tailed

α	2α	$\nu=1$	2	3	4	6	8	10	15	20	30	60	∞
0.05	0.10	6.34	2.92	2.35	2.13	1.94	1.86	1.81	1.75	1.72	1.70	1.67	1.64
0.025	0.05	12.71	4.30	3.18	2.78	2.45	2.31	2.23	2.13	2.09	2.04	2.00	1.96
0.01	0.02	31.82	6.96	4.54	3.75	3.14	2.90	2.76	2.60	2.53	2.46	2.39	2.33
0.005	0.01	63.66	9.92	5.84	4.60	3.71	3.36	3.17	2.95	2.84	2.75	2.66	2.58

value $t_{\alpha,\nu}$ such that $\Pr(t > t_{\alpha,\nu}) = \alpha$, for the t-distribution with ν degrees of freedom. Corresponding to the second column value 2α, the table gives the two-tailed value $t_{\alpha,\nu}$ such that $\Pr(|t| > t_{\alpha,\nu}) = 2\alpha$. Interpolation may be used for values of ν not included.

techniques of integration *See* INTEGRATION.

telescoping series A series $\{a_r\}$ which can be expressed as a sum of differences of successive terms of another series $\{b_r\}$, i.e. that

$a_r = b_r - b_{r-1}$ so $\displaystyle\sum_{r=a}^{n} a_r = b_n - b_0$. For example, if

$a_r = \dfrac{1}{r(r+1)} = \dfrac{1}{r} - \dfrac{1}{r+1}$, $\displaystyle\sum_{r=1}^{n} a_r$ can be written as

$$\sum_{r=1}^{n}\left(\frac{1}{r} - \frac{1}{r+1}\right) = 1 - \frac{1}{n+1} = \frac{n}{n+1}.$$

tend to To have as a limit. For example, the sequence $a_n = \dfrac{1}{n}$ tends to 0 as n tends to infinity.

tension An internal force that exists at each point of a string or spring. If the string or spring were cut at any particular point, then equal and opposite forces applied to the two cut ends would be needed to maintain the illusion that it remains uncut. The magnitude of these applied forces is the magnitude of the tension at that point. When a particle is suspended from a fixed point by a light string, the tension in the string exerts a force vertically upwards on the particle, and it exerts an equal and opposite force on the point of suspension.

tensor The generalization of scalars, which have no index, vectors with one index and matrices with two indices to n indices in m-dimensional space. The components of tensors obey certain transformation rules.

tera- Prefix used with *SI units to denote multiplication by 10^{12}.

term *See* SEQUENCE and SERIES.

terminal speed When an object falls to Earth from a great height, its speed, in certain *mathematical models, tends to a value called its terminal speed. One possible mathematical model gives the equation $m\ddot{\mathbf{r}} = -mg\mathbf{j} - c\dot{\mathbf{r}}$, where m is the mass of the particle and $\mathbf{j}$ the unit vector in the direction vertically upwards. The second term on the right-hand side, in which c is a positive constant, arises from the *air resistance. The velocity corresponding to $\ddot{\mathbf{r}} = 0$ is equal to $(-mg/c)\mathbf{j}$, which is called the terminal velocity. The terminal speed is the magnitude mg/c of this velocity. If, instead, the equation $m\ddot{\mathbf{r}} = -mg\mathbf{j} - c|\dot{\mathbf{r}}|\dot{\mathbf{r}}$ is used, the terminal speed is $\sqrt{mg/c}$ and the terminal velocity is $-\sqrt{mg/c}\,\mathbf{j}$. As the particle falls, its velocity tends to the terminal velocity as t increases, irrespective of the initial conditions.

terminating decimal See DECIMAL REPRESENTATION.

ternary relation See RELATION.

ternary representation The representation of a number to *base 3.

tessellation In its most general form, a tessellation is a covering of the plane with shapes. Often the shapes are polygons and the pattern is in some sense repetitive. A tessellation is regular if it consists of congruent regular polygons. There are just the three possibilities shown here: the polygon is either an equilateral triangle, a square or a regular hexagon.

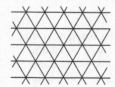  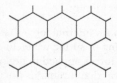

A tessellation is semi-regular if it consists of regular polygons, not all congruent. It can be shown that there are just eight of these, one of which has two forms that are mirror-images of each other. They use triangles, squares, hexagons, octagons and dodecagons. For example, one consisting of octagons and squares and another consisting of hexagons and triangles, shown here, may be familiar as patterns of floor-coverings.

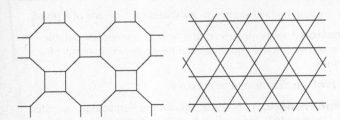

The possibilities are unlimited if tessellations in which the polygons are not regular are considered. For example, there are many interesting tessellations using congruent non-regular pentagons.

⊕ SEE WEB LINKS

• Tessellations with various grids and interactive design tools.

test statistic A *statistic used in *hypothesis testing that has a known distribution if the null hypothesis is true.

tetrahedral number An integer of the form $\frac{1}{6}n(n+1)(n+2)$, where n is a positive integer. This number equals the sum of the first n *triangular numbers.

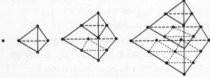

The first few tetrahedral numbers are 1, 4, 10 and 20, and the reason for the name can be seen from the figure.

tetrahedron (tetrahedra) A solid figure bounded by four triangular faces, with four vertices and six edges. A regular tetrahedron has equilateral triangles as its faces, and so all its edges have the same length.

t-formulae A range of trigonometric identities that are useful as substitutions for integrations, often as half-angle formulae.
If $t = \tan\frac{1}{2}A$, then $\sin A = \dfrac{2t}{1+t^2}$, $\quad \cos A = \dfrac{1-t^2}{1+t^2}$.

Thales of Miletus (about 585 BC) Greek philosopher frequently regarded as the first mathematician in the sense of being the one to whom particular discoveries have been attributed. These include a number of geometrical propositions, including the theorem that the angle in a semicircle is a right angle, and methods of measuring heights by means of shadows and calculating the distances of ships at sea.

theorem A mathematical statement established by means of a proof.

theta function A range of special functions of a complex variable. The roots of all polynomial equations can be expressed in terms of theta functions.

third derivative *See* HIGHER DERIVATIVE.

Thom, René Frédéric (1923–2002) French mathematician, awarded the *Fields Medal in 1958 for work in *topology, but best known for his development of the mathematics in *catastrophe theory, where continuous or incremental change brings about some discontinuity as in the occurrence of earthquakes.

Thomson, William *See* KELVIN, LORD.

thrust The force of compression in a rod. Any force which is pushing rather than pulling.

tie A rod in tension, usually in a *light framework.

tiling = TESSELLATION.

time In the real world, the passage of time is a universal experience which clocks of various kinds have been designed to measure. In a *mathematical model, time is represented by a real variable, usually denoted by t, with $t = 0$ corresponding to some suitable starting-point. An *observer associated with a *frame of reference has the capability to measure the duration of time intervals between events occurring in the problem being investigated.

Time has the dimension T, and the SI unit of measurement is the *second.

time series A sequence of observations taken over a period of time, usually at equal intervals. Time series analysis is concerned with identifying the factors that influence the variation in a time series, perhaps with a view to predicting what will happen in the future. For many time series, two important components are seasonal variation, which is periodic change usually in yearly cycles, and trend, which is the long-term change in the seasonally adjusted figures.

ton The mass of 20 hundredweights or 2240 pounds (lbs). It is the imperial near equivalent of the tonne (approximately 2200 lbs).

tonne The mass of 1000 kilograms.

topologically distinguishable In a *topological space X, points x and y in X are topologically indistinguishable if and only if they have exactly the same *neighbourhoods. Otherwise, they are distinguishable. Generally, 'distinguishable' is stronger than 'distinct' but not as strong as being *separated.

topological space A set of x with an associated set T of all its subsets that contains the empty set and the whole set x, and which is closed under union and intersection. The members of T are the open sets of X. Any set of points making up a geometrical figure and satisfying these constraints is a topological space and its topology is determined by the properties of its open sets.

topological vector space A space which is both a *topological space and a *vector space in which addition and scalar multiplication are continuous. These spaces are the most general spaces used in *functional analysis. They include all normed vector spaces, so, in particular, *Hilbert spaces and *Banach spaces are examples of topological vector spaces.

topology The area of mathematics concerned with the general properties of shapes and space, and in particular with the study of properties that are not changed by continuous distortions such as stretching.

toppling *See* SLIDING–TOPPLING CONDITION.

torque = MOMENT.

torus (tori) Suppose that a circle of radius a is rotated through one revolution about a line, in the plane of the circle, a distance b from the centre of the circle, where $b > a$. The resulting surface or solid is called a torus, the shape of a 'doughnut' or 'anchor ring'. The surface area of such a torus is equal to $4\pi^2 ab$ and its volume equals $2\pi^2 a^2 b$.

total differential = EXACT DIFFERENTIAL.

total force The (vector) sum of all the forces acting on a particle, a system of particles or a rigid body.

totally bounded (of a space) Describing a *metric space which, for any $\varepsilon > 0$, can be presented as the union of a finite number of sets each with *diameter smaller than ε. Essentially, it can be covered by a finite number of sets which are arbitrarily small. The real numbers in the interval $(0, 1)$ are totally bounded since, once ε is stated, the number of intervals of width ε needed to cover a unit interval can be calculated and is finite. However, the set of rationals in the interval $(0, 1)$ is not totally bounded.

totally disconnected (of a space) Describing a *topological space in which the only connected subsets are the empty set and the one-point sets. Since these are connected in every topological space, 'totally disconnected' means that the space has no non-trivial connected subsets. The *Cantor set is an example of a totally disconnected space.

total probability law Let $A_1, A_2, \ldots, A_n$ be mutually exclusive events whose union is the whole sample space of an experiment. Then the total probability law is the following formula for the probability of an event B:

$$\Pr(B) = \Pr(B|A_1)\Pr(A_1) + \Pr(B|A_2)\Pr(A_2) + \cdots + \Pr(B|A_n)\Pr(A_n).$$

Tower of Brahma *See* TOWER OF HANOI.

Tower of Hanoi Imagine three poles with a number of discs of different sizes initially all on one of the poles in decreasing order. The problem is to transfer all the discs to one of the other poles, moving the discs individually from pole to pole, so that one disc is never placed above another of smaller diameter. A version with 8 discs, known as the Tower of Hanoi, was invented by Edouard Lucas and sold as a toy in 1883, but the idea may be much older. The toy referred to a legend about the Tower of Brahma which has 64 discs being moved from pole to pole in the same way by a group of priests, the prediction being that the world will end when they have finished their task.

If t_n is the number of moves it takes to transfer n discs from one pole to another, then $t_1 = 1$ and $t_{n+1} = 2t_n + 1$. It can be shown that the solution of this *difference equation is $t_n = 2^n - 1$, so the original Tower of Hanoi puzzle takes 255 moves, and the Tower of Brahma takes $2^{64} - 1$.

(⊕) SEE WEB LINKS

• An interactive applet for different numbers of discs, with animated solutions.

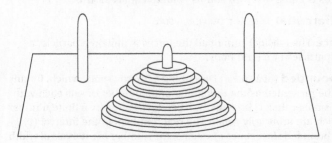

trace The trace of a square *matrix is the sum of the entries in the *main diagonal.

trail A *walk in which no edge is repeated.

trajectory The path traced out by a *projectile.

transcendental number A real number that is not a root of a polynomial equation with integer coefficients. In other words, a number is transcendental if it is not an *algebraic number. In 1873, Hermite showed that e is transcendental; and it was shown by Lindemann, in 1882, that π is transcendental.

transfinite number A *cardinal number relating to an *infinite set such as *aleph-null and the cardinality of the *continuum.

transformation (of the plane) Let S be the set of points in the plane. A transformation of the plane is a *one-to-one mapping from S to S. The most important transformations of the plane are the linear transformations, which are those that, in terms of Cartesian coordinates, can be represented by linear equations. For a linear transformation T, there are constants a, b, c, d, h and k such that T maps the point P with coordinates (x, y) to the point P' with coordinates (x', y'), where

$$x' = ax + by + h,$$
$$y' = cx + dy + k.$$

When $h = k = 0$, the origin O is a fixed point, since T maps O to itself, and then the transformation can be written $\mathbf{x'} = \mathbf{Ax}$, where

$$\mathbf{x} = \begin{bmatrix} x \\ y \end{bmatrix}, \quad \mathbf{x'} = \begin{bmatrix} x' \\ y' \end{bmatrix}, \quad \mathbf{A} = \begin{bmatrix} a & b \\ c & d \end{bmatrix}.$$

Examples of such transformations are *rotations about O, *reflections in lines through O and *dilatations from O. *Translations are examples of linear transformations in which O is not a fixed point.

transformation group A group with elements that are transformations with the binary operation of composition of transformations. For example, the set of rotations about a fixed point form a transformation group, but the set of reflections in axes through a fixed point do not, as the set is not closed under composition. For example, successive reflections in lines at right angles to one another is equivalent to a half-turn about the point of intersection of the axes of reflection.

transformation matrix A linear transformation can be represented by a transformation matrix T which contains the coefficients as its elements. So the transformation in two dimensions given by the equations $x' = a_{11}x + a_{12}y$, $y' = a_{21}x + a_{22}y$ can be written as

$$\begin{pmatrix} x' \\ y' \end{pmatrix} = \mathbf{T} \begin{pmatrix} x \\ y \end{pmatrix}, \quad \text{where} \quad \mathbf{T} = \begin{pmatrix} a_{11} & a_{12} \\ a_{21} & a_{22} \end{pmatrix}.$$

transition matrix, transition probability See MARKOV CHAIN.

transitive relation A *binary relation $\sim$ on a set S is transitive if, for all a, b and c in S, whenever $a \sim b$ and $b \sim c$ then $a \sim c$.

translation (of the plane) A *transformation of the plane in which a point P with coordinates (x, y) is mapped to the point P' with coordinates (x', y'), where $x' = x + h$, $y' = y + k$. Thus the origin O is mapped to the point O' with coordinates (h, k), and the point P is mapped to the point P', where the *directed line segment $\overrightarrow{PP'}$ has the same direction and length as $\overrightarrow{OO'}$.

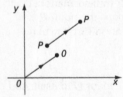

translation of axes (in the plane) Suppose that a Cartesian coordinate system has a given x-axis and y-axis with origin O and given unit length, so that a typical point P has coordinates (x, y). Let O' be the point with coordinates (h, k), and consider a new coordinate system with X-axis and Y-axis parallel and similarly directed to the x-axis and y-axis, with the same unit length, with origin at O'. With respect to the new coordinate system, the point P has coordinates (x, y). The old and new coordinates in such a translation of axes are related by $x = X + h$, $y = Y + k$, or, put another way, $X = x - h$, $Y = y - k$.

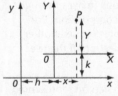

This procedure is useful for investigating, for example, a curve with equation $9x^2 + 4y^2 - 18x + 16y - 11 = 0$. Completing the square in x and y gives $9(x-1)^2 + 4(y+2)^2 = 36$, and so, with respect to a new coordinate system with origin at the point $(1, -2)$, the curve has the simple equation $9X^2 + 4Y^2 = 36$.

translation of axes (in 3-dimensional space) Just like a translation of axes in the plane, a translation of axes in 3-dimensional space can be made, to a new origin with coordinates (h, k, l), with new axes parallel and similarly directed to the old. The old coordinates (x, y, z) and new coordinates (X, Y, Z) are related by $x = X + h$, $y = Y + k$, $z = Z + l$, or, to put it another way, $X = x - h$, $Y = y - k$, $Z = z - l$.

transportation problem A problem in which units of a certain product are to be transported from a number of factories to a number of retail outlets in a way that minimizes the total cost. For example, suppose that there are m factories and n retail outlets, and that the transportation costs are specified by an $m \times n$ matrix $[c_{ij}]$, where c_{ij} is the cost, in suitable units, of transporting one unit of the product from the i-th factory to the j-th retail outlet. Suppose also that the maximum number of units that each factory can supply and the minimum number of units that each outlet requires are specified. By introducing suitable variables, the problem of minimizing the total cost can be formulated as a *linear programming problem.

transpose The transpose of an $m \times n$ matrix is the $n \times m$ matrix obtained by interchanging the rows and columns. The transpose of $\mathbf{A}$ is denoted by $\mathbf{A}^T$, $\mathbf{A}^t$ or $\mathbf{A}'$. Thus if $\mathbf{A} = [a_{ij}]$, then $\mathbf{A}^T = [a'_{ij}]$, where $a'_{ij} = a_{ji}$; that is,

$$
\mathbf{A} = \begin{bmatrix} a_{11} & a_{12} & \cdots & a_{1n} \\ a_{21} & a_{22} & \cdots & a_{2n} \\ \vdots & \vdots & \ddots & \vdots \\ a_{m1} & a_{m2} & \cdots & a_{mn} \end{bmatrix}, \qquad \mathbf{A}^T = \begin{bmatrix} a_{11} & a_{12} & \cdots & a_{m1} \\ a_{21} & a_{22} & \cdots & a_{m2} \\ \vdots & \vdots & \ddots & \vdots \\ a_{1n} & a_{2n} & \cdots & a_{mn} \end{bmatrix}.
$$

The following properties hold, for matrices $\mathbf{A}$ and $\mathbf{B}$ of appropriate orders:
  (i) $(\mathbf{A}^T)^T = \mathbf{A}$.
  (ii) $(\mathbf{A} + \mathbf{B})^T = \mathbf{A}^T + \mathbf{B}^T$.
  (iii) $(k\mathbf{A})^T = k\mathbf{A}^T$.
  (iv) $(\mathbf{A}\mathbf{B})^T = \mathbf{B}^T\mathbf{A}^T$.

transversable graph One that can be drawn without removing pen from paper or going over the same edge twice.

transversal A straight line that intersects a given set of two or more straight lines in the plane. When a transversal intersects a given pair of lines, eight angles are formed; the four angles between the pair of lines are interior angles and the four others are exterior angles.

An interior angle between the transversal and one line and an exterior angle between the transversal and the other line, such that these two angles are on the same side of the transversal, are **corresponding angles**. An interior angle between the transversal and one line and an interior angle between the transversal and the other line, such that these two angles are on opposite sides of the transversal, are **alternate angles**. The given two lines are parallel if and only if corresponding angles are equal: they are also parallel if and only if alternate angles are equal.

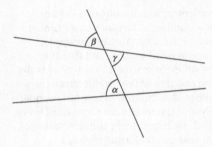

The figure shows a transversal intersecting two lines. The angles α and β are corresponding angles and α and γ are alternate angles.

transverse axis *See* HYPERBOLA.

transverse component *See* RADIAL AND TRANSVERSE COMPONENTS.

trapezium (trapezia) A quadrilateral with two parallel sides. If the parallel sides have lengths a and b, and the distance between them is h, the area of the trapezium equals $\frac{1}{2}h(a+b)$.

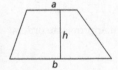

trapezium rule An approximate value can be found for the definite integral

$$\int_a^b f(x)\,dx,$$

using the values of $f(x)$ at equally spaced values of x between a and b, as follows. Divide $[a, b]$ into n equal subintervals of length h by the partition

$$a = x_0 < x_1 < x_2 < \cdots < x_{n-1} < x_n = b,$$

where $x_{i-1} - x_i = h = (b-a)/n$. Denote $f(x_i)$ by f_i, and let P_i be the point (x_i, f_i). If the line segment $P_i P_{i+1}$ is used as an approximation to the curve $y = f(x)$ between x_i and x_{i+1}, the area under that part of the curve is approximately the area of the trapezium shown in the figure, which equals $\frac{1}{2} h(f_i + f_{i+1})$. By adding up the areas of such trapezia between a and b, the resulting trapezium rule gives

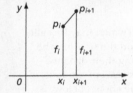

$$\frac{1}{2} h(f_0 + 2f_1 + 2f_2 + \cdots + 2f_{n-1} + f_n)$$

as an approximation to the value of the integral. This approximation has an *error that is roughly proportional to $1/n^2$; when the number of subintervals is doubled, the error is roughly divided by 4. *Simpson's rule is significantly more accurate.

trapezoidal rule = TRAPEZIUM RULE.

travelling salesman problem (TSP) (in graph theory) This is a situation similar to the *minimum connector problem but the salesman wishes to return to the starting point (home) at the end, and so essentially the problem is to find a closed walk which visits every vertex, and which minimizes the total distance travelled. In practice there can be unusual situations in which the most efficient route is to go $A \to B \to A$ because to get to B from any other vertex, without going through A, is difficult. However, the problem is much easier to analyse if the assumption is made that every vertex is visited exactly once, and the problem reduces to finding the *Hamiltonian cycle of minimum length. There is no algorithm to solve the TSP but an upper limit is easy to identify—the length of any route satisfying the conditions is immediately an upper bound. A minimum bound can be constructed by the following argument: if a minimum connected path is found for a subset of $n-1$ vertices, and the length of the two shortest edges from the n-th vertex to any two of the $n-1$ vertices are added, then the total length provides a lower bound which will be the minimum distance if and only if the graph constructed in this way is a Hamiltonian cycle, i.e. there is a route round it where no retracing would be required.

tree A *connected graph with no *cycles. It can be shown that a connected simple graph with n vertices is a tree if and only if it has $n-1$ edges. The figure shows all the trees with up to five vertices.

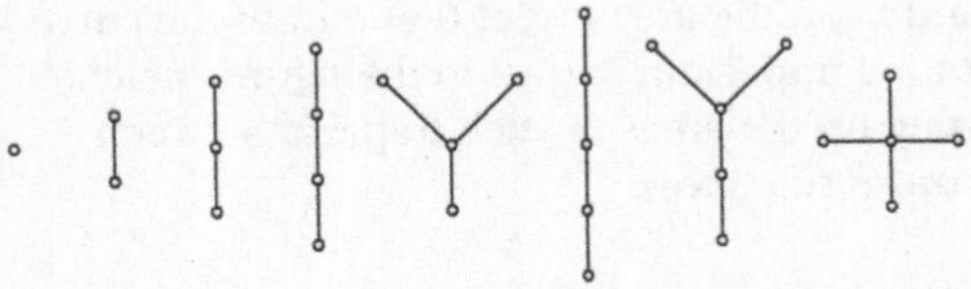

Particularly in applications, one of the vertices (which may be called nodes) of a tree may be designated as the root, and the tree may be drawn with the vertices at different levels indicating their distance from the root. A rooted tree in which every vertex (except the root, of degree 2) has degree either 1 or 3, such as the one shown below, is called a binary tree.

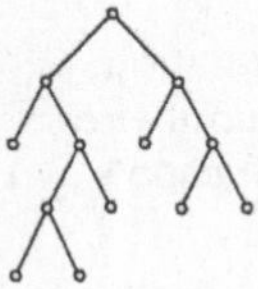

trend *See* TIME SERIES.

tri- Prefix denoting three, for example as in triangle.

trial (in statistics) A single observation or experiment.

trial and improvement *See* FALSE POSITION.

triangle Using the properties of angles made when one line cuts a pair of parallel lines, it is proved, as illustrated on the left, that the angles of a triangle ABC add up to 180°. By considering separate areas, illustrated on the right, it can be shown that the area of the triangle is half that of the rectangle shown, and so the area of a triangle is 'half base times height'.

If now A, B and C denote the angles of the triangle, and a, b and c the lengths of the sides opposite them, the following results hold:

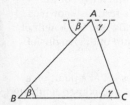 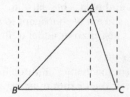

(i) The area of the triangle equals $\frac{1}{2}bc \sin A$.

(ii) The sine rule:

$$\frac{a}{\sin A} = \frac{b}{\sin B} = \frac{c}{\sin C} = 2R,$$

where R is the radius of the *circumcircle.

(iii) The cosine rule: $a^2 = b^2 + c^2 - 2bc \cos A$, or, in another form,

$$\cos A = \frac{b^2 + c^2 - a^2}{2bc}.$$

(iv) The tangent rule:

$$\tan \frac{B - C}{2} = \frac{b - c}{b + c} \cot \frac{A}{2}.$$

(v) Hero's formula: Let $s = \frac{1}{2}(a + b + c)$. Then the area of the triangle equals $\sqrt{s(s-a)(s-b)(s-c)}$.

triangle inequality (for complex numbers) If z_1 and z_2 are complex numbers, then $|z_1 + z_2| \leq |z_1| + |z_2|$. This result is known as the triangle inequality because it follows from the fact that $|OQ| \leq |OP_1| + |P_1 Q|$, where P_1, P_2 and Q represent z_1, z_2 and $z_1 + z_2$ in the *complex plane.

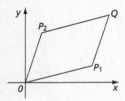

triangle inequality (for points in the plane) For points A, B and C in the plane, $|AC| \leq |AB| + |BC|$. This result, the triangle inequality, says that the length of one side of a triangle is less than or equal to the sum of the lengths of the other two sides.

triangle inequality (for vectors) Let $|\mathbf{a}|$ denote the length of the vector $\mathbf{a}$. For vectors $\mathbf{a}$ and $\mathbf{b}$, $|\mathbf{a} + \mathbf{b}| \leq |\mathbf{a}| + |\mathbf{b}|$. This result is known as the triangle inequality, since it is equivalent to saying that the length of one side of a triangle is less than or equal to the sum of the lengths of the other two sides.

triangle of forces (in mechanics) If three forces act at a point on a body in equilibrium their vector sum must be zero, and consequently the forces can be drawn in a closed triangle.

triangular matrix A square matrix that is either lower triangular or upper triangular. It is lower triangular if all the entries above the main diagonal are zero, and upper triangular if all the entries below the main diagonal are zero.

triangular number An integer of the form $\frac{1}{2}n(n+1)$, where n is a positive integer. The first few triangular numbers are 1, 3, 6, 10 and 15, and the reason for the name can be seen from the figure.

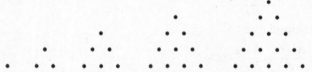

triangulation The method of fixing a point by using directions from two known points to construct a triangle.

tridiagonal matrix A square matrix with zero entries everywhere except on the main diagonal and the neighbouring diagonal on either side.

trigonometric function Though the distinction tends to be overlooked, each of the trigonometric functions has two forms, depending upon whether degrees or radians are used.

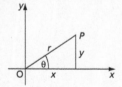

Using degrees
The basic trigonometric functions, cosine, sine and tangent, are first introduced by using a right-angled triangle, but $\cos \theta°$, $\sin \theta°$ and $\tan \theta°$ can also be defined when θ is larger than 90 and when θ is negative. Let

P be a point (not at O) with Cartesian coordinates (x, y). Suppose that OP makes an angle of $\theta°$ with the positive x-axis and that $|OP| = r$. Then the following are the definitions: $\cos \theta° = x/r$, $\sin \theta° = y/r$, and (when $x \neq 0$) $\tan \theta° = y/x$. It follows that $\tan \theta° = \sin \theta°/\cos \theta°$, and that $\cos^2 \theta° + \sin^2 \theta° = 1$. Some of the most frequently required values are given in the table.

θ	0	30	45	60	90
$\cos \theta°$	1	$\dfrac{\sqrt{3}}{2}$	$\dfrac{1}{\sqrt{2}}$	$\dfrac{1}{2}$	0
$\sin \theta°$	0	$\dfrac{1}{2}$	$\dfrac{1}{\sqrt{2}}$	$\dfrac{\sqrt{3}}{2}$	1
$\tan \theta°$	0	$\dfrac{1}{\sqrt{3}}$	1	$\sqrt{3}$	not defined

2 sin positive	1 all positive
3 tan positive	4 cos positive

The point P may be in any of the four quadrants. By considering the signs of x and y, the quadrants in which the different functions take positive values can be found, and are shown in the figure. The following are useful for calculating values, when P is in quadrant 2, 3 or 4:

$$\cos(180 - \theta)° = -\cos \theta°, \qquad \sin(180 - \theta)° = \sin \theta°,$$
$$\cos(180 + \theta)° = -\cos \theta°, \qquad \sin(180 + \theta)° = -\sin \theta°,$$
$$\cos(-\theta)° = \cos \theta°, \qquad \sin(-\theta)° = -\sin \theta°.$$

The functions cosine and sine are periodic, of period 360; that is to say, $\cos(360 + \theta)° = \cos \theta°$ and $\sin(360 + \theta)° = \sin \theta°$. The function tangent is periodic, of period 180; that is, $\tan(180 + \theta)° = \tan \theta°$.

Using radians

In more advanced work, it is essential that angles are always measured in *radians. It is necessary to introduce new trigonometric functions $\cos x$, sin x and tan x, where now x is a real number. These agree with the former functions, if x is treated as being the measure in radians of the angle formerly measured in degrees. For example, since π radians $= 180°$,

$$\cos \pi = \cos 180° = -1, \qquad \sin \frac{\pi}{3} = \sin 60° = \frac{\sqrt{3}}{2},$$

$$\tan \frac{\pi}{4} = \tan 45° = \frac{1}{\sqrt{2}}.$$

The functions cos and sin are periodic with period 2π, and tan has period π.

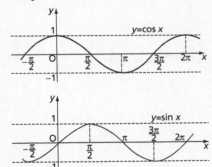

It is important to distinguish between the functions $\cos x$ and $\cos x°$. They are different functions, but are related since $\cos x° = \cos (\pi x/180)$. The same applies to sin and tan. Sometimes authors do not make the distinction and use the notation $\cos A$, for example, where A is an angle measured in degrees or radians. This is what has been done in the Table of Trigonometric Formulae (Appendix 12).

The other trigonometric functions, cotangent, secant and cosecant, are defined as follows:

$$\cot x = \frac{\cos x}{\sin x}, \qquad \sec x = \frac{1}{\cos x}, \qquad \operatorname{cosec} x = \frac{1}{\sin x},$$

where, in each case, values of x that make the denominator zero must be excluded from the domain. The basic identities satisfied by the trigonometric functions will be found in the Table of Trigonometric Formulae (Appendix 12).

In order to find the derivatives of the trigonometric functions, it is first necessary to show that

$$\lim_{x \to 0} \frac{\sin x}{x} = 1.$$

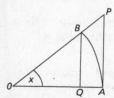

In view of the essentially geometric definition of sin x, a geometric method has to be used. By considering (when $0 < x < \pi/2$) the areas of $\triangle OBQ$, sector OAB and $\triangle OAP$, where $\angle OAB$ measures x radians, illustrated above, it can be shown that

$$\cos x < \frac{\sin x}{x} < \frac{1}{\cos x}.$$

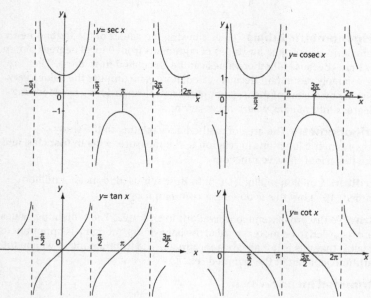

After dealing also with $x < 0$, the required limit can be deduced. Hence it can be shown that

$$\frac{d}{dx}(\sin x) = \cos x, \qquad \frac{d}{dx}(\cos x) = -\sin x.$$

The derivatives of the other trigonometric functions are found from these, by using the rules for differentiation, and are given in the Table of Derivatives (Appendix 6).

trigonometric series expansions For x measured in radians, the basic trigonometric functions can be expressed by the following series expansions, derived as Taylor series:

$$\sin x = x - \frac{x^3}{3!} + \frac{x^5}{5!} + \cdots + (-1)^r \frac{x^{2r+1}}{(2r+1)!} + \ldots \text{ for all } x,$$

$$\cos x = 1 - \frac{x^2}{2!} + \frac{x^4}{4!} + \cdots + (-1)^r \frac{x^{2r}}{(2r)!} + \ldots \text{ for all } x,$$

$$\tan^{-1}x = x - \frac{x^3}{3} + \frac{x^5}{5} + \cdots + (-1)^r \frac{x^{2r+1}}{(2r+1)} + \ldots \text{ for } -1 \le x \le 1.$$

trigonometric tables Tables showing the values of the trigonometric functions, usually just for values of arguments from 0 to 90 degrees or 0 to $\pi/2$ rads because all other values can be calculated from these. They are now largely redundant because calculators can compute the values very quickly by means of the trigonometric series expansions—transforming degrees into radians where necessary first.

trigonometry The area of mathematics relating to the study of trigonometric functions in relation to the measurements in triangles, and the behaviour of wave functions.

trillion A million million (10^{12}). In Britain it used to mean a million cubed (10^{18}) but this is no longer common usage.

trim To discard extreme observations in a sample. The remaining values are regarded to be more typical of the behaviours, though it is important to realize that this is not always appropriate. The trimmed mean will be the arithmetic mean of the trimmed set.

trimmed mean *See* TRIM.

trinomial Consisting of three terms, usually in the form $(a + b + c)$ or $ax^2 + bx + c$.

triple A triple consists of 3 objects normally taken in a particular order, denoted, for example, by (x_1, x_2, x_3).

triple product (of vectors) *See* SCALAR TRIPLE PRODUCT and VECTOR TRIPLE PRODUCT.

triple root *See* ROOT.

trisect To divide into three equal parts.

trisection of an angle One of the problems that the Greek geometers attempted (like the *duplication of the cube and the *squaring of the circle) was to find a construction, with ruler and compasses, to trisect any angle; that is, to divide it into three equal parts. (The construction for bisecting an angle is probably familiar.) Now constructions of the kind envisaged can give only lengths belonging to a class of numbers obtained, essentially, by addition, subtraction, multiplication, division and the taking of square roots. It can be shown that the trisection of certain angles is equivalent to the construction of numbers that do not belong to this class. So, in general, the trisection of an angle is impossible.

trisector The trisectors of an angle are the two lines that divide the angle into three equal angles.

trivial solution The solution of a *homogeneous set of linear equations in which all the unknowns are equal to zero.

trivial subgroup The subset of a group which contains only the identity element. It will always be a group itself.

truncated cube One of the *Archimedean solids, with 6 octagonal faces and 8 triangular faces. It can be formed by cutting off the corners of a cube in such a way that the original square faces become regular octagons.

truncated tetrahedron One of the *Archimedean solids, with 4 hexagonal faces and 4 triangular faces. It can be formed by cutting off the corners of a (regular) *tetrahedron in such a way that the original triangular faces become regular hexagons.

truncation Suppose that a number has more digits than can be conveniently handled or stored. In truncation (as opposed to *rounding), the extra digits are simply dropped; for example, when truncated to 1 decimal place, the numbers 1.875 and 1.845 both become 1.8. *See also* DECIMAL PLACES and SIGNIFICANT FIGURES.

truncation error The error resulting in using an approximation obtained by truncation.

truth table The *truth value of a *compound statement can be determined from the truth values of its components. A table that gives, for

all possible truth values of the components, the resulting truth values of
the compound statement is a truth table. The truth table for ¬*p* is

p	¬*p*
T	*F*
F	*T*

and combined truth tables for *p* ∧ *q*, *p* ∨ *q* and *p* ⇒ *q* are as follows:

p	*q*	*p*∧*q*	*p*∨*q*	*p*⇒*q*
T	*T*	*T*	*T*	*T*
T	*F*	*F*	*T*	*F*
F	*T*	*F*	*T*	*T*
F	*F*	*F*	*F*	*T*

From these, any other truth table can be completed. For example, the
final column below, giving the truth table for the compound statement
(*p* ∧ *q*) ∨ (¬*r*), is found by first completing columns for *p* ∧ *q* and ¬*r*:

p	*q*	*r*	*p*∧*q*	¬*r*	(*p*∧*q*)∨(¬*r*)
T	*T*	*T*	*T*	*F*	*T*
T	*T*	*F*	*T*	*T*	*T*
T	*F*	*T*	*F*	*F*	*F*
T	*F*	*F*	*F*	*T*	*T*
F	*T*	*T*	*F*	*F*	*F*
F	*T*	*F*	*F*	*T*	*T*
F	*F*	*T*	*F*	*F*	*F*
F	*F*	*F*	*F*	*T*	*T*

truth value A term whose meaning is apparent from the following
usage: if a statement is true, its truth value is *T* (or TRUE); if the statement
is false, its truth value is *F* (or FALSE).

t-test A test to determine whether or not a sample of size *n* with mean $\bar{x}$
comes from a *normal distribution with mean μ. The *statistic *t* given by

$$t = \frac{\sqrt{n}(\bar{x} - \mu)}{s}$$

has a *t*-distribution with *n*−1 degrees of freedom, where s^2 is the
(unbiased) sample *variance. The *t*-test can also be used to test whether a
sample mean differs significantly from a population mean, and to test
whether two samples have been drawn from the same population.

See T-DISTRIBUTION for a table for use in the *t*-test.

Tukey, John Wilder (1915–2000) American mathematician and statistician who published work in topology and on the fast Fourier transform but it is for his work in statistics that he will be best remembered. He worked initially with time series, and made contributions to the *analysis of variance and resolving difficulties about making inferences about a set of parameter values from a single sample. He is perhaps best known for his 1977 book *Exploratory Data Analysis* which changed the basis on which data is analysed, and accelerated the move away from purely parametric statistics.

Turing, Alan Mathison (1912–54) British mathematician and logician who conceived the notion of the *Turing machine. During the Second World War, he was involved in cryptanalysis, the breaking of codes, and afterwards worked on the construction of some of the early digital computers and the development of their programming systems.

Turing machine A theoretical machine which operates according to extremely simple rules, invented by Turing with the aim of obtaining a mathematically precise definition of what is 'computable'. It has been generally agreed that the machine can calculate or compute anything for which there is an 'effective' *algorithm. The resulting understanding of computability has been shown to be equivalent to other attempts at defining the concept.

turning point A point on the graph $y = f(x)$ at which $f'(x) = 0$ and $f'(x)$ changes sign. A turning point is either a *local maximum or a *local minimum. Some authors use 'turning point' as equivalent to *stationary point.

twin primes A pair of prime numbers that differ by 2. For example, 29 and 31, 71 and 73, and 10 006 427 and 10 006 429, are twin primes. A conjecture that there are infinitely many such pairs has been neither proved nor disproved.

((()) SEE WEB LINKS

• More information on twin primes, including a list of the twenty largest pairs identified.

two-person zero-sum game A game with two players in which the total *payoff is zero, i.e. anything which one player gains is directly at the expense of the other player.

two-sample tests (in statistics) Any test which is to be applied to two independent samples, in contrast to paired-sample tests when the two samples are combined before applying a one-sample test to the resulting sample.

two-sided test *See* HYPOTHESIS TESTING.

two-tailed test *See* HYPOTHESIS TESTING.

Tychonoff space A *topological space that is both *completely regular and *Hausdorff.

Type I error Occurs when the null hypothesis is rejected by a statistical test when the null hypothesis was true. Pr{Type I error} = significance level of the test.

Type II error Occurs when the null hypothesis is not rejected when it should have been, i.e. the null hypothesis was not true. Pr{Type II error $|\theta\} = 1 -$ power of the test when θ is a value of the parameter occurring in the alternative hypothesis and it is a function of θ. Ideally a test would have power $= 1$ for all θ in the alternative hypothesis and this provides a basis for comparing alternative tests. For example, a paired-sample test, if it can be constructed in a given context, will usually be more powerful than a two-sample test because a major source of variability has been removed.

unary operation A unary operation on a set S is a rule that associates with any element of S a resulting element. If this resulting element is always also in S, then it is said that S is closed under the operation. The following are examples of unary operations: the rule that associates with each integer a its negative $-a$; the rule that associates with each non-zero real number a its inverse $1/a$; and the rule that associates with any subset A of a universal set E its complement A'.

unbiased estimator *See* ESTIMATOR.

unbounded function f(x) Not possessing both an upper and a lower bound. So for all positive real values V there is a value of the independent variable x for which $|f(x)| > V$. For example, $f(x) = x^2$ is unbounded because $f(x) \geq 0$ but $f(x) \to \infty$ as $x \to \pm\infty$, i.e. it is bounded below but not above, while $f(x) = x^3$ has neither upper nor lower bound.

unconditional inequality Always true, irrespective of the values taken by the variables so $x^2 + 2x + 3 > 0$ is unconditional since it can be expressed as $(x+1)^2 + 2 > 0$ which is always > 0, but $x^2 + 2x - 3 > 0$ is a conditional inequality because it can be expressed as $(x+1)^2 > 4$ which is only true for $x > 1$ or $x < -3$.

uncountable Not *countable, so the elements cannot be put into one-to-one correspondence with the natural numbers, or a subset of them.

under-determined A set of equations for which there are more variables than there are equations.

uniform A quantity in mechanics may be called uniform when it is, in some sense, constant. For example, a particle may be moving with uniform velocity or with uniform acceleration. A uniform rod, lamina or rigid body is one whose density is uniform; that is, whose density is the same at every point.

uniform convergence Describing a sequence of functions on a given set D which *converge at the same rate for all points in D. If, for every $\varepsilon > 0$ and for every x in D, there is an integer N for which

$|f_m(x) - f_n(x)| < \varepsilon$ for all m, $n > N$, the sequence is said to have uniform convergence.

uniform distribution The uniform distribution on the interval $[a, b]$ is the continuous probability *distribution whose *probability density function f is given by $f(x) = 1/(b - a)$, where $a \leq x \leq b$. It has mean $(a + b)/2$ and variance $(b - a)^2/12$. There is also a discrete form: on the set 1, 2, ..., n, it is the probability distribution whose *probability mass function is given by $\Pr(X = r) = 1/2$, for $r = 1, 2, ..., n$. For example, the random variable for the winning number in a lottery has a uniform distribution on the set of all the numbers entered in the lottery.

uniform gravitational force A *gravitational force, acting on a particular body, that is independent of the position of the body. The gravitational force on a body in a limited region near the surface of a planet is approximately uniform.

For example, the gravitational force acting on a particle of mass m near the Earth's surface, assumed to be a horizontal plane, can be taken to be $-mg\mathbf{k}$, where $\mathbf{k}$ is a unit vector directed vertically upwards and g is the magnitude of the acceleration due to gravity.

uniformly continuous function If the domain and range of a function are both metric spaces in which, for every $\varepsilon > 0$ in the domain there exists a single $\delta > 0$ in the range so that for all x in the domain $|f(x + \delta) - f(x)| < \varepsilon$. For continuity only, the value of δ can depend on x as well as ε.

uniform space A generalization of the notion of a *metric space where the metric provides a distance between two points and therefore a numerical measure of how close the two points are to one another. In a uniform space some notions of closeness of points are kept, without having the numerical measure provided by a metric, but providing more than just what the topological structure would provide.

union The union of sets A and B (subsets of a *universal set) is the set consisting of all objects that belong to A or B (or both), and it is denoted by $A \cup B$ (read as 'A union B'). Thus the term 'union' is used for both the resulting set and the operation, a *binary operation on the set of all subsets of a universal set. The following properties hold:

(i) For all A, $A \cup A = A$ and $A \cup \varnothing = A$.
(ii) For all A and B, $A \cup B = B \cup A$; that is, $\cup$ is commutative.
(iii) For all A, B and C, $(A \cup B) \cup C = A \cup (B \cup C)$; that is, $\cup$ is associative.

In view of (iii), the union $A_1 \cup A_2 \cup \cdots \cup A_n$ of more than two sets can be written without brackets, and it may also be denoted by

$$\bigcup_{i=1}^{n} A_i.$$

For the union of two events, *see* EVENT.

unique factorization theorem The process of writing any positive integer as the product of its prime factors is probably familiar; it may be taken as self-evident that this can be done in only one way. Known as the unique factorization theorem, this result of elementary number theory can be proved from basic axioms about the integers.

Theorem:

Any positive integer ($\neq 1$) can be expressed as a product of primes. This expression is unique except for the order in which the primes occur.

Thus, any positive integer $n(\neq 1)$ can be written as $p_1^{\alpha_1} p_2^{\alpha_2} \ldots p_r^{\alpha_r}$, where $p_1, p_2, \ldots, p_r$ are primes, satisfying $p_1 < p_2 < \cdots < p_r$, and $\alpha_1, \alpha_2, \ldots, \alpha_r$ are positive integers. This is the prime decomposition of n. For example, writing $360 = 2^3 \times 3^2 \times 5$ shows the prime decomposition of 360.

unit *See* SI UNITS.

unitary ratio *See* RATIO.

unit circle In the plane, the unit circle is the circle of radius 1 with its centre at the origin. In Cartesian coordinates, it has equation $x^2 + y^2 = 1$. In the *complex plane, it represents those complex numbers z such that $|z| = 1$.

unit cube A cube of side 1 unit. The unit cube defined by the points $(0, 0, 0)$, $(1, 0, 0)$, $(0, 1, 0)$, $(0, 0, 1)$ can be used in 3-dimensional matrix transformations in an analogous manner to the *unit square.

unit matrix = IDENTITY MATRIX.

unit square A square of side 1 unit. Specifically the unit square defined by $O(0, 0)$, $I(1, 0)$, $K(1, 1)$, $J(0, 1)$ is of interest in matrix transformations in the plane. It can be used to identify the transformation from a matrix, or find the matrix if the transformation is known because the images of points I and J form the columns of the matrix which performs that transformation. For example, in a rotation of $90°$ clockwise (about the origin), the image of I is $(0, -1)$ and the image of J is $(1, 0)$ so the matrix is $\begin{pmatrix} 0 & 1 \\ -1 & 0 \end{pmatrix}$. If a matrix of a transformation is $\begin{pmatrix} 1 & 3 \\ 0 & 1 \end{pmatrix}$ then the image of I is just the first column, i.e. $(1, 0)$, so it has not moved, and the image of J

is (3, 1), so the transformation is a shear parallel to the x-axis which moves (0, 1) to (3, 1).

unit vector A vector with magnitude, or length, equal to 1. For any non-zero vector **a**, a unit vector in the direction of **a** is **a**/|**a**|. The row vector $[a_1 \ a_2 \ \ldots \ a_n]$ or the column vector

$$\begin{bmatrix} a_1 \\ a_2 \\ \vdots \\ a_n \end{bmatrix}$$

may be called a unit vector if $\sqrt{a_1^2 + a_2^2 + \cdots + a_n^2} = 1$.

unity The number or digit 1.

universal gravitational constant = GRAVITATIONAL CONSTANT.

Universal Product Code (UPC) A standardized system of product bar coding introduced in 1973.

(🌐) SEE WEB LINKS
• More details of how UPC works, including a video case study of their use in tracking parcels.

universal quantifier *See* QUANTIFIER.

universal set In a particular piece of work, it may be convenient to fix the universal set E, a set to which all the objects to be discussed belong. Then all the sets considered are subsets of E.

unknown A variable or function whose value is to be found.

unknown constant A constant whose value is not currently known. For example, in the general equation of a straight line $y = mx + c$, x and y are variables and m and c are unknown constants which are determined by the gradient and intercept of the particular line. In the indefinite integral $\int x \, dx = \dfrac{x^2}{2} + c$ the constant of integration c is an unknown constant which requires a *boundary condition for it to be found.

unstable equilibrium *See* EQUILIBRIUM.

upper bound *See* BOUND.

upper limit *See* LIMIT OF INTEGRATION.

upper triangular matrix *See* TRIANGULAR MATRIX.

URL Abbreviation for 'Uniform Resource Locator', though 'Identifier' would be a more accurate descriptor of its function. This is a character string which gives the address at which a resource can be found.

utility A measure of the total perceived value resulting from an outcome or course of action. This may be negative, for example if an oil exploration company undertakes a survey showing that no oil can be extracted, the costs of undertaking the survey will be reflected in a negative utility for that outcome.

utility function A function which defines the *utility for the range of possible outcomes. If a probability distribution is known or can be estimated for those outcomes, then the *expected utility can be calculated.

V The Roman *numeral for 5.

valency = DEGREE (of a vertex of a graph).

validation (of a simulation model) The process of ensuring a simulation is an adequate representation of the reality it is attempting to model.

Vallée-Poussin, Charles-Jean De La (1866–1962) Belgian mathematician who in 1896 proved the *prime number theorem independently of *Hadamard.

value *See* CONSTANT FUNCTION, FUNCTION and INFINITE PRODUCT.

value (of a matrix game) *See* FUNDAMENTAL THEOREM OF GAME THEORY.

Vandermonde's convolution formula The following relationship between *binomial coefficients:

$$\binom{m+n}{r} = \binom{m}{0}\binom{n}{r} + \binom{m}{1}\binom{n}{r-1} + \cdots + \binom{m}{r}\binom{n}{0}.$$

The formula may be proved by equating the coefficients of x^r on both sides of the identity $(1+x)^{m+n} = (1+x)^m(1+x)^n$.

vanish To become zero or to tend to zero.

variable An expression, usually denoted by a letter, that is defined for values within a given set. Can be used to represent elements of sets which are not numbers but frequently it relates to numerical quantities and functions defined in them together with the relationship between them.

variance A measure of the *dispersion of a random variable or of a sample. For a random variable X, the population variance is the second moment about the population mean μ and is equal to $E((X-\mu)^2)$ (*see* EXPECTED VALUE). It is usually denoted by σ^2 or $\mathrm{Var}(X)$. For a sample, the sample variance, denoted by s^2, is the second moment of the data about the sample mean $\bar{x}$ but the denominator is usually taken as $n-1$ rather than n in order to make it an *unbiased estimator of the population variance. So

$$s^2 = \frac{\sum (x_i - \bar{x})^2}{n-1}.$$

For computational purposes, notice that $\sum(x_i - \bar{x})^2 = \sum x_i^2 - n\bar{x}^2$.

variance, analysis of *See* ANALYSIS OF VARIANCE.

variate = RANDOM VARIABLE.

varies directly, varies inversely *See* PROPORTION.

vector In physics or engineering, the term 'vector' is used to describe a physical quantity like velocity or force that has a magnitude and a direction. Sometimes there may also be a specified point of application, but generally in mathematics that is not of concern. Thus a vector is defined to be 'something' that has magnitude and direction.

One approach is to define a vector to be an *ordered pair consisting of a positive real number, the magnitude or length, and a direction in space. The vectors **a** and **b** are said to be equal if they have the same magnitude and the same direction. The zero vector **0** that has magnitude 0 and no direction is also allowed.

Another way is to make use of the well-defined notion of *directed line segment and define a vector to be the collection of all directed line segments with a given length and a given direction. If $\overrightarrow{AB}$ is a directed line segment in the collection that is the vector **a**, it is said that $\overrightarrow{AB}$ Represents **a**. If $\overrightarrow{AB}$ and $\overrightarrow{CD}$ represent the same vector, then $\overrightarrow{AB}$ and $\overrightarrow{CD}$ are parallel and have the same length. The magnitude, or length, |**a**| of the vector **a** is the length of any of the directed line segments that represent **a**. According to this definition, two vectors are equal if they are the same collection of directed line segments. It is also necessary to permit directed line segments of length zero and define the zero vector **0** to be the collection of all these.

In the first approach, the connection with directed line segments is made by saying that a directed line segment $\overrightarrow{AB}$ represents the vector **a** if the length and direction of $\overrightarrow{AB}$ are equal to the magnitude and direction of **a**. Some authors write $\overrightarrow{AB} = \mathbf{a}$ if $\overrightarrow{AB}$ represents the vector **a**, and then, if $\overrightarrow{AB} = \mathbf{a}$ and $\overrightarrow{CD} = \mathbf{a}$, they go on to write $\overrightarrow{AB} = \overrightarrow{CD}$. Some authors actually use the word 'vector' for what we prefer to call a directed line segment. They then write $\overrightarrow{AB} = \overrightarrow{CD}$ if $\overrightarrow{AB}$ and $\overrightarrow{CD}$ have the same length and the same direction.

vector analysis The area of mathematics concerned with the application of calculus to functions of vectors. For example, the extension of the rectilinear differential equations of motion such as $v = \dfrac{dx}{dt}$ to the vector form $\mathbf{v} = \dfrac{d\mathbf{r}}{dt}$ where $\mathbf{r}(t)$ is the position vector.

vector equation (of a line) Given a line in space, let **a** be the *position vector of a point A on the line, and **u** any vector with direction along the line. Then the line consists of all points P whose position vector **p** is given by $\mathbf{p} = \mathbf{a} + t\mathbf{u}$ for some value of t. This is a vector equation of the line. It is established by noting that P lies on the line if and only if $\mathbf{p} - \mathbf{a}$ has the direction of **u** or its negative, which is equivalent to $\mathbf{p} - \mathbf{a}$ being a scalar multiple of **u**. If, instead, the line is specified by two points A and B on it, with position vectors **a** and **b**, then the line has vector equation $\mathbf{p} = (1 - t)\mathbf{a} + t\mathbf{b}$. This is obtained by setting $\mathbf{u} = \mathbf{b} - \mathbf{a}$ in the previous form.

vector equation (of a plane) Given a plane in 3-dimensional space, let **a** be the *position vector of a point A in the plane, and **n** a *normal vector to the plane. Then the plane consists of all points P whose position vector **p** satisfies $(\mathbf{p} - \mathbf{a}) \cdot \mathbf{n} = 0$. This is a vector equation of the plane. It may also be written $\mathbf{p} \cdot \mathbf{n} = $ constant. By supposing that **p** has components x, y, z, that **a** has components x_1, y_1, z_1, and that **n** has components l, m, n, the first form of the equation becomes $l(x - x_1) + m(y - y_1) + n(z - z_1) = 0$, and the second form becomes the standard linear equation $lx + my + nz = $ constant.

vector norm For an n-dimensional *vector **x** a vector norm is a non-negative function which satisfies:

$|\mathbf{x}| > 0$ when $\mathbf{x} \neq 0$ and $|\mathbf{x}| = 0$ when $\mathbf{x} = 0$,

$|k\mathbf{x}| = k \times |\mathbf{x}|$ for any scalar k,

$|\mathbf{x} + \mathbf{y}| \leq |\mathbf{x}| + |\mathbf{y}|$.

vector product Let **a** and **b** be non-zero non-parallel vectors, and let θ be the angle between them (θ in radians, with $0 < \theta < \pi$). The vector product $\mathbf{a} \times \mathbf{b}$ of **a** and **b** is defined as follows. Its magnitude equals $|\mathbf{a}||\mathbf{b}| \sin \theta$, and its direction is perpendicular to **a** and **b** such that **a**, **b** and $\mathbf{a} \times \mathbf{b}$ form a right-handed system. So, viewed from a position facing the direction of $\mathbf{a} \times \mathbf{b}$, the vector **a** has to be rotated clockwise through the angle θ to have the direction of **b**. If **a** is parallel to **b** or if one of them is the zero vector **0**, then $\mathbf{a} \times \mathbf{b}$ is defined to be **0**. The notation $\mathbf{a} \wedge \mathbf{b}$ is also used for $\mathbf{a} \times \mathbf{b}$. The following properties hold, for all vectors **a**, **b** and **c**:

   (i) $\mathbf{b} \times \mathbf{a} = -(\mathbf{a} \times \mathbf{b})$.

  (ii) $\mathbf{a} \times (k\mathbf{b}) = (k\mathbf{a}) \times \mathbf{b} = k(\mathbf{a} \times \mathbf{b})$.

 (iii) The magnitude $|\mathbf{a} \times \mathbf{b}|$ is equal to the area of the parallelogram with sides determined by **a** and **b**.

 (iv) $\mathbf{a} \times (\mathbf{b} + \mathbf{c}) = \mathbf{a} \times \mathbf{b} + \mathbf{a} \times \mathbf{c}$, the distributive law.

(v) If the vectors, in terms of components (with respect to standard vectors **i**, **j**, **k**), are $\mathbf{a} = a_1\mathbf{i} + a_2\mathbf{j} + a_3\mathbf{k}$, $\mathbf{b} = b_1\mathbf{i} + b_2\mathbf{j} + b_3\mathbf{k}$, then $\mathbf{a} \times \mathbf{b} = (a_2 b_3 - a_3 b_2)\mathbf{i} + (a_3 b_1 - a_1 b_3)\mathbf{j} + (a_1 b_2 - a_2 b_1)\mathbf{k}$. This can be written, with an abuse of 3×3 determinant notation, as

$$\mathbf{a} \times \mathbf{b} = \begin{vmatrix} \mathbf{i} & \mathbf{j} & \mathbf{k} \\ a_1 & a_2 & a_3 \\ b_1 & b_2 & b_3 \end{vmatrix}.$$

vector projection (of a vector on a vector) Given non-zero vectors **a** and **b**, let $\overrightarrow{OA}$ and $\overrightarrow{OB}$ be *directed line segments representing **a** and **b**, and let θ be the angle between them (θ in radians, with $0 \leq \theta \leq \pi$). Let C be the *projection of B on the line OA. The vector projection of **b** on **a** is the vector represented by $\overrightarrow{OC}$ Since $|OC| = |OB|\cos\theta$, this vector projection is equal to $|\mathbf{b}|\cos\theta$ times the unit vector $\mathbf{a}/|\mathbf{a}|$. Thus the vector projection of **b** on **a** equals

$$\left(\frac{\mathbf{a} \cdot \mathbf{b}}{\mathbf{a} \cdot \mathbf{a}} \right) \mathbf{a}.$$

The scalar projection of **b** on **a** is equal to $(\mathbf{a} \cdot \mathbf{b})/|\mathbf{a}|$, which equals $|\mathbf{b}|\cos\theta$. It is positive when the vector projection of **b** on **a** is in the same direction as **a**, and negative when the vector projection is in the opposite direction to **a**; its absolute value gives the length of the vector projection of **b** on **a**.

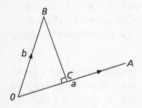

vector space (linear space) The mathematical structure defined by two sets and associated operations with the following properties.

(i) The set A, of vectors, must form an Abelian group, i.e. it requires an operation of addition to be defined on the set, to be closed under addition, to be commutative and to have an identity and inverse elements.

(ii) The other set B, of scalars, must form a field, i.e. be an *abelian group with respect to addition, an abelian group with respect to multiplication when the zero element is removed, and for multiplication to be distributed over addition.

(iii) A multiplicative operation has to be defined on the two sets with the following properties:

(iv) If p, q are scalars in set B and x, y are vectors in set A then

$$p(\mathbf{x} + \mathbf{y}) = p\mathbf{x} + p\mathbf{y}.$$
$$(p + q)\mathbf{x} = p\mathbf{x} + q\mathbf{x}$$
$$p(q\mathbf{x}) = (pq)\mathbf{x}$$

(v) if 1 is the multiplicative identity in the set B, then $1.\mathbf{x} = \mathbf{x}$.

vector subspace A subset of a *vector space which is itself a vector space.

vector sum The binary operation that returns a vector of which the length and direction are those of the diagonal of the parallelogram found by using the two vectors for the sides, i.e. to add vectors $\mathbf{p}$ and $\mathbf{q}$ shown below, if $\overrightarrow{OA} = \mathbf{p} = \overrightarrow{CB}$ and $\overrightarrow{OC} = \mathbf{q} = \overrightarrow{AB}$ then the vector $\mathbf{p} + \mathbf{q} = \overrightarrow{OB}$

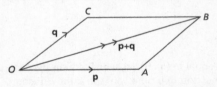

vector triple product For vectors $\mathbf{a}$, $\mathbf{b}$ and $\mathbf{c}$, the vector $\mathbf{a} \times (\mathbf{b} \times \mathbf{c})$, being the *vector product of $\mathbf{a}$ with the vector $\mathbf{b} \times \mathbf{c}$, is called a vector triple product. The use of brackets here is essential, since $(\mathbf{a} \times \mathbf{b}) \times \mathbf{c}$, another vector triple product, gives, in general, quite a different result. The vector $\mathbf{a} \times (\mathbf{b} \times \mathbf{c})$ is perpendicular to $\mathbf{b} \times \mathbf{c}$ and so lies in the plane determined by $\mathbf{b}$ and $\mathbf{c}$. In fact, $\mathbf{a} \times (\mathbf{b} \times \mathbf{c}) = (\mathbf{a} . \mathbf{c})\mathbf{b} - (\mathbf{a} . \mathbf{b})\mathbf{c}$.

velocity Suppose that a particle is moving in a straight line, with a point O on the line taken as origin and one direction taken as positive. Let x be the *displacement of the particle at time t. The velocity of the particle is equal to $\dot{x}$ or dx/dt, the *rate of change of x with respect to t. The velocity is positive when the particle is moving in the positive direction and negative when it is moving in the negative direction.

In the preceding paragraph, a common convention has been followed in which the unit vector $\mathbf{i}$ in the positive direction along the line has been suppressed. Velocity is in fact a vector quantity, and in the 1-dimensional case above is equal to $\dot{x}\mathbf{i}$.

When the motion is in two or three dimensions, vectors are used explicitly. The velocity $\mathbf{v}$ of a particle P with position vector $\mathbf{r}$ is given by $\mathbf{v} = d\mathbf{r}/dt = \dot{\mathbf{r}}$. When Cartesian coordinates are used, $\mathbf{r} = x\mathbf{i} + y\mathbf{j} + z\mathbf{k}$, and then $\dot{\mathbf{r}} = \dot{x}\mathbf{i} + \dot{y}\mathbf{j} + \dot{z}\mathbf{k}$.

Velocity has the dimensions LT^{-1}, and the SI unit of measurement is the metre per second, abbreviated to 'm s^{-1}'.

velocity ratio *See* MACHINE.

velocity–time graph A graph that shows velocity plotted against time for a particle moving in a straight line. Let $x(t)$ and $v(t)$ be the displacement and velocity, respectively, of the particle at time t. The velocity–time graph is the graph $y = v(t)$, where the t-axis is horizontal and the y-axis is vertical with the positive direction upwards. When the graph is above the horizontal axis, the particle is moving in the positive direction, when it is below the horizontal axis, the particle is moving in the negative direction.

The gradient of the velocity–time graph at any point is equal to the acceleration of the particle at that time. Also,

$$\int_{t_1}^{t_2} v(t) \ dt = x(t_2) - x(t_1).$$

So, with the convention that any area below the horizontal axis is negative, the area under the graph gives the change in position during the time interval concerned. Note that this is not necessarily the distance travelled.

(Here a common convention has been followed in which the unit vector **i** in the positive direction along the line has been suppressed. The displacement, velocity and acceleration of the particle are in fact vector quantities equal to $x(t)\mathbf{i}$, $v(t)\mathbf{i}$ and $a(t)\mathbf{i}$, where $v(t) = \dot{x}(t)$ and $a(t) = \dot{v}(t)$.)

Venn, John (1834–1923) British logician who, in his work *Symbolic Logic* of 1881, introduced what are now called *Venn diagrams.

Venn diagram A method of displaying relations between subsets of some *universal set. The universal set E is represented by the interior of a rectangle, say, and subsets of E are represented by regions inside this, bounded by simple closed curves. For instance, two sets A and B can be represented by the interiors of overlapping circles and then the sets $A \cup B$, $A \cap B$ and $A\backslash B = A \cap B'$, for example, are represented by the shaded regions shown in the figures.

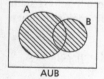

AUB

A∩B

A\B

Given one set, A, the universal set is divided into two disjoint subsets A and A', which can be clearly seen in a simple Venn diagram. Given two sets A and B, the universal set E is divided into four disjoint subsets $A \cap B$, $A' \cap B$, and $A \cap B'$ and $A' \cap B'$. A Venn diagram drawn with two overlapping circles for A and B clearly shows the four corresponding regions. Given three sets A, B and C, the universal set E is divided into eight disjoint subsets $A \cap B \cap C$, $A' \cap B \cap C$, $A \cap B' \cap C$, $A \cap B \cap C'$, $A \cap B' \cap C'$, $A' \cap B \cap C'$, $A' \cap B' \cap C$ and $A' \cap B' \cap C'$, and these can be illustrated in a Venn diagram as shown here.

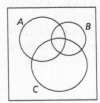

Venn diagrams can be used with care to prove properties such as *De Morgan's laws, but some authors prefer other proofs because a diagram may only illustrate a special case. Four general sets, for example, should not be represented by four overlapping circles because they cannot be drawn in such a way as to make apparent the 16 disjoint subsets into which E should be divided.

verification (of a simulation model) The process of checking that the program output is consistent with the outcomes which should result from the random numbers generated in the simulation.

vertex *See* ELLIPSE, HYPERBOLA and PARABOLA; and also CONE.

vertex (of a graph), **vertex-set** *See* GRAPH.

vertex method For a *linear programming problem where the decision variables are not required to take integer values, the optimal solutions will occur at one or more of the extreme points, i.e. vertices of the feasible region. The method of solution is to calculate the value of the objective function at each vertex, and the maximum or minimum (as required) of these values identifies any optimal solutions. Where more than one such vertex is found, all points on the boundary of the feasible region between them will also be an optimal solution.

vertical angles (vertical opposite angles) A pair of non-adjacent angles formed by the intersection of two straight lines.

Viète, François (1540–1603) French mathematician who moved a step towards modern algebraic notation by using vowels to represent unknowns and consonants to represent known numbers. This practice, together with other improvements in notation, facilitated the handling of equations and enabled him to make considerable advances in algebra. He also developed the subject of trigonometry.

virtual work (in mechanics) The total work done by a system in an infinitesimal displacement, which is in a direction where motion is physically possible.

virtual work principle (in mechanics) A static system is in equilibrium if and only if the *virtual work at that position is zero.

volume A measure of the 3-dimensional space enclosed by a solid.

volume of a solid of revolution Let $y = f(x)$ be the graph of a function f, a *continuous function on $[a, b]$, and such that $f(x) \geq 0$ for all x in $[a, b]$. The volume V of the solid of revolution obtained by rotating, through one revolution about the x-axis, the region bounded by the curve $y = f(x)$, the x-axis and the lines $x = a$ and $x = b$, is given by

$$V = \int_b^a \pi y^2 \, dx = \int_b^a \pi (f(x))^2 \, dx.$$

Parametric form

For the curve $x = x(t)$, $y = y(t)$ ($t \in [\alpha, \beta]$), the volume V is given by

$$V = \int_\alpha^\beta \pi y^2 \frac{dx}{dt} dt = \int_\alpha^\beta \pi (y(t))^2 x'(t) \, dt.$$

Von Neumann, John (1903–57) Mathematician who made important contributions to a wide range of areas of pure and applied mathematics. He was born in Budapest but lived in the United States from 1930. In applied mathematics, he was one of the founders of optimization theory and the theory of games, with applications to economics. Within pure mathematics, his work in functional analysis is important. His lifelong interest in mechanical devices led to his being involved crucially in the initial development of the modern electronic computer and the important concept of the stored program.

vulgar fraction A *fraction written as a/b where a and b are positive integers, as opposed to a *decimal fraction. For example, 0.75 is a decimal fraction; $\frac{3}{4}$ is a vulgar fraction.

W Symbol of *watt.

walk (in graph theory) The general name given to a sequence of edges and vertices in a *graph. Walks with certain properties are of particular interest and are given specific names. So a walk in which no edge is repeated is a *trail, and if no vertex is revisited in the course of a trail it is a *path. The special case where the path finishes at the vertex where it started is an exception to this, and is termed a *cycle or *loop.

Wallis, John (1616–1703) The leading English mathematician before *Newton. His most important contribution was in the results he obtained by using infinitesimals, developing Cavalieri's method of indivisibles. He also wrote on mechanics. Newton built on Wallis's work in his development of calculus and his laws of motion. It was Wallis who first made use of negative and fractional indices. He was instrumental in the founding of the Royal Society in 1662.

Wallis's Product *See* PI.

warning limits The inner limits set on a *control chart in a production process. If the observed value falls between the warning and *action limits, then it is taken as a signal that the process may be off target, and another sample is taken immediately. If it is also outside the warning limits, action will be taken, but if the second observation is within limits, the production is assumed to be on target. For the means of samples of size n in a process with standard deviation σ with target mean μ, the warning limits will be set at $\mu \pm 1.96 \dfrac{\sigma}{\sqrt{n}}$.

watt The SI unit of *power, abbreviated to 'W'. One watt is equal to one *joule per second.

weak law of large numbers The sequence $\{m_n\}$ generated by calculating the average of the first n observations of a random variable X with finite variance, i.e. $m_n = \dfrac{1}{n}\sum_{i=1}^{n} x_i$ tends to the mean $\mu = E(X)$

with probability 1 as $n \to \infty$, i.e. for any $\varepsilon > 0$ and any $\delta > 0$ there exists an N for which $\Pr\{|m_n - \mu| > \varepsilon\} < \delta$ for all $n > N$.

weakly hereditary property (of spaces) *See* HEREDITARY PROPERTY (OF SPACES).

Weierstrass, Karl (Theodor Wilhelm) (1815–97) German mathematician who was a leading figure in the field of mathematical analysis. His work was concerned with providing the subject with the necessary rigour as it developed out of 18th-century calculus. One particular area in which he made significant contributions was in the expansion of functions in *power series. To show that intuition is not always reliable, he gave an example of a function that is continuous at every point but not differentiable at any point. Some of his work was done while he was a provincial school teacher, having little contact with the world of professional mathematicians. He was promoted directly to professor of mathematics in Berlin at the age of 40.

weight The magnitude of the force acting on a body due to gravity. It is generally assumed that when a body of mass m is near the Earth's surface it is acted upon by a *uniform gravitational force equal to $-mg\mathbf{k}$, where $\mathbf{k}$ is a unit vector directed vertically upwards and g is the magnitude of the acceleration due to gravity. Thus the weight of the body equals mg.

Weight has the dimensions MLT^{-2}, the same as force, and the SI unit of measurement is the *newton. In common usage, when an object is weighed the measurement is recorded in kilograms. In fact, the weighing machine measures the weight, but the reading gives the corresponding mass.

weighted mean *See* MEAN.

well–conditioned problem A problem which is not *ill-conditioned.

Whitehead, Alfred North (1861–1947) British mathematician and philosopher who produced *Principia Mathematica* with Bertrand *Russell in 1913.

whole number An integer, though sometimes it is taken to mean only non-negative integers, or just the positive integers.

Wiener, Norbert (1899–1969) American mathematician and logician who founded cybernetics, and worked on guided missiles during the Second World War.

Wilcoxon paired sample test If paired observations from two distributions are available and it is known, or can be assumed, that the difference in value between the two distributions is symmetrically distributed about the median difference, then the *Wilcoxon signed rank test can be applied to the differences. In particular, looking for a shift in the population median is achieved by testing the *null hypothesis of zero difference.

Wilcoxon rank–sum test A non-parametric test (*see* NON-PARAMETRIC METHODS) for testing the null hypothesis that two independent samples of size n and m are from the same population. The observations are combined and ranked. If the null hypothesis is true, the sum U of the ranks of the observations in the sample of size n has mean $n(n+m+1)/2$ and variance $nm(n+m+1)/12$. For even relatively small values of n and m, the value of U can be tested against a *normal distribution. This is also known as the Mann–Whitney U test.

Wilcoxon signed rank test A non-parametric test of the null hypothesis that the median is a specified value. It requires the assumption or knowledge that the distribution being sampled is symmetric. It is based on ranking the observations by their distance above and below the median and comparing the total of the rankings above and below. It is a more powerful test than the *sign test which only counts the numbers of observations above and below the median and therefore discards quite a lot of information. The procedure is to rank the observations not equal to the hypothesized median by their distance from the median, and to sum the ranks above the median and below the median. There are tables of critical values for different values of n (the number of observations not equal to the hypothesized median) to conduct both one-tailed and two-tailed tests, but for even relatively small values of n, the distribution

$$N\left(\frac{1}{4}n(n+1),\ \frac{1}{24}n(n+1)(2n+1)\right)$$ is a good approximation.

Wiles, Andrew John FRS (1953–   ) British mathematician famous for the proof of one of mathematics most famous problems—*Fermat's Last Theorem. He announced he had the proof in 1993 and was widely tipped to receive a *Fields Medal as a result. However, that work was found to contain an error and it was not until 1995 that a complete proof was published, by which time Wiles was too old to be considered for a Fields Medal. However, the Fields Institute honoured his extraordinary achievement by a special award in 1998 of a silver plate. He also won the *Wolf Prize in 1996. Winner of the Abel Prize for 2016, 'for his stunning proof of Fermat's Last Theorem by way of the modularity conjecture for semistable elliptic curves, opening a new era in number theory.'

within-subjects design A designed experiment where the same group of subjects (often people) are measured under different experimental conditions.

Witten, Edward (1957–) American mathematician and theoretical physicist working on superstring theory. Winner of a *Fields Medal in 1990 for his work relating knot theory and quantum theory.

Wolf Prize Made by the Wolf Foundation in Israel which awards five or six yearly prizes across a number of scientific fields including mathematics. The prizes started in 1978.

word (in computer science) The basic unit of data which is used by a particular system—a fixed-size group of *bits that are handled as a block by the hardware and/or the software. The number of bits in a word is known as the word size or width or length and is an important feature of any particular computer architecture.

work done The work done by a force **F** during the time interval from $t = t_1$ to $t = t_2$ is equal to

$$\int_{t_1}^{t_2} \mathbf{F}.\mathbf{v}\, dt,$$

where **v** is the velocity of the point of application of **F**.

From the equation of motion $m\mathbf{a} = \mathbf{F}$ for a particle of mass m moving with acceleration **a**, it follows that $m\mathbf{a} . \mathbf{v} = \mathbf{F} . \mathbf{v}$, and this then gives $(d/dt)(\frac{1}{2}\, m\mathbf{v}.\mathbf{v}) = \mathbf{F} . \mathbf{v}$. By integration, it follows that the change in *kinetic energy is equal to the work done by the force.

Suppose that a particle, moving along the x-axis in the positive direction, with displacement $x(t)\mathbf{i}$ and velocity $v(t)\mathbf{i}$ at time t, is acted on by a constant force $F\mathbf{i}$ in the same direction. Then the work done by this constant force during the time interval from $t = t_1$ to $t = t_2$ is equal to

$$\int_{t_1}^{t_2} F\, v(t)\, dt,$$

which equals $F(x(t_2) - x(t_1))$. This is usually interpreted as 'work = force × distance'.

The work done against a force **F** should be interpreted as the work done by an applied force equal and opposite to **F**. When the force is *conservative, this is equal to the change in *potential energy. When a person lifts an object of mass m from ground level to a height z above the ground, work is done against the *uniform gravitational force and the work done equals the increase mgz in potential energy.

Work has the dimensions $ML^2\, T^{-2}$, the same as energy, and the SI unit of measurement is the *joule.

work–energy principle The principle that the change in *kinetic energy of a particle during some time interval is equal to the *work done by the total force acting on the particle during the time interval.

wrt Abbreviation for with respect to. In particular to identify the variable in an integration or differentiation process so $f'(x) = \dfrac{dy}{dx}$ is the differential of $f(x)$ wrt x, and $\int \pi(x^2 + k)\, dx$ is an integral wrt x.

X The Roman *numeral for 10.

***x*-axis** One of the axes in a Cartesian coordinate system.

Yates' correction A continuity correction applied to a *chi-squared contingency table test. Normally only applied for a 2×2 table, where the effect is greatest and by a happy coincidence the arithmetic required is the simplest. It involves diminishing the numerical size of the difference between observed and expected values by 0.5 before squaring, i.e. the chi-square statistic calculated is $\sum \dfrac{(|O - E| - 0.5)^2}{E}$.

y-axis One of the axes in a Cartesian coordinate system.

Young's modulus of elasticity The modulus of elasticity of elastic strings or springs made of the same material, but of different cross-sectional area will vary and Young's modulus of elasticity states the relationship as $E = \dfrac{\lambda}{A}$ giving Hooke's law as $T = \dfrac{EA}{l}x$ where $E =$ Young's modulus of elasticity, $\lambda =$ modulus of elasticity, $A =$ cross-sectional area, $l =$ natural length and $x =$ extension from natural length.

Z *See* INTEGER.

z-axis One of the axes in a Cartesian coordinate system.

Zeeman, Sir Erik Christopher FRS (1925–2016) British mathematician whose work in *topology and *catastrophe theory found wide applicability in areas as diverse as physics, the social sciences and economics.

Zeno of Elea (5th century BC) Greek philosopher whose paradoxes, known through the writings of Aristotle, may have significantly influenced the Greeks' perception of magnitude and number. The four paradoxes of motion are concerned with whether or not space and time are fundamentally made up of minute indivisible parts. If not, the first two paradoxes, the Dichotomy and the Paradox of Achilles and the Tortoise, appear to lead to absurdities. If so, the third and fourth, the Paradox of the Arrow and the Paradox of the Stadium, apparently give contradictions. *See* ACHILLES PARADOX and ARROW PARADOX.

Zermelo, Ernst (Friedrich Ferdinand) (1871–1953) German mathematician considered to be the founder of axiomatic set theory. In 1908, he formulated a set of axioms for set theory, which attempted to overcome problems such as that posed by *Russell's paradox. These, with modifications, have been the foundation on which much subsequent work in the subject has been built.

zero The real number 0, which is the additive identity, i.e. $x + 0 = 0 + x = x$ for any real number x. The element in any ring with the property that $x \cdot 0 = 0 \cdot x = 0$.

zero (of a function) *See* ROOT.

zero-dimensional (of a space) Describing a *topological space which has a base of sets that are *clopen. Every *discrete space is zero-dimensional, but the converse is not true: the set of rational numbers is zero-dimensional but does have *isolated points. Being zero-dimensional is a hereditary property of a space.

zero divisors Non-zero elements in a ring whose product is zero. There are no zero divisors within the real or complex numbers, but there are in other systems. For example, if $\mathbf{A} = \begin{pmatrix} 1 & 0 \\ 0 & 0 \end{pmatrix}$ and $\mathbf{B} = \begin{pmatrix} 0 & 0 \\ 0 & 1 \end{pmatrix}$ then $\mathbf{AB} = \mathbf{BA} = \begin{pmatrix} 0 & 0 \\ 0 & 0 \end{pmatrix}$ so $\mathbf{A}$ and $\mathbf{B}$ are zero divisors within the ring of 2×2 matrices, defined with the usual matrix addition and multiplication.

zero element An element z is a zero element for a *binary operation $\circ$ on a set S if, for all a in S, $a \circ z = z \circ a = z$. Thus the real number 0 is a zero element for multiplication since, for all a, $a0 = 0a = 0$. The term 'zero element', also denoted by 0, may be used for an element such that $a + 0 = 0 + a = a$ for all a in S, when S is a set with a binary operation $+$ called addition. Strictly speaking, this is a *neutral element for the operation $+$.

zero function In real analysis, the zero function is the *real function f such that $f(x) = 0$ for all x in $\mathbf{R}$.

zero matrix The $m \times n$ zero matrix $\mathbf{O}$ is the $m \times n$ matrix with all its entries zero. A zero column matrix or row matrix may be denoted by $\mathbf{0}$.

zero measure A set of points capable of being enclosed in intervals whose total length is arbitrarily small.

zero-sum game See MATRIX GAME.

zero vector See VECTOR.

zeta function The function $\zeta(s) = \sum_{n=1}^{\infty} \dfrac{1}{n^s}$ defined on the complex variables s. The sum is convergent, and gives an *analytic function, when the real part of s is >1, and in particular $\zeta(2) = \dfrac{\pi^2}{6}$; $\zeta(4) = \dfrac{\pi^4}{90}$; $\zeta(6) = \dfrac{\pi^6}{945}$. While the exact distribution of primes is unknown, *Riemann observed that it approximates to the zeta function.

$\mathbf{Z}_n$ See RESIDUE CLASS (modulo n).

zone A zone of a sphere is the part between two parallel planes. If the sphere has radius r and the distance between the planes is h, the area of the curved surface of the zone equals $2\pi rh$.

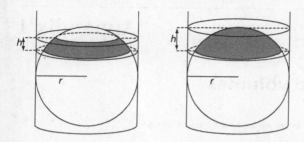

A circumscribing cylinder of a sphere is a right-circular cylinder with the same radius as the sphere and with its axis passing through the centre of the sphere. Suppose that a zone is formed by two parallel planes. Take the circumscribing cylinder with its axis perpendicular to the two planes as shown. It is an interesting fact that the area of the curved surface of the zone is equal to the area of the part of the circumscribing cylinder between the same two planes. The special case when one of the planes touches the sphere gives the area of a spherical cap as $2\pi rh$.

z

Areas and volumes

(For unexplained notation, see under the relevant reference.)

*Rectangle, length a, width b:
Area $= ab$.

*Parallelogram:
Area $= bh = ab \sin \theta$.

*Triangle:
Area $= \frac{1}{2}$base $\times$ height $= \frac{1}{2}bc \sin A$.

*Trapezium:
Area $= \frac{1}{2}h(a + b)$.

*Circle, radius r:
Area $= \pi r^2$.
Length of circumference $= 2\pi r$.

*Right-circular *cylinder, radius r, height h:
Volume $= \pi r^2 h$,
Curved surface area $= 2\pi rh$.

*Right-circular *cone:
Volume $= \frac{1}{3}\pi r^2 h$,
Curved surface area $= \pi rl$.

*Frustum of a *cone:
Volume $= \frac{1}{3}\pi h(a^2 + ab + b^2)$,
Curved surface area $= \pi(a + b)l$.

*Sphere, radius r:
Volume $= \frac{4}{3}\pi r^3$,
Surface area $= 4\pi r^2$.

Surface area of a *cylinder:
Curved surface area $= 2\pi rh$.

Appendix 2

Centres of mass

Triangular *lamina:
> 2/3 of the way along any median from its vertex.

Solid *hemisphere, radius r:
> 3/8r from the centre, along the radius which is perpendicular to the base.

Hemispherical shell, radius r:
> ½ r from the centre, along the radius which is perpendicular to the base.

*Cone or *pyramid:
> ¾ of the way from the vertex to the centroid of the base.

Conical shell:
> 2/3 of the way from the vertex to the centre of the base.

Circular *sector, radius r and angle 2θ:
> $\dfrac{2r \sin \theta}{3\theta}$ from the centre of the circle.

Circular *arc, radius r and angle 2θ:
> $\dfrac{r \sin \theta}{\theta}$ from the centre of the circle.

Moments of inertia

For uniform bodies of mass M.

Thin *rod, length 2l, about perpendicular axis through centre:
$\frac{1}{3}Ml^2$.

Rectangular *lamina, about axis in plane bisecting edges of length 2l:
$\frac{1}{3}Ml^2$.

Thin rod, length 2l, about perpendicular axis through end:
$\frac{4}{3}Ml^2$.

Rectangular lamina, about edge perpendicular to edges of length 2l:
$\frac{4}{3}Ml^2$.

Rectangular lamina, sides 2l and 2b, about perpendicular axis through centre:
$\frac{1}{3}M(l^2 + b^2)$.

Hoop or cylindrical shell, radius r, about axis:
Mr^2.

Hoop, radius r, about a diameter:
$\frac{1}{2}Mr^2$.

*Disc or solid *cylinder, radius r, about axis:
$\frac{1}{2}Mr^2$.

Disc, radius r, about a diameter:
$\frac{1}{4}Mr^2$.

Solid *sphere, radius r, about a diameter:
$\frac{2}{5}Mr^2$.

Spherical shell, radius r, about a diameter:
$\frac{2}{3}Mr^2$.

Elliptical lamina, axes 2a and 2b, about axis of length 2a:
$\frac{1}{4}Mb^2$.

Elliptical lamina, axes 2a and 2b, about perpendicular axis through centre:
$\frac{1}{4}M(a^2 + b^2)$.

Solid *ellipsoid, axes 2a, 2b, and 2c, about axis of length 2c:
$\frac{1}{5}M(a^2 + b^2)$.

Ellipsoidal shell, axes 2a, 2b, and 2c, about axis of length 2c:
$\frac{1}{3}M(a^2 + b^2)$.

*Right-circular *cone, base radius r, about axis of cone:
$\frac{3}{10}Mr^2$.

Appendix 4

Geometry: equations of lines and planes

Distance from a point to a line

The distance s from $ax + by + c = 0$ to $P(x_1, y_1)$ is $s = \dfrac{|ax_1 + by_1 + c|}{\sqrt{a^2 + b^2}}$.

Equation of a plane

If a plane contains the point (x_1, y_1, z_1) and the vector $\begin{pmatrix} a \\ b \\ c \end{pmatrix}$ is normal to the plane, then

the plane is given by $\dfrac{x - x_1}{a} = \dfrac{y - y_1}{b} = \dfrac{z - z_1}{c}$, or $ax + by + cz = d$ (where $d = ax_1 + by_1 + cz_1$).

Distance from a point to a plane

The distance s from $ax + by + cz = d$ to $P(x_1, y_1, z_1)$ is $s = \dfrac{|ax_1 + by_1 + cz_1 - d|}{\sqrt{a^2 + b^2 + c^2}}$.

Angle between two planes

The angle between the planes $a_1 x + b_1 y + c_1 z = d_1$ and $a_2 x + b_2 y + c_2 z = d_2$ is

$\arccos\left(\dfrac{a_1 a_2 + b_1 b_2 + c_1 c_2}{\sqrt{a_1^2 + b_1^2 + c_1^2}\sqrt{a_2^2 + b_2^2 + c_2^2}} \right)$; the planes are parallel if and only if

$\dfrac{a_1}{a_2} = \dfrac{b_1}{b_2} = \dfrac{c_1}{c_2}$, and the planes are perpendicular if and only if $a_1 a_2 + b_1 b_2 + c_1 c_2 = 0$.

The circle

The equation of a circle with centre (a, b) and radius r is $(x - a)^2 + (y - b)^2 = r^2$.
In parametric form, $x = a + r \cos t$, $y = b + r \sin t$.

*Conic sections

	*Ellipse	*Parabola	*Hyperbola	*Rectangular hyperbola
Cartesian form	$\dfrac{x^2}{a^2} + \dfrac{y^2}{b^2} = 1$	$y^2 = 4ax$	$\dfrac{x^2}{a^2} - \dfrac{y^2}{b^2} = 1$	$xy = c^2$
Parametric form	$(a\cos\theta,\ b\sin\theta)$	$(at^2,\ 2at)$	$(a\sec\theta,\ b\tan\theta)$ $(\pm a\cosh\theta,\ b\sinh\theta)$	$\left(ct, \dfrac{c}{t}\right)$
*Eccentricity	$e < 1$, $b^2 = a^2(1 - e^2)$	$e = 1$	$e > 1$, $b^2 = a^2(e^2 - 1)$	$e = \sqrt{2}$
*Foci	$(\pm ae, 0)$	$(a, 0)$	$(\pm ae, 0)$	$(\pm\sqrt{2}c, \pm\sqrt{2}c)$
*Directrices	$x = \pm\dfrac{a}{e}$	$x = -a$	$x = \pm\dfrac{a}{e}$	$x + y = \pm\sqrt{2}c$
*Asymptotes	None	None	$\dfrac{x}{a} = \pm\dfrac{y}{b}$	$x = 0,\ y = 0$

Basic algebra

*Arithmetic series (first term a, common difference d, last term l with n terms)
$u_n = a + (n-1)d$, $S_n = \frac{1}{2}n(a+l) = \frac{1}{2}n\{2a + (n-1)d\}$.

*Geometric series (first term a, common ratio r, with n terms)

$u_n = ar^{n-1}$, $S_n = \frac{a(1-r^n)}{1-r}$, $S_\infty = \frac{a}{1-r}$ if $|r| < 1$.

*Logarithms and exponentials

$\log_a x = \frac{\log_b x}{\log_b a}$, $e^{x \ln a} = a^x$.

*Numerical integration
*Trapezium rule:

$$\int_a^b y\,dx \approx \frac{1}{2}h\{(y_0 + y_n) + 2(y_1 + y_2 + \cdots + y_{n-1})\}, \quad \text{where } h = \frac{b-a}{n}.$$

*Simpson's rule, for even n:

$$\int_a^b y\,dx \approx \frac{1}{3}h\{(y_0 + y_n) + 4 \times (y_2 + \cdots + y_{n-2}) + 2 \times (y_1 + \cdots + y_{n-1})\},$$

$$\text{where } h = \frac{b-a}{n}.$$

Summations

$$\sum_{r=1}^{n} r = \frac{1}{2}n(n+1),$$

$$\sum_{r=1}^{n} r^2 = \frac{1}{6}n(n+1)(2n+1),$$

$$\sum_{r=1}^{n} r^3 = \frac{1}{4}n^2(n+1)^2.$$

*Newton–Raphson iterative method of solution

To solve $f(x) = 0$: $x_{n+1} = x_n - \dfrac{f(x_n)}{f'(x_n)}$.

*Complex numbers

$e^{i\theta} = \cos\theta + i\sin\theta$,

$\{r(\cos\theta + i\sin\theta)\}^n = r^n(\cos n\theta + i\sin n\theta)$.

The roots of $z^n = 1$ are given by $z = e^{\frac{i2\pi k}{n}}$, for $k = 0, 1, 2, \ldots, n-1$.

Derivatives

$f(x)$	$f'(x)$
k (constant)	0
x	1
x^k	kx^{k-1}
$\sin x$	$\cos x$
$\cos x$	$-\sin x$
$\tan x$	$\sec^2 x$
$\sec x$	$\sec x \tan x$
$\operatorname{cosec} x$	$-\operatorname{cosec} x \cot x$
$\cot x$	$-\operatorname{cosec}^2 x$
e^{kx}	ke^{kx}
$\ln x$	$1/x$
$a^x (a > 0)$	$a^x \ln a$
$\sin^{-1} x$	$\dfrac{1}{\sqrt{1-x^2}}$
$\cos^{-1} x$	$-\dfrac{1}{\sqrt{1-x^2}}$
$\tan^{-1} x$	$\dfrac{1}{1+x^2}$
$\sinh x$	$\cosh x$
$\cosh x$	$\sinh x$
$\tanh x$	$\operatorname{sech}^2 x$
$\coth x$	$-\operatorname{cosech}^2 x$
$\sinh^{-1} x$	$\dfrac{1}{\sqrt{x^2+1}}$
$\cosh^{-1} x$	$\dfrac{1}{\sqrt{x^2-1}}$
$\tanh^{-1} x$	$\dfrac{1}{\sqrt{1-x^2}}$

Integrals

Notes:

(i) The table gives, for each function f, an *antiderivative ϕ. The function ϕ_1 given by $\phi_1(x) = \phi(x) + c$, where c is an arbitrary constant, is also an antiderivative of f.

(ii) In certain cases, for example when $f(x) = 1/x$, $\tan x$, $\cot x$, $\sec x$, $\operatorname{cosec} x$, and $\sqrt{x^2 - a^2}$, the function f is not continuous for all x. In these cases, a definite integral $\int_a^b f(x)$ can be evaluated as $\phi(b) - \phi(a)$ only if a and b both belong to an interval in which f is continuous.

$f(x)$	$\phi(x) = \int f(x)dx$
$x^k \; (k \neq -1)$	$\dfrac{x^{k+1}}{k+1}$
$1/x$	$\ln|x|$
$\sin x$	$-\cos x$
$\cos x$	$\sin x$
$\tan x$	$-\ln|\cos x| = \ln|\sec x|$
$\sec x$	$\ln|\sec x + \tan x|$
$\operatorname{cosec} x$	$\ln|\operatorname{cosec} x - \cot x| = \ln|\tan \tfrac{1}{2}x|$
$\cot x$	$\ln|\sin x|$
$\sin^2 x$	$\tfrac{1}{2}\left(x - \tfrac{1}{2}\sin 2x\right)$
$\cos^2 x$	$\tfrac{1}{2}\left(x + \tfrac{1}{2}\sin 2x\right)$
$\sinh x$	$\cosh x$
$\cosh x$	$\sinh x$
$e^{kx} \; (k \neq 0)$	e^{kx}/k
$e^{ax} \sin bx$	$\dfrac{e^{ax}}{a^2 + b^2}\left(a \sin bx - b \cos bx\right)$
$e^{ax} \cos bx$	$\dfrac{e^{ax}}{a^2 + b^2}\left(a \cos bx + b \sin bx\right)$
$a^x \; (a>0, \; a \neq 1)$	$a^x/\ln a$
$\ln x$	$x \ln x - x$

$\dfrac{1}{\sqrt{a^2 - x^2}}\,(a > 0)$	$\sin^{-1}\dfrac{x}{a}$
$\dfrac{1}{\sqrt{a^2 + x^2}}\,(a > 0)$	$\dfrac{1}{a}\tan^{-1}\dfrac{x}{a}$
$\dfrac{1}{x^2 - a^2}\,(a > 0)$	$\dfrac{1}{2a}\ln\left|\dfrac{x-a}{x+a}\right|$
$\dfrac{1}{\sqrt{x^2 + a^2}}\,(a > 0)$	$\sinh^{-1}\dfrac{x}{a}$ or $\ln(x + \sqrt{x^2 + a^2})$
$\dfrac{1}{\sqrt{x^2 - a^2}}\,(a > 0)$	$\cosh^{-1}\dfrac{x}{a}$ or $\ln(x + \sqrt{x^2 - a^2})$
$\sqrt{x^2 + a^2}$	$\dfrac{1}{2}x\sqrt{x^2 + a^2} + \dfrac{1}{2}a^2\,\ln(x + \sqrt{x^2 + a^2})$
$\sqrt{x^2 - a^2}$	$\dfrac{1}{2}x\sqrt{x^2 - a^2} - \dfrac{1}{2}a^2\,\ln\left|x + \sqrt{x^2 - a^2}\right|$
$\sqrt{x^2 + x^2}\,(a > 0)$	$\dfrac{1}{2}x\sqrt{x^2 - x^2} + \dfrac{1}{2}a^2\,\sin^{-1}\dfrac{x}{a}$

Common ordinary differential equations and solutions

Type	Form	Solution
1st order, separable variables (*see* SEPARABLE FIRST-ORDER DIFFERENTIAL EQUATION)	Can be written as $f(y) . \frac{dy}{dx} = g(x)$	$\int f(y) . dy = \int g(x) . dx$
1st order, *integrating factor	$\frac{dy}{dx} + P(x) . y = Q(x)$	$y e^{\int P(x) . dx} = \int (Q(x) e^{P(x) . dx}) . dx$
Homogeneous 1st order (*see* HOMOGENEOUS FIRST-ORDER DIFFERENTIAL EQUATION)	$dy/dx = f(x, y)$ in which the function f, of two variables, has the property that $f(kx, ky) = f(x, y)$ for all k.	Let $y = vx$ so that $dy/dx = x \, dv/dx + v$. The differential equation for v as a function of x that is obtained is always *separable*.
*Simple harmonic motion	$\frac{d^2 x}{dt^2} = -k^2 x$	$x = a\cos(kt + \varepsilon)$ where a and ε are found from boundary conditions.
2nd order with constant coefficients	$a\frac{d^2 y}{dx^2} + b\frac{dy}{dx} + cy = 0$ where $am^2 + bm + cm = 0$ has 2 real roots, m_1 and m_2.	$y = Ae^{m_1 x} + Be^{m_2 x}$
	$a\frac{d^2 y}{dx^2} + b\frac{dy}{dx} + cy = 0$ where $am^2 + bm + cm = 0$ has 2 complex roots, m_1 and $m_2 = \alpha \pm \beta i$	$y = Ae^{m_1 x} + Be^{m_2 x}$ $= e^{ax}(C \cos(\beta x) + D \sin(\beta x))$
	$a\frac{d^2 y}{dx^2} + b\frac{dy}{dx} + cy = 0$ where $am^2 + bm + cm = 0$ has 1 repeated root m.	$y = (Ax + B)e^{mx}$
	$a\frac{d^2 y}{dx^2} + b\frac{dy}{dx} + cy = f(x)$	If $y = h(x)$ is a solution to $a\frac{d^2 y}{dx^2} + b\frac{dy}{dx} + cy = f(x)$ and $y = g(x)$ is a solution to $a\frac{d^2 y}{dx^2} + b\frac{dy}{dx} + cy = 0$ then the general solution is $y = h(x) + \lambda g(x)$.

Series

$$e^x = 1 + \frac{x}{1!} + \frac{x^2}{2!} + \cdots + \frac{x^n}{n!} + \cdots \quad \text{(for all } x\text{)}$$

$$\sin x = x - \frac{x^3}{3!} + \frac{x^5}{5!} - \frac{x^7}{7!} + \cdots + (-1)^n \frac{x^{2n+1}}{(2n+1)!} + \cdots \quad \text{(for all } x\text{)}$$

$$\cos x = 1 - \frac{x^2}{2!} + \frac{x^4}{4!} + \frac{x^6}{6!} + \cdots + (-1)^n \frac{x^{2n}}{(2n)!} + \cdots \quad \text{(for all } x\text{)}$$

$$\sinh x = x + \frac{x^3}{3!} + \frac{x^5}{5!} + \frac{x^7}{7!} + \cdots + \frac{x^{2n+1}}{(2n+1)!} + \cdots \quad \text{(for all } x\text{)}$$

$$\cosh x = 1 + \frac{x^2}{2!} + \frac{x^4}{4!} + \frac{x^6}{6!} + \cdots + \frac{x^{2n}}{(2n)!} + \cdots \quad \text{(for all } x\text{)}$$

$$\ln(1+x) = x - \frac{x^2}{2} + \frac{x^3}{3} - \frac{x^4}{4} + \cdots + (-1)^{n-1}\frac{x^n}{n} + \cdots \quad (-1 < x \leq 1)$$

$$\tan^{-1}x = x - \frac{x^3}{3} + \frac{x^5}{5} - \frac{x^7}{7} + \cdots + (-1)^n \frac{x^{2n+1}}{2n+1} + \cdots \quad (-1 \leq x \leq 1)$$

$$\sin^{-1}x = x + \frac{1}{2}\frac{x^3}{3} + \frac{1\times3}{2\times4}\frac{x^5}{5} + \frac{1\times3\times5}{2\times4\times6}\frac{x^7}{7} + \cdots$$
$$+ \binom{2n}{n} \frac{x^{2n+1}}{2^{2n}(2n+1)} + \cdots \quad (-1 \leq x \leq 1)$$

$$(1+x)^\alpha = 1 + \frac{\alpha}{1!}x + \frac{\alpha(\alpha-1)}{2!}x^2 + \cdots$$
$$+ \frac{\alpha(\alpha-1)\cdots(\alpha-n+1)}{n!}x^n + \cdots (-1 < x < 1)$$

(This is a binomial expansion; when α is a non-negative integer, the expansion is a finite series and is then valid for all x.)

Convergence tests for series

*Absolute convergence	If the series $\sum_{i=1}^{\infty} |a_i|$ converges then $\sum_{i=1}^{\infty} |a_i|$ also converges.
*Alternating series test	If $\{a_n\}$ is a sequence of non-negative real numbers, with $a_{i+1} \geq a_i$ and $a_i \to 0$ as $i \to \infty$, then $\sum_{i=1}^{\infty} a_i(-1)$ converges.
*Abel's test	If $\sum a_n$ is a convergent infinite sequence, and $\{b_n\}$ is monotonically decreasing, i.e. $b_{n+1} \leq b_n$ for all n, then $\sum a_n b_n$ is also convergent.
Comparison test	If $0 \leq a_i \leq b_i$ for all $i >$ some positive integer N, then: if $\sum_{i=1}^{\infty} b_i$ converges $\sum_{i=1}^{\infty} a_i$ also converges and if $\sum_{i=1}^{\infty} a_i$ diverges $\sum_{i=1}^{\infty} b_i$ also diverges.
*Dirichlet's test	If $\{a_n\}$ is a series which has bounded partial sums, i.e. $|\sum_{n=1}^{m} a_n| < K$ for all values of m, and $\{b_n\}$ is decreasing and converges to zero, i.e. $b_n < b_{n-1}$ and $\lim_{n \to \infty} b_n = 0$ then $\sum_{n=1}^{m} a_n n_n$ converges.
*Geometric series	If $\{a_n\}$ is a sequence of real numbers, with $a_{i+1} = r \times a_i$ then $\sum_{i=1}^{\infty} a_i$ converges $\left(\text{to } \frac{a_1}{1-r}\right)$ if $|r| < 1$ and diverges otherwise.
*Ratio test	For an infinite series $\{a_n\}$, if the ratio $\left|\dfrac{a_{n+1}}{a_n}\right| \to k$ as $n \to \infty$ then when k is < 1 the series converges absolutely, and for $k > 1$ the series diverges. If $k = 1$, the ratio test is not sufficient to determine whether it converges.

Common inequalities

*Triangle inequality	$\lvert\lvert x\rvert - \lvert y\rvert\rvert \leq \lvert x+y\rvert \leq \lvert x\rvert + \lvert y\rvert$
	$\lvert x + x_2 + \ldots + x_n\rvert \leq \lvert x_1\rvert + \lvert x_2\rvert + \ldots + \lvert x_n\rvert$
*Means	arithmetic mean $\geq$ geometric mean $\geq$ harmonic mean
	$\dfrac{x_1 + x_2 + \ldots + x_n}{n} \geq n\sqrt{x_1 . x_2 + \ldots + x_n}$
	$\geq \dfrac{n}{\dfrac{1}{x_1} + \dfrac{1}{x_2} + \ldots + \dfrac{1}{x_n}}$ where $x_1, x_2, \ldots, x_n$ are
	non-negative real numbers
Abel's	$\left\lvert \sum_{i=1}^{n} a_i f_i \right\rvert \leq A f_1$ where $\{f_i\}$ is a sequence of real numbers with $f_{i+1} \geq f_i$ and $\{a_i\}$ a sequence of real or complex numbers, and $A = \max\{\lvert a_1\rvert, \lvert a_2\rvert, \ldots \lvert a_n\rvert\}$.
Bernoulli's	$(1+x)^n \geq 1 + nx$ for integers $n > 1$ and any real number $x > -1$.
*Cauchy–Schwarz inequality for integrals	If $f(x)$, $g(x)$ are real functions then $\left\{ \int [f(x)g(x)]dx \right\}^2$ $\leq \left\{ \int [f(x)]^2 dx \right\} \left\{ \int [g(x)]^2 dx \right\}$ if all these integrals exist.
*Cauchy–Schwarz inequality for sums	If a_i and b_i are real numbers, $i = 1, 2, \ldots, n$ then $\sum_{i=1}^{n} a_i b_i \leq \sqrt{(\sum_{i=1}^{n} a_i^2)(\sum_{i=1}^{n} b_i^2)}$.
*Chebyshev's inequalities	$\Pr\{\lvert X - \mu_x\rvert > k\sigma\} \leq \dfrac{1}{k^2}$
	If X is a *random variable* and $g(X)$ is always ≥ 0 then $\Pr\{g(X) \geq k\} \leq \dfrac{E\{g(X)\}}{K}$
Holder's for integrals	$\int_a^b \lvert f(x).g(x)\rvert dx \leq \left\lvert \int_a^b \lvert f(x)\rvert^p dx \right\rvert^{\frac{1}{p}} \times \left\lvert \int_a^b \lvert g(x)\rvert^q dx \right\rvert^{\frac{1}{q}}$ for $\dfrac{1}{p} + \dfrac{1}{q} = 1$
Holder's for sums	$\sum_{i=1}^{n} \lvert a_i b_i\rvert \leq \left(\sum_{i=1}^{n} \lvert a_i\rvert^p \right)^{\frac{1}{p}} \times \left(\sum_{i=1}^{n} \lvert b_i\rvert^q \right)^{\frac{1}{q}}$ for $\frac{1}{p} + \frac{1}{q} = 1$
*Isoperimetric inequality	If p is the perimeter of a closed curve in a plane and the area enclosed by the curve is A, then $p^2 \leq 4\pi A$

Trigonometric formulae

$$\tan A = \frac{\sin A}{\cos A}, \quad \cot A = \frac{\cos A}{\sin A} = \frac{1}{\tan A},$$

$$\sec A = \frac{1}{\cos A}, \quad \operatorname{cosec} A = \frac{1}{\sin A},$$

$$\cos^2 A + \sin^2 A = 1, \quad \sec^2 A = 1 + \tan^2 A, \quad \operatorname{cosec}^2 A = 1 + \cot^2 A.$$

*Addition formulae

$$\sin(A + B) = \sin A \cos B + \cos A \sin B,$$
$$\sin(A - B) = \sin A \cos B - \cos A \sin B,$$
$$\cos(A + B) = \cos A \cos B - \sin A \sin B,$$
$$\cos(A - B) = \cos A \cos B + \sin A \sin B,$$
$$\tan(A + B) = \frac{\tan A + \tan B}{1 - \tan A \tan B},$$
$$\tan(A - B) = \frac{\tan A - \tan B}{1 + \tan A \tan B}.$$

*Double-angle formulae

$$\sin 2A = 2\sin A \cos A,$$

$$\cos 2A = \cos^2 A - \sin^2 A,$$

$$\cos 2A = 1 - 2\sin^2 A, \quad \sin^2 A = \frac{1}{2}(1 - \cos 2A),$$

$$\cos 2A = 2\cos^2 A - 1, \quad \cos^2 A = \frac{1}{2}(1 + \cos 2A),$$

$$\tan 2A = \frac{2\tan A}{1 - \tan^2 A}.$$

Tangent-of-half-angle formulae

Let $t = \tan \frac{1}{2}A$; then

$$\sin A = \frac{2t}{1 + t^2}, \quad \cos A = \frac{1 - t^2}{1 + t^2}, \quad \tan A = \frac{2t}{1 - t^2}.$$

Product formulae

$$\sin A \cos B = \frac{1}{2}\Big(\sin (A + B) + \sin (A - B)\Big),$$

$$\cos A \sin B = \frac{1}{2}\Big(\sin (A + B) - \sin (A - B)\Big),$$

$$\cos A \cos B = \frac{1}{2}\Big(\cos (A + B) + \cos (A - B)\Big),$$

$$\sin A \sin B = \frac{1}{2}\Big(\cos (A - B) - \cos (A + B)\Big).$$

Sums and differences

$$\sin C + \sin D = 2\sin \frac{1}{2}(C + D)\cos \frac{1}{2}(C - D),$$

$$\sin C - \sin D = 2\cos \frac{1}{2}(C + D)\sin \frac{1}{2}(C - D),$$

$$\cos C + \cos D = 2\cos \frac{1}{2}(C + D)\cos \frac{1}{2}(C - D),$$

$$\cos C - \cos D = -2\sin \frac{1}{2}(C + D)\sin \frac{1}{2}(C - D).$$

Trigonometric values for some special angles

$\sin 0 = 0$, $\cos 0 = 1$, $\tan 0 = 0$;

$\sin 30° = \frac{1}{2}$, $\cos 30° = \frac{\sqrt{3}}{2}$, $\tan 30° = \frac{1}{\sqrt{3}}$;

$\sin 45° = \frac{1}{\sqrt{2}}$, $\cos 45° = \frac{1}{\sqrt{2}}$, $\tan 45° = 1$;

$\sin 60° = \frac{\sqrt{3}}{2}$, $\cos 60° = \frac{1}{2}$, $\tan 60° = \sqrt{3}$;

$\sin 90° = 1$, $\cos 90° = 0$; $\tan 90°$ is undefined.

Symmetry

$\sin (-\theta) = -\sin \theta$, $\cos (-\theta) = \cos \theta$, $\tan (-\theta) = -\tan \theta$;

$\sin (90° - \theta) = \cos \theta$, $\cos (90° - \theta) = \sin \theta$, $\tan (90° - \theta) = \cot \theta$;

$\sin (180° - \theta) = \sin \theta$, $\cos (180° - \theta) = -\cos \theta$, $\tan (180° - \theta) = -\tan \theta$.

Shifts and periodicity

A shift by 360° for sin and cos and a shift by 180° for tan, or by multiples of these values, leave the function value unchanged because these are the periods of those functions. In addition, the following hold:

$\sin (90° + \theta) = \cos \theta$, $\cos (90° + \theta) = -\sin \theta$, $\tan (90° + \theta) = -\cot \theta$;

$\sin (180° + \theta) = -\sin \theta$, $\cos (180° + \theta) = -\cos \theta$.

Appendix 13

Probability distributions

Discrete probability distributions

*Distribution	*Parameters	*Probability mass function	*Mean	*Variance	*Moment generating function
*Binomial	n, p	$P(X = r) = \binom{n}{r} p^r (1 - p)^{n-r}$, $r = 0, 1, 2, \ldots, n$	$\mu = np$	$\sigma^2 = np(1 - p)$	$M(t) = (1 + p(e^t - 1))^n$
*Geometric	p	$P(X = r) = p(1 - p)^{r-1}$, $r = 1, 2, \ldots$	$\mu = \dfrac{1}{p}$	$\sigma^2 = \dfrac{(1 - p)}{p^2}$	$M(t) = \dfrac{pe^t}{1 - (1 - p)e^t}$, $t < -\ln(1 - p)$
*Negative binomial	k, p	$p(X = r) = \binom{r - 1}{k - 1} p^k (1 - p)^{r-k}$, $r = k, k + 1, k + 2, \ldots$	$\mu = \dfrac{k(1 - p)}{p}$	$\sigma^2 = \dfrac{k(1 - p)}{p^2}$	$M(t) = \dfrac{(pe^t)^k}{\{1 - (1 - p)e^t\}^k}$, $t < -\ln(1 - p)$
*Hypergeometric	n, N, M	$P(X = r) = \dfrac{\binom{M}{r}\binom{N - M}{n - r}}{\binom{N}{n}}$, for r with $\max(0, n - N + M) \leq r \leq \min(n, M)$	$\mu = n \cdot \dfrac{M}{N}$	$\sigma^2 = n \cdot \dfrac{M}{N}\left(1 - \dfrac{M}{N}\right)\left(\dfrac{N - n}{N - 1}\right)$	Not useful
*Poisson	λ	$P(X = r) = e^{-\lambda}\dfrac{\lambda^r}{r!}$, $r = 0, 1, 2, \ldots$	$\mu = \lambda$	$\sigma^2 = \lambda$	$M(t) = \exp\{\lambda(e^t - 1)\}$

Continuous probability distributions

*Distribution	*Parameters	*Probability density function	*Mean	*Variance	*Moment generating function
*Uniform on $[a, b]$	a, b	$f(x) = \dfrac{1}{b-a}, a \leq x \leq b$	$\mu = \dfrac{(b+a)}{2}$	$\sigma^2 = \dfrac{(b-a)^2}{12}$	$M(t) = \dfrac{e^{tb} - e^{ta}}{(b-a)t}, t \neq 0$
*Normal	μ, σ	$f(x) = \dfrac{1}{\sigma\sqrt{(2\pi)}} e^{\dfrac{-(x-\mu)^2}{2\sigma^2}},$ $-\infty \leq x \leq \infty$	μ	σ^2	$M(t) = \exp\left(\mu t + \dfrac{1}{2}\sigma^2 t^2\right)$
*Exponential	λ	$f(x) = \lambda e^{-\lambda x}, x \geq 0$	$\mu = \dfrac{1}{\lambda}$	$\sigma^2 = \dfrac{1}{\lambda^2}$	$M(t) = \dfrac{\lambda}{\lambda - t}, t < \lambda$
*Gamma	λ, r	$f(x) = \dfrac{\lambda^r}{\Gamma(r)} x^{r-1} e^{-\lambda x}, x > 0$	$\mu = \dfrac{r}{\lambda}$	$\sigma^2 = \dfrac{r}{\lambda^2}$	$M(t) = \left(\dfrac{\lambda}{\lambda - t}\right)^r, t < \lambda$
*Chi-squared	n	$f(x) = \dfrac{1}{2^{\frac{n}{2}} \Gamma\left(\frac{n}{2}\right)} x^{\frac{n}{2}-1} e^{-\frac{x}{2}}, x > 0$	$\mu = n$	$\sigma^2 = 2n$	$M(t) = \left(\dfrac{\frac{1}{2}}{\frac{1}{2} - t}\right)^{\frac{n}{2}} = \dfrac{1}{(1 - 2t)^{\frac{n}{2}}}, t < \frac{1}{2}$

Appendix 14

Symbols

Symbol	Reference
$\neg$	*negation
$\wedge$	*conjunction
$\vee$	*disjunction
$\Rightarrow, \Leftrightarrow$	*implication
$\sim$	*equivalence relation
$\exists, \forall$	*quantifier
$\in, \notin$	*belongs to, does not belong to
$\subseteq, \supseteq$	*subset
$\subset, \supset$	*proper subset
$\cup, \bigcup$	*union
$\cap, \bigcap$	*intersection
$A', \bar{A}$	*complement
$\varnothing$	*empty set
$A \times B$	*Cartesian product
$A \setminus B, A - B$	*difference set
$A + B, A \,\Delta\, B$	*symmetric difference
$n(A), \# (A), |A|$	*cardinality
$\mathscr{P}(A)$	*power set
$n!$	*factorial
$[a, b]$	*closed interval, *least common multiple
(a, b)	*open interval, *greatest common factor
$[a, b), (a, b]$	*interval, open at one end
$\binom{n}{r}$	*binomial coefficient
$[x]$	*integer part
$\{x\}$	*fractional part
$|x|$	*absolute value
$|z|$	*modulus
$\bar{z}$	*conjugate
$\Re z, \operatorname{Re} z$	*real part
$\Im z, \operatorname{Im} z$	*imaginary part
$\overrightarrow{AB}$	*directed line segment

$\lvert AB \rvert, \lvert \overrightarrow{AB} \rvert$	*length
$\lVert P \rVert$	*norm
$\sqrt{}$	*square root
$\neq, <, \leq, >, \geq$	*inequality
$\approx$	*approximation
$\equiv$	*congruence
$\propto$	*proportion
$\sum, \displaystyle\sum$	*summation notation
$\Pi, \displaystyle\prod$	*product notation
π	*pi
$f: x \mapsto y$	*function, *mapping
$f: S \to T$	*function, *mapping
$\to$	*limit
$\nearrow, \searrow$	*limit from the left and right
$f \circ g$	*composition
f^{-1}	*inverse function, *inverse mapping
$f', \dfrac{df}{dx}, y', \dfrac{dy}{dx}$	*derivative, *derived function
$f'', f''' \ldots, f^{(n)}, \dfrac{d^2f}{dx^2}, \ldots, \dfrac{d^nf}{dx^n}$	*higher derivative
$y'', y''' \ldots, y^{(n)}, \dfrac{d^2y}{dx^2}, \ldots, \dfrac{d^ny}{dx^n}$	*higher derivative
$f_x, f_y, f_1, f_2, \dfrac{\partial f}{\partial x}, \dfrac{\partial f}{\partial y}$	*partial derivative
$f_{xx}, f_{xy}, \ldots, f_{11}, f_{12}, \ldots$	*higher-order partial derivative
$\dfrac{\partial^2 f}{\partial x^2}, \dfrac{\partial^2 f}{\partial x \partial y}, \ldots$	*higher-order partial derivative
$\dot{x}, \ddot{x}$	*rate of change
$\displaystyle\int$	*integral, *antiderivative
$\mathbf{a.b}$	*scalar product
$\mathbf{a} \times \mathbf{b}, \mathbf{a} \wedge \mathbf{b}$	*vector product
$\mathbf{a}.(\mathbf{b} \times \mathbf{c}), [\mathbf{a}, \mathbf{b}, \mathbf{c}]$	*scalar triple product
$\mathbf{a} \times (\mathbf{b} \times \mathbf{c})$	*vector triple product
$\mathbf{A}^T, \mathbf{A}^t, \mathbf{A}'$	*transpose
$\mathbf{A}^{-1}$	*inverse matrix
$\lvert \mathbf{A} \rvert$	*determinant
$\langle G, \circ \rangle$	*group
$\langle R, +, \times \rangle$	*ring
$\mathrm{E}(X)$	*expected value
$\mathrm{Var}(X)$	*variance
$\mathrm{Cov}(X, Y)$	*covariance
$\mathrm{Pr}(A)$	*probability
$\mathrm{Pr}(A \mid B)$	*conditional probability

Greek letters

Name	Lower case	Capital
Alpha	α	A
Beta	β	B
Gamma	γ	Γ
Delta	δ	Δ
Epsilon	ε	E
Zeta	ζ	Z
Eta	η	H
Theta	θ	Θ
Iota	ι	I
Kappa	κ	K
Lambda	λ	Λ
Mu	μ	M
Nu	ν	N
Xi	ξ	Ξ
Omicron	o	O
Pi	π	Π
Rho	ρ	P
Sigma	σ	Σ
Tau	τ	T
Upsilon	υ	Y
Phi	φ	Φ
Chi	χ	X
Psi	ψ	Ψ
Omega	ω	Ω

Roman numerals

If smaller numbers follow larger numbers, then the numbers are added, but if a smaller number comes before a larger number it is subtracted, so CM is 900 and MC is 1 100.

The basic values are I = 1, V = 5, X = 10, L = 50, C = 100, D = 500 and M = 1 000.

	Roman		Roman
1	I	50	L
2	II	51	LI
3	III	55	LV
4	IV	60	LX
5	V	70	LXX
6	VI	80	LXXX
7	VII	90	XC
8	VIII	95	XCV
9	IX	99	XCIX
10	X	100	C
11	XI	101	CI
12	XII	105	CV
13	XIII	110	CX
14	XIV	150	CL
15	XV	200	CC
16	XVI	300	CCC
17	XVII	400	CD
18	XVIII	500	D
19	XIX	600	DC
20	XX	700	DCC
21	XXI	800	DCCC
22	XXII	900	CM
23	XXIII	1000	M
24	XXIV	1100	MC
25	XXV	1500	MD
30	XXX	1900	MCM
35	XXXV	2000	MM
40	XL	2009	MMIX
45	XLV		
49	XLIX		

Fields Medal winners

Year	Location	Winners
2018	Rio de Janeiro, Brazil	Caucher Birkar, Alessio Figalli, Peter Scholze, Akshay Venkatesh
2014	Seoul, South Korea	Artur Avila, Manjul Bhargava, Martin Hairer, Marvam Mirzakhani
2010	Hyderabad, India	Elon Lindenstrauss, Ngô Bao Châu, Stanislav Smiirnov, Cedric Vilani
2006	Madrid, Spain	Andrei Okounkov, Grigori Perelman (who declined the award), Terence Tao, Wendelin Werner
2002	Beijing, China	Laurent Lafforgue, Vladimir Voevodsky
1998	Berlin, Germany	Richard Ewen Borcherds, William Timothy *Gowers, Maxim Kontsevich, Curtis T. McMullen
		A silver plate was awarded to Andrew *Wiles as a special tribute.
1994	Zürich, Switzerland	Efim Isakovich Zelmanov, Jacques-Louis Lions, Jean Bourgain, Jean-Christophe Yoccoz
1990	Kyoto, Japan	Vladimir Drinfeld, Vaughan Frederick Randal Jones, Shigefumi Mori, Edward *Witten
1986	Berkeley, California, USA	Simon Donaldson, Gerd Faltings, Michael *Freedman
1982	Warsaw, Poland	Alain Connes, William Thurston, Shing-Tung Yau
1978	Vancouver, British Columbia, Canada	Pierre Deligne, Charles Fefferman, Grigory Margulis, Daniel *Quillen
1974	Helsinki, Finland	Enrico Bombieri, David Mumford
1970	Nice, France	Alan Baker, Heisuke Hironaka, Sergei Petrovich Novikov, John Griggs Thompson
1966	Moscow, Russia	Michael Francis *Atiyah, Paul Joseph Cohen, Alexander Grothendieck, Stephen Smale
1962	Stockholm, Sweden	Lars Hörmander, John *Milnor
1958	Edinburgh, Scotland, United Kingdom	Klaus Roth, René *Thom
1954	Amsterdam, Netherlands	Kunihiko Kodaira, Jean-Pierre Serre
1950	Cambridge, Massachusetts, USA	Laurent Schwartz, Atle Selberg
1936	Oslo, Norway	Lars Ahlfors, Jesse Douglas

Millennium Prize problems

1. P versus NP

The class of problems for which an algorithm can find a solution quickly (in polynomial time) is termed P. The class of problems for which an algorithm can verify a solution quickly is termed NP. The question is whether all problems in NP are also in P. Most mathematicians and computer scientists expect that they are not, but there has been no proof one way or the other by 2013.

2. The Hodge conjecture

The Hodge conjecture is a major unsolved problem in algebraic geometry. It arises out of attempts to approximate the shape of a given object by putting together simple geometric building blocks of increasing dimension.

The Hodge conjecture is that for projective algebraic varieties, Hodge cycles are rational linear combinations of algebraic cycles.

3. The *Riemann hypothesis

The distribution of prime numbers does not follow any regular pattern, but Riemann observed that it is closely related to the (Riemann) zeta function

$$\zeta(s) = \sum_{n=1}^{\infty} \frac{1}{n^s} = 1 + \frac{1}{2^s} + \frac{1}{3^s} + \frac{1}{4^s} + \cdots.$$

The Riemann hypothesis is the conjecture that the real part of any non-trivial solution to $\zeta(s) = 0$ is $\frac{1}{2}$, i.e. it is in the form $s = \frac{1}{2} + iy$ for some real value y.

This conjecture was first stated in 1859 and was included in Hilbert's list of 23 unsolved problems published in 1900. While the statement has been confirmed for the first one and half billion cases, it has resisted a general proof for over one hundred and fifty years.

4. Yang–Mills existence and mass gap

Yang–Mills theory is now the foundation of most elementary-particle theory in quantum mechanics, generalizing the work of Maxwell in electromagnetism. Its predictions in relation to the strong interactions of elementary particles have been tested and verified experimentally, and confirmed by computer simulations, but the mathematics remains unexplained. Classical theory says that electromagnetism travels in waves at the speed of light with zero mass, while the Yang–Mills theory depends on the quantum particles having positive masses (hence the 'mass gap').

The problem is to establish rigorously the Yang–Mills quantum theory and that the mass of the least massive particle of the force field is strictly positive, that is, that each type of particle has a lower mass bound.

5. Navier–Stokes existence and smoothness

The Navier–Stokes equations describe the motion of fluids. Since 'fluids' is a collective term for any substance which is not solid, it embraces liquids and gases, meaning that everything from the behaviour of the contents of a glass of tap water, through waves in the oceans, to the turbulence created by the motion of a Formula 1 car taking a corner at 170 mph is governed by these equations. As with the *Riemann hypothesis, the statement of these equations dates back to the 19th century, being first published by Claude Louis Navier in 1821 and 1822, and developed by Sir George Gabriel Stokes over the following years. Again, although these equations have been the subject of a huge amount of effort by the best mathematicians and physicists over an extended period, the complete solutions have yet to be found.

It has not yet been proven that solutions to the Navier–Stokes equations always exist in three dimensions (the existence), nor that if they do exist, then they do not contain any singularities (the smoothness). These two aspects comprise the Navier–Stokes existence and smoothness problem.

6. The *Birch and Swinnerton-Dyer conjecture

Much of mathematics is engaged with trying to produce generalizations of what can be done for particular cases; so, the quadratic formula generates solutions for all quadratic equations, and we know there is nothing left to be resolved in the case of equations of that type. Still valuable in terms of mathematics is the knowledge of when it is not possible to produce a generalization—so the negative proof in the early 1970s of Hilbert's tenth problem represented a very significant piece of mathematics. It was proved that there is no algorithm to determine whether a given polynomial Diophantine equation with integer coefficients has a solution in integers. However, it remains possible that there are some special cases for which it is possible to say something.

The Birch and Swinnerton-Dyer conjecture is that when the solutions are the points of an abelian variety, the size of the group of rational points is related to the behaviour of an associated zeta function $\zeta(s)$ near $s = 1$. In particular, it asserts that if $\zeta(1) = 0$ then there are an infinite number of rational points or solutions, and that if $\zeta(1) \neq 0$ then there are only a finite number of such points.

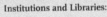